Association for Women in Mathematics Series

Volume 6

Series editor

Kristin Lauter, Redmond, WA, USA

Focusing on the groundbreaking work of women in mathematics past, present, and future, Springer's Association for Women in Mathematics Series presents the latest research and proceedings of conferences worldwide organized by the Association for Women in Mathematics (AWM). All works are peer-reviewed to meet the highest standards of scientific literature, while presenting topics at the cutting edge of pure and applied mathematics. Since its inception in 1971, The Association for Women in Mathematics has been a non-profit organization designed to help encourage women and girls to study and pursue active careers in mathematics and the mathematical sciences and to promote equal opportunity and equal treatment of women and girls in the mathematical sciences. Currently, the organization represents more than 3000 members and 200 institutions constituting a broad spectrum of the mathematical community in the United States and around the world.

More information about this series at http://www.springer.com/series/13764

Gail Letzter
Editor-in-Chief

Kristin Lauter · Erin Chambers
Nancy Flournoy · Julia Elisenda Grigsby
Carla Martin · Kathleen Ryan
Konstantina Trivisa
Editors

Advances in the Mathematical Sciences

Research from the 2015 Association for Women in Mathematics Symposium

Springer

Editor-in-Chief
Gail Letzter
National Security Agency
Fort Meade, MD
USA

Editors
Kristin Lauter
Microsoft Research
Redmond, WA
USA

Erin Chambers
Department of Mathematics and Computer Science
Saint Louis University
St. Louis, MO
USA

Nancy Flournoy
Department of Statistics
University of Missouri-Columbia
Columbia, MO
USA

Julia Elisenda Grigsby
Department of Mathematics
Boston College
Chestnut Hill, MA
USA

Carla Martin
National Security Agency
Fort Meade, MD
USA

Kathleen Ryan
Department of Mathematics and Computer Science
DeSales University
Center Valley, PA
USA

Konstantina Trivisa
Department of Mathematics
University of Maryland
College Park, MD
USA

ISSN 2364-5733 ISSN 2364-5741 (electronic)
Association for Women in Mathematics Series
ISBN 978-3-319-34137-8 ISBN 978-3-319-34139-2 (eBook)
DOI 10.1007/978-3-319-34139-2

Library of Congress Control Number: 2016940319

Gail Letzter is the Editor-in-Chief for this volume.

Printed on acid-free paper

This Springer imprint is published by Springer Nature
The registered company is Springer International Publishing AG Switzerland

Dedicated to our fellow editor
Carla Dee Martin
1972–2015

Preface

Participants of the 2015 AWM research symposium

In 2011, the Association for Women in Mathematics marked its 40th anniversary with the organization's first major research symposium "40 Years and Counting: AWM's Celebration of Women in Mathematics." The extensive scientific program was a tribute to the depth and breadth of the technical contributions made by today's women mathematicians. Inspired by its success, AWM's leadership decided to make major research gatherings a regular event, thus beginning a new tradition of AWM biennial research symposia. The second conference was held at Santa Clara University in 2013. This volume is based on the third of the series, the 2015 AWM Research Symposium held at the University of Maryland.

Table 1 2015 AWM research symposium: the plenary lectures

Ingrid Daubechies	Applied Mathematics helping Art Historians and Conservators: Digital Cradle Removal
Maria Chudnovsky	Coloring square-free perfect graphs
Jill Pipher	Dyadic Analysis: From Fourier to Haar to Wavelets, and back
Katrin Wehrheim	String diagrams in Topology, Geometry and Analysis

The 2015 AWM Research Symposium attracted over 330 participants and showcased research of women mathematicians from academia, industry, and government. With four exciting plenary talks (Table 1) and fourteen special sessions (Table 2) on topics from across the mathematical spectrum, this conference had something for everyone. There was an air of excitement throughout the meeting as participants learned about all sorts of new advances and novel applications to other fields such as art history, biology, and computer science. Plenary lectures and special sessions had packed audiences; attendees wandered through the exhibit hall or spent time at poster sessions featuring results of recent PhDs in between talks. This volume of the new AWM Springer series commemorates the conference through a set of invited peer-reviewed contributions from special session speakers and special session organizers as well as one of the plenary speakers (Katrin Wehrheim). The volume, which includes papers based on talks from eight out of the fourteen special sessions (those starred in Table 2), reflects the broad range of mathematics presented at the conference.

Table 2 2015 AWM research symposium: the special sessions

Research from the "Cutting EDGE"*
Many facets of Probability*
Topics in Computational Topology and Geometry
Low-dimensional Topology*
Number Theory
Mathematics at Government Labs and Centers*
Symplectic Topology/Geometry
Recent mathematical advancements empowering signal/image processing
Algebraic Geometry
Statistics*
PDEs in Continuum Mechanics
Discrete Math (and Theoretical Computer Science)*
Mathematical Biology*
Sharing the Joy: Engaging Undergraduate Students in Mathematics*

AWM presidential award winners: Rhonda Hughes (*left*), Sylvia Bozeman (*standing up*) with first AWM president Mary Gray (*right*)

Speakers and organizers for the special session "Research from the 'Cutting EDGE'", from *left* to *right*: Kathleen Ryan, Carmen Wright, Ami Radunskaya, Candice Price, Sarah Bryant, Rhonda Hughes, Amy Buchmann, and Ulrica Wilson

A highlight of the symposium was the presentation of the first AWM Presidential Award to the co-founders Sylvia Bozeman and Rhonda Hughes of the Enhancing Diversity in Graduate Education (EDGE) program. This highly successful venture, established in 1998, was designed to increase the number of women and minorities who complete graduate school in the mathematical sciences. Participants attend an initial intensive summer program and receive extensive mentoring that continues through their first year of graduate school and beyond. To date, a total of 67 EDGE participants have obtained their doctorates in the mathematical sciences. The AWM was also honored to host a special session "Research from the 'Cutting EDGE'" at the 2015 conference; all eight speakers in this session attended the EDGE program and have finished or are in the process of finishing their PhDs. We are pleased to have a significant EDGE representation in this volume with three papers based on talks by EDGE special session speakers Candice Price, Sarah Bryant, and Raegan Higgins, and a fourth paper whose authors include the EDGE session speaker Amy Buchmann and the current EDGE co-director Ami Radunskaya. Our editorial group benefited greatly by having EDGE graduate and special session presenter Kathleen Ryan, as part of the team.

In addition to the special session and plenary talks, the AWM symposium featured keynote speaker Shirley Malcolm, Head of Education and Human Resources Programs at the American Association for the Advancement of Science and author of "The Double Bind: The Price of Being a Minority Women in Science." A long-time advocate for underrepresented groups in STEM, Dr. Malcolm spoke eloquently about the challenges faced by women and minorities in the mathematical sciences. Her moving speech affirmed the importance of the AWM as well as initiatives such as the EDGE program in providing the support women need as they pursue mathematical careers. More information about her presentation as well as other symposium events can be found on the symposium blog:

https://sites.google.com/site/awmmath/home/announcements/awmsymposiumblog

For the full symposium schedule with a list of all the talks, poster sessions, presenters, and other activities, please follow the program link on the conference's website:

https://sites.google.com/site/awmmath/home/awm-research-symposium-2015

Organizer Talitha Washington (*bottom left*), keynote speaker Shirley Malcolm (*bottom middle*) and AWM presidential award winner Sylvia Bozeman (*top left*) with AWM presidents Kristin Lauter (*top middle*), Jill Pipher (*top right*), Ruth Charney (*bottom right*)

This volume opens with a part entitled *From the Plenary Talks* and consists of a survey of Floer field theory written by Katrin Wehrheim, one of the symposium plenary speakers. After that, the papers are grouped together in parts based on subject areas. For the most part, the parts correspond to special sessions with some modifications. Each paper from the special session "Research from the 'Cutting EDGE'", as well as a contribution from the "Mathematics at Government Labs and Centers" and the contributions by special session organizers have been placed in a part based on its content.

Part II, *Low-Dimensional Topology*, consists of three papers based on talks in the special session of the same name and a contribution by the special session organizer Elisenda Grigsby. Part III, *Mathematical Biology*, contains three papers corresponding to presentations in the "Mathematical Biology" special session and two papers written by participants from "Research from the 'Cutting EDGE'" (one by Candice Price and the other by Amy Buchmann, Ami Radunskaya and others). Part IV, *Probability and Stochastic Processes*, is a combination of a paper co-written by Kavita Ramanan, a speaker from the "Many facets of probability" special session and a paper by "Research from the 'Cutting EDGE'" speaker Sarah Bryant. Part V, *Statistics*, consists of three papers based on talks from the "Statistics" special session. This is followed by a *Differential Equations* part

(Part VI) that has been created for the paper by Raegan Higgins, a speaker from the "Research from the 'Cutting EDGE'" session. The volume includes two papers from the special session "Sharing the Joy: Engaging Undergraduate Students in Mathematics" in a part of the same name (Part VII). The last part (Part VIII), *Discrete Mathematics and Computer Science*, contains three papers, each related to a different special session: a paper by Shari Wiley who talked in the special session of the same name, a paper co-authored by Carol Woodward who spoke in the "Mathematics at Government Labs and Centers" special session, and a paper co-authored by Erin Chambers, who organized the special session "Topics in Computational Topology and Geometry".

Symposium organizers with plenary speaker and past AWM president Jill Pipher from *left* to *right*: Konstantina Trivisa, Magnhild Lien, Shelly Harvey, Talitha Washington, Kristin Lauter, Jill Pipher, Ruth Charney (missing: Gail Letzter)

The editors would like to express their gratitude to the National Science Foundation and the National Security Agency for funding the symposium through their respective grant programs and to the AWM Research Symposium's corporate sponsors Microsoft Research, Google, Springer, Elsevier, Wolfram, and INTECH for their generous financial support. We also give special thanks to the University of Maryland, College Park for hosting the event and providing us with appropriate rooms and other infrastructure necessary to make the meeting go smoothly. Finally, we would like to acknowledge the organizers of the conference (Ruth Charney, Shelly Harvey, Kristin Lauter, Gail Letzter, Magnhild Lien, Konstantina Trivisa,

Talitha Washington), Jay Popham from the Springer staff, the referees who reviewed the papers, and the AWM staff including AWM Managing Director Jennifer Lewis whose hard work made both the symposium and this volume possible.

January 2016

Gail Letzter
Kristin Lauter
Erin Chambers
Nancy Flournoy
Julia Elisenda Grigsby
Carla Martin
Kathleen Ryan
Konstantina Trivisa

Contents

Part I
From the Plenary Talks

Floer Field Philosophy

Katrin Wehrheim

Abstract Floer field theory is a construction principle for example, 3-manifold invariants via decomposition in a bordism category and a functor to the symplectic category, and is conjectured to have natural four-dimensional extensions. This survey provides an introduction to the categorical language for the construction and extension principles and provides the basic intuition for two gauge theoretic examples which conceptually frame Atiyah–Floer type conjectures in Donaldson theory as well as the relations of Heegaard Floer homology to Seiberg–Witten theory.

Keywords Bordism bicategories · Floer theory · Topological field theory · Quilted 2-categories · Quilted Atiyah–Floer conjectures

Mathematics Subject Classification 57R56 · 53R57 · 58D29 · 81T45

1 Introduction

In the 1980s, the areas of low dimensional topology and symplectic geometry both saw important progress arise from the study of moduli spaces of solutions of nonlinear elliptic PDEs. In the study of smooth 4-manifolds, Donaldson [14] introduced the use of ASD Yang–Mills instantons,[1] which were soon followed by Seiberg–Witten

[1] A smooth four manifold can be thought of as a curved *four*-dimensional space-time. ASD (anti-self-dual) instantons in this space-time satisfy a reduction of Maxwell's equations for the electromagnetic potential in vacuum, which has an infinite dimensional gauge symmetry.

K. Wehrheim (✉)
Department of Mathematics, University of California, Berkeley, CA 94720, USA
e-mail: katrin@math.berkeley.edu

G. Letzter et al. (eds.), *Advances in the Mathematical Sciences*, Association for Women in Mathematics Series 6, DOI 10.1007/978-3-319-34139-2_1

equations [51]—another gauge theoretic[2] PDE. In the study of symplectic manifolds, Gromov [28] introduced pseudoholomorphic curves[3] In both subjects Floer [22, 23] then introduced a new approach to infinite dimensional Morse theory[4] based on the respective PDEs. This sparked the construction of various algebraic structures—such as the Fukaya A_∞-category of a symplectic manifold [63], a Chern–Simons field theory for 3-manifolds and 4-cobordisms [16], and analogous Seiberg–Witten 3-manifold invariants [33]—from these and related PDEs, which encode significant topological information on the underlying manifolds. Chern–Simons field theory in particular comprises the Donaldson invariants of 4-manifolds, together with algebraic tools to calculate these by decomposing a closed 4-manifold into 4-manifolds whose common boundary is given by a three-dimensional submanifold. This strategy of decomposition into simpler pieces inspired the new topic of "topological (quantum) field theory" [2, 42, 62, 85], in which the properties of such theories are described and studied.

In trying to extend the field-theoretic strategy to the decomposition of 3-manifolds along two-dimensional submanifolds, Floer and Atiyah [3] realized a connection to symplectic geometry: A degeneration of the ASD Yang–Mills equation on a 4-manifold with two-dimensional fibers Σ yields the Cauchy–Riemann equation on a (singular) symplectic manifold M_Σ given by the flat connections on Σ modulo gauge symmetries. Along with this, 3-dimensional handlebodies H with boundary $\partial H = \Sigma$ induce Lagrangian submanifolds $L_H \subset M_\Sigma$ given by the boundary restrictions of flat connections on H. Now Lagrangians[5] are the most fundamental topological object studied in symplectic geometry. They are often studied by means of the Floer homology $HF(L_0, L_1)$ of pairs of Lagrangians, which arises from a complex that is generated by the intersection points $L_0 \cap L_1$ and whose homology is invariant under Hamiltonian deformations of the Lagrangians. For the pair $L_{H_0}, L_{H_1} \subset M_\Sigma$ arising from the splitting $Y = H_0 \cup_\Sigma H_1$ of a 3-manifold into two handlebodies H_0, H_1, these generators are naturally identified with the generators of the instanton Floer homology $HF_{\text{inst}}(Y)$, given by flat connections on Y modulo gauge symmetries. (Indeed, restricting the latter to $\Sigma \subset Y$ yields a flat connection on Σ that extends to both H_0 and H_1—in other words, an intersection point of L_{H_0} with L_{H_1}.)

[2]In mathematics, "gauge theory" refers to the study of connections on principal bundles, where "gauge symmetries" arise from the pullback action by bundle isomorphisms; see, e.g., [67, Appendix A].

[3]Symplectic manifolds can be thought of as the configuration spaces of classical mechanical systems, with the position-momentum pairing providing the symplectic structure as well as a class of almost complex structures J. Pseudoholomorphic curves can then be thought of as *two*-dimensional surfaces in a $2n$-dimensional symplectic ambient space, which can be locally described as the image of $2n$ real-valued functions $\underline{u}$ of a complex variable $z = x + iy$ that satisfy a generalized Cauchy–Riemann equation $\partial_x \underline{u} = J(u)\, \partial_y \underline{u}$.

[4]Morse theory captures the topological shape of a space by studying critical points of a function and flow lines of its gradient vector field. In finite dimensions it yields a complex whose homology is independent of choices (e.g., of function) and in fact equals the singular homology of the space.

[5]Throughout this paper, the term "Lagrangian" refers to a half-dimensional isotropic submanifold of a symplectic manifold—corresponding to fixing the integrals of motion, e.g. the momentums.

These observations inspired the **Atiyah–Floer conjecture**

$$HF_{\text{inst}}(H_0 \cup_\Sigma H_1) \simeq HF(L_{H_0}, L_{H_1}),$$

which asserts an equivalence between the differentials on the Floer complexes—arising from ASD instantons on $\mathbb{R} \times Y$ and pseudoholomorphic maps $\mathbb{R} \times [0, 1] \to M_\Sigma$ with boundary values on L_{H_0}, L_{H_1}, respectively. While this conjecture is not well defined due to singularities in the symplectic manifolds M_Σ, and the proof of a well-defined version by Dostoglou–Salamon [18] required hard adiabatic limit analysis, the underlying ideas sparked inquiry into relationships between low-dimensional topology and symplectic geometry. At this point, the two fields are at least as tightly intertwined as algebraic and symplectic geometry (via mirror symmetry), most notably through the **Heegaard–Floer invariants for 3- and 4-manifolds** (as well as knots and links), which were discovered by Ozsvath–Szabo [52] by following the line of argument of Atiyah and Floer in the case of Seiberg–Witten theory. In both cases, the concept for the construction of an invariant of 3-manifolds Y is the same:

1. Split $Y = H_0 \cup_\Sigma H_1$ along a surface Σ into two handlebodies H_i with $\partial H_i = \Sigma$.
2. Represent the dividing surface Σ by a symplectic manifold M_Σ and the two handle bodies by Lagrangians $L_{H_i} \subset M_\Sigma$ arising from dimensional reductions of a gauge theory which is known to yield topological invariants.
3. Take the Lagrangian Floer homology $HF(L_{H_0}, L_{H_1})$ of the pair of Lagrangians.
4. Argue that different splittings yield isomorphic Floer homology groups—due to an isomorphism to a gauge theoretic invariant of Y or by direct symplectic isomorphisms $HF(L_{H_0}, L_{H_1}) \simeq HF(L_{\widetilde{H}_0}, L_{\widetilde{H}_1})$ for different splittings $Y = \widetilde{H}_0 \cup_{\widetilde{\Sigma}} \widetilde{H}_1$.

Floer field theory is an extension of this approach to more general decompositions of 3-manifolds, by phrasing Step 4 above as the existence of a functor between topological and symplectic categories that extends the association

$$\Sigma \mapsto M_\Sigma, \qquad H \mapsto L_H, \qquad \partial H = \Sigma \;\Rightarrow\; L_H \subset M_\Sigma.$$

It gives a conceptual explanation for Step 4 invariance proofs such as [52] which bypass a comparison to the gauge theory by directly relating the Floer homologies of Lagrangians $L_{H_0}, L_{H_1} \subset M_\Sigma$ and $L_{\widetilde{H}_0}, L_{\widetilde{H}_1} \subset M_{\widetilde{\Sigma}}$. Since these can arise from surfaces $\Sigma \not\simeq \widetilde{\Sigma}$ of different genus, the comparison between pseudoholomorphic curves in symplectic manifolds $M_\Sigma \not\simeq M_{\widetilde{\Sigma}}$ of different dimension must crucially use the fact that the Lagrangian boundary conditions encode different splittings of the same 3-manifold. Floer field theory encodes this as an isomorphism between algebraic compositions of the Lagrangians, which in turn yields isomorphic Floer homologies (a strategy that we elaborate on in Sects. 2.4 and 3.5),

$$\begin{aligned} H_0 \cup_\Sigma H_1 \simeq \widetilde{H}_0 \cup_{\widetilde{\Sigma}} \widetilde{H}_1 \quad &\Longrightarrow \quad L_{H_0} \# L_{H_1} \sim L_{\widetilde{H}_0} \# L_{\widetilde{H}_1} \\ &\Longrightarrow \quad HF(L_{H_0}, L_{H_1}) \simeq HF(L_{\widetilde{H}_0}, L_{\widetilde{H}_1}). \end{aligned}$$

Floer field theory, in particular its key isomorphism of Floer homologies [78] hinted at above, was discovered by the author and Woodward [80, 81] when attempting to formulate well-defined versions of the Atiyah–Floer conjecture. While the isomorphism of Floer homologies in [78] is usually formulated in terms of strip-shrinking in a new notion of quilted Floer homology [75], it can be expressed purely in terms of Floer homologies of pairs of Lagrangians, which lie in different products of symplectic manifolds. In this language, strip shrinking then is a degeneration of the Cauchy–Riemann operator to a limit in which the curves in one factor of the product of symplectic manifolds become trivial. (For more details, see Sect. 3.5.) This relation between pseudoholomorphic curves in different symplectic manifolds then provides a purely symplectic analogue of the adiabatic limit in [18], which relates ASD instantons to pseudoholomorphic curves.

The value of Floer field theory to 3-manifold topology is mostly of philosophical nature—giving a conceptual understanding for invariance proofs and a general construction principle for 3-manifold invariants (and similarly for knots and links), which has since been applied in a variety of contexts [5, 36, 44, 56, 80, 81]. One main purpose of this paper and the content of Sect. 2 is to explain this philosophy and cast the construction principle into rigorous mathematical terms. For that purpose Sect. 2.1 gives brief expositions of the notions of categories and functors, the category Cat, and bordism categories Bor_{d+1}. After introducing the symplectic category Symp in Sect. 2.2, the categorical structure in symplectic geometry that can be related to low dimensional topology, in Sect. 2.3 we cast the concept of Cerf decompositions (cutting manifolds into simple cobordisms) into abstract categorical terms that apply equally to bordism categories and our construction of the symplectic category. We then exploit the existence of Cerf decompositions in Bor_{d+1} and Symp together with a Yoneda functor Symp $\to$ Cat (see Lemma 3.5.6) to formulate a general construction principle for Floer field theories. This notion of Floer field theory is defined in Sect. 2.4 as a functor $\mathrm{Bor}_{d+1} \to$ Cat that factors through Symp. This construction is exemplified in Sect. 2.5 by naive versions of two gauge theoretic examples related to Yang–Mills–Donaldson resp. Seiberg–Witten theory in dimensions $2 + 1$. Finally, Sect. 2.6 explains how this yields conjectural symplectic versions of the gauge theoretic 3-manifold invariants, as predicted by Atiyah and Floer.

The second purpose of this paper and content of Sect. 3 is to lay some foundations for an **extension of Floer field theories to dimension 4**. Our goal here is to provide a rigorous exposition of the algebraic language in which this extension principle can be formulated—at a level of sophistication that is easily accessible to geometers while sufficient for applications. Thus, we review in detail the notions of 2-categories and bicategories in Sect. 3.1, including the 2-category Cat of (categories, functors, natural transformations), explicitly construct a bordism bicategory Bor_{2+1+1} in Sect. 3.2, and summarize notions and Yoneda constructions of 2-functors between these higher categories in Sect. 3.3. Moreover, Sect. 3.5 outlines the construction of symplectic 2-categories, based on abstract categorical notions of adjoints and quilt diagrams that we develop in Sect. 3.4. The latter transfers notions of adjunction and spherical string diagrams from monoidal categories into settings without natural monoidal structure.

This provides sufficient language to at least advertise an extension principle which we further discuss in [74]:

Any Floer field theory $\mathrm{Bor}_{2+1} \to \mathrm{Symp} \to \mathrm{Cat}$ *which satisfies a quilted naturality axiom has a natural extension to a 2-functor* $\mathrm{Bor}_{2+1+1} \to \mathrm{Symp} \to \mathrm{Cat}$.

This says in particular that any 3-manifold invariant which is constructed along the lines of the Atiyah–Floer conjecture naturally induces a 4-manifold invariant. While it does seem surprising, such a result could be motivated from the point of view of gauge theory, since the Atiyah–Floer conjecture and Heegaard–Floer theory were inspired by dimensional reductions of $3+1$ field theories $\mathrm{Bor}_{3+1} \to \mathscr{C}$. It also can be viewed as a pedestrian version of the cobordism hypothesis [42],[6] saying that a functor $\mathrm{Bor}_{2+1+\varepsilon} \to \mathrm{Symp}$ (where the ε stands for compatibility with diffeomorphisms of 3-manifolds) has a canonical extension $\mathrm{Bor}_{2+1+1} \to \mathrm{Symp}$.

Finally, the extensions of Floer field theory to dimension 4 are again expected to be isomorphic to the associated gauge theoretic 4-manifold invariants, in a way that is compatible with decomposition into 3- and 2-manifolds. We phrase these expectations in Sect. 3.6 as **quilted Atiyah–Floer conjectures**, which identify field theories $\mathrm{Bor}_{2+1+1} \to \mathscr{C}$. The last section Sect. 3.6 also demonstrates the construction principle for 2-categories via associating elliptic PDEs to quilt diagrams in several more gauge theoretic examples, which provide not only the proper context for stating all the generalized Atiyah–Floer conjectures, but also yield conceptually clear contexts for the various approaches to their proofs.

While this $2+1$ field theoretic circle of ideas has been and used in various publications, its rigorous abstract formulation in terms of a notion of "category with Cerf decompositions" is new to the best of the author's knowledge. Similarly, the notions of bordism bicategories, the symplectic 2-category, generalized string diagrams, and field theoretic proofs of Floer homology isomorphisms have been known and (at least implicitly) used in similar contexts, but are here cast into a new concept of "quilted bicategories" which will be central to the extension principle—both of which seem significantly beyond the known circle of ideas. Finally, note that Floer field theory should not be confused with the symplectic field theory (SFT) introduced by [21], in which another symplectic category—given by contact-type manifolds and symplectic cobordisms—is the domain, not the target of a functor.

We end this introduction by a more detailed explanation of the notion of an "invariant" as it applies to the study of topological or smooth compact manifolds, and a very brief introduction to the resulting classification of manifolds.

---

[6]Lurie's constructions involve the canonical extension of a functor $\mathrm{Bor}_{0+1+\ldots+\varepsilon} \to \mathscr{C}$ to $\mathrm{Bor}_{0+1+\ldots+1} \to \mathscr{C}$. However, this requires an extension of the field theory to dimensions 1 and 0 (which we do not even have ideas for) as well as a monoidal structure on the target category $\mathscr{C}$ (which is lacking at present because the gauge theoretic functors are well defined only on the connected bordism category). On the other hand, we have other categorical structures at our disposal, which we formalize in Sect. 3.4 as the notion of a quilted 2-category (akin to a spherical 2-category as described in [43]). In that language, the diagram of a Morse 2-function as in [26] expresses a 4-manifold as a quilt diagram in the bordism bicategory Bor_{2+1+1}. Now the key idea for [74] is that a functor $\mathrm{Bor}_{2+1} \to \mathrm{Symp}$ translates the diagram of a 4-manifold into a quilt diagram in the symplectic 2-category in Sect. 3.5, where it is reinterpreted in terms of pseudoholomorphic curves.

1.1 A Brief Introduction to Invariants of Manifolds

In order to classify manifolds of a fixed dimension n up to diffeomorphism, one would ideally like to have a complete invariant $I : \mathrm{Man}_n \to \mathscr{C}$. Here Man_n is the category of n-manifolds and diffeomorphisms between them (see Example 2.1.4), and $\mathscr{C}$ is a category such as $\mathscr{C} = \mathbb{Z}$ with trivial morphisms or the category $\mathscr{C} = \mathrm{Gr}$ of groups and homomorphisms. Such I is an invariant if it is a functor (see Definition 2.1.3), since this guarantees that diffeomorphic manifolds are mapped to isomorphic objects of $\mathscr{C}$ (e.g., the same integer or isomorphic groups). In other words, functoriality guarantees that I induces a well-defined map $|I| : |\mathrm{Man}_n| \to |\mathscr{C}|$ from diffeomorphism classes of manifolds to, e.g., $\mathbb{Z}$ or isomorphism classes of groups. Such an invariant lets us distinguish manifolds: If $I(X), I(Y)$ are not isomorphic (i.e. $|I|([X]) \neq |I|([Y])$) then X and Y cannot be diffeomorphic. Moreover, an invariant is called "complete" if an isomorphism $I(X) \simeq I(Y)$ implies the existence of a diffeomorphism $X \simeq Y$, i.e., $|I|([X]) = |I|([Y]) \Rightarrow [X] = [Y]$.

Simple examples of invariants—when restricting Man_n to compact oriented manifolds—are the homology groups $H_k : \mathrm{Man}_n \to$ Groups for fixed $k \in \mathbb{N}_0$ or their rank, i.e., the Betti numbers $\beta_k : \mathrm{Man}_n \to \mathbb{Z}$. These are in fact topological—rather than smooth—invariants since homeomorphic—rather than just diffeomorphic—manifolds have isomorphic homology groups. The 0th Betti number β_0 is complete for $n = 0, 1$ since it determines the number of connected components, and there is only one compact, connected manifold of dimension 0 (the point) or 1 (the circle). The first more nontrivial complete invariant—now also restricting to connected manifolds—is the first Betti number $\beta_1 : \mathrm{Man}_2 \to \mathbb{Z}$, since compact, connected, oriented 2-manifolds are determined by their genus $g = \frac{1}{2}\beta_1$.

The fundamental group $\pi_1 : \mathrm{Man}_n \to \mathrm{Gr}$ is not strictly well-defined since it requires the choice of a base point and thus is a functor on the category of manifolds with a marked point. However, for connected manifolds it still induces a well defined map $|\pi_1| : |\mathrm{Man}_n| \to |\mathrm{Gr}|$ from manifolds modulo diffeomorphism to groups modulo isomorphism, since change of base point induces an isomorphism of fundamental groups. Viewing this as an invariant, it is complete for $n = 0, 1, 2$. In dimension $n = 3$, completeness would mean that the isomorphism type of the fundamental group of a (compact, connected) 3-manifold determines the 3-manifold up to diffeomorphism. This is true in the case of the trivial fundamental group: By the Poincaré conjecture, any simply connected 3-manifold "is the 3-sphere," i.e., is diffeomorphic to S^3. It is also true for a large class (irreducible, nonspherical) of 3-manifolds, but there are plenty of groups that can be represented by many non-diffeomorphic 3-manifolds, e.g., lens spaces and connected sums with them (see [1, 30] for surveys). Thus $|\pi_1|$ is a useful but incomplete invariant of closed, connected 3-manifolds. In dimension $n \geq 4$ however, the classification question should be posed for fixed $|\pi_1|$ since on the one hand any finitely presented group appears as the fundamental group of a closed, connected n-manifold, and on the other hand the classification of finitely presented groups is a wide open problem itself.[7]

Moreover, while in dimension $n \leq 3$, the classifications up to homeomorphism and up to diffeomorphism coincide (i.e., topological n-manifolds can be equipped with a unique smooth structure), these differ in dimensions $n \geq 4$. In dimension $n \geq 5$, both classifications can be undertaken with the help of surgery theory introduced by Milnor [49]. In dimension 4, the classification of smooth 4-manifolds differs drastically from that of topological manifolds (see [61] for a survey). Here, gauge theory—starting with the work of Donaldson, and continuing with Seiberg–Witten theory—is the main source of invariants which can differentiate between different smooth structures on the same topological manifold. In particular, Donaldson's first results using ASD Yang–Mills instantons [14] showed that a large number of topological manifolds (those with nondiagonalizable definite intersection form $H_2(X;\mathbb{Z}) \times H_2(X;\mathbb{Z}) \to \mathbb{Z}$) in fact do not support any smooth structure.

2 Floer Field Theory

2.1 Categories and Functors

Definition 2.1.1 A **category**[8] $\mathscr{C}$ consists of

- a set $\mathrm{Obj}_{\mathscr{C}}$ of **objects**,
- for each pair $x_1, x_2 \in \mathrm{Obj}_{\mathscr{C}}$ a set of **morphisms** $\mathrm{Mor}_{\mathscr{C}}(x_1, x_2)$,
- for each triple $x_1, x_2, x_3 \in \mathrm{Obj}_{\mathscr{C}}$ a **composition map**

$$\mathrm{Mor}_{\mathscr{C}}(x_1, x_2) \times \mathrm{Mor}_{\mathscr{C}}(x_2, x_3) \to \mathrm{Mor}_{\mathscr{C}}(x_1, x_3), \quad (f_{12}, f_{23}) \mapsto f_{12} \circ f_{23},$$

such that

- composition is *associative*, i.e., we have $(f_{12} \circ f_{23}) \circ f_{34} = f_{12} \circ (f_{23} \circ f_{34})$ for any triple of composable morphisms f_{12}, f_{23}, f_{34},
- composition has *identities*, i.e., for each $x \in \mathrm{Obj}_{\mathscr{C}}$ there exist a unique[9] morphism $\mathrm{id}_x \in \mathrm{Mor}_{\mathscr{C}}(x, x)$ such that $\mathrm{id}_x \circ f = f$ and $g \circ \mathrm{id}_x = g$ hold for any $f \in \mathrm{Mor}_{\mathscr{C}}(x, y)$ and $g \in \mathrm{Mor}_{\mathscr{C}}(y, x)$.

---

[7] As a matter of curiosity: The group isomorphism problem—determining whether different finite group presentations define isomorphic groups—is undecidable, i.e., cannot be solved for all general presentations by an algorithm; see, e.g., [32].

[8] Throughout, all categories are meant to be small, i.e., consist of sets of objects and morphisms. However, we will usually neglect to specify constructions in sufficient detail—e.g., require manifolds to be submanifolds of some $\mathbb{R}^N$—in order to obtain sets.

[9] Note that uniqueness follows immediately from the defining properties: If id'_x is another identity morphism then we have $\mathrm{id}'_x = \mathrm{id}'_x \circ \mathrm{id}_x = \mathrm{id}_x$.

The very first example of a category consists of objects which are sets (possibly with extra structure such as a linear structure, metric, or smooth manifold structure), morphisms that are maps (preserving the extra structure), composition given by composition of maps, and identities given by the identity maps. The following bordism categories contain more general morphisms, which are more rigorously constructed in Remark 3.2.2.

Example 2.1.2 The **bordism category $\mathbf{Bor_{d+1}}$** in dimension $d \geq 0$ is roughly defined as follows; see Fig. 1 for illustration.

- Objects are the closed, oriented, d-dimensional manifolds Σ.
- Morphisms in $\mathrm{Mor}(\Sigma_1, \Sigma_2)$ are the compact, oriented, $(d+1)$-dimensional cobordisms Y with identification of the boundary $\partial Y \simeq \Sigma_1^- \sqcup \Sigma_2$, modulo diffeomorphisms relative to the boundary.
- Composition of morphisms $[Y_{12}] \in \mathrm{Mor}(\Sigma_1, \Sigma_2)$ and $[Y_{23}] \in \mathrm{Mor}(\Sigma_2, \Sigma_3)$ is given by gluing $[Y_{12}] \circ [Y_{23}] := [Y_{12} \cup_{\Sigma_2} Y_{23}] \in \mathrm{Mor}(\Sigma_1, \Sigma_3)$ along the common boundary.

Here one needs to be careful to include the choice of boundary identifications in the notion of morphism. Thus, a diffeomorphism $\phi : \Sigma_0 \to \Sigma_1$ can be cast as a morphism

$$Z_\phi := \big[\big([0,1] \times \Sigma_1, \{0\} \times \phi, \{1\} \times \mathrm{id}_{\Sigma_1}\big)\big] \in \mathrm{Mor}_{\mathrm{Bor}_{d+1}}(\Sigma_0, \Sigma_1) \qquad (2.1.1)$$

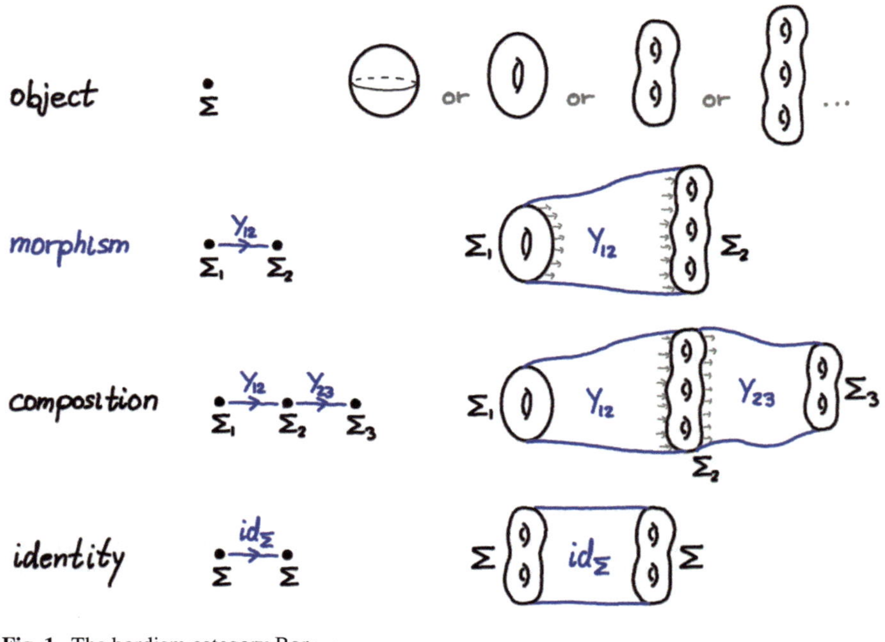

Fig. 1 The bordism category Bor_{2+1}

given by the cobordism $[0,1] \times \Sigma_1$ with boundary identifications $\{0\} \times \phi : \Sigma_0 \to \{0\} \times \Sigma_1$ and $\{1\} \times \mathrm{id}_{\Sigma_1} : \Sigma_1 \to \{1\} \times \Sigma_1$, as illustrated in Fig. 2. In that sense, the identity morphisms $\mathrm{id}_\Sigma = Z_{\mathrm{id}_\Sigma}$ are given by the identity maps $\mathrm{id}_\Sigma : \Sigma \to \Sigma$.

Equipping the composed morphism $[Y_{12}] \circ [Y_{23}]$ with a smooth structure moreover requires a choice of tubular neighborhoods of Σ_2 in the gluing operation. The good news is that gluing with respect to different choices yields diffeomorphic results, so that composition is well defined. The interesting news is that this ambiguity in the composition precludes the extension to a 2-category; see Example 3.2.3.

The notion of categories becomes most useful in the notion of a functor relating two categories, since preservation of various structures (composition and identities) can be expressed efficiently as "functoriality."

Definition 2.1.3 A **functor** $\mathscr{F} : \mathscr{C} \to \mathscr{D}$ between two categories $\mathscr{C}$, $\mathscr{D}$ consists of

- a map $\mathscr{F} : \mathrm{Obj}_\mathscr{C} \to \mathrm{Obj}_\mathscr{D}$ between the sets of objects,
- for each pair $x_1, x_2 \in \mathrm{Obj}_\mathscr{C}$ a map $\mathscr{F}_{x_1,x_2} : \mathrm{Mor}_\mathscr{C}(x_1, x_2) \to \mathrm{Mor}_\mathscr{D}(\mathscr{F}(x_1), \mathscr{F}(x_2))$,

that are compatible with identities and composition in the sense that

$$\mathrm{id}_{\mathscr{F}(x)} = \mathscr{F}_{x,x}(\mathrm{id}_x), \qquad \mathscr{F}_{x_1,x_3}(f_{12} \circ f_{23}) = \mathscr{F}_{x_1,x_2}(f_{12}) \circ \mathscr{F}_{x_2,x_3}(f_{23}).$$

For example, the inclusion of diffeomorphisms into the bordism category in Example 2.1.2 can be phrased as a functor as follows.

Example 2.1.4 Let Man_d be the category consisting of the same objects as Bor_{d+1}, morphisms given by diffeomorphisms, and composition given by composition of maps. Then, there is a functor $\mathrm{Man}_d \to \mathrm{Bor}_{d+1}$ given by

- the identity map between the sets of objects,
- for each pair Σ_0, Σ_1 of diffeomorphic d-manifolds the map $\mathrm{Mor}_{\mathrm{Man}_d}(\Sigma_0, \Sigma_1) \to \mathrm{Mor}_{\mathrm{Bor}_{d+1}}(\Sigma_0, \Sigma_1)$ that associates to a diffeomorphism ϕ the cobordism Z_ϕ defined in (2.1.1).

A more algebraic example of a category is given by categories and functors.

Example 2.1.5 The **category of categories** Cat consists of

- objects given by categories $\mathscr{C}$,
- morphisms in $\mathrm{Mor}_{\mathrm{Cat}}(\mathscr{C}_1, \mathscr{C}_2)$ given by functors $\mathscr{F}_{12} : \mathscr{C}_1 \to \mathscr{C}_2$,
- composition of morphisms given by composition of functors—i.e. composition of the maps on both object and morphism level.

2.2 The Symplectic Category

The vision of Alan Weinstein [82] was to construct a symplectic category along the following lines. (See [11, 47] for introductions to symplectic topology.)

- Objects are the symplectic manifolds $M := (M, \omega)$.
- Morphisms are the Lagrangian submanifolds[10] $L \subset M_1^- \times M_2$, where we denote by $M_1^- := (M_1, -\omega_1)$ the same manifold with reversed symplectic structure.
- Composition of morphisms $L_{12} \subset M_1^- \times M_2$ and $L_{23} \subset M_2^- \times M_3$ is defined by the **geometric composition** (where $\Delta_M \subset M \times M^-$ denotes the diagonal)

$$L_{12} \circ L_{23} := \mathrm{pr}_{M_1^- \times M_3}\big(L_{12} \times L_{23} \cap M_1^- \times \Delta_{M_2} \times M_3\big) \subset M_1^- \times M_3.$$

This notion includes symplectomorphisms $\phi : M_1 \to M_2, \phi^*\omega_2 = \omega_1$ as morphisms given by their graph $\mathrm{gr}(\phi) = \{(x, \phi(x)) \,|\, x \in M_1\} \subset M_1^- \times M_2$. Also, geometric composition is defined exactly so as to generalize the composition of maps. That is, we have $\mathrm{gr}(\phi) \circ \mathrm{gr}(\psi) = \mathrm{gr}(\psi \circ \phi)$. On the other hand, this more generalized notion allows one to view pretty much all constructions in symplectic topology as morphisms—for example, symplectic reduction from $\mathbb{CP}^2$ to $\mathbb{CP}^1$ is described by a Lagrangian 3-sphere $\Lambda \subset (\mathbb{CP}^2)^- \times \mathbb{CP}^1$; see [29, 75, 82] for details and more examples.

Unfortunately, geometric composition generally—even after allowing for perturbations (e.g., isotopy through Lagrangians)—at best yields immersed or multiply covered Lagrangians.[11] However, Floer homology[12] is at most expected to be invariant under **embedded geometric composition**, i.e., when the intersection in

$$\mathrm{pr}_{M_1^- \times M_3} : L_{12} \times_{M_2} L_{23} := L_{12} \times L_{23} \cap M_1^- \times \Delta_{M_2} \times M_3 \longrightarrow M_1^- \times M_3 \tag{2.2.1}$$

is transverse, and the projection is an embedding. In the linear case—for symplectic vector spaces and linear Lagrangian subspaces—this issue was resolved in [29] by observing that linear composition, even if not transverse, always yields another Lagrangian subspace. In higher generality, and compatible with Floer homology, a symplectic category $\mathrm{Symp} = \mathrm{Symp}^\# / \sim$ was constructed in [76] by the following general algebraic completion construction for a partially defined composition.

---

[10]Other terms for a Lagrangian, viewed as a morphism $M_1 \to M_2$, are "Lagrangian relation" or "Lagrangian correspondence," but we will largely avoid such distinctions in this paper.

[11]Even the question of finding a Lagrangian $L \subset \mathbb{CP}^2$ with embedded composition $L \circ \Lambda \subset \mathbb{CP}^1$ was open until the recent construction of a new Lagrangian embedding $\mathbb{RP}^2 \hookrightarrow L \subset \mathbb{CP}^2$ in [12].

[12]Floer homology is a central tool in symplectic topology introduced by Floer [23] in the 1980s, inspired by Gromov [28] and Witten [84]. It has been extended to a wealth of algebraic structures such as Fukaya categories; see, e.g., [63]. It can be thought of as the Morse homology of a symplectic action functional on the space of paths connecting two Lagrangians, and recasts the ill posed gradient flow ODE as a Cauchy–Riemann PDE (whose solutions are pseudoholomorphic curves).

Definition 2.2.1 The **extended symplectic category** $\mathrm{Symp}^{\#}$ is defined as follows.

- Objects are the symplectic manifolds (M, ω).
- Simple morphisms $L_{12} \in \mathrm{SMor}(M_1, M_2)$ are the Lagrangian submanifolds $L_{12} \subset M_1^- \times M_2$.
- General morphisms $\underline{L} = (L_{01}, \ldots, L_{(k-1)k}) \in \mathrm{Mor}_{\mathrm{Symp}^{\#}}(M, N)$ are the composable chains of simple morphisms $L_{ij} \in \mathrm{SMor}(M_i, M_j)$ between symplectic manifolds $M = M_0, M_1, \ldots, M_k = N$.
- Composition of morphisms $\underline{L} = (L_{01}, \ldots, L_{(k-1)k}) \in \mathrm{Mor}_{\mathrm{Symp}^{\#}}(M, N)$ and $\underline{L}' = (L'_{01}, \ldots, L'_{(k'-1)k'}) \in \mathrm{Mor}_{\mathrm{Symp}^{\#}}(N, P)$ is given by algebraic concatenation $\underline{L}\#\underline{L}' := (L_{01}, \ldots, L_{(k-1)k}, L'_{01}, \ldots, L'_{(k'-1)k'})$.

For this to form a strict category, we include trivial chains $(\,) \in \mathrm{Mor}(M, M)$ of length $k = 0$ as identity morphisms.

While this is a welldefined category, its composition notion is not related to geometric composition yet. However, the following quotient construction ensures that composition is given by geometric composition when the result is embedded.

Definition 2.2.2 The **symplectic category** Symp is defined as follows.

- Objects are the symplectic manifolds (M, ω).
- Morphisms are the equivalence classes in $\mathrm{Mor}_{\mathrm{Symp}}(M, N) := \mathrm{Mor}_{\mathrm{Symp}^{\#}}(M, N)/\sim$.
- Composition $[\underline{L}] \circ [\underline{L}'] := [\underline{L}\#\underline{L}']$ is induced by the composition in $\mathrm{Symp}^{\#}$.

Here, the composition-compatible equivalence relation $\sim$ on the morphism spaces of $\mathrm{Symp}^{\#}$ is obtained as follows.

- The subset of **geometric composition moves** $\mathrm{Comp} \subset \mathrm{Mor}_{\mathrm{Symp}^{\#}} \times \mathrm{Mor}_{\mathrm{Symp}^{\#}}$ consists of all pairs $\big((L_{12}, L_{23}), L_{12} \circ L_{23}\big)$ and $\big(L_{12} \circ L_{23}, (L_{12}, L_{23})\big)$ for which the geometric composition $L_{12} \circ L_{23}$ is embedded as in (2.2.1).
- The equivalence relation $\sim$ on $\mathrm{Mor}_{\mathrm{Symp}^{\#}}$ is defined by $\underline{L} \sim \underline{\tilde{L}}$ if there is a finite sequence of moves $\underline{L} \leadsto \underline{L}' \leadsto \underline{L}'' \ldots \leadsto \underline{L}^{(N)} = \underline{\tilde{L}}$ in which each move replaces one subchain of simple morphisms by another,

$$\begin{aligned} \underline{L}^{(k)} = \big(\ldots, L_{ij}, L_{jl}, \ldots\big) &\leadsto \underline{L}^{(k+1)} = \big(\ldots, L_{ij} \circ L_{jl}, \ldots\big) \\ \text{resp.}\;\; \underline{L}^{(k)} = \big(\ldots, L_{ij} \circ L_{jl}, \ldots\big) &\leadsto \underline{L}^{(k+1)} = \big(\ldots, L_{ij}, L_{jl}, \ldots\big) \end{aligned}$$

  according to a geometric composition move $\big((L_{ij}, L_{jl}), L_{ij} \circ L_{jl}\big) \in \mathrm{Comp}$ resp. $\big(L_{ij} \circ L_{jl}, (L_{ij}, L_{jl})\big) \in \mathrm{Comp}$.

The result of this quotient construction is that the composition of morphisms is given by geometric composition $[L_{12}] \circ [L_{23}] = [L_{12} \circ L_{23}]$ if the latter is embedded. We will later recast this construction in terms of an extension of the symplectic category $\mathrm{Symp}^{\#}$ to a 2-category in which the equivalence relation $\sim$ is obtained from 2-isomorphisms; see Example 3.1.6 and Sect. 3.5.

Remark 2.2.3 The present equivalence relation does not identify a Lagrangian $L \subset M^- \times N$ with its image $\phi_H(L) \subset M^- \times N$ under a Hamiltonian symplectomorphism ϕ_H. Indeed, any morphism $\underline{L}$ in $\mathrm{Mor}_{\mathrm{Symp}^\#}(M, N)$ induces a (Lagrangian where immersed) subset of $M^- \times N$ by complete geometric composition, and this subset is invariant under geometric composition moves. However, such equivalences under Hamiltonian deformation can also be cast as 2-isomorphisms; see Example 3.5.1.

2.3 Categories with Cerf Decompositions

The basic idea of Cerf decompositions is to decompose a $(d+1)$-manifold $Y = Y_{01} \cup_{\Sigma_1} Y_{12} \ldots \cup_{\Sigma_{k-1}} Y_{(k-1)k}$ into simpler pieces $Y_{ij} = f^{-1}([b_i, b_j])$ by cutting at regular level sets $\Sigma_i = f^{-1}(b_i)$ of a Morse function $Y \to \mathbb{R}$ as illustrated in Fig. 4 below. By viewing Y as a cobordism between empty sets, i.e., as a morphism in $\mathrm{Mor}_{\mathrm{Bor}_{d+1}}(\emptyset, \emptyset)$, this can be seen as a factorization $[Y] = [Y_{01}] \circ [Y_{12}] \circ \ldots \circ [Y_{(k-1)k}]$ in Bor_{d+1}. Here, the Morse function f and regular levels b_i can be chosen such that each piece $Y_{i(i+1)}$ contains either none or one critical point, and thus is either a cylindrical cobordism—diffeomorphic to the product cobordism $Z_\phi = [0, 1] \times \Sigma_j$ as in (2.1.1)—or a handle attachment as in the following remark. These "simple cobordisms" are illustrated in Figs. 2 and 3.

Remark 2.3.1 A **k-handle attachment** Y_α of index $0 \leq k \leq d+1$ is a $(d+1)$-dimensional cobordism, which is obtained by attaching to a cylinder $[0, 1] \times \Sigma$ a handle $B^k \times B^{d+1-k}$ along an attaching cycle $S^{k-1} \hookrightarrow \alpha \subset \{1\} \times \Sigma$, as illustrated in Fig. 3. Here, B^k denotes a k-dimensional ball with boundary $\partial B^k = S^{k-1}$.

By reversing the orientation and boundary identifications of any k-handle attachment Y_α from Σ to Σ', we obtain a cobordism Y_α^- from Σ' to Σ. This reversed cobordism is also a $d+1-k$-handle attachment $Y_\alpha^- = Y_{\alpha^*}$ for an attaching cycle $S^{d-k} \hookrightarrow \alpha^* \subset \{1\} \times \Sigma'$. It moreover is the adjoint of Y_α in the sense of Remark 2.4.3 and will become useful in the formulation of Cerf moves below.

Specifying to dimension $d = 2$ and the connected bordism category, it will suffice to consider 2-handle attachments (and their adjoints) with attaching circles

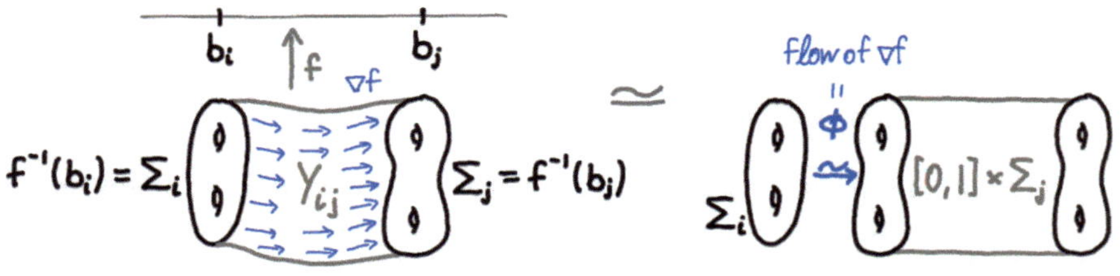

Fig. 2 A cylindrical cobordism supports a Morse function without critical points, whose gradient flow induces a diffeomorphism to the product cobordism $[0, 1] \times \Sigma_j$ with natural identification $\Sigma_j \cong \{1\} \times \Sigma_j$ and boundary identification $\phi : \Sigma_i \xrightarrow{\sim} \{0\} \times \Sigma_j$ arising from the flow as well

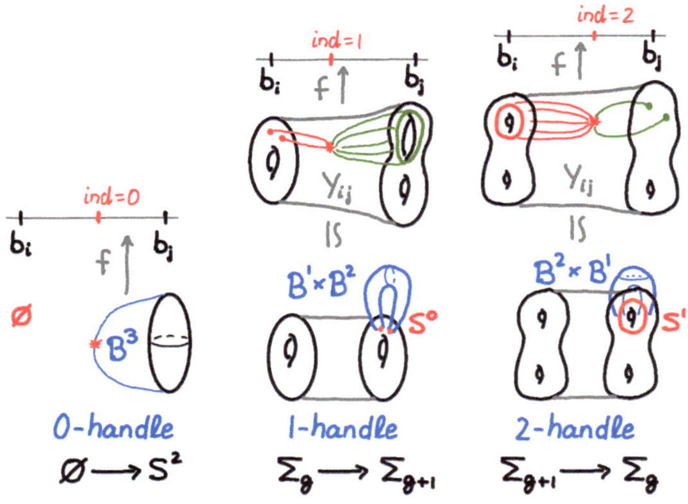

Fig. 3 Handle attachments in dimension $d = 2$ are "simple cobordisms" which support a Morse function f with a single critical point of index $0 \leq k \leq 3$. The attaching cycles are given by intersection of the unstable manifold (in *red*) with the boundary. Index $k = 0$ and $k = 3$ handle attachments are adjoint via orientation reversal and only appear between the empty set and sphere S^2. Index $k = 1$ and $k = 2$ handle attachments are adjoint via orientation reversal (which interchanges unstable and stable manifolds) and appear between surfaces Σ_g, Σ_{g+1} of adjacent genus $g, g+1 \in \mathbb{N}_0$

that are homologically nontrivial and thus do not disconnect the surface. More precisely, any attaching circle $S^1 \simeq \alpha \subset \Sigma$ in a closed surface Σ determines a 2-handle attachment as follows: Replacing an annulus neighborhood of α by two disks specifies a lower genus surface $\Sigma' = \Sigma_\alpha$ together with a diffeomorphism $\pi_\alpha : \Sigma \setminus \alpha \to \Sigma' \setminus \{2 \text{ points}\}$. Given this construction, the 2-handle attaching cobordism Y_α from Σ to Σ' is unique up to diffeomorphism fixing the boundary.

More detailed introductions to Cerf theory can be found in e.g. [13, 27, 50]. Here, we concentrate on the algebraic structure that it equips the bordism categories with. To describe this structure, we may think of Cerf decompositions as a prime decomposition of $(d+1)$-manifolds, and more generally of $(d+1)$-cobordisms: A decomposition into simple cobordisms (cylindrical cobordisms and handle attachments) always exist and simple cobordisms have no further simplifying decomposition. And while these Cerf decompositions are not unique, any two choices of decomposition are related via just a few moves, some of which are shown in Fig. 4. These moves reflect changes in the Morse function (critical point cancelations and critical point switches), cutting levels (cylinder cancellation), and the ways in which pieces are glued together (diffeomorphism equivalences which in particular encode handle slides). All of these Cerf moves are local in the sense[13] that they replace only

[13] While a diffeomorphism equivalence is not local, it decomposes into a sequence of local moves.

one or two consecutive cobordisms by one or two consecutive cobordisms with the same composition. That is, the moves are of one of three forms:

$$\begin{aligned}
\ldots \cup_{\Sigma_i} Y_{ij} \cup_{\Sigma_j} Y_{jl} \cup_{\Sigma_l} \ldots &= \ldots \cup_{\Sigma_1} \tilde{Y}_{ij} \cup_{\tilde{\Sigma}_j} \tilde{Y}_{jl} \cup_{\Sigma_l} \ldots && \text{for } Y_{ij} \cup_{\Sigma_j} Y_{jl} = \tilde{Y}_{ij} \cup_{\tilde{\Sigma}_j} \tilde{Y}_{jl}, \\
\ldots \cup_{\Sigma_1} Y_{ij} \cup_{\Sigma_j} Y_{jl} \cup_{\Sigma_l} \ldots &= \ldots \cup_{\Sigma_1} \tilde{Y}_{il} \cup_{\Sigma_l} \ldots && \text{for } Y_{ij} \cup_{\Sigma_j} Y_{jl} = \tilde{Y}_{il}, \\
\ldots \cup_{\Sigma_1} Y_{il} \cup_{\Sigma_l} \ldots &= \ldots \cup_{\Sigma_1} \tilde{Y}_{ij} \cup_{\tilde{\Sigma}_j} \tilde{Y}_{jl} \cup_{\Sigma_l} \ldots && \text{for } Y_{il} = \tilde{Y}_{ij} \cup_{\tilde{\Sigma}_j} \tilde{Y}_{jl}.
\end{aligned}$$

In the following, we will cast this notion—decompositions into simple pieces that are unique up to a set of moves—into more formal terms. For that purpose, we denote the union of all morphisms of a category $\mathscr{C}$ by $\mathrm{Mor}_{\mathscr{C}} := \bigcup_{x_1, x_2 \in \mathrm{Obj}_{\mathscr{C}}} \mathrm{Mor}_{\mathscr{C}}(x_1, x_2)$, and we denote all relations between composable chains[14] of morphisms by

$$\mathrm{Rel}_{\mathscr{C}} := \bigcup_{k, \ell \in \mathbb{N}} \big\{ \big((f_i), (g_j) \big) \in (\mathrm{Mor}_{\mathscr{C}})^k \times (\mathrm{Mor}_{\mathscr{C}})^\ell \, \big| f_1 \circ \ldots \circ f_k = g_1 \circ \ldots \circ g_\ell \big\}.$$

Definition 2.3.2 A **category with Cerf decompositions** is a category $\mathscr{C}$ together with

- a subset $\mathrm{SMor} \subset \mathrm{Mor}_{\mathscr{C}}$ of **simple morphisms,**
- a subset $\mathrm{Cerf} \subset \mathrm{Rel}_{\mathscr{C}}$ of **local Cerf moves,** which is symmetric (under exchanging the factors) and consists of pairs of composable chains of simple morphisms $f_{12}, \ldots, f_{(k-1)k} \in \mathrm{SMor}$, $g_{12}, \ldots, g_{(\ell-1)\ell} \in \mathrm{SMor}$ whose compositions are equal,

such that

- the simple morphisms generate all morphisms, i.e., for any $m \in \mathrm{Mor}_{\mathscr{C}}$ there exist $h_{12}, \ldots, h_{(n-1)n} \in \mathrm{SMor}$ such that $m = h_{12} \circ \ldots \circ h_{(n-1)n}$,
- the presentation in terms of simple morphisms is unique up to **Cerf moves**, i.e., any two presentations of the same morphism in terms of $h_{12}, \ldots, h_{(n-1)n} \in \mathrm{SMor}$ and $\tilde{h}_{12}, \ldots, \tilde{h}_{(\tilde{n}-1)\tilde{n}} \in \mathrm{SMor}$ are related by a finite sequence[15] of identities

  $$h_{12} \circ \ldots \circ h_{(n-1)n} = h'_{12} \circ \ldots \circ h'_{(n'-1)n'} = \ldots = \tilde{h}_{12} \circ \ldots \circ \tilde{h}_{(\tilde{n}-1)\tilde{n}}$$

  in which each equality replaces one subchain of simple morphisms by another,

  $$\ldots \circ f_{12} \circ \ldots \circ f_{(k-1)k} \circ \ldots = \ldots \circ g_{12} \circ \ldots \circ g_{(\ell-1)\ell} \circ \ldots$$

  according to a local Cerf move $\big((f_{12}, \ldots, f_{(k-1)k}), (g_{12}, \ldots, g_{(\ell-1)\ell}) \big) \in \mathrm{Cerf}$.

The bordism categories Bor_{d+1} are the motivating example of categories with Cerf decompositions, with SMor and Cerf given by the simple cobordisms and Cerf

[14]Throughout, we will use the term "composable chain" to denote ordered tuples of morphisms, in which each consecutive pair is composable, so that the entire tuple—by associativity of composition—has a well defined composition.

[15]Throughout, we will use the term "sequence" to denote a finite totally ordered set.

moves as discussed above (for a more detailed exposition see [27]). However, in the examples arising from gauge theory, we consider the $2+1$-dimensional connected bordism category, the $d=2$ case of the following general notion for $d \geq 2$.[16]

Example 2.3.3 The **connected bordism category** $\mathbf{Bor}^{\mathbf{conn}}_{\mathbf{d+1}}$ is defined as follows.

- Objects are the closed, connected, oriented d-dimensional manifolds.
- Morphisms are the compact, connected, oriented $d+1$-dimensional cobordisms with identification of the boundary, and modulo diffeomorphisms as in Bor_{d+1}.
- Composition is by gluing via boundary identifications as in Bor_{d+1}.

If we allow $\Sigma = \emptyset$ as object, then closed, connected, oriented $d+1$-manifolds are contained in this category as morphisms from $\emptyset$ to $\emptyset$.

In this language, the Cerf decomposition theorem for 3-manifolds—in the connected case proven in [26] and reviewed in [27]—can be stated as in the following theorem, and is illustrated in Fig. 4 and further explained in Remark 2.3.5. Here, in strict categorical language, a 3-cobordism from Σ_- to Σ_+ is an equivalence class $[(Y, \iota^-, \iota^+)]$ of 3-cobordisms and embeddings $\iota^\pm : \Sigma_\pm \to \partial Y$ modulo diffeomorphisms relative to the boundary identifications $\iota^\pm$. However, the decomposition and boundary identifications are actually induced by a decomposition of representatives, and thus, we drop the brackets and embeddings—see [27] and Sect. 3.2 for more deliberations on this. Moreover, we may again generalize to dimension $d \geq 2$.

Theorem 2.3.4 $\mathrm{Bor}^{\mathrm{conn}}_{d+1}$ *is a category with Cerf decompositions as follows.*

- *The set of simple morphisms* $\mathrm{SMor} \subset \mathrm{Mor}_{\mathrm{Bor}^{\mathrm{conn}}_{d+1}}$ *consists of*
  - *cylindrical cobordisms Z_ϕ for diffeomorphisms $\phi : \Sigma \to \Sigma'$ as in* (2.1.1),
  - *k-handle attachments $Y_\alpha \in \mathrm{Mor}(\Sigma, \Sigma')$ for $1 \leq k \leq d$ as in Remark 2.3.1.*
- *The Cerf moves* $\mathrm{Cerf} \subset \mathrm{Rel}_{\mathrm{Bor}^{\mathrm{conn}}_{d+1}}$ *are the following and their transpositions:*
  - *Cylinder cancellations $\big((Z_\phi, Z_\psi), Z_{\psi\circ\phi}\big)$ for all composable pairs of diffeomorphisms ϕ, ψ.*
  - *Cylinder cancellations $\big((Z_\phi, Y), Y'\big)$ resp. $\big((Y, Z_\phi), Y'\big)$ in which Y' is the same cobordism as Y (up to diffeomorphism), but with incoming resp. outgoing boundary inclusion pre- resp. post-composed with a diffeomorphism ϕ.*
  - *Critical point cancellations $\big((Y_\alpha^-, Y_\beta), Z_\phi\big)$ occur for attaching cycles $\alpha, \beta \subset \Sigma$ with transverse intersection in a single point; these give rise to a pair of cobordisms $Y_\alpha^- \in \mathrm{Mor}(\Sigma', \Sigma)$, $Y_\beta \in \mathrm{Mor}(\Sigma, \Sigma'')$ whose composition is a cylindrical cobordism representing a diffeomorphism $\phi : \Sigma' \to \Sigma''$.*
  - *Critical point switches $\big((Y_\alpha, Y'_\beta), (Y_\beta, Y'_\alpha)\big)$ and $\big((Y_\alpha^-, Y_\beta), (Y'_\beta, Y'^-_\alpha)\big)$ occur for disjoint attaching cycles $\alpha, \beta \subset \Sigma$; these give rise to a pair of cobordisms*[17]

[16] We restrict to dimension $d \geq 2$ when discussing connected bordisms since the handle attachments in dimension $d = 1$ are morphisms between generally disconnected 1-manifolds, so that $\mathrm{Bor}^{\mathrm{conn}}_{1+1}$ does not have useful connected Cerf decompositions.

[17] See Remark 2.5.1 for more details on the notation used here.

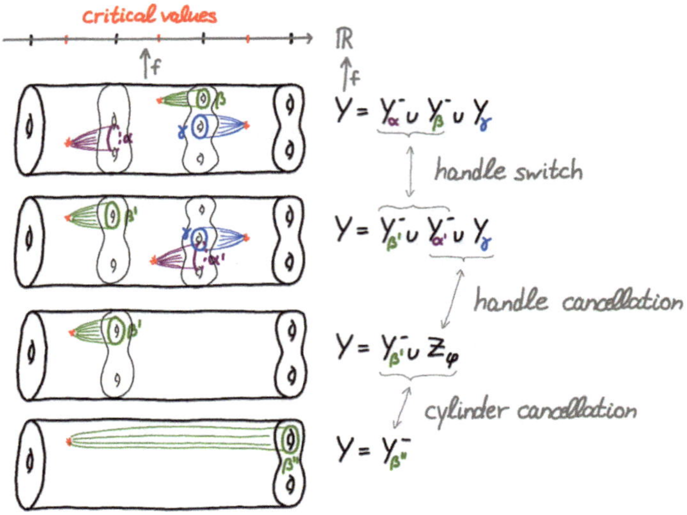

Fig. 4 Cerf decompositions of a 3-cobordism Y and Cerf moves between them

$Y_\alpha \in \mathrm{Mor}(\Sigma, \Sigma'_\alpha)$, $Y'_\beta \in \mathrm{Mor}(\Sigma'_\alpha, \Sigma'')$ *whose composition is the same as that of the pair* $Y_\beta \in \mathrm{Mor}(\Sigma, \Sigma'_\beta)$, $Y'_\alpha \in \mathrm{Mor}(\Sigma_\beta,' \Sigma'')$.

Remark 2.3.5 For $d = 2$ the objects of $\mathrm{Mor}_{\mathrm{Bor}^{\mathrm{conn}}_{2+1}}$—closed, connected, oriented surfaces—can be classified up to diffeomorphism by their genus. Moreover, the simple morphisms $\mathrm{SMor} \subset \mathrm{Mor}_{\mathrm{Bor}^{\mathrm{conn}}_{2+1}}$ can be further specified:

- Cylindrical cobordisms Z_ϕ represent diffeomorphisms $\phi : \Sigma \to \Sigma'$ between surfaces of the same genus as in (2.1.1).
- 2-Handle attachments $Y_\alpha \in \mathrm{Mor}(\Sigma, \Sigma')$ specified by a homologically nontrivial circle $S^1 \simeq \alpha \subset \Sigma$ are simple morphisms[18] from a surface Σ of genus g to a surface Σ' of genus $g - 1$.
- 1-Handle attachments are 2-handle attachments with reversed orientation, i.e., the simple morphisms $Y_\alpha^- \in \mathrm{Mor}(\Sigma', \Sigma)$ from a surface Σ' of genus $g - 1$ to a surface Σ of genus g.

The structural similarities between the symplectic and bordism categories can now be phrased in terms of abstract Cerf decompositions.

Lemma 2.3.6 *The symplectic category* Symp *from Definition 2.2.2 is a category with Cerf decompositions as follows:*

- *The set of simple morphisms* $\mathrm{SMor} \subset \mathrm{Mor}_{\mathrm{Symp}}$ *consists of the equivalence classes* $[L_{12}]$ *of Lagrangian submanifolds* $L_{12} \subset M_1^- \times M_2$.

[18]More precisely, Y_α is obtained by attaching to the cylindrical cobordism $[0, 1] \times \Sigma$ a 2-handle $B^2 \times [-\varepsilon, \varepsilon]$ along a thickening $[-\varepsilon, \varepsilon] \times S^1 \subset \{1\} \times \Sigma$ of the attaching circle.

- *The set of local Cerf moves* $\mathrm{Cerf} \subset \mathrm{Rel}_{\mathrm{Symp}}$ *consists of the relations*

$$\big([(L_{01}, L_{12})]\,,\,[L_{01} \circ L_{12}]\big) \quad \text{and} \quad \big([L_{01} \circ L_{12}]\,,\,[(L_{01}, L_{12})]\big)$$

for embedded geometric compositions $L_{01} \circ L_{12}$ *as in* (2.2.1).

Proof To check that the simple morphisms generate all morphisms, consider a general morphism $\underline{L} \in \mathrm{Mor}_{\mathrm{Symp}}(M, N)$ and pick a representative $(L_{01}, \ldots, L_{(k-1)k})$, given by a composable chain of Lagrangian submanifolds $L_{ij} \subset M_i^- \times M_j$ from $M_0 = M$ to $M_k = N$. The definition of composition in Symp yields the identity

$$\underline{L} = \big[(L_{01}, \ldots, L_{(k-1)k})\big] = \big[L_{01}\# \ldots \#L_{(k-1)k}\big] = [L_{01}] \circ \ldots \circ [L_{(k-1)k}].$$

Since each $[L_{ij}]$ is a simple morphism, this is the required decomposition of $\underline{L}$ into simple morphisms. To show that these decompositions are unique up to the given Cerf moves, note that an equality

$$[L_{01}] \circ \ldots \circ [L_{(k-1)k}] = [L'_{01}] \circ \ldots \circ [L'_{(k'-1)k'}]$$

in $\mathrm{Mor}_{\mathrm{Symp}}$ means by definition that the corresponding morphisms in $\mathrm{Symp}^{\#}$ are equivalent

$$(L_{01}, \ldots, L_{(k-1)k}) \sim (L'_{01}, \ldots, L'_{(k'-1)k'})$$

under the equivalence relation $\sim$ given in Definition 2.2.2. Recall that this relation is generated by the geometric composition moves $\mathrm{Comp} \subset \mathrm{Mor}_{\mathrm{Symp}^{\#}} \times \mathrm{Mor}_{\mathrm{Symp}^{\#}}$, so that there is a sequence of moves from $(L_{01}, \ldots, L_{(k-1)k})$ to $(L'_{01}, \ldots, L'_{(k'-1)k'})$ in which adjacent pairs are replaced by their embedded geometric composition. Our definition of $\mathrm{Cerf} \subset \mathrm{Rel}_{\mathrm{Symp}}$ by moves on equivalence classes encoded by Comp translates this into a sequence of Cerf moves from $[L_{01}] \circ \ldots [L_{(k-1)k}]$ to $[L'_{01}] \circ \ldots [L'_{(k'-1)k'}]$. □

2.4 Construction Principle for Floer Field Theories

The algebraic background of Floer field theory is the following construction principle for functors between categories with Cerf decompositions.

Lemma 2.4.1 *Let* $\mathscr{C}, \mathscr{D}$ *be two categories with Cerf decompositions and a* **Cerf-compatible partial functor** $\mathscr{F} : (\mathrm{Obj}_{\mathscr{C}}, \mathrm{SMor}_{\mathscr{C}}) \to (\mathrm{Obj}_{\mathscr{D}}, \mathrm{SMor}_{\mathscr{D}})$ *consisting of*

- *a map* $\mathrm{Obj}_{\mathscr{C}} \to \mathrm{Obj}_{\mathscr{D}}$,
- *a map* $\mathrm{SMor}_{\mathscr{C}} \to \mathrm{SMor}_{\mathscr{D}}$ *which induces a map* $\mathrm{Cerf}_{\mathscr{C}} \to \mathrm{Cerf}_{\mathscr{D}}$ *given by*

$$\big((f_{(i-1)i})_{i=1,\ldots,k}, (g_{(j-1)j})_{j=1,\ldots,\ell}\big) \mapsto \big((\mathscr{F}(f_{(i-1)i}))_{i=1,\ldots,k}, (\mathscr{F}(g_{(j-1)j}))_{j=1,\ldots,\ell}\big).$$

Then $\mathscr{F}$ *has a unique extension to a functor* $\overline{\mathscr{F}} : \mathscr{C} \to \mathscr{D}$ *which restricts to* $\mathscr{F}$ *on* $\mathrm{Obj}_{\mathscr{C}}$ *and* $\mathrm{SMor}_{\mathscr{C}} \subset \mathrm{Mor}_{\mathscr{C}}$.

Proof Compatibility of $\overline{\mathscr{F}}$ with composition requires its value on a general morphism $f \in \mathrm{Mor}_{\mathscr{C}}$ to be $\overline{\mathscr{F}}(f) = \mathscr{F}(f_{01}) \circ \ldots \circ \mathscr{F}(f_{(k-1)k})$ for any Cerf decomposition $f = f_{01} \circ \ldots \circ f_{(k-1)k}$ into simple morphisms $f_{ij} \in \mathrm{SMor}_{\mathscr{C}}$. The induced map $\mathrm{Cerf}_{\mathscr{C}} \to \mathrm{Cerf}_{\mathscr{D}}$ guarantees that this definition of $\overline{\mathscr{F}}(f)$ is independent of the choice of decomposition, thus yields a well defined map $\mathrm{Mor}_{\mathscr{C}} \to \mathrm{Mor}_{\mathscr{D}}$. Moreover, this map is compatible with composition by construction. Thus, a well-defined functor $\overline{\mathscr{F}}$ is uniquely determined by $\mathscr{F}$. □

The next Lemma specializes this abstract construction principles to $\mathscr{C} = \mathrm{Bor}^{\mathrm{conn}}_{d+1}$ and $\mathscr{D} = \mathrm{Symp}$ and is illustrated in Fig. 5. It can be read in two ways: In the strictly categorical sense, a partial functor should assign to a class $[Y]$ of simple cobordisms modulo diffeomorphisms relative to the boundary identifications a class $[L_Y]$ of Lagrangian submanifolds modulo embedded geometric composition. In practice, this will be achieved by assigning to each simple cobordism Y a Lagrangian submanifold L_Y in a way that is compatible with diffeomorphisms.

Strictly speaking, the following is a mild generalization of Lemma 2.4.1 because critical point switches really correspond to two Cerf moves in Symp, for example,

$$Y_{01} \cup_{\Sigma_1} Y_{12} \simeq Z'_{01} \cup_{\Sigma'_1} Y'_{12} \quad \Longrightarrow \quad (L_{Y_{01}}, L_{Y_{12}}) \sim L_{Y_{01}} \circ L_{Y_{12}} = L_{Y'_{01}} \circ L_{Y'_{12}} \sim (L_{Y'_{01}}, L_{Y'_{12}}).$$

Examples of Floer field theories constructed in this way will be discussed in Sect. 2.5.

Lemma 2.4.2 *Let* $\mathscr{F} : (\mathrm{Obj}_{\mathrm{Bor}^{\mathrm{conn}}_{d+1}}, \mathrm{SMor}_{\mathrm{Bor}^{\mathrm{conn}}_{d+1}}) \to (\mathrm{Obj}_{\mathrm{Symp}}, \mathrm{SMor}_{\mathrm{Symp}})$ *be a* Cerf-*compatible partial functor consisting of the following:*

- *Symplectic manifolds* M_Σ *for each d-manifold* $\Sigma \in \mathrm{Obj}_{\mathrm{Bor}^{\mathrm{conn}}_{d+1}}$;
- *Lagrangian submanifolds* $L_{[Y]} \in \mathrm{SMor}_{\mathrm{Symp}}(M_\Sigma, M_{\Sigma'})$ *for each simple* $d+1$-*cobordism* $[Y] \in \mathrm{SMor}_{\mathrm{Bor}^{\mathrm{conn}}_{d+1}}(\Sigma, \Sigma')$;
  (*More precisely, this requires the following:*

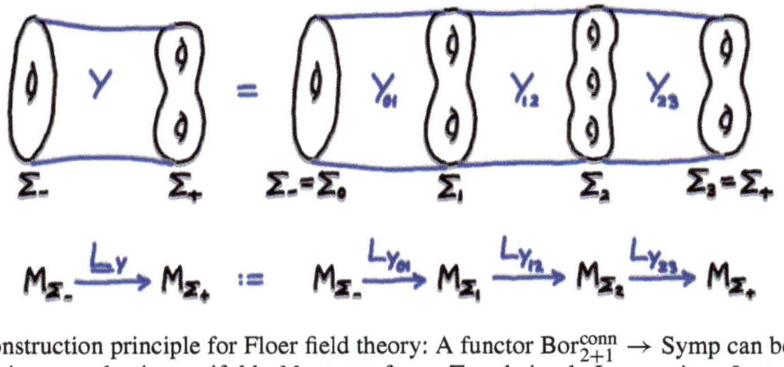

Fig. 5 Construction principle for Floer field theory: A functor $\mathrm{Bor}^{\mathrm{conn}}_{2+1} \to \mathrm{Symp}$ can be specified by associating symplectic manifolds M_Σ to surfaces Σ and simple Lagrangians L_Y to simple 3-cobordisms Y in a way that is compatible with Cerf moves

- *Lagrangian submanifolds* $L_Y \subset M^-_{\partial^- Y} \times M_{\partial^+ Y}$ *for each handle attachment* Y *with partitioned boundary* $\partial Y = \partial^- Y \sqcup \partial^+ Y$,
- *symplectomorphisms* $M_\Sigma \to M_{\Sigma'}$ *denoted by their graphs* $L_\phi \subset M^-_\Sigma \times M_{\Sigma'}$ *for the cylindrical cobordisms* Z_ϕ *representing each diffeomorphism* $\phi : \Sigma \to \Sigma'$,
- *identities*[19] $L_{\psi\circ\phi} = L_\phi \circ L_\psi$ *for diffeomorphisms* $\phi : \Sigma \to \Sigma'$, $\psi : \Sigma' \to \Sigma''$ *and* $L_{\Psi(Y)} = L_{\Psi|^{-1}_{\partial^- Y}} \circ L_Y \circ L_{\Psi|_{\partial^+ Y}}$ *for* $\Psi : Y \to Z$.

Then any choice of a representative cobordism Y *with orientation preserving diffeomorphisms* $\iota^- : \Sigma^- \to \partial^- Y$, $\iota^+ : \Sigma' \to \partial^+ Y$ *induces well-defined morphism*

$$L_{[Y]} := \left[\left(L_{\iota^-}, L_Y, L_{(\iota^+)^{-1}}\right)\right] = \left[\, \left(L_{\iota^-}^{-1} \times L_{\iota^+}^{-1}\right)(L_Y) \subset M^-_\Sigma \times M_{\Sigma'} \,\right],$$

where in the last equality we view $L_{\iota^\pm} : M_{\Sigma^\pm} \to M_{\partial^\pm Y}$ *as maps.*)

- *identities of Lagrangians for each local Cerf move*

$$\begin{aligned} \big((X, Y), Z\big) \in \mathrm{Cerf}_{\mathrm{Bor}^{\mathrm{conn}}_{d+1}} \quad &\Rightarrow \quad L_X \circ L_Y = L_Z, \\ \big(X, (Y, Z)\big) \in \mathrm{Cerf}_{\mathrm{Bor}^{\mathrm{conn}}_{d+1}} \quad &\Rightarrow \quad L_X = L_Y \circ L_Z, \\ \big((V, W), (X, Y)\big) \in \mathrm{Cerf}_{\mathrm{Bor}^{\mathrm{conn}}_{d+1}} \quad &\Rightarrow \quad L_V \circ L_W = L_X \circ L_Y, \end{aligned}$$

where all geometric compositions on the right-hand side are embedded as in (2.2.1).

Then $\mathcal{F}$ *has a unique extension to a functor* $\overline{\mathcal{F}} : \mathrm{Bor}^{\mathrm{conn}}_{d+1} \to \mathrm{Symp}$.

Moreover, if $\mathcal{F}$ *takes values in an exact or monotone symplectic category* Symp^τ *(see Remark 3.5.4), then* $\mathcal{F}$ *induces a functor* $\mathrm{Bor}^{\mathrm{conn}}_{d+1} \to \mathrm{Cat}$.

Proof To check that the construction of simple morphisms in the second bullet point is well-defined we need to consider a diffeormorphism $\Psi : Y \to Z$ which preserves the partition of boundary components, i.e., $\Psi_{\partial^\pm Y}$ maps $\partial^\pm Y$ to $\partial^\pm Z$. Then Y and $Z = \Psi(Y)$ with the corresponding boundary identifications yields the same Lagrangian submanifold $L_{[\Psi(Y)]} = L_{[Y]}$ since we have

$$\begin{aligned} &\left(L_{\Psi|_{\partial^- Y}\circ\iota^-}\,,\; L_{\Psi(Y)}\,,\; L_{(\Psi|_{\partial^+ Y}\circ\iota^+)^{-1}}\right) \\ &= \left(L_{\iota^-} \circ L_{\Psi|_{\partial^- Y}}\,,\; L_{\Psi|^{-1}_{\partial^- Y}} \circ L_Y \circ L_{\Psi|_{\partial^+ Y}}\,,\; L_{(\Psi|_{\partial^+ Y})^{-1}} \circ L_{(\iota^+)^{-1}}\right) \\ &\sim \left(L_{\iota^-} \circ L_{\Psi|_{\partial^- Y}} \circ L_{(\Psi|_{\partial^- Y})^{-1}}\,,\; L_Y\,,\; L_{\Psi|_{\partial^+ Y}} \circ L_{(\Psi|_{\partial^+ Y})^{-1}} \circ L_{(\iota^+)^{-1}}\right) \\ &= \left(L_{(\Psi|_{\partial^- Y})^{-1}\circ\Psi|_{\partial^- Y}\circ\iota^-}\,,\; L_Y\,,\; L_{(\iota^+)^{-1}\circ(\Psi|_{\partial^+ Y})^{-1}\circ\Psi|_{\partial^+ Y}}\right) = \left(L_{\iota^-}\,,\; L_Y\,,\; L_{(\iota^+)^{-1}}\right). \end{aligned}$$

Now on objects Σ, the functor $\overline{\mathcal{F}}$ is determined by the symplectic manifolds M_Σ. For a morphism $[Y] \in \mathrm{Bor}^{\mathrm{conn}}_{d+1}(\Sigma, \Sigma')$, pick a representative cobordism Y with orientation preserving embeddings $\iota^-_Y : \Sigma^- \to \partial Y$, $\iota^+_Y : \Sigma' \to \partial Y$ to the respective boundary components. By the Cerf decomposition Theorem 2.3.4, there exist a decomposition $Y = Y_{01} \cup_{\Sigma_1} Y_{12} \cup \ldots \cup_{\Sigma_{n-1}} Y_{(n-1)n}$ into simple morphisms which are either

[19] In these identities L_ϕ is the graph of a map so that $\circ$ is geometric composition of Lagrangians. Viewing L_ϕ as map, they could be rewritten as $L_{\psi\circ\phi} = L_\psi \circ L_\phi$ and $L_{\Psi(Y)} = \left(L_{\Psi|_{\partial^- Y}} \times L_{\Psi|_{\partial^+ Y}}\right)(L_Y)$.

handle attachments $Y_{i(i+1)}$ with boundary identifications $\iota_i^- : \Sigma_i^- \to \partial Y_{i(i+1)}$, $\iota_{i+1}^+ : \Sigma_{i+1} \to \partial Y_{i(i+1)}$ or cylindrical cobordisms $Y_{i(i+1)} = Z_{\phi_i}$ representing a diffeomorphism $\phi_i : \Sigma_i \to \Sigma_{i+1}$. As in Lemma 2.4.1, functoriality then requires

$$\overline{\mathscr{F}}([Y]) = \big[\big(L_{\iota_Y^-}, L_{Y_{01}}, L_{Y_{12}}, \ldots, L_{Y_{(n-1)n}}, L_{(\iota_Y^+)^{-1}}\big)\big]$$

to be given by the algebraic composition in Symp of the corresponding Lagrangian submanifolds. This fully determines $\overline{\mathscr{F}}$, but to see that it is well defined we need to consider not just another Cerf decomposition of Y—for which the proof is exactly as in Lemma 2.4.1—but also allow for a diffeomorphism $\Psi : Y \to Z$ that intertwines boundary identifications, $\Psi \circ \iota_Y^\pm = \iota_Z^\pm$. The latter induces a Cerf decomposition $Z = \Psi(Y_{01}) \cup_{\Psi(\Sigma_1)} \Psi(Y_{12}) \cup \ldots \cup_{\Psi(\Sigma_{n-1})} \Psi(Y_{(n-1)n})$ with $\Sigma_i := Y_{(i-1)i} \cap Y_{i(i+1)} \subset Y$, whose value under $\overline{\mathscr{F}}$ is

$$\begin{aligned}\overline{\mathscr{F}}([Z]) &= \big[\big(L_{\Psi|_{\partial^- Y}\circ\iota_Y^-}, L_{\Psi(Y_{01})}, L_{\Psi(Y_{12})}, \ldots, L_{\Psi(Y_{(n-1)n})}, L_{(\Psi|_{\partial^+ Y}\circ\iota_Y^+)^{-1}}\big)\big]\\ &= \big[\big(L_{\iota_Y^-} \circ L_{\Psi|_{\partial^- Y}}, L_{(\Psi|_{\partial^- Y})^{-1}} \circ L_{Y_{01}} \circ L_{\Psi|_{\Sigma_1}}, L_{(\Psi|_{\Sigma_1})^{-1}} \circ L_{Y_{12}} \circ L_{\Psi|_{\Sigma_2}}, \ldots\\ &\qquad \ldots L_{(\Psi|_{\Sigma_{n-1}})^{-1}} \circ L_{Y_{(n-1)n}} \circ L_{\Psi|_{\partial^+ Y}}, L_{(\Psi|_{\partial^+ Y})^{-1}} \circ L_{(\iota_Y^+)^{-1}}\big)\big]\\ &= \big[\big(L_{\iota_Y^-}, L_{Y_{01}}, L_{Y_{12}}, \ldots, L_{Y_{(n-1)n}}, L_{(\iota_Y^+)^{-1}}\big)\big] \;=\; \overline{\mathscr{F}}([Y]).\end{aligned}$$

This finishes the proof that the unique extension $\overline{\mathscr{F}}$ is a well-defined functor.

Finally, if $\overline{\mathscr{F}} : \mathrm{Bor}_{d+1}^{\mathrm{conn}} \to \mathrm{Symp}^\tau$ takes values in a monotone symplectic category (for a monotonicity constant $\tau \geq 0$; see Remark 3.5.4), then it can be composed with the Yoneda functor $\mathrm{Symp}^\tau \to \mathrm{Cat}$ constructed in [76] and Lemma 3.5.6 below to induce a functor $\mathrm{Bor}_{d+1}^{\mathrm{conn}} \to \mathrm{Cat}$, as claimed. Here, the existence of the Yoneda functor follows from the fact that Symp^τ extends to a 2-category. □

A formal notion of $d+1$ Floer field theory should also include a notion of duality. However, the abstract categorical notion of duality requires a monoidal structure—roughly speaking, an associative multiplication of objects that extends to a bifunctor. While in the bordism category Bor_{d+1} a monoidal structure is naturally given by disjoint unions of objects and morphisms, an extension of the gauge theoretic examples in Sect. 2.5 to disconnected bordisms remains elusive; see Remark 2.5.7. Instead, we work with the following practical notion of adjunctions, which will be part of an abstract notion of quilted 2-categories in Definition 3.4.2.

Remark 2.4.3 The **adjoint of a cobordism** $[Y] \in \mathrm{Mor}_{\mathrm{Bor}_{d+1}}(\Sigma_0, \Sigma_1)$ with boundary embeddings $\iota_Y^\pm : \Sigma_i \to \partial Y$ is the cobordism $[Y^-] \in \mathrm{Mor}_{\mathrm{Bor}_{d+1}}(\Sigma_1, \Sigma_0)$ obtained by reversing the orientation and boundary embeddings $\iota_Y^+ : \Sigma_1^- \to \partial Y^-$, $\iota_Y^- : \Sigma_0 \to \partial Y^-$. In particular, the adjoint of a k-handle attachment is a $d+1-k$-handle attachment.

The **adjoint of a Lagrangian** $L \subset M_0^- \times M_1$ is $L^T := \tau(L) \subset M_1^- \times M_0$ obtained by transposition $\tau(p_0, p_1) := (p_1, p_0)$. For very simple morphisms—cylindrical cobordisms and graphs of symplectomorphisms—these adjoints are also inverse morphisms, but not in general.

In the category of categories, not every functor may have an adjoint, but there also is a notion of two functors $f : \mathscr{C} \to \mathscr{D}$ and $f^T : \mathscr{D} \to \mathscr{C}$ being adjoint; see Definition 3.4.2.

With this we can somewhat formalize our notion of connected Floer field theories. We will keep the definition flexible to allow for current progress toward constructing more general symplectic 2-categories as discussed in Example 2.5.2 and Remark 3.5.5.

Definition 2.4.4 A **d+1 connected Floer field theory** is an adjunction preserving functor $\mathrm{Bor}_{d+1}^{\mathrm{conn}} \to \mathscr{C}$ to an algebraic category (such as $\mathscr{C} = \mathrm{Cat}$) that arises as composition of a functor $\mathscr{F} : \mathrm{Bor}_{d+1}^{\mathrm{conn}} \to \mathscr{S}$ to a symplectic category (i.e. a category such as $\mathscr{S} = \mathrm{Symp}^\tau$ whose objects are symplectic manifolds) with a Yoneda-type functor arising from a 2-categorical structure on $\mathscr{S}$ that encodes Floer theory (such as the functor $\mathrm{Symp}^\tau \to \mathrm{Cat}$ constructed in Lemma 3.5.6).

Here, the Yoneda functor $\mathrm{Symp}^\tau \to \mathrm{Cat}$ arises from a quilted generalization of Floer homology which was developed in [75, 76, 79] within a mononote symplectic category (see Remark 3.5.4) that guarantees well-behaved moduli spaces of pseudoholomorphic quilts; see Sect. 3.5. Since the composition with this functor is automatic (if it exists), we will sometimes also refer to a functor $\mathrm{Bor}_{d+1}^{\mathrm{conn}} \to \mathrm{Symp}$ (even if it does not take values in a monotone subcategory) as a Floer field theory—because it reduces the question of constructing a functor $\mathrm{Bor}_{d+1}^{\mathrm{conn}} \to \mathrm{Cat}$ to ensuring that quilted Floer homology is well defined on its image. One might be tempted to call a functor $\mathrm{Bor}_{d+1} \to \mathrm{Symp}$ a "d+1 symplectic field theory," but the label of SFT = symplectic field theory was given by [21] to a theory in which another symplectic category—given by contact-type manifolds and symplectic cobordisms—is the domain, not the target of a functor.

2.5 $2+1$ *Floer Field Theories Arising from Gauge Theory*

Working more specifically in dimensions 2+1, and making use of the adjunctions in Remark 2.4.3, we can specialize Lemma 2.4.2 even further to observe that a $2+1$ connected Floer field theory $\mathrm{Bor}_{2+1}^{\mathrm{conn}} \to \mathrm{Cat}$ in the sense of Definition 2.4.4 can be obtained by essentially just fixing symplectic data for one surface of each genus and attaching circles in these. Here, we will be somewhat cavalier about diffeomorphisms that are isotopic to the identity. These do not affect the representation spaces in Example 2.5.4, but in general, e.g., in Example 2.5.6, more vigilance such as in [26, 27, 53] is required.

Remark 2.5.1 In order to construct a $2+1$ connected Floer field theory $\mathrm{Bor}_{2+1}^{\mathrm{conn}} \to \mathrm{Cat}$, it suffices to construct a functor $\mathscr{F} : \mathrm{Bor}_{2+1}^{\mathrm{conn}} \to \mathrm{Symp}^\tau$ that preserves adjunctions. The latter can be obtained as in Lemma 2.4.2 by the following constructions.

1. To a closed, connected, oriented surface Σ, associate a symplectic manifold M_Σ (that is compact and τ-monotone for a fixed $\tau \geq 0$; see Remark 3.5.4).
2. To a diffeomorphism $\phi : \Sigma_0 \to \Sigma_1$ associate a symplectomorphism $L_\phi : M_{\Sigma_0} \to M_{\Sigma_1}$ such that $L_\phi \circ L_\psi = L_{\phi\circ\psi}$ (as maps) when ϕ, ψ are composable.
3. To a 2-handle attaching cobordism $Y_\alpha \in \mathrm{Mor}_{\mathrm{Bor}_{2+1}^{\mathrm{conn}}}(\Sigma, \Sigma')$ between connected surfaces as in Remark 2.3.1 associate a Lagrangian submanifold $L_\alpha \subset M_\Sigma^- \times M_{\Sigma'}$ (that is compact and τ-monotone).

3.′ To the reversed 1-handle attachment $Y_\alpha^- \in \mathrm{Mor}_{\mathrm{Bor}_{2+1}^{\mathrm{conn}}}(\Sigma', \Sigma)$ associate the transposed Lagrangian $L_\alpha^T \subset M_{\Sigma'}^- \times M_\Sigma$.

4. For attaching circles $\alpha, \phi(\alpha) \subset \Sigma$ related by a diffeomorphism $\phi : \Sigma \to \Sigma$, there is a diffeomorphism $\phi' : \Sigma_\alpha \to \Sigma_{\phi(\alpha)}$ determined by $\phi' \circ \pi_\alpha = \pi_{\phi(\alpha)} \circ \phi$ such that the 3-cobordisms $Y_\alpha \simeq Y_{\phi(\alpha)}$ are diffeomorphic relative to ϕ, ϕ' on the boundary. Ensure that this is reflected by an identity of Lagrangians $(L_\phi \times L_{\phi'})(L_\alpha) = L_{\phi(\alpha)}$ via the symplectomorphisms given in 2.
5. For disjoint attaching circles $\alpha, \beta \subset \Sigma$, denote by $\beta' := \pi_\alpha(\beta) \subset \Sigma_\alpha$ and $\alpha' := \pi_\beta(\alpha) \subset \Sigma_\beta$ the attaching circles in the outgoing boundary of Y_α resp. Y_β that are obtained from β resp. α. Then, there is a diffeomorphism $\phi'' : (\Sigma_\alpha)_{\beta'} \to (\Sigma_\beta)_{\alpha'}$ between the outgoing boundaries of $Y_{\beta'}, Y_{\alpha'}$, determined by $\phi'' \circ \pi_{\beta'} \circ \pi_\alpha = \pi_{\alpha'} \circ \pi_\beta$, such that the 3-cobordisms[20] $Y_\alpha^- \cup_\Sigma Y_\beta \simeq Y_{\beta'} \cup_{\phi''} Y_{\alpha'}^-$ are diffeomorphic with fixed boundary, and the 3-cobordisms $Y_\alpha \cup_{\Sigma_\alpha} Y_{\beta'} \simeq Y_\beta \cup_{\Sigma_\beta} Y_{\alpha'}$ are diffeomorphic relative to $\mathrm{id}_\Sigma, \phi''$ on the boundary. Ensure that this is reflected by embedded geometric compositions $L_\alpha^T \circ L_\beta$, $(\mathrm{id} \times \phi'')(L_{\beta'}) \circ L_{\alpha'}^T$, $L_\alpha \circ L_{\beta'}$, $L_\beta \circ L_{\alpha'}$ and identities

$$(\mathrm{id} \times L_{\phi''})(L_\alpha \circ L_{\beta'}) = L_\beta \circ L_{\alpha'}, \qquad L_\alpha^T \circ L_\beta = (\mathrm{id} \times \phi'')(L_{\beta'}) \circ L_{\alpha'}^T. \quad (2.5.1)$$

6. For attaching circles $\alpha, \beta \subset \Sigma$ with transverse intersection in a single point, the composition $Y_\alpha^- \cup_\Sigma Y_\beta \simeq Z_\phi$ is diffeomorphic with fixed boundary to the cylindrical cobordism of a diffeomorphism $\phi : \Sigma_\alpha \to \Sigma_\beta$ determined by $\phi \circ \pi_\alpha = \pi_\beta$ on $\Sigma \setminus (\alpha \cup \beta)$ and $\phi(\pi_\alpha(\beta)) = \pi_\beta(\alpha)$. Ensure that this is reflected by an embedded geometric composition $L_\alpha^T \circ L_\beta = \mathrm{gr}(L_\phi)$.

While step 1 fixes the functor $\mathcal{F}$ on all objects, steps 2 and 3 fix explicit Lagrangians $\mathcal{F}([Y]) = \underline{L}_Y$ only for simple morphisms Y as $L_{Z_\phi} = L_\phi$ for cylindrical cobordisms, $L_{Y_\alpha} = L_\alpha$ for 2-handle attachments, and $L_{Y_\alpha^-} = L_\alpha^T$ for their adjoint 1-handle attachments. To determine the value of the functor $\mathcal{F}([Y]) = [\underline{L}_Y]$ on a general cobordism $Y \in \mathrm{Mor}_{\mathrm{Bor}_{2+1}}(\Sigma, \Sigma')$, we choose a Cerf decomposition $Y = Y_{01} \cup_{\Sigma_1} Y_{12} \ldots \cup_{\Sigma_{k-1}} Y_{(k-1)k}$ into a composable chain of simple morphisms $Y_{ij} \in \mathrm{Mor}_{\mathrm{Bor}_{2+1}}(\Sigma_i, \Sigma_j)$ from $\Sigma_0 = \Sigma$ to $\Sigma_k = \Sigma'$. Then functoriality requires

$$\begin{aligned}[\underline{L}_Y] = \mathcal{F}([Y]) &= \mathcal{F}([Y_{01}]) \circ \mathcal{F}([Y_{12}]) \ldots \circ \mathcal{F}([Y_{(k-1)k}]) \\ &= [L_{Y_{01}}] \circ [L_{Y_{12}}] \ldots \circ [L_{Y_{(k-1)k}}],\end{aligned}$$

[20] Here $\cup_{\phi''}$ denotes a gluing of the boundaries of $Y_{\beta'}, Y_{\alpha'}^-$ via the diffeomorphism ϕ''.

and this is well-defined since different Cerf decompositions of $[Y]$ are related by Cerf moves, which steps 4–6 guarantee to correspond to embedded geometric compositions, i.e., yield the same morphisms in the symplectic category. More precisely, steps 2,3 associate to a cobordism with Cerf decomposition (a factorization in $\mathrm{Bor}_{2+1}^{\mathrm{conn}}$) a morphism in the extended symplectic category of Definition 2.2.1,

$$Y = Y_{01} \cup_{\Sigma_1} Y_{12} \ldots \cup_{\Sigma_{k-1}} Y_{(k-1)k} \quad \mapsto \quad \underline{L}_Y = L_{Y_{01}} \# L_{Y_{12}} \ldots \# L_{Y_{(k-1)k}}.$$

Then Cerf moves can be viewed as isomorphisms between different factorizations in $\mathrm{Bor}_{2+1}^{\mathrm{conn}}$, and steps 4–6 relate these to isomorphisms in $\mathrm{Symp}^{\#}$ given by the relation used in Definition 2.2.2 of the symplectic category as the quotient of $\mathrm{Symp}^{\#}$. This could more precisely be phrased as a 2-functor between extensions of $\mathrm{Bor}_{2+1}^{\mathrm{conn}}$ to a bicategory as in Example 3.2.1 and of $\mathrm{Symp}^{\#}$ to a 2-category as in Example 3.1.6.

Since its first announcement in [80], this Floer field philosophy has been applied to obtain various proposals for $2+1$ field theories, which are inspired from various gauge theories. Unfortunately, these are still preprints [80, 81], work in progress [38], or published [5, 36, 44, 56] but hinging on generalizations of the crucial isomorphism in Floer homology under geometric compositions beyond the (compact monotone) setting in which it was proven in [78]; see Remarks 3.5.4–3.5.8. Instead of discussing the technicalities and possible obstructions, this section focusses on the motivations, and thus presents both intuitive and naive reasonings why theories along these lines are to be expected.

The intuitive reason for an intimate connection between symplectic geometry and gauge theory in dimensions $2+1$ is the following example of a partial functor from $\mathrm{Bor}_{2+1}^{\mathrm{conn}}$ to a category of infinite dimensional symplectic Banach spaces and Lagrangian Banach-submanifolds. It provides the basic data from which one expects a 2+1+1 field theory which comprises Donaldson invariants and instanton Floer homology[21] for certain 4- and 3-manifolds, as discussed in Sect. 2.6.

Example 2.5.2 (Infinite dimensional Floer field theory from spaces of connections). Fix a compact, connected, simply connected Lie group G, and let $\langle\cdot,\cdot\rangle$ be a G-invariant inner product on the Lie algebra $\mathfrak{g}$. (The main and first nontrivial examples are $G = SU(r)$ for $r \geq 2$.) The following constructions will use some basic notations from gauge theory, which can be found in, e.g., [67]. These constructions also have natural extensions to nontrivial bundles—such as the unique nontrivial $SO(3)$-bundles over surfaces and handle attachments used in [80], which also serve to avoid issues of reducible connections.

1. To each closed, connected, oriented surface Σ, we associate the space of connections $\mathscr{A}(\Sigma) := \Omega^1(\Sigma, \mathfrak{g})$ on the trivial G-bundle over Σ. It has a natural

[21] Donaldson invariants and instanton Floer homology are invariants for smooth 4- and 3-manifolds that were developed in the 1980s [15, 22]; see [16, 17] for introductions. Similar to the symplectic versions of Floer homology, the 3-manifold invariant can be viewed as the Morse homology of the Chern–Simons functional on a space of connections (modulo gauge) on the 3-manifold, with the gradient flow recast as the ASD Yang–Mills PDE (whose stationary solutions are the flat connections).

symplectic structure given by $\omega(a_1, a_2) = \int_\Sigma \langle a_1 \wedge a_2\rangle$ for $a_i \in \mathscr{A}(\Sigma)$; see [4, 57, 71]. Indeed, ω is bilinear and alternating (recall that $\alpha_1 \wedge \alpha_2 = -\alpha_2 \wedge \alpha_1$ for real-valued 1-forms), and it is nondegenerate since the Hodge star operator for any choice of metric on Σ induces an L^2-metric $g(a_1, a_2) = \omega(a_1, *a_2)$ on $\mathscr{A}(\Sigma)$.

Note here that reversing the orientation of Σ corresponds to reversing the sign of the symplectic form, i.e. $\mathscr{A}(\Sigma^-) = \mathscr{A}(\Sigma)^-$. Moreover, $*|_{\mathscr{A}(\Sigma)}$ is in fact an ω-compatible complex structure since $*^2 = -\mathrm{id}$.

2. To each diffeomorphism $\phi : \Sigma_0 \to \Sigma_1$, we associate the push forward $L_\phi := \phi_* : \mathscr{A}(\Sigma_0) \to \mathscr{A}(\Sigma_1)$ given by $(\phi_* a)(v) := a(\mathrm{d}\phi^{-1}(v))$. This is a symplectomorphism since for $a_1, a_2 \in \mathscr{A}(\Sigma_0)$ we have

$$\begin{aligned}\big(L_\phi^* \omega_{\mathscr{A}(\Sigma_1)}\big)(a_1, a_2) &= \textstyle\int_{\Sigma_1} \langle \phi_*(a_1) \wedge \phi_*(a_2)\rangle = \int_{\Sigma_1} \phi_* \langle a_1 \wedge a_2\rangle \\ &= \textstyle\int_{\phi^{-1}(\Sigma_1)} \langle a_1 \wedge a_2\rangle = \omega_{\mathscr{A}(\Sigma_0)}(a_1, a_2).\end{aligned}$$

Moreover we have $L_\phi \circ L_\psi = \phi_* \circ \psi_* = (\phi \circ \psi)_* = L_{\phi\circ\psi}$ as required when ϕ, ψ are composable.

3. To each 2-handle attachment $Y_\alpha \in \mathrm{Mor}_{\mathrm{Bor}^{\mathrm{conn}}_{2+1}}(\Sigma, \Sigma')$, we associate the space of restrictions of flat connections on Y_α to the boundary components $\partial Y_\alpha = \Sigma^- \sqcup \Sigma'$,

$$\mathscr{L}(Y_\alpha) := \big\{\big(\tilde{A}|_\Sigma, \tilde{A}|_{\Sigma'}\big) \,|\, \tilde{A} \in \mathscr{A}(Y_\alpha), F_{\tilde{A}} = 0\big\} \subset \mathscr{A}(\Sigma)^- \times \mathscr{A}(\Sigma').$$

This yields an isotropic of $\mathscr{A}(\Sigma)^- \times \mathscr{A}(\Sigma') \cong \mathscr{A}(\Sigma^- \sqcup \Sigma') = \mathscr{A}(Y_\alpha)$ since the linearization of curvature $\frac{\mathrm{d}}{\mathrm{d}t}\big|_{t=0} F_{\tilde{A}+t\tilde{a}} = \mathrm{d}_{\tilde{A}}\tilde{a}$ at a connection $\tilde{A}$ is the associated differential, so that $\omega(\tilde{a}_1|_{\partial Y}, \tilde{a}_2|_{\partial Y}) = \int_Y \langle \mathrm{d}_{\tilde{A}}\tilde{a}_1 \wedge \tilde{a}_2\rangle - \langle \tilde{a}_1 \wedge \mathrm{d}_{\tilde{A}}\tilde{a}_2\rangle = 0$ by Stokes' theorem. In appropriate Banach space completions, one can also show that $\mathscr{L}(Y_\alpha)$ is a Banach submanifold and coisotropic, hence, a Lagrangian submanifold of $\mathscr{A}(\Sigma)^- \times \mathscr{A}(\Sigma')$. (This is a direct generalization of [68, Lemma 4.6] which proves these claims for Y_α replaced by a handlebody.)

3'. The analogous construction for the 1-handle attachment $Y_\alpha^- \in \mathrm{Mor}_{\mathrm{Bor}^{\mathrm{conn}}_{2+1}}(\Sigma', \Sigma)$ yields the transposed Lagrangian

$$\mathscr{L}(Y_\alpha^-) := \big\{\big(\tilde{A}|_{\Sigma'}, \tilde{A}|_\Sigma\big) \,|\, \tilde{A} \in \mathscr{A}(Y_\alpha^-), F_{\tilde{A}} = 0\big\} = \mathscr{L}(Y_\alpha)^T.$$

4. To check $(\phi_* \times \phi'_*)(\mathscr{L}(Y_\alpha)) = \mathscr{L}(Y_{\phi(\alpha)})$ for a diffeomorphism $\phi : \Sigma \to \Sigma$, recall that ϕ, ϕ' are the boundary restrictions of a diffeomorphism $\widetilde{\phi} : Y_\alpha \to Y_{\phi(\alpha)}$. Then the relation between the Lagrangians follows from the fact that the spaces of flat connections on Y_α and $Y_{\phi(\alpha)}$ are identified by pullback with $\widetilde{\phi}$.

5, 6. For any composable pair of cobordisms $Y_{ij} \in \mathrm{Mor}_{\mathrm{Bor}^{\mathrm{conn}}}(\Sigma_i, \Sigma_j)$, we have the Lagrangian for the composition of cobordisms given by the geometric composition of the Lagrangians for the separate cobordisms,

$$
\begin{aligned}
&\mathscr{L}(Y_{01} \cup_{\Sigma_1} Y_{12}) \\
&= \big\{ (A_0, A_2) \mid \exists \tilde{A}_{\text{flat}} \in \mathscr{A}(Y_{01} \cup_{\Sigma_1} Y_{12}), \tilde{A}|_{\Sigma_i} = A_i \big\} \\
&= \big\{ (A_0, A_2) \mid \exists \tilde{A}_{ij} \in \mathscr{A}_{\text{flat}}(Y_{ij}), \tilde{A}_{01}|_{\Sigma_1} = \tilde{A}_{12}|_{\Sigma_1}, \tilde{A}_{01}|_{\Sigma_0} = A_0, \tilde{A}_{12}|_{\Sigma_2} = A_2 \big\} \\
&= \pi_{\mathscr{A}(\Sigma_0)\times\mathscr{A}(\Sigma_2)}\big(\big(\mathscr{L}(Y_{01}) \times \mathscr{L}(Y_{12})\big) \cap \big(\mathscr{A}(\Sigma_0) \times \Delta_{\mathscr{A}(\Sigma_1)} \times \mathscr{A}(\Sigma_2)\big)\big) \\
&= \mathscr{L}(Y_{01}) \circ \mathscr{L}(Y_{12}),
\end{aligned}
$$

where we denote the sets of flat connections by $\mathscr{A}_{\text{flat}}(Y) := \{\tilde{A} \in \mathscr{A}(Y) \mid F_{\tilde{A}} = 0\}$. This proves all required identities of geometric compositions. However, these geometric compositions are never embedded since all restrictions of the connections to Σ_1 are flat, thus cannot span the complement of the diagonal.

While, these constructions do not yield a functor $\text{Bor}_{2+1}^{\text{conn}} \to \text{Symp}$ via the principle of Remark 2.5.1, we will explain in Example 3.6.2 how one might use quilts (see Sect. 3.5) made up of ASD instantons in place of pseudoholomorphic curves to extend this partial functor to a Floer field theory $\text{Bor}_{2+1}^{\text{conn}} \to \text{Cat}$ that factors through a symplectic instanton 2-category whose objects are symplectic Banach spaces of connections.

The beginning of an instanton Floer field theory given above is the natural intermediate step in an expected relation between Chern–Simons theory on 3-manifolds and symplectic invariants arising from a choice of decomposition of the 3-manifold as formulated by Atiyah [3] in terms of Floer homologies [22, 23]; see also [57, 71] and Sect. 2.6. This symplectic invariant uses Heegaard splittings as explained before Example 2.5.6 and finite dimensional symplectic quotients of the above spaces of connections, as explained in the following remark. Moreover, the Chern–Simons theory on 3-manifolds is naturally coupled with Donaldson–Yang–Mills theory on 4-manifolds; see [15–17]. Thus, the subsequent sketch of Floer field theories arising from representation spaces should be viewed as the beginning of a symplectic categorification of Donaldson–Yang–Mills theory (in various versions, depending on choice of group and twisting). It also serves as a purely symplectic explanation of the conjecture that the Floer homology arising from a decomposition of the 3-manifold is in fact a 3-manifold invariant, i.e. independent of the choice of decomposition; see Sect. 2.6 for details.

Remark 2.5.3 (Finite dimensional reduction of instanton Floer field theory). While the spaces of connections in Example 2.5.2 are infinite dimensional and tend to have a smooth structure, a symplectic reduction by the Hamiltonian action of the gauge group yields finite dimensional but generally singular spaces. Here, the gauge group $\mathscr{G}(\Sigma) = \mathscr{C}^\infty(\Sigma, G)$ acts on $\mathscr{A}(\Sigma)$ by pulling back connections with bundle isomorphisms, and its moment map is the curvature; see [4, 57, 71]. The symplectic quotient $M_\Sigma := \mathscr{A}(\Sigma)/\!\!/\mathscr{G}(\Sigma)$ can thus be understood topologically as the space of representations of the fundamental group $\pi_1(\Sigma)$ in the Lie group G—given by the holonomies of flat connections—modulo gauge symmetries represented by simultaneous conjugation of the holonomies. The quotient[22] of the Lagrangian $L_{Y_\alpha} := \mathscr{L}(Y_\alpha)/\mathscr{G}(\Sigma) \times \mathscr{G}(\Sigma') \subset M_\Sigma^- \times M_{\Sigma'}$ is given by those representations that arise as the restriction of a representation of $\pi_1(Y_\alpha)$, i.e., are trivial on loops in $\partial Y_\alpha = \Sigma^- \sqcup \Sigma'$ that are contractible in Y_α.

Singularities in these spaces are due to reducible connections, corresponding to representations $\rho : \pi_1(\Sigma) \to G$ on which conjugation by G acts with nondiscrete stabilizer $G_\rho = \{g \in G \mid g^{-1}\rho g = \rho\}$ (e.g., the stabilizer of the trivial representation is the whole group G). These can be avoided by working on appropriately twisted bundles or making holonomy requirements around punctures in Σ resp. tangles[23] in Y_α. (The latter usually yields field theories for cobordisms with tangles, but there are specific—central in G—holonomy requirements for which the position of puncture resp. tangle is irrelevant.) Then, the symplectic quotient by the gauge group $\mathscr{G}(\Sigma)$ yields a finite dimensional Lagrangian submanifold $L_{Y_\alpha} \subset M_\Sigma^- \times M_{\Sigma'}$.

Instead of discussing possible twisting constructions to avoid the reducibles noted above, the following example gives an idea of a finite dimensional Floer field theory in terms of sets rather than manifolds. For abelian groups G, this will actually yield smooth symplectic and Lagrangian manifolds, but a field theory based on these would only capture homological information of the bordism category.

Example 2.5.4 (Naive Floer field theory from representation spaces). We will go through the Floer field theory construction outlined in Remark 2.5.1 in the example of representations of a compact, connected, simply connected Lie group G, such as $G = SU(2)$, which arise from trivial G-bundles in Example 2.5.2 and Remark 2.5.3.

1. To each closed, connected, oriented surface Σ, associate the representation space

$$M_\Sigma := \{\rho \in \mathrm{Hom}(\pi_1(\Sigma), G)\}/_\sim \quad \text{with} \quad \rho \sim \rho' :\Leftrightarrow \exists g \in G : \rho' = g^{-1}\rho g.$$

Any standard basis $(\alpha_1, \beta_1, \ldots, \alpha_g, \beta_g)$ for $\pi_1(\Sigma)$, i.e., loops that are disjoint except for single transverse intersection points $\alpha_i \pitchfork \beta_i$ and whose concatenation $\prod_{i=1}^g \alpha_i\beta_i\alpha_i^{-1}\beta_i^{-1}$ is homotopic to the constant loop, yields an identification

$$M_\Sigma \simeq \{(a_1, b_1, \ldots, a_g, b_g) \in G^{2g} \mid \textstyle\prod_{i=1}^g a_i b_i a_i^{-1} b_i^{-1} = \mathrm{id}\}/_\sim$$

(with $\mathrm{id} \in G$ denoting the identity), modulo simultaneous conjugation

$$(a_i, b_i)_{i=1,\ldots,g} \sim (g^{-1}a_i g, g^{-1}b_i g)_{i=1,\ldots,g} \qquad \forall g \in G.$$

[22] The fact that we can take the quotient by the product of gauge groups is due to the identification $\mathscr{G}(\Sigma) \times \mathscr{G}(\Sigma') = \mathscr{C}^\infty(\partial Y_\alpha, G) = \mathscr{C}^\infty(Y_\alpha, G)|_{\partial Y_\alpha}$ with the boundary values of the gauge group $\mathscr{G}(Y_\alpha)$, which uses the assumption of G being connected and simply connected.

[23] A tangle in a cobordism is an embedded submanifold whose boundary coincides with given punctures on the boundary of the cobordism.

2. To each diffeomorphism $\phi : \Sigma_0 \to \Sigma_1$ associate the map $L_\phi : M_{\Sigma_0} \to M_{\Sigma_1}$ which maps $\rho \in M_{\Sigma_0}$ to the representation $L_\phi(\rho) \in M_{\Sigma_1}$ given by $[\gamma] \mapsto \rho([\phi^{-1} \circ \gamma])$ for any circle $\gamma : S^1 \to \Sigma_1$. Observe that $L_\phi \circ L_\psi = L_{\psi\circ\phi}$ when ϕ, ψ are composable.
3. For each attaching circle $\alpha \subset \Sigma$ we use the bijection $\pi_\alpha : \Sigma \setminus \alpha \to \Sigma' \setminus \{2 \text{ points}\}$ and a deformation of any loop $\gamma : S^1 \to \Sigma'$ to avoid the special points to construct

$$L_\alpha := \left\{ ([\rho], [\rho']) \in M_\Sigma^- \times M_{\Sigma'} \,\middle|\, \rho([\alpha]) = \mathrm{id}, \forall \gamma : \rho'([\gamma]) = \rho([\pi_\alpha^{-1} \circ \gamma]) \right\}.$$

Note that this construction is independent of the choice of a parametrization $\alpha : S^1 \to \Sigma$ of the attaching circle (and deformation to $\alpha(1) = z$). In the identification obtained from a standard basis $(\alpha_i, \beta_i)_{i=1,\dots,g}$ for Σ with $[\alpha_1] = [\alpha]$ and the induced basis $(\pi_\alpha \circ \alpha_i, \pi_\alpha \circ \beta_i)_{i=2,\dots,g}$ for Σ' we have

$$L_\alpha = \left\{ ([(a_i, b_i)_{i=1,\dots,g}], [(a_i', b_i')_{i=2,\dots,g}]) \,\middle|\, a_1 = \mathrm{id}, \forall i \geq 2 : a_i' = a_i, b_i' = b_i \right\}.$$

3'. The analogous construction for the adjoint cobordism $Y_\alpha^- \in \mathrm{Mor}_{\mathrm{Bor}_{2+1}^{\mathrm{conn}}}(\Sigma', \Sigma)$ yields the transposed Lagrangian $L_\alpha^T \subset M_{\Sigma'}^- \times M_\Sigma$.

4. For any attaching circle $\alpha : S^1 \to \Sigma$ and diffeomorphism $\phi : \Sigma \to \Sigma$ we can rewrite $\rho'([\gamma]) = \rho([\pi_{\phi(\alpha)}^{-1} \circ \gamma])$ in the construction of $L_{\phi(\alpha)}$ equivalently as $\rho'([\phi' \circ \tilde{\gamma}]) = \rho([\phi \circ \pi_\alpha^{-1} \circ \tilde{\gamma}])$ for all loops $\tilde{\gamma}$ since $\pi_{\phi(\alpha)} \circ \phi = \phi' \circ \pi_\alpha$, and thus

$$\begin{aligned} L_{\phi(\alpha)} &= \left\{ ([\rho], [\rho']) \in M_\Sigma^- \times M_{\Sigma'} \,\middle|\, \rho([\phi \circ \alpha]) = \mathrm{id}, \rho'([\phi' \circ \gamma]) = \rho([\phi \circ \pi_\alpha^{-1} \circ \gamma]) \right\} \\ &= \left\{ (L_\phi([\tilde{\rho}]), L_{\phi'}([\tilde{\rho}'])) \in M_\Sigma^- \times M_{\Sigma'} \,\middle|\, \tilde{\rho}([\alpha]) = \mathrm{id}, \tilde{\rho}'([\gamma]) = \tilde{\rho}([\pi_\alpha^{-1} \circ \gamma]) \right\} \\ &= (L_\phi \times L_{\phi'})(L_\alpha). \end{aligned}$$

5. For disjoint attaching circles $\alpha \cap \beta = \emptyset$ we calculate the geometric composition

$$\begin{aligned} L_\alpha \circ L_{\beta'} &= \left\{ ([\rho], [\rho'']) \,\middle|\, \exists [\rho'] \in M_{\Sigma_\alpha} : ([\rho], [\rho']) \in L_\alpha, ([\rho'], [\rho'']) \in L_{\beta'} \right\} \\ &= \left\{ ([\rho], [\rho'']) \,\middle|\, \rho([\alpha]) = \rho([\beta]) = \mathrm{id}, \rho''([\gamma]) = \rho([(\pi_{\beta'}\pi_\alpha)^{-1} \circ \gamma]) \right\} \end{aligned}$$

by noting that $[\rho'']$ is determined from $[\rho]$ by

$$\rho''([\gamma]) = \rho'([\pi_{\beta'}^{-1} \circ \gamma]) = \rho([\pi_\alpha^{-1} \circ \pi_{\beta'}^{-1} \circ \gamma]) \qquad \forall\, \gamma : S^1 \to \Sigma'' := (\Sigma_\alpha)_{\beta'}$$

and the additional requirement $\mathrm{id} = \rho'([\beta']) = \rho([\pi_\alpha^{-1} \circ \beta'])$, where we have $\pi_\alpha^{-1}(\beta') = \beta$ because the attaching circles are disjoint. Analogously, in the composition $L_\beta \circ L_{\alpha'} = \left\{ ([\rho], [\rho'']) \,\middle|\, \dots \right\}$ we have $\rho''([\gamma]) = \rho([(\pi_{\alpha'}\pi_\beta)^{-1} \circ \gamma])$ for all $\gamma : S^1 \to (\Sigma_\beta)_{\alpha'}$. Using, the diffeomorphism ϕ'' given by $\phi'' \circ \pi_{\beta'}\pi_\alpha = \pi_{\alpha'}\pi_\beta$, we rewrite this as $\rho''([\phi'' \circ \tilde{\gamma}]) = \rho([(\pi_{\beta'}\pi_\alpha)^{-1} \circ \tilde{\gamma}])$ for all $\tilde{\gamma} = (\phi'')^{-1} \circ \gamma$ so that we obtain the first identity in (2.5.1),

$$L_\beta \circ L_{\alpha'} = \big\{ \big([\rho],[\rho'']\big) \,\big|\, \rho([\beta]) = \rho([\alpha]) = \mathrm{id}, \rho''([\phi'' \circ \gamma]) = \rho([(\pi_{\beta'}\pi_\alpha)^{-1} \circ \gamma]) \big\}$$
$$= (\mathrm{id} \times L_{\phi''})\big(L_\alpha \circ L_{\beta'}\big).$$

The second identity between geometric compositions of Lagrangians is similar:

$$\begin{aligned}
&(\mathrm{id} \times L_{\phi''})(L_{\beta'}) \circ L^T_{\alpha'} \\
&= \big\{ \big([\rho'],[\sigma']\big) \,\big|\, \exists [\rho''] \in M_{(\Sigma_\alpha)_{\beta'}} : \big([\rho'],[\rho'']\big) \in L_{\beta'}, \big([\sigma'], L_{\phi''}([\rho''])\big) \in L_{\alpha'} \big\} \\
&= \big\{ \big([\rho'],[\sigma']\big) \,\big|\, \rho'([\beta']) = \sigma'([\alpha']) = \mathrm{id}, \rho'([\pi_{\beta'}^{-1} \circ \gamma]) = \sigma'([\pi_{\alpha'}^{-1} \circ \phi'' \circ \gamma]) \;\forall \gamma \big\} \\
&= \big\{ \big([\rho'],[\sigma']\big) \,\big|\, \rho([\alpha]) = \rho([\beta]) = \mathrm{id}, \rho' = \rho([\pi_\alpha^{-1} \circ \ldots]), \sigma' = \rho([\pi_\beta^{-1} \circ \ldots]) \big\} \\
&= \big\{ \big([\rho'],[\sigma']\big) \,\big|\, \exists [\rho] \in M_\Sigma : \big([\rho],[\rho']\big) \in L_\alpha, \big([\rho],[\sigma']\big) \in L_\beta \big\} \;=\; L^T_\alpha \circ L_\beta,
\end{aligned}$$

where the first composition requires $\rho'([\beta']) = \mathrm{id} = \sigma'([\alpha'])$ in addition to

$$\rho'([\pi_{\beta'}^{-1} \circ \gamma]) = \rho''([\gamma]) = \sigma'([\pi_{\alpha'}^{-1} \circ \phi'' \circ \gamma]) \qquad \forall\, \gamma : S^1 \to (\Sigma_\alpha)_{\beta'}.$$

Using $\phi'' \circ \pi_{\beta'} \circ \pi_\alpha = \pi_{\alpha'} \circ \pi_\beta$ we can rewrite this as

$$\rho'([\tilde\gamma]) = \sigma'([\pi_{\alpha'}^{-1} \circ \phi'' \circ \pi'_\beta \circ \tilde\gamma]) = \sigma'([\pi_\beta \circ \pi_\alpha^{-1} \circ \tilde\gamma]) \qquad \forall\, \tilde\gamma : S^1 \to \Sigma_\alpha \setminus \beta',$$

i.e., the conditions in $L^T_\alpha \circ L_\beta$ for these loops, which also correspond to the loops in $\Sigma_\beta \setminus \alpha'$. In addition, this second geometric composition requires $\rho'([\beta']) = \rho([\beta]) = \mathrm{id}$, $\sigma'([\alpha']) = \rho([\alpha]) = \mathrm{id}$, which identifies it with the first composition.

Note here that either one of the representations $\big([\rho'],[\sigma']\big) \in L'^T_\beta \circ L_{\alpha'}$ of $\pi_1(\Sigma_\beta)$ or $\pi_1(\Sigma_\alpha)$ fully determines the intermediate representation $[\rho'']$ of $(\Sigma_\alpha)_{\beta'}$. This can also be seen from the fact that $\pi_{\beta'}$ (as well as $\pi_{\alpha'}$) acts surjectively on fundamental groups, in fact any loop in $Y_{\beta'}$ (not just in $(\Sigma_\alpha)_{\beta'} \subset \partial Y_{\beta'}$) can be homotoped into the boundary component Σ_α of higher genus. This uniqueness of the intermediate representations proves injectivity of the projection in the geometric compositions, and—if there was a smooth structure—the corresponding infinitesimal fact would also prove transversality of the intersection, thus embeddedness of the geometric composition $L^T_{\beta'} \circ L_{\alpha'}$.

Embeddedness of $L_\beta \circ L_{\alpha'}$ resp. $L_\alpha \circ L_{\beta'}$ analogously follows from π_1-surjectivity of π_β resp. π_α. For the last geometric composition corresponding to the gluing of cobordisms $Y^-_\alpha \cup_\Sigma Y_\beta$ at the highest genus surface Σ, the fact that the intermediate representation $[\rho]$ on Σ is determined by the representations $\big([\rho'],[\sigma']\big) \in L^T_\alpha \circ L_\beta$ on the two lower genus surfaces Σ_α, Σ_β, is not evident from the formulas. In fact, it is false if we allow α, β to be homologous. However, this is excluded by the assumption of all surfaces, in particular $(\Sigma_\alpha)_{\beta'} \simeq (\Sigma_\beta)_{\alpha'}$ being connected. Thus, we can choose a standard basis $(\alpha_1, \beta_1, \ldots, \alpha_g, \beta_g)$ for $\pi_1(\Sigma)$ with $\alpha_1 = \alpha$ and $\beta_g = \beta$ to see that points in $L^T_\alpha \circ L_\beta$ have the form $\big([(a_i, b_i)_{i=2,\ldots,g}], [(a_i, b_i)_{i=1,\ldots,g-1}]$, which determines the indermediate $[(a_i, b_i)_{i=1,\ldots,g}] \in M_\Sigma$ uniquely.

6. For attaching circles $\alpha, \beta \subset \Sigma$ with unique transverse intersection point we can choose a standard basis $(\alpha_1, \beta_1, \ldots, \alpha_g, \beta_g)$ for $\pi_1(\Sigma)$ with $\alpha_1 = \alpha$ and $\beta_1 = \beta$. Then $L_\alpha^T \circ L_\beta$ is given by pairs $\big([(a_i, b_i)_{i=2,\ldots g}], [(a_i', b_i')_{i=2,\ldots g}]\big) \in M_{\Sigma_\alpha}^- \times M_{\Sigma_\beta}$ for which—after conjugation of the representative $(a_i', b_i')_{i=2,\ldots g}$—there exists $[(a_i, b_i)]_{i=1,\ldots g} \in M_\Sigma$ such that $a_1 = b_1 = \mathrm{id}$ and $a_i = a_i', b_i = b_i'$ for $i \geq 2$. That is, in this basis $L_\alpha^T \circ L_\beta$ is identified with the diagonal over the identified representation spaces $M_{\Sigma_\alpha} \simeq M_{\Sigma_\beta}$. Since this identification is by the map L_ϕ, it shows the identity $L_\alpha^T \circ L_\beta = \mathrm{gr}(L_\phi)$. Moreover, in the presence of a smooth structure, the geometric composition $L_\alpha^T \circ L_\beta$ would be embedded since the intermediate point $[(a_i, b_i)]_{i=1,\ldots g} \in M_\Sigma$ is uniquely determined.

Remark 2.5.5 (Rigorous Floer field theories from representation spaces). Even for the simplest nonabelian group $G = SU(2)$, the representation space for the torus $\Sigma = T^2$ in Example 2.5.4 is the pillowcase $M_{T^2} \simeq S^1 \times S^1/\mathbb{Z}_2$ (here $\mathbb{Z}_2$ acts on each factor S^1 by reflection with two fixed points), and more complicated representation spaces may not even be orbifolds. In some simple cases, e.g., in [31] for knots represented by Lagrangians in the pillowcase, one can deal explicitly with these singularities. To obtain a full Floer field theory, [80] replaces moduli spaces of flat G-connections with moduli spaces of central-curvature connections on unitary bundles with fixed determinant and coprime rank r and degree d. For $r = 2, d = 1$ this corresponds to flat connections on nontrivial $SO(3)$-bundles, which can also be viewed as taking the above representation spaces for $G = SU(2)$ on a punctured surface $\Sigma\setminus\mathrm{pt}$, and instead of holonomy id requiring $-\mathrm{id}$ around the puncture. This yields monotone symplectic manifolds

$$\widehat{M}_\Sigma \simeq \left\{(a_i, b_i)_{i=1,\ldots,g} \in SU(2)^{2g} \,\middle|\, \textstyle\prod_{i=1}^g a_i b_i a_i^{-1} b_i^{-1} = -\mathrm{id}\right\}\big/_{\sim}.$$

If instead of $-\mathrm{id}$ we replace id with a noncentral element $k \in G$, then the representation spaces for the cobordisms are no longer independent of the choice of paths connecting the punctures on the surface (around which the holonomy is required to be conjugate to k). The corresponding Floer field theory in [81] thus yields invariants for pairs of cobordisms with embedded tangles (though invariance under isotopies of the embedding is not yet discussed, so the field theory falls short of yielding knot or link invariants).

Just as dimensional reductions of Donaldson–Yang–Mills theory give rise to the Atiyah–Floer conjecture, the Seiberg–Witten theory for 4-manifolds motivated the development of Heegaard–Floer homology by Ozsváth-Szabó [52]. Since a two-dimensional reduction of the Seiberg–Witten equations gives rise to vortex equations, whose moduli spaces of solutions can be identified with symmetric products of the ambient space [25], they arrived at a 3-manifold invariant that on a given 3-manifold Y is constructed by choosing a so-called Heegaard splitting $Y = H_0^- \cup_\Sigma H_1$

into two handlebodies,[24] representing the handlebodies by Lagrangians $L_{H_i} \subset M_\Sigma = \mathrm{Sym}^g(\Sigma)$ in the symmetric product of the dividing surface Σ, and taking Floer theoretic invariants of the pair $L_{H_0}, L_{H_1} \subset M_\Sigma$. Here and throughout, g will denote the genus of the present surface Σ. Since Heegaard splittings are not unique by any means, Ozsváth-Szabó had to explicitly compare holomorphic curves in symmetric products of different surfaces to prove that the Heegaard–Floer homology groups $HF(L_{H_0}, L_{H_1})$ (with "plus/minus/hat" decorations arising from keeping track of intersections with a marked point in Σ) are in fact 3-manifold invariants, i.e., independent of the choice of splitting.

There are several more conceptual explanations of this independence. First, [35] recently proved an Atiyah–Floer type identification of $HF(L_{H_0}, L_{H_1})$ with monopole Floer homology—the 3-manifold invariant arising directly from Seiberg–Witten gauge theory [33]. Second, as explained in Sect. 2.6, an extension of Heegaard–Floer homology to a $2+1$ Floer field theory would also reproduce the Heegaard–Floer 3-manifold invariant. In addition, this would provide a symplectic categorification of Seiberg–Witten theory. Perutz established the basics of such a theory by constructing Lagrangian matching invariants [55] for 4-manifolds equipped with broken Lefshetz fibrations, which are expected to be equal to the Seiberg–Witten invariants, in particular independent of the choice of broken fibration. The core of this approach is a construction in [53] of Lagrangians in symmetric products associated to simple 3-cobordisms, whose basic structure we explain in the following.

Example 2.5.6 (Naive Floer field theory from symmetric products). We will use the steps in Remark 2.5.1 to outline the extension of Heegaard–Floer homology to a Floer field theory as proposed in [36, 38, 53] for any fixed $n \geq 0$ (or $n < 0$ with surfaces restricted to genus $g \geq -n$). To avoid dealing with complex geometry, we will work with a naive version of symmetric products in which they are constructed as sets rather than smooth algebraic varieties. The smooth, symplectic, and Lagrangian structures are discussed in [53].

1. To each closed, connected, oriented surface Σ, associate the symmetric product

$$M_\Sigma := \mathrm{Sym}^{g+n}\Sigma \;=\; {\Sigma^{g+n}}\big/{S_{g+n}} \;=\; {\Sigma \times \ldots \times \Sigma}\big/{(z_1,\ldots,z_{g+n}) \sim (z_{\sigma(1)},\ldots,z_{\sigma(g+n)})},$$

   where g is the genus of Σ, and S_{g+n} is the symmetric group acting by permutations $\sigma : \{1, \ldots, g+n\} \to \{1, \ldots, g+n\}$. On the complement of the diagonal $\Delta \subset \Sigma^{g+n}$ (where two or more points coincide) this is a smooth quotient, but to obtain a global smooth structure it has to be viewed as the symmetric product of an algebraic curve. This requires the choice of a complex structure on Σ, and the symplectic structure is an additional choice—induced by the broken fibration in [53]—all of which we suppress here.

[24] A handlebody is a 3-manifold H with boundary $\partial H = \Sigma$ (i.e., a cobordism from Σ to the empty set), which is obtained from handle attachments along a maximal number of disjoint attaching circles $\alpha_1, \ldots, \alpha_g \subset \Sigma$ that are homologically independent.

2. To each diffeomorphism $\phi : \Sigma_0 \to \Sigma_1$ associate the map

$$L_\phi : M_{\Sigma_0} \to M_{\Sigma_1}, \quad \big[(z_1, \ldots, z_{g+n})\big] \mapsto \big[(\phi(z_1), \ldots, \phi(z_{g+n}))\big]$$

and observe that $L_\phi \circ L_\psi = L_{\phi\circ\psi}$ when ϕ, ψ are composable. This yields a smooth map when ϕ is holomorphic in the chosen complex structures on Σ_i, but in general this naive construction only yields the correct map outside of the diagonal.

3. To each attaching circle $\alpha \subset \Sigma$ associate $L_\alpha \subset M_\Sigma^- \times M_{\Sigma'}$ given by

$$L_\alpha := \big\{\big(\big[(z_1, \ldots, z_{g+n})\big], \big[(\pi_\alpha(z_2), \ldots, \pi_\alpha(z_{g+n}))\big]\big) \,\big|\, z_1 \in \alpha, z_2, \ldots, z_{g+n} \in \Sigma\setminus\alpha\big\}.$$

Note that this naively constructed subset is not even closed, let alone a smooth submanifold. However, [53] rigorously constructs Lagrangian submanifolds $\widehat{V}_\alpha$ that are smoothly isotopic to L_α on the subset $U_0 \cup U_1 \subset \mathrm{Sym}^{g+n}\Sigma$ given by tuples with up to one point in a given tubular neighborhood $\widetilde{\alpha} \subset \Sigma$ of α. Thus, it makes some sense to discuss the field theory construction in this model.

3'. To the adjoint 1-handle attachment $Y_\alpha^- \in \mathrm{Mor}_{\mathrm{Bor}_{2+1}^{\mathrm{conn}}}(\Sigma', \Sigma)$ we associate the transposed Lagrangian $L_\alpha^T \subset M_{\Sigma'}^- \times M_\Sigma$.

4. For an attaching circle $\alpha \subset \Sigma$ and diffeomorphism $\phi : \Sigma \to \Sigma$ note that we have $(L_\phi \times L_{\phi'})(L_\alpha) = L_{\phi(\alpha)}$ because $z_i' := \phi(z_i)$ yields an identification

$$\begin{aligned}&\big\{\big(\big[(\phi(z_i))_{i=1,\ldots g+n}\big], \big[(\phi'(\pi_\alpha(z_i)))_{i=2,\ldots g+n}\big]\big) \,\big|\, z_1 \in \alpha, z_{i\geq 2} \in \Sigma\setminus\alpha\big\}\\ &= \big\{\big(\big[(z_i')_{i=1,\ldots g+n}\big], \big[(\pi_{\phi(\alpha)}(z_i'))_{i=2,\ldots g+n}\big]\big) \,\big|\, z_1' \in \phi(\alpha), z_{i\geq 2}' \in \Sigma\setminus\phi(\alpha)\big\}.\end{aligned}$$

In [53, 2.3.1], actual symplectomorphisms are associated to diffeomorphisms ϕ that arise from parallel transport in a broken fibration.

5. For disjoint attaching circles $\alpha \cap \beta = \emptyset$ the bijectivity of $\pi_\beta : \Sigma\setminus\beta \to \Sigma_\beta \setminus\{2 \text{ points}\}$ implies $x_1 \in \alpha \Leftrightarrow \pi_\beta(x_1) \in \alpha'$ and analogously $x_2 \in \beta \Leftrightarrow \pi_\alpha(x_2) \in \beta'$. Thus

$$\begin{aligned}L_\alpha \circ L_{\beta'} &= \big\{([\underline{x}], [\underline{z}]) \,\big|\, \exists[\underline{y}] \in M_{\Sigma_\alpha} : ([\underline{x}], [\underline{y}]) \in L_\alpha, ([\underline{y}], [\underline{z}]) \in L_{\beta'}\big\}\\ &= \big\{\big(\big[(x_i)_{i=1,\ldots g+n}\big], \big[(z_i)_{i=3,\ldots g+n}\big]\big) \,\big|\, x_1 \in \alpha, \pi_\alpha(x_2) \in \beta', z_i = \pi_{\beta'}(\pi_\alpha(x_i))\big\},\\ L_\beta \circ L_{\alpha'} &= \big\{\big(\big[(x_i)_{i=1,\ldots g+n}\big], \big[(z_i)_{i=3,\ldots g+n}\big]\big) \,\big|\, x_2 \in \beta, \pi_\beta(x_1) \in \alpha', z_i = \pi_{\alpha'}(\pi_\beta(x_i))\big\}\end{aligned}$$

are related via $\mathrm{id} \times L_{\phi''}$ by the defining property $\phi'' \circ \pi_{\beta'} \circ \pi_\alpha = \pi_{\alpha'} \circ \pi_\beta$ of ϕ''. The second identity between geometric compositions of Lagrangians is similar:

$$\begin{aligned}&(\mathrm{id} \times L_{\phi''})(L_{\beta'}) \circ L_{\alpha'}^T\\ &= \big\{([\underline{x}], [\underline{z}]) \,\big|\, \exists[\underline{v}] \in M_{(\Sigma_\alpha)_{\beta'}} : ([\underline{x}], [\underline{v}]) \in L_{\beta'}, ([\underline{z}], L_{\phi''}([\underline{v}])) \in L_{\alpha'}\big\}\\ &= \big\{\big(\big[(x_i)_{i=2,\ldots g+n}\big], \big[(z_i)_{i=2,\ldots g+n}\big]\big) \,\big|\, x_2 \in \beta', z_2 \in \alpha', \phi''(\pi_{\beta'}(x_i)) = \pi_{\alpha'}(z_i)\ \forall i \geq 3\big\}\\ &= \big\{([\underline{x}], [\underline{z}]) \,\big|\, \exists[\underline{y}] \in M_\Sigma : ([\underline{y}], [\underline{x}]) \in L_\alpha, ([\underline{y}], [\underline{z}]) \in L_\beta\big\} \;=\; L_\alpha^T \circ L_\beta.\end{aligned}$$

Indeed, we have $[\underline{y}] = [(y_1, \tilde{x}_2, \dots, \tilde{x}_{g+n})] = [(y_1', \tilde{z}_2, \dots, \tilde{z}_{g+n})]$ for $\tilde{x}_i = \pi_\alpha^{-1}(x_i)$, $\tilde{z}_i = \pi_\beta^{-1}(z_i)$ and some $y_1 \in \alpha$, $y_1' \in \beta$. Since α, β are disjoint, this implies $y_1 = \tilde{z}_i$ and $y_1' = \tilde{x}_j$ for some $i, j \geq 2$ which we can permute to $i = j = 2$ to obtain $z_2 = \pi_\beta(y_1) \in \alpha'$ and $x_2 = \pi_\alpha(y_1') \in \beta'$. Permutation also achieves $\tilde{x}_i = \tilde{z}_i$ for $i \geq 3$ and hence $x_i = \pi_\alpha(y_i)$, $z_i = \pi_\beta(y_i)$ for some $y_i \in \Sigma \setminus (\alpha \cup \beta)$, which can be rewritten as $\phi''(\pi_{\beta'}(x_i)) = \pi_{\alpha'}(z_i)$ by the defining property of ϕ'' applied to y_i.

While transversality cannot be discussed at the level of sets, note that the intermediate points $[\underline{y}]$ resp. $[\underline{v}]$ in the four geometric compositions above are uniquely determined by $([\underline{x}], [\underline{z}])$. This proves injectivity of the projection in the geometric composition, and the same infinitesimal fact in the presence of a smooth structure also proves transversality of the intersection, thus embeddedness of the geometric compositions. For the true Lagrangian submanifolds, the corresponding identities—up to Hamiltonian isotopy—are conjectured in [53, 3.6.1].

6. For attaching circles $\alpha, \beta \subset \Sigma$ with unique transverse intersection point we have

$$\begin{aligned} L_\alpha^T \circ L_\beta &= \{([\underline{x}], [\underline{z}]) \mid \exists [\underline{y}] \in M_\Sigma : ([\underline{y}], [\underline{x}]) \in L_\alpha, ([\underline{y}], [\underline{z}]) \in L_\beta\} \\ &= \{([(x_i)_{i=2,\dots g+n}], [(z_i)_{i=2,\dots g+n}]) \mid (2.5.2), \phi(x_i) = z_i \; \forall i \geq 3\} \simeq \mathrm{gr}(L_\phi), \end{aligned}$$

where the intermediate point $[\underline{y}] = [(y_1, \tilde{x}_2, \dots, \tilde{x}_{g+n})] = [(y_1', \tilde{z}_2, \dots, \tilde{z}_{g+n})]$ after permutation satisfies either $\tilde{z}_2 = y_1 \in \alpha$, $\tilde{x}_2 = y_1' \in \beta$ or $y_1 = y_1' \in \alpha \pitchfork \beta$, $\pi_\alpha^{-1}(x_2) = \pi_\beta^{-1}(z_2)$. In both cases $\pi_\alpha^{-1}(x_i) = \pi_\beta^{-1}(z_i)$ for $i \geq 3$ can be rewritten as $\phi(x_i) = z_i$ by $\phi \circ \pi_\alpha = \pi_\beta$. For $i = 2$ we have

$$x_2 \in \pi_\alpha(\beta), z_2 \in \pi_\beta(\alpha) \qquad \text{or} \qquad \pi_\alpha^{-1}(x_2) = \pi_\beta^{-1}(z_2) \in \Sigma \setminus (\alpha \cup \beta). \quad (2.5.2)$$

In view of the additional property $\phi(\pi_\alpha(\beta)) = \pi_\beta(\alpha)$ of the diffeomorphism $\phi : \Sigma_\alpha \to \Sigma_\beta$ the expectation is that (2.5.2) is equivalent (up to Hamiltonian isotopy of the Lagrangian) to $\phi(x_2) = z_2$.

Note moreover that the intermediate point $[\underline{y}]$ is uniquely determined by $([\underline{x}], [\underline{z}])$, which as before would proves embeddedness of the geometric composition $L_\alpha^T \circ L_\beta$ if the same fact holds after adjustment to achieve a smooth structure.

In the true Lagrangian setting of [53], this move has not been addressed yet.

Remark 2.5.7 (Monoidal structures and gauge theory for disconnected surfaces). Note that the functor arising from infinite dimensional gauge theory in Example 2.5.2 can equally be applied to disconnected surfaces and cobordisms and intertwines the disjoint union $\sqcup$ on Bor_{2+1} with a natural monoidal structure on the symplectic category—the Cartesian product:

$$\mathscr{A}(\Sigma \sqcup \Sigma') = \mathscr{A}(\Sigma) \times \mathscr{A}(\Sigma'), \qquad \mathscr{L}(Y \sqcup Y') = \mathscr{L}(Y) \times \mathscr{L}(Y').$$

The same can be said for the representation spaces in Example 2.5.4, but it no longer holds in the gauge theoretic settings in which we actually obtain smooth, finite dimensional symplectic manifolds and Lagrangians. While the symmetric product of a disconnected surface at least is given by a union of Cartesian products, e.g.,

$$\mathrm{Sym}^2(\Sigma \sqcup \Sigma') = \mathrm{Sym}^2(\Sigma) \sqcup \mathrm{Sym}^1(\Sigma) \times \mathrm{Sym}^1(\Sigma') \sqcup \mathrm{Sym}^2(\Sigma'),$$

the representation spaces of Remark 2.5.5 become singular on disconnected surfaces. Indeed, a puncture $\mathrm{pt} \in \Sigma \sqcup \Sigma'$ lies on only one of the connected components, w.l.o.g. $\mathrm{pt} \in \Sigma$, so that the holonomy of a flat connection yields an element of

$$\mathrm{Hom}\big(\pi_1((\Sigma \sqcup \Sigma')\setminus\mathrm{pt}), SU(2)\big) = \mathrm{Hom}\big(\pi_1(\Sigma\setminus\mathrm{pt}), SU(2)\big) \times \mathrm{Hom}\big(\pi_1(\Sigma'), SU(2)\big),$$

and thus the moduli space of flat connections is $\widehat{M}_{\Sigma\sqcup\Sigma'} = \widehat{M}_\Sigma \times M_{\Sigma'}$, where the second factor is the singular representation space from Example 2.5.4.

Moreover, adding a requirement of compatibility with monoidal structures to our notion of Floer field theory, such as $\Sigma \sqcup \Sigma' \mapsto M_{\Sigma\sqcup\Sigma'} = M_\Sigma \times M_{\Sigma'}$ for a functor $\mathrm{Bor}_{2+1} \to \mathrm{Symp}$, only makes sense if we also have compatibility such as $M \times M' \mapsto \mathscr{C}_{M\times M'} = \mathscr{C}_M \otimes \mathscr{C}_{M'}$ for the functor Symp $\to$ Cat, i.e., a natural factorization of the category $\mathscr{C}_{M\times M'}$ that is associated to a Cartesian product of symplectic manifolds. However, our construction of the symplectic 2-category and the induced functor in Sect. 3.5 is such that the objects of $\mathscr{C}_{M\times M'}$ are general Lagrangian submanifolds of $M \times M'$, not just split Lagrangians $L \times L' \subset M \times M'$ arising from objects $L \in \mathrm{Obj}_{\mathscr{C}_M}$ and $L' \in \mathrm{Obj}_{\mathscr{C}_{M'}}$ in the categories associated to the factors of the Cartesian product.

Homological algebra allows one to formulate a sense in which refined versions of these categories may be equivalent, $\widetilde{\mathscr{C}}_{M\times M'} \sim \widetilde{\mathscr{C}}_M \otimes \widetilde{\mathscr{C}}_{M'}$, but it would likely require significant restrictions on the geometry of the symplectic manifolds M, M'.

2.6 *Atiyah–Floer Type Conjectures for 3-Manifold Invariants*

This section discusses the invariants of 3-manifolds in the sense of Sect. 1.1 which arise abstractly from 2 + 1 connected Floer field theories as in Definition 2.4.4, and in the more specific examples surveyed in Sect. 2.5. The notion of field theories originated with the idea of obtaining invariants for manifolds by decomposing them into simpler pieces. This also motivated the Atiyah–Floer conjecture in the context of Example 2.5.4 and Heegaard–Floer homology, in which a (conjectural) invariant $|I| : |Man_3| \to |\mathrm{Gr}|$ takes values in isomorphism classes of groups and is constructed roughly as follows (c.f. the outline before Example 2.5.6).

1. Choose a representative of $[Y]$ and a Heegaard splitting $Y = H_0^- \cup_\Sigma H_1$ along a surface Σ into two handlebodies H_i with $\partial H_i = \Sigma^-$. This is a special case of a decomposition of the morphism $[Y] \in \mathrm{Mor}_{2+1}(\emptyset, \emptyset)$ given by $[Y] = [H_0^-] \circ [H_1]$.

2. Represent the dividing surface Σ by a symplectic manifold M_Σ and the two handlebodies by Lagrangians $L_{H_i} \subset M_\Sigma$, e.g., as follows in the Examples:

   Example 2.5.4: Using the map $\pi_1(\Sigma) \to \pi_1(H)$ induced by inclusion $\Sigma \hookrightarrow H$ we set
   $L_H := \{\rho \mid \rho(\gamma) = \text{id } \forall[\gamma] = 0 \in \pi_1(H)\} \subset M_\Sigma := \text{Hom}(\pi_1(\Sigma), G)/\sim$.
   Example 2.5.6: For g the genus of Σ and disjoint generators $\alpha_1, \ldots, \alpha_g \subset \Sigma \subset H$ of $\pi_1(H)$ (an additional choice that the invariant may depend on) we set
   $L_H := T_{\underline{\alpha}} := \{[(z_1, \ldots, z_g)] \mid z_i \in \alpha_i\} \subset M_\Sigma := \text{Sym}^g \Sigma$.

3. Take $|I|([Y])$ to be the isomorphism class of the Floer homology $HF(L_{H_0}, L_{H_1})$.
4. Check that different choices of representatives and Heegaard splittings yield isomorphic Floer homology groups.

The last step of this program is a major challenge. In the context of Example 2.5.4, this step would follow from the Atiyah–Floer conjecture below, since instanton Floer homology arises from the ASD Yang–Mills equation on $\mathbb{R} \times Y$, thus does not depend on the choice of a Heegaard splitting, and in fact yields a 3-manifold invariant, i.e., is independent—up to isomorphism—from other choices involved in the construction. In the context of Example 2.5.6, the invariance in Step 4 was proven as part of the construction [52], but also follows from the analogue of the Atiyah–Floer conjecture established in [35], which identifies the three flavors of Heegaard Floer homology with three flavors of monopole Floer homology. The latter arise from the Seiberg–Witten equation on $\mathbb{R} \times Y$ and were proven to be a 3-manifold invariant in [33].

Conjecture 2.6.1 (Atiyah–Floer type conjectures for Heegaard splittings). For any Heegaard splitting $Y = H_0^- \cup_\Sigma H_1$ of a closed 3-manifold Y there are isomorphisms

Example 2.5.4: $HF(L_{H_0}, L_{H_1}) \simeq HF_{\text{inst}}(Y)$, if Y is a homology 3-sphere,[25]
Example 2.5.6: $HF^{\cdots}(L_{H_0}, L_{H_1}) \simeq HF^{\cdots}_{\text{mon}}(Y)$ for the three versions HF^+, HF^-, $\widehat{HF}$.

In the context of Example 2.5.4, a well defined part of this conjecture (equality of Euler characteristics) was proven by Taubes [65]. Defining the full Lagrangian Floer homology would require a notion of pseudoholomorphic curves in the singular representation space M_Σ, which has not yet been approached. Aside from this, a proof approach was outlined in [57, 71], extending the proof in [18] of the well posed Atiyah–Floer type conjecture described in Remark 2.6.4. Another well-defined version of the original Atiyah–Floer conjecture for trivial $SU(2)$-bundles was formulated by Salamon [57] in the context of the infinite dimensional Floer field theory outlined in Example 2.5.2. For Heegaard splittings of homology 3-spheres $Y = H_0^- \cup_\Sigma H_1$ it asserts the existence of an isomorphism that involves an instanton Floer homology for the pair of infinite dimensional Lagrangians $\mathcal{L}_{H_0}, \mathcal{L}_{H_1}$ and was

[25] A closed 3-manifold is called (integral) homology 3-sphere if its homology groups with $\mathbb{Z}$-coefficients $H_*(Y; \mathbb{Z}) \simeq H_*(S^3; \mathbb{Z})$ coincide with those of the 3-sphere. This assumption guarantees the absence of nontrivial reducible connections on Y.

recently proven in [60] with field theoretic methods. Roughly speaking, the existence of a 2-category which comprises both handlebodies H and their associated Lagrangians $\mathscr{L}_H$ as 1-morphisms allows us to express the notion of a "local" isomorphism $H \sim \mathscr{L}_H$, which—once proven—implies more "global" isomorphisms (see Remark 3.4.5) such as

$$HF_{\text{inst}}([0,1] \times \Sigma, \mathscr{L}_{H_0} \times \mathscr{L}_{H_1}) \simeq HF_{\text{inst}}(Y). \tag{2.6.1}$$

In particular, this proves that the above Steps 1–3 applied to Example 2.5.2 yield a well-defined invariant for homology 3-spheres, i.e., the left-hand side of (2.6.1) is independent of the choice of Heegaard splitting, as required in Step 4.

A more conceptual reason[26] for the invariance in Step 4 would be given by an extension of the constructions in Steps 1–3 to a 3-manifold invariant resulting from a (connected) 2 + 1 Floer field theory as outlined below, together with an extension of the symplectic category to a 2-category as in Example 3.5.1.

1. The Heegaard splitting $Y = H_0^- \cup_\Sigma H_1$ is a decomposition of the morphism $[Y] \in \mathrm{Mor}_{\mathrm{Bor}_{2+1}^{\mathrm{conn}}}(\emptyset, \emptyset)$ given by $[Y] = [H_0^-] \circ [H_1]$.
2. The representation by symplectic data can be viewed as determining parts of a functor $\mathscr{F} : \mathrm{Bor}_{2+1}^{\mathrm{conn}} \to \mathrm{Symp}$ by associating to the empty set $\emptyset \in \mathrm{Obj}_{\mathrm{Bor}_{2+1}^{\mathrm{conn}}}$ the trivial symplectic manifold given by a point $\mathscr{F}(\emptyset) := \mathrm{pt} \in \mathrm{Obj}_{\mathrm{Symp}}$, to nonempty surfaces $\Sigma \in \mathrm{Obj}_{\mathrm{Bor}_{2+1}^{\mathrm{conn}}}$ the given symplectic manifolds $\mathscr{F}(\Sigma) := M_\Sigma$, and to handlebodies $H \in \mathrm{Mor}_{\mathrm{Bor}_{2+1}^{\mathrm{conn}}}(\emptyset, \Sigma)$ the Lagrangian $\mathscr{F}(H_i) := L_{H_i} \subset \mathrm{pt}^- \times M_\Sigma$.
3. The Floer homology $HF(L_{H_0}, L_{H_1}) = \mathrm{Mor}^2_{\mathrm{Symp}}(L_{H_0}, L_{H_1})$ is the 2-morphism space for $L_{H_0}, L_{H_1} \in \mathrm{Mor}^1_{\mathrm{Symp}}(\mathrm{pt}, M_\Sigma)$ in the symplectic 2-category.
4. Check that the construction in 2. extends to a functor $\mathscr{F} : \mathrm{Bor}_{2+1}^{\mathrm{conn}} \to \mathrm{Symp}$.

Here, the functoriality in Step 4 guarantees in particular that different Heegaard decompositions $\widetilde{H}_0^- \cup_{\widetilde{\Sigma}} \widetilde{H}_1 = Y = H_0^- \cup_\Sigma H_1$ of the same 3-manifold, i.e., different factorizations $[\widetilde{H}_0^-] \circ [\widetilde{H}_1] = [Y] = [H_0^-] \circ [H_1] \in \mathrm{Mor}_{\mathrm{Bor}_{2+1}^{\mathrm{conn}}}(\emptyset, \emptyset)$ are mapped to equivalent composable chains of Lagrangians

$$[L_{\widetilde{H}_0^-}] \circ [L_{\widetilde{H}_1}] = [\underline{L}_Y] = [L_{H_0^-}] \circ [L_{H_1}] \ \in\ \mathrm{Mor}_{\mathrm{Symp}}(\mathrm{pt}, \mathrm{pt}).$$

Within the symplectic 2-category, this corresponds to isomorphic 1-morphisms

$$L_{\widetilde{H}_0^-} \circ L_{\widetilde{H}_1} \ \sim\ \underline{L}_Y \ \sim\ L_{H_0^-} \circ L_{H_1} \qquad \text{in } \mathrm{Mor}^1_{\mathrm{Symp}}(\mathrm{pt}, \mathrm{pt}).$$

Now the symplectic 2-morphism spaces extend to tuples using quilted Floer homology as explained in Remark 3.4.4 and Sect. 3.5. These cyclic morphism spaces have

[26] This reasoning is based on noting that Heegaard splittings $Y = H_0^- \cup_\Sigma H_1$ arise from special Cerf decompositions $Y = Y_{\alpha_1}^- \cup_{\Sigma_1} \dots Y_{\alpha_n}^- \cup_\Sigma Y_{\beta_n} \dots \cup_{\Sigma_1'} Y_{\beta_1}$ in which all handles of the same index are grouped together. Composing the handles of equal index yields the corresponding handlebodies $H_1 = Y_{\beta_n} \dots \cup_{\Sigma_1'} Y_{\beta_1}$ and $H_0 = (Y_{\alpha_1}^- \cup_{\Sigma_1} \dots Y_{\alpha_n}^-)^- = Y_{\alpha_n} \dots \cup_{\Sigma_1} Y_{\alpha_1}$, and the moves between Heegaard splittings can be expressed in terms of Cerf moves.

a cyclic symmetry that in particular induces identifications $\mathrm{Mor}^2_{\mathrm{Symp}}(L_{H_0}, L_{H_1}) = \mathrm{Mor}^2_{\mathrm{Symp}}(\mathrm{id}_{\mathrm{pt}}, L_{H_0^-} \circ L_{H_1})$ where $\mathrm{id}_{\mathrm{pt}} \in \mathrm{Mor}^1_{\mathrm{Symp}}(\mathrm{pt}, \mathrm{pt})$ is the identity element given by the diagonal. With that, the isomorphism between 1-morphisms induces an isomorphism between the Floer homologies viewed as 2-morphism spaces,

$$\begin{aligned} HF(L_{H_0}, L_{H_1}) &= \mathrm{Mor}^2_{\mathrm{Symp}}(\mathrm{id}_{\mathrm{pt}}, L_{H_0^-} \circ L_{H_1}) \\ &\simeq \mathrm{Mor}^2_{\mathrm{Symp}}(\mathrm{id}_{\mathrm{pt}}, L_{\widetilde{H}_0^-} \circ L_{\widetilde{H}_1}) = HF(L_{\widetilde{H}_0}, L_{\widetilde{H}_1}). \end{aligned} \tag{2.6.2}$$

While, this is a more conceptual explanation of the invariance of Heegaard Floer homology than the direct proof by Ozsvath–Szabo in [52], it is yet to be completed in the setting of Example 2.5.6. In the above language, [53, Lemma 3.17] shows that the geometric composition $L_{Y_{\alpha_n}} \circ \ldots \circ L_{Y_{\alpha_1}}$ is smoothly isotopic to the Heegaard torus $L_{H_0} = T_{\underline{\alpha}}$ used in [52], and [36] announces this to be a Hamiltonian isotopy, but a result along these lines is so far only proven for handle slides (changing the order between handle attachments and diffeomorphisms) in [54].

These 2-categorical considerations do however lead to natural extensions of the Atiyah–Floer type conjectures to Cerf decompositions.

Conjecture 2.6.2 (Atiyah–Floer type conjectures for (cyclic) Cerf decompositions). For any Cerf decomposition $[Y] = [Y_{01}] \circ \ldots [Y_{(k-1)k}]$ of a closed 3-manifold Y into simple cobordisms $Y_{i(i+1)}$ there are isomorphisms

Example 2.5.4: $HF(L_{Y_{01}}, \ldots, L_{Y_{(k-1)k}}) \simeq HF_{\mathrm{inst}}(Y)$,
Example 2.5.6: $HF^{\cdots}(L_{Y_{01}}, \ldots, L_{Y_{(k-1)k}}) \simeq HF^{\cdots}_{\mathrm{mon}}(Y)$.

Another version of this conjecture is an expected isomorphism between link invariants that arise from a Floer field theory for tangle categories in [80] (in which invariance under isotopy of the link embedding remains to be proven) and Floer homology invariants defined from singular instantons in [34].

Here, the Cerf decomposition $Y = Y_{01} \cup_{\Sigma_1} \ldots \cup_{\Sigma_{k-1}} Y_{(k-1)k}$ arises from a Morse function $f : Y \to \mathbb{R}$ and choices of regular level sets $\Sigma_j = f^{-1}(b_j)$ which separate the critical points of f. Analogous conjectures are obtained by working with cyclic Cerf decompositions $Y = \big(Y_{01} \cup_{\Sigma_1} \ldots \cup_{\Sigma_{k-1}} Y_{(k-1)k}\big)/\Sigma_0 \simeq \Sigma_k$ which arise from S^1-valued Morse functions $f : Y \to S^1$ and regular level sets $\Sigma_j = f^{-1}(b_j)$ with a cyclic identification $\partial Y_{01} \supset \Sigma_0 = f^{-1}(b_0 = b_k) = \Sigma_k \subset \partial Y_{(k-1)k}$. In this setting we can work as in Example 2.5.6 with symmetric products $\mathrm{Sym}^{g+n}(\Sigma)$, where g is the genus of the surface Σ and $n \in \mathbb{Z}$ is any fixed integer, by restricting consideration to Morse functions whose regular fibers have genus $\geq -n$ as laid out in [36, 53, 66]. Similarly, cyclic Cerf decompositions in the context of Example 2.5.4 allow us to work with nontrivial bundles as in Remark 2.5.5, and thus obtain well-defined Atiyah–Floer type conjectures, most notably for nontrivial $SO(3)$ bundles as discussed below.

Example 2.6.3 (Cyclic Atiyah–Floer type conjecture for nontrivial $SO(3)$ bundles). The Floer field theory outlined in Example 2.5.4 is made rigorous in [80] by replacing trivial $SU(2)$-bundles with nontrivial $SO(3)$-bundles. However, this excludes the empty set (over which any bundle is trivial) from the objects and thus does not allow for handlebodies (cobordisms from the empty set to a surface) as morphisms. So in this context we cannot rigorously formulate an Atiyah–Floer type Conjecture 2.6.1 for Heegaard splittings, but Conjecture 2.6.2 for cyclic Cerf decompositions does have a well-defined meaning: An identification between the instanton Floer homology of a 3-manifold Y with cyclic Cerf decomposition and the Floer homology of a cyclic sequence of Lagrangians $\underline{L}_Y$ arising from this Cerf decomposition,

$$HF\big(\underline{L}_Y = (L_{Y_{(j-1)j}})_{j\in\mathbb{Z}_k}\big) \simeq HF_{\text{inst}}\big(Y = (Y_{01} \cup_{\Sigma_1} \ldots \cup_{\Sigma_{k-1}} Y_{(k-1)k})/\Sigma_0 \simeq \Sigma_k\big).$$

Here $HF(\underline{L}_Y)$ is defined in terms of quilted pseudoholomorphic cylinders in [75], but is also directly identical to the standard Floer homology

$$HF(\underline{L}_Y) = HF\big((L_{Y_{01}} \times \ldots L_{Y_{(k-1)k}})^T, \Delta_{M_{\Sigma_0}} \times \ldots \Delta_{M_{\Sigma_{k-1}}}\big)$$

for a pair of Lagrangians in $M_{\Sigma_0} \times M^-_{\Sigma_0} \times \ldots M_{\Sigma_{k-1}} \times M^-_{\Sigma_{k-1}}$, where the first requires a permutation of factors $(\ldots)^T : M^-_{\Sigma_0} \times \ldots M^-_{\Sigma_{k-1}} \times M_{\Sigma_0} \to M_{\Sigma_0} \times M^-_{\Sigma_0} \times \ldots M^-_{\Sigma_{k-1}}$.

Remark 2.6.4 (Approaches to Atiyah–Floer conjecture for nontrivial $SO(3)$ bundles). The cyclic version of Conjecture 2.6.2 in Example 2.6.3 was proven by Dostoglou–Salamon [18] for 3-manifolds equipped with a Morse function $f : Y \to S^1$ without critical points, so that all fibers $\Sigma_i \simeq \Sigma$ are diffeomorphic and thus $Y = ([0, 1] \times \Sigma)/\phi$ is the mapping cylinder of a diffeomorphism $\phi : \Sigma \to \Sigma$ on a regular level set $\Sigma = f^{-1}(b_0)$ arising from the flow of ∇f on $Y \setminus f^{-1}(b_0)$. This gives rise to a symplectomorphism $L_\phi : M_\Sigma \to M_\Sigma$, for which the Floer homology $HF(L_\phi) = HF(\text{graph}\, L_\phi, \Delta_{M_\Sigma})$ can be constructed without boundary conditions. Then, the proof of the Atiyah–Floer type isomorphism $HF(L_\phi) \simeq HF_{\text{inst}}(([0, 1] \times \Sigma)/\phi)$ directly identifies the Floer complexes by an adiabatic limit in which the metric on Σ is scaled by $\varepsilon^2 \to 0$.

In the presence of critical points this argument is expected to generalize by partitioning $Y = (Y_{01} \cup_{\Sigma_1} \ldots \cup_{\Sigma_{k-1}} Y_{(k-1)k})/\Sigma_0 \simeq \Sigma_k$ into thickenings of the surfaces $[0, 1] \times \Sigma_j$ with metric $\mathrm{d}s^2 + \varepsilon^2 g_{\Sigma_j}$ and handle attachments Y_{ij} with metric $\varepsilon^2 g_{Y_{ij}}$. As a result, the volumes $\text{Vol}([0, 1] \times \Sigma_j) \sim \varepsilon^2$ and $\text{Vol}(Y_{ij}) \sim \varepsilon^3$ scale differently, which indicates different degenerations on these types of pieces. Here, the absence of reducibles guarantees linear bounds $d\big(A, \mathscr{A}_{\text{flat}}\big) \leq C\|F_A\|$ of the distance of a connection (on Σ_j or Y_{ij}) to the flat connections in terms of its curvature, so that the adiabatic limit analysis [18] can be combined with the analytic setup for boundary conditions in gauge theory [69, 70] to obtain a compactness result: Solutions of the ASD equation defining the instanton Floer differential converge for $\varepsilon \to 0$ to pseudoholomorphic curves in the M_{Σ_j}, whose boundaries match up via the $L_{Y_{ij}}$, thus giving rise to a contribution to the (quilted) Lagrangian Floer differential. This initial motivation for the conjecture was fleshed out with analytic details in [72] and recently explicitly put into quilted settings in [20], but the proof of a 1-1 correspondence between the differentials remains to be completed.

An alternative approach closer to completion is based on a strategy outlined in [57, 71] via an intermediate instanton Floer homology $HF_{\text{inst}}\big(\bigsqcup_j[0,1]\times\Sigma_j, (\mathscr{L}_{Y_{ij}})\big)$ associated to the manifold with boundary $\bigsqcup_j[0,1]\times\Sigma_j$ and boundary conditions given by infinite dimensional Lagrangians $\mathscr{L}_{Y_{ij}}$ as in Example 2.5.2. Its identification with $HF\big(\underline{L}_Y = (L_{Y_{ij}})\big)$ should follow from a direct generalization of the adiabatic limit analysis [18] to Lagrangian boundary conditions. On the other hand, an isomorphism as in (2.6.1) between the "closed" and "open" instanton Floer homologies $HF_{\text{inst}}\big(\bigsqcup_j[0,1]\times\Sigma_j, (\mathscr{L}_{Y_{ij}})\big) \simeq HF_{\text{inst}}(Y)$ could also be approached by degenerating only the metric on the handle attachments Y_{ij}. Instead of using a metric degeneration, this can also be approached by a "local to global" field theoretic argument as explained above and in Remark 3.4.5, where the strategy is to prove an isomorphism $Y_{ij} \sim \mathscr{L}_{Y_{ij}}$ in a 2-categorical setting, which then implies the desired isomorphism of Floer homologies similar to the argument for equation (2.6.2) above. In the case of trivial SU(2)-bundles, this approach is implemented in [60] and directly transfers to any setting in which the fundamental classes $[L_{Y_{ij}}]$ of the associated finite dimensional Lagrangians induce well defined classes in the "2-morphism spaces" resp. "localized Floer theories" $HF_{\text{inst}}(Y_{ij}, \mathscr{L}_{Y_{ij}})$.

Finally, a third approach pioneered by Fukaya [24] is to avoid adiabatic limit analysis and construct a direct chain map between the instanton and Lagrangian Floer chain complexes. This should again be understood as a "local to global" field theoretic argument, based on an implicit isomorphism $Y_{ij} \sim L_{Y_{ij}}$. In this case, an appropriate 2-categorical setting needs to combine the ASD and Cauchy–Riemann equation. Aside from metric degeneration proposals in [24], this can be achieved by Lagrangian seam conditions as discussed in Example 3.6.7 and [41].

3 Extensions of Floer Field Theories

3.1 2-Categories and Bicategories

In the construction of both the bordism and symplectic categories we introduced an equivalence relation between the morphisms in order to obtain a geometrically meaningful notion of composition. An algebraically cleaner way of phrasing the requirements on this relation—in particular compatibility with the desired notion of composition—is in terms of 2-morphisms between the morphisms, forming either a 2-category or the slightly weaker notion of bicategory. Once these notions and some examples are established, we will cast the construction of bordism and symplectic categories in these 2-categorical terms.

Definition 3.1.1 A **2-category** $\mathscr{C}$ is a category enriched in categories, i.e., consists of

- a set $\text{Obj}_{\mathscr{C}}$ of **objects**,
- for each pair $x_1, x_2 \in \text{Obj}_{\mathscr{C}}$ a category of **morphisms** $\text{Mor}_{\mathscr{C}}(x_1, x_2)$, i.e.,

- a set $\mathrm{Mor}^1_{\mathscr{C}}(x_1, x_2)$ of **1-morphisms**,
- for each pair $f, g \in \mathrm{Mor}^1_{\mathscr{C}}(x_1, x_2)$ a set of **2-morphisms** $\mathrm{Mor}^2_{\mathscr{C}}(f, g)$,
- for each triple $f, g, h \in \mathrm{Mor}^1_{\mathscr{C}}(x_1, x_2)$ an associative **vertical composition**

$$\mathrm{Mor}^2_{\mathscr{C}}(f, g) \times \mathrm{Mor}^2_{\mathscr{C}}(g, h) \to \mathrm{Mor}^2_{\mathscr{C}}(g, h), \quad (\alpha_{12}, \beta_{12}) \mapsto \alpha_{12} \circ_{\mathrm{v}} \beta_{12},$$

- for each $f \in \mathrm{Mor}^1_{\mathscr{C}}(x_1, x_2)$ an identity $\mathrm{id}_f \in \mathrm{Mor}^2_{\mathscr{C}}(f, f)$ for $\circ_{\mathrm{v}}$,

- a **composition functor** $\mathrm{Mor}_{\mathscr{C}}(x_1, x_2) \times \mathrm{Mor}_{\mathscr{C}}(x_2, x_3) \to \mathrm{Mor}_{\mathscr{C}}(x_1, x_3)$ for each triple $x_1, x_2, x_3 \in \mathrm{Obj}_{\mathscr{C}}$, i.e.,
  - an associative **horizontal composition on 1-morphisms**

$$\mathrm{Mor}^1_{\mathscr{C}}(x_1, x_2) \times \mathrm{Mor}^1_{\mathscr{C}}(x_2, x_3) \to \mathrm{Mor}^1_{\mathscr{C}}(x_1, x_3), \quad (f_{12}, f_{23}) \mapsto f_{12} \circ_{\mathrm{h}} f_{23},$$

  - for each $x \in \mathrm{Obj}_{\mathscr{C}}$ an identity $1_x \in \mathrm{Mor}^1_{\mathscr{C}}(x, x)$ for $\circ_{\mathrm{h}}$, that is $1_x \circ_{\mathrm{h}} f = f$ for any $f \in \mathrm{Mor}^1_{\mathscr{C}}(x, y)$ and $g \circ_{\mathrm{h}} 1_x = g$ for any $g \in \mathrm{Mor}^1_{\mathscr{C}}(w, x)$.
  - for any $(f_{12}, f_{23}), (g_{12}, g_{23}) \in \mathrm{Mor}^1_{\mathscr{C}}(x_1, x_2) \times \mathrm{Mor}^1_{\mathscr{C}}(x_2, x_3)$ an associative **horizontal composition on 2-morphisms**,

$$\begin{aligned} \mathrm{Mor}^2_{\mathscr{C}}(f_{12}, g_{12}) \times \mathrm{Mor}^2_{\mathscr{C}}(f_{23}, g_{23}) &\longrightarrow \mathrm{Mor}^2_{\mathscr{C}}(f_{12} \circ_{\mathrm{h}} f_{23}, g_{12} \circ_{\mathrm{h}} g_{23}) \\ (\alpha_{12}, \alpha_{23}) &\longmapsto \alpha_{12} \circ_{\mathrm{h}} \alpha_{23}, \end{aligned}$$

  that is compatible with identities, i.e. for $f_{12} = g_{12} \in \mathrm{Mor}^1_{\mathscr{C}}(x_1, x_2)$ and $f_{23} = g_{23} \in \mathrm{Mor}^1_{\mathscr{C}}(x_2, x_3)$ we have

$$\mathrm{id}_{f_{12}} \circ_{\mathrm{h}} \mathrm{id}_{f_{23}} = \mathrm{id}_{f_{12} \circ_{\mathrm{h}} f_{23}},$$

  and is compatible with vertical composition, i.e., for $\alpha_{12}, \beta_{12} \in \mathrm{Mor}^2_{\mathscr{C}}(f_{12}, g_{12})$ and $\alpha_{23}, \beta_{23} \in \mathrm{Mor}^2_{\mathscr{C}}(f_{23}, g_{23})$ we have

$$(\alpha_{12} \circ_{\mathrm{v}} \beta_{12}) \circ_{\mathrm{h}} (\alpha_{23} \circ_{\mathrm{v}} \beta_{23}) = (\alpha_{12} \circ_{\mathrm{h}} \alpha_{23}) \circ_{\mathrm{v}} (\beta_{12} \circ_{\mathrm{h}} \beta_{23}).$$

A graphical representation of the structure and axioms of 2-categories is by string diagrams, as discussed in Sect. 3.4. In [73] these were motivated as a natural visualization of four-dimensional manifolds with boundary and corners, as they appear in the extension of $2+1$ bordism categories. However, we will see in Sect. 3.2 that instead of a 2-category this yields the following notion of bicategory, in which the horizontal unital and associativity requirements on 1-morphisms are relaxed.

Definition 3.1.2 A **bicategory** $\mathscr{C}$ consists of a set $\mathrm{Obj}_{\mathscr{C}}$ of objects and categories of morphisms $\mathrm{Mor}_{\mathscr{C}}(x_1, x_2)$ as in Definition 3.1.1 (i.e., 1-morphisms, 2-morphisms, vertical composition $\circ_{\mathrm{v}}$, and units id_f), and for each triple $x_1, x_2, x_3 \in \mathrm{Obj}_{\mathscr{C}}$ a **horizontal bifunctor** $\circ_{\mathrm{h}} : \big(\mathrm{Mor}_{\mathscr{C}}(x_1, x_2), \mathrm{Mor}_{\mathscr{C}}(x_2, x_3)\big) \to \mathrm{Mor}_{\mathscr{C}}(x_1, x_3)$ consisting of

- a **horizontal map on 1-morphisms**

$$\circ_h^1 : \mathrm{Mor}^1_{\mathscr{C}}(x_1, x_2) \times \mathrm{Mor}^1_{\mathscr{C}}(x_2, x_3) \to \mathrm{Mor}^1_{\mathscr{C}}(x_1, x_3),$$

- for any composable pairs (f_{12}, f_{23}), (g_{12}, g_{23}) a **horizontal map on 2-morphisms**

$$\circ_h^2 : \mathrm{Mor}^2_{\mathscr{C}}(f_{12}, g_{12}) \times \mathrm{Mor}^2_{\mathscr{C}}(f_{23}, g_{23}) \to \mathrm{Mor}^2_{\mathscr{C}}\big(\circ_h(f_{12}, f_{23}), \circ_h(g_{12}, g_{23})\big),$$

that make $\circ_h$ into a bifunctor and **horizontal composition up to 2-isomorphism**, as follows:

- $\circ_h = (\circ_h^1, \circ_h^2)$ is compatible with vertical identities, i.e., $\circ_h^2(\mathrm{id}_{f_{12}}, \mathrm{id}_{f_{23}}) = \mathrm{id}_{\circ_h^1(f_{12}, f_{23})}$,
- $\circ_h^2$ is associative and compatible with vertical composition, i.e.,

$$\circ_h^2(\circ_h^2(\alpha_{12}, \alpha_{23}), \alpha_{34}) = \circ_h^2(\alpha_{12}, \circ_h^2(\alpha_{23}, \alpha_{34})),$$
$$\circ_h^2(\alpha_{12} \circ_v \beta_{12}, \alpha_{23} \circ_v \beta_{23}) = \circ_h^2(\alpha_{12}, \alpha_{23}) \ \circ_v \ \circ_h^2(\beta_{12}, \beta_{23}),$$

- $\circ_h^1$ is associative up to 2-isomorphism, i.e., $\circ_h^1\big(f, \circ_h^1(g, h)\big) \sim \circ_h^1\big(\circ_h^1(f, g), h\big)$, where the relation $\sim$ on $\mathrm{Mor}^1_{\mathscr{C}}$ is defined by

$$k \sim k' \iff \exists \alpha, \beta \in \mathrm{Mor}^2_{\mathscr{C}} : \ \mathrm{id}_k = \alpha \circ_v \beta, \ \mathrm{id}_{k'} = \beta \circ_v \alpha,$$

- $\circ_h^1$ is unital up to 2-isomorphism, i.e., for each $y \in \mathrm{Obj}_{\mathscr{C}}$ there exist a (not necessarily unique) **weak identity 1-morphism** $1_y \in \mathrm{Mor}^1_{\mathscr{C}}(y, y)$ such that

$$\circ_h^1(f, 1_y) \sim f \quad \forall f \in \mathrm{Mor}^1_{\mathscr{C}}(x, y), \qquad \circ_h^1(1_y, g) \sim g \quad \forall g \in \mathrm{Mor}^1_{\mathscr{C}}(y, z).$$

An instructive non-example of a bicategory (or 2-category) is the attempt to extend the category of sets and maps by a notion of conjugacy as 2-morphisms.

Example 3.1.3 (Categorical structure of sets, maps, and conjugacy). The category of sets consists of sets as objects and maps $f_{12} : S_1 \to S_2$ as 1-morphisms from S_1 to S_2, with horizontal composition $f_{12} \circ_h f_{23}$ given by composition of maps. One might want to add further structure to the sets and ask it to be preserved by the maps—e.g., forming the linear category of vector spaces and homomorphisms—but the underlying form of many categories is given by sets and maps. In the linear category, a natural relation between homomorphisms arises from a change of basis, which is formalized as the conjugation with an isomorphism. Generally, for each pair of maps $f_{12}, g_{12} \in \mathrm{Mor}^1_{\mathscr{C}}(S_1, S_2)$ between the same two sets S_1, S_2, one would like to let 2-morphisms be given by conjugation with bijections,

$$\mathrm{Mor}^2_{\mathscr{C}}(f_{12}, g_{12}) := \big\{\alpha_{12} = (\alpha_1, \alpha_2) \,\big|\, \alpha_i : S_i \to S_i \text{ bijections}, \alpha_2^{-1} \circ f_{12} \circ \alpha_1 = g_{12}\big\}.$$

This defines a category $\mathrm{Mor}_{\mathscr{C}}(S_1, S_2)$ since conjugations have a well-defined (and associative, unital) vertical composition:

$$\alpha_2^{-1} \circ f_{12} \circ \alpha_1 = g_{12}, \quad \beta_2^{-1} \circ g_{12} \circ \beta_1 = h_{12} \quad \Rightarrow \quad (\alpha_2 \circ \beta_2)^{-1} \circ f_{12} \circ (\alpha_1 \circ \beta_1) = h_{12}.$$

That is, setting $(\alpha_1, \alpha_2) \circ_{\mathrm{v}} (\beta_1, \beta_2) := (\alpha_1 \circ \beta_1, \alpha_2 \circ \beta_2)$ composes two conjugacies from f_{12} to g_{12} and from g_{12} to h_{12} to a conjugacy from f_{12} to h_{12}. In other words, conjugacy is an equivalence relation—its transitivity corresponds to a well defined vertical composition. Next, a well-defined horizontal bifunctor would require conjugacy to be compatible with composition of maps, that is

$$\alpha_2^{-1} \circ f_{12} \circ \alpha_1 = g_{12}, \quad \tilde{\alpha}_3^{-1} \circ f_{23} \circ \tilde{\alpha}_2 = g_{23} \quad \Rightarrow \quad \gamma_3^{-1} \circ (f_{23} \circ f_{12}) \circ \gamma_1 = g_{23} \circ g_{12}$$

for some bijections $(\gamma_1, \gamma_3) := \circ_{\mathrm{h}}\big((\alpha_1, \alpha_2), (\tilde{\alpha}_2, \tilde{\alpha}_3)\big)$. However, this implication generally only holds if we have $\alpha_2 = \tilde{\alpha}_2$ in

$$\tilde{\alpha}_3^{-1} \circ f_{23} \circ \tilde{\alpha}_2 \circ \alpha_2^{-1} \circ f_{12} \circ \alpha_1 = g_{23} \circ g_{12}.$$

Thus, our constructions do not yield a 2-category or bicategory. This corresponds to the fact that composition of maps does not descend to a well-defined composition of conjugacy classes. In those terms, the above discussion shows that $[f_{12}] = [g_{12}]$ and $[f_{23}] = [g_{23}]$ does generally not imply $[f_{12} \circ f_{23}] = [g_{12} \circ g_{23}]$. In fact, we will see in Remark 3.1.7 that a bicategorical structure is exactly what is needed to obtain a well defined composition on the level of equivalence classes of 1-morphisms.

A better behaved notion of conjugacy-type equivalence between map-type objects is the following notion of natural transformations between functors, which are an equivalence relation (define a category) and are compatible with composition (fit into a horizontal composition functor) as we show in the following Lemma. Here we moreover review the category of functors and a composition functor on it.

Lemma 3.1.4 *The* **category of functors** $\mathrm{Fun}(\mathscr{C}, \mathscr{D})$ *is well defined as follows.*

- *Objects are functors* $\mathscr{F} : \mathscr{C} \to \mathscr{D}$.
- *Morphisms* $\eta \in \mathrm{Mor}_{\mathrm{Fun}(\mathscr{C},\mathscr{D})}(\mathscr{F}, \mathscr{G})$ *are natural transformations* $\eta : \mathscr{F} \Rightarrow \mathscr{G}$ *given by a map* $\eta : \mathrm{Obj}_{\mathscr{C}} \to \mathrm{Mor}_{\mathscr{D}}$ *which takes each* $x \in \mathrm{Obj}_{\mathscr{C}}$ *to a morphism* $\eta(x) \in \mathrm{Mor}_{\mathscr{D}}(\mathscr{F}(x), \mathscr{G}(x))$ *such that we have*

$$k \in \mathrm{Mor}_{\mathscr{C}}(x, y) \quad \Longrightarrow \quad \mathscr{F}(k) \circ \eta(y) = \eta(x) \circ \mathscr{G}(k).$$

- *Composition of natural transformations* $\eta : \mathscr{F} \Rightarrow \mathscr{G}$ *and* $\zeta : \mathscr{G} \Rightarrow \mathscr{H}$ *is given by* $(\eta \circ_{\mathrm{v}} \zeta) : x \mapsto \eta(x) \circ \zeta(x)$ *as in Fig. 6, giving rise to a map*

$$\circ_{\mathrm{v}} : \mathrm{Mor}_{\mathrm{Fun}(\mathscr{C},\mathscr{D})}(\mathscr{F}, \mathscr{G}) \times \mathrm{Mor}_{\mathrm{Fun}(\mathscr{C},\mathscr{D})}(\mathscr{G}, \mathscr{H}) \to \mathrm{Mor}_{\mathrm{Fun}(\mathscr{C},\mathscr{D})}(\mathscr{F}, \mathscr{H}).$$

- *The identity natural transformations* $\mathrm{id}_{\mathscr{F}} : \mathscr{F} \Rightarrow \mathscr{F}$ *are given by* $\mathrm{id}_{\mathscr{F}}(x) = \mathrm{id}_{\mathscr{F}(x)}$ *for all* $x \in \mathrm{Obj}_{\mathscr{C}}$.

Moreover, for any triple of categories $\mathscr{C}_0, \mathscr{C}_1, \mathscr{C}_2$, *the* **horizontal composition functor** $\circ_{\mathrm{h}} : \mathrm{Fun}(\mathscr{C}_0, \mathscr{C}_1) \times \mathrm{Fun}(\mathscr{C}_1, \mathscr{C}_2) \to \mathrm{Fun}(\mathscr{C}_0, \mathscr{C}_2)$ *is well defined as follows.*

- *Composition of functors* $(\mathscr{F}_{01}, \mathscr{F}_{12}) \mapsto \mathscr{F}_{01} \circ_{\mathrm{h}} \mathscr{F}_{12}$ *is given by composition of the maps* $\mathrm{Obj}_{\mathscr{C}_i} \to \mathrm{Obj}_{\mathscr{C}_{i+1}}$ *and* $\mathrm{Mor}_{\mathscr{C}_i} \to \mathrm{Mor}_{\mathscr{C}_{i+1}}$ *which make up* $\mathscr{F}_{i(i+1)}$ *for* $i = 0, 1$.

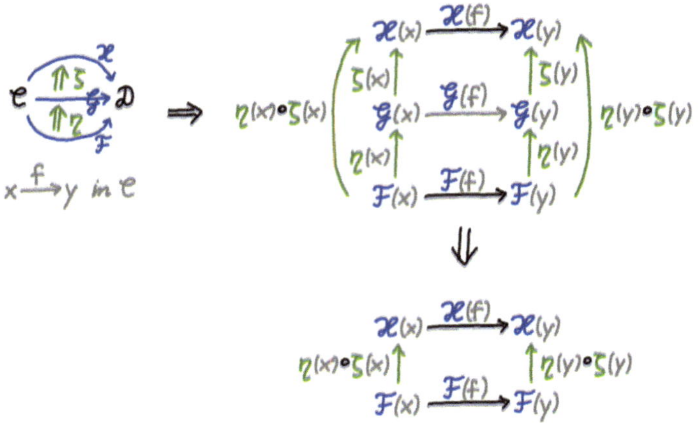

Fig. 6 Vertical composition of natural transformations

- *The identities for this horizontal composition are given by the identity functors* $1_{\mathscr{C}} \in \mathrm{Fun}(\mathscr{C}, \mathscr{C})$.
- *For each pair of objects* $(\mathscr{F}_{01}, \mathscr{F}_{12}), (\mathscr{G}_{01}, \mathscr{G}_{12}) \in \mathrm{Fun}(\mathscr{C}_0, \mathscr{C}_1) \times \mathrm{Fun}(\mathscr{C}_1, \mathscr{C}_2)$ *in the product category, the horizontal composition of natural transformations, as illustrated in Fig. 7 is*

$$\begin{aligned} \circ_{\mathrm{h}} : \ & \mathrm{Mor}_{\mathrm{Fun}(\mathscr{C}_0,\mathscr{C}_1)}(\mathscr{F}_{01}, \mathscr{G}_{01}) \times \mathrm{Mor}_{\mathrm{Fun}(\mathscr{C}_1,\mathscr{C}_2)}(\mathscr{F}_{12}, \mathscr{G}_{12}) \\ & \qquad \to \mathrm{Mor}_{\mathrm{Fun}(\mathscr{C}_0,\mathscr{C}_2)}(\mathscr{F}_{01} \circ \mathscr{F}_{12}, \mathscr{G}_{01} \circ \mathscr{G}_{12}) \\ & (\eta_{01} \circ_{\mathrm{h}} \eta_{12})(x) := \eta_{12}\big(\mathscr{F}_{01}(x)\big) \circ \mathscr{G}_{12}\big(\eta_{01}(x)\big) = \mathscr{F}_{12}\big(\eta_{01}(x)\big) \circ \eta_{12}\big(\mathscr{G}_{01}(x)\big). \end{aligned}$$

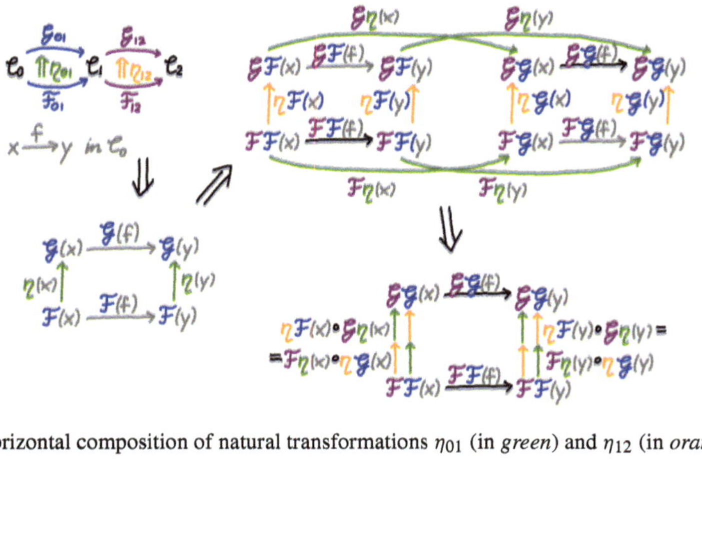

Fig. 7 Horizontal composition of natural transformations η_{01} (in *green*) and η_{12} (in *orange*)

Proof The (vertical) composition of natural transformations in $\mathrm{Fun}(\mathscr{C},\mathscr{D})$ is well-defined since for all $k \in \mathrm{Mor}_{\mathscr{C}}(x,y)$ we have

$$\begin{aligned}\mathscr{F}(k)\circ(\eta\circ\zeta)(y) = \mathscr{F}(k)\circ\eta(y)\circ\zeta(y) &= \eta(x)\circ\mathscr{G}(k)\circ\zeta(y)\\ &= \eta(x)\circ\zeta(x)\circ\mathscr{H}(k) = (\eta\circ\zeta)(x)\circ\mathscr{H}(k).\end{aligned}$$

It is associative by associativity of the composition $\circ$ in $\mathscr{D}$, and it is unital with $1_{\mathscr{F}} : x \mapsto \mathrm{id}_{\mathscr{F}(x)}$. The (horizontal) composition of functors is well defined in the same way in which composition of maps is well defined. The horizontal composition of natural transformations $\eta_{01} : \mathscr{F}_{01} \Rightarrow \mathscr{G}_{01}$ and $\eta_{12} : \mathscr{F}_{12} \Rightarrow \mathscr{G}_{12}$ is well defined since for all $k \in \mathrm{Mor}_{\mathscr{C}}(x,y)$ we have

$$\begin{aligned}(\mathscr{F}_{01}\circ\mathscr{F}_{12})(k)\circ(\eta_{01}\circ_{\mathrm{h}}\eta_{12})(y) &= \mathscr{F}_{12}\big(\mathscr{F}_{01}(k)\big)\circ\eta_{12}\big(\mathscr{F}_{01}(y)\big)\circ\mathscr{G}_{12}\big(\eta_{01}(y)\big)\\ &= \eta_{12}\big(\mathscr{F}_{01}(x)\big)\circ\mathscr{G}_{12}\big(\mathscr{F}_{01}(k)\big)\circ\mathscr{G}_{12}\big(\eta_{01}(y)\big)\\ &= \eta_{12}\big(\mathscr{F}_{01}(x)\big)\circ\mathscr{G}_{12}\big(\mathscr{F}_{01}(k)\circ\eta_{01}(y)\big)\\ &= \eta_{12}\big(\mathscr{F}_{01}(x)\big)\circ\mathscr{G}_{12}\big(\eta_{01}(x)\circ\mathscr{G}_{01}(k)\big)\\ &= \eta_{12}\big(\mathscr{F}_{01}(x)\big)\circ\mathscr{G}_{12}\big(\eta_{01}(x)\big)\circ\mathscr{G}_{12}\big(\mathscr{G}_{01}(k)\big)\\ &= (\eta_{01}\circ_{\mathrm{h}}\eta_{12})(x)\circ(\mathscr{G}_{01}\circ\mathscr{G}_{12})(k).\end{aligned}$$

Moreover, the identity $\eta_{12}\big(\mathscr{F}_{01}(x)\big)\circ\mathscr{G}_{12}\big(\eta_{01}(x)\big) = \mathscr{F}_{12}\big(\eta_{01}(x)\big)\circ\eta_{12}\big(\mathscr{G}_{01}(x)\big)$ follows from applying η_{12} to the morphism $\eta_{01}(x) : \mathscr{F}_{01}(x) \to \mathscr{G}_{01}$ in $\mathscr{C}_1$.

This horizontal composition is compatible with identities since for $\mathscr{F}_{ij} : \mathscr{C}_i \to \mathscr{C}_j$ and $x \in \mathrm{Obj}_{\mathscr{C}_0}$ we have

$$\big(1_{\mathscr{F}_{01}}\circ_{\mathrm{h}} 1_{\mathscr{F}_{12}}\big)(x) = \mathrm{id}_{\mathscr{F}_{12}(\mathscr{F}_{01}(x))}\circ\mathscr{F}_{12}\big(\mathrm{id}_{\mathscr{F}_{01}(x)}\big) = \mathrm{id}_{\mathscr{F}_{12}(\mathscr{F}_{01}(x))} = 1_{\mathscr{F}_{12}\circ_{\mathrm{h}}\mathscr{F}_{01}}(x),$$

and it is compatible with vertical composition since for $\eta_{ij} : \mathscr{F}_{ij} \Rightarrow \mathscr{G}_{ij}$ and $\zeta_{ij} : \mathscr{G}_{ij} \Rightarrow \mathscr{H}_{ij}$ and $x \in \mathrm{Obj}_{\mathscr{C}_0}$ we have

$$\begin{aligned}\big(\eta_{01}\circ_{\mathrm{v}}\zeta_{01}\big)\circ_{\mathrm{h}}\big(\eta_{12}\circ_{\mathrm{v}}\zeta_{12}\big)(x) &= (\eta_{12}\circ_{\mathrm{v}}\zeta_{12})\big(\mathscr{F}_{01}(x)\big)\circ\mathscr{H}_{12}\big((\eta_{01}\circ_{\mathrm{v}}\zeta_{01})(x)\big)\\ &= \eta_{12}\big(\mathscr{F}_{01}(x)\big)\circ\zeta_{12}\big(\mathscr{F}_{01}(x)\big)\circ\mathscr{H}_{12}\big(\eta_{01}(x)\big)\circ\mathscr{H}_{12}\big(\zeta_{01}(x)\big)\\ &= \eta_{12}\big(\mathscr{F}_{01}(x)\big)\circ\mathscr{G}_{12}\big(\eta_{01}(x)\big)\circ\zeta_{12}\big(\mathscr{G}_{01}(x)\big)\circ\mathscr{H}_{12}\big(\zeta_{01}(x)\big)\\ &= \big(\eta_{01}\circ_{\mathrm{h}}\eta_{12}\big)\circ_{\mathrm{v}}\big(\zeta_{01}\circ_{\mathrm{h}}\zeta_{12}\big)(x).\end{aligned}$$

□

The well-defined category of functors and horizontal composition functor now yield an extension of the category of categories in Example 2.1.5 to a 2-category.

Example 3.1.5 The **2-category of categories** Cat consists of

- objects given by categories $\mathscr{C}$,
- the morphism category for any pair of categories $\mathscr{C}_1, \mathscr{C}_2$ given by the category of functors $\mathrm{Fun}(\mathscr{C}_1, \mathscr{C}_2)$,

- the horizontal composition functor $\circ_h$: $\mathrm{Fun}(\mathscr{C}_0, \mathscr{C}_1) \times \mathrm{Fun}(\mathscr{C}_1, \mathscr{C}_2) \to \mathrm{Fun}(\mathscr{C}_0, \mathscr{C}_2)$ for each triple of categories $\mathscr{C}_0, \mathscr{C}_1, \mathscr{C}_2$.

Leading toward the next example of 2-categories, the construction of the symplectic category in Definition 2.2.2 can be understood as reconstructing a category from its simple morphisms and Cerf moves. For that purpose it is useful to express this Cerf data as the following "resolution" of the original category.

Example 3.1.6 (Resolution of a category with Cerf decompositions). Given a category $\mathscr{C}$ with Cerf decompositions into simple morphisms unique up to Cerf moves as in Definition 2.3.2, the 2-category $\mathscr{C}^{\#}$ is defined by

- the set of objects $\mathrm{Obj}_{\mathscr{C}^{\#}} := \mathrm{Obj}_{\mathscr{C}}$,
- the set of 1-morphisms $\mathrm{Mor}^1_{\mathscr{C}^{\#}}$ given by finite composable chains of simple morphisms, with horizontal composition given by concatenation of chains,
- the set of 2-morphisms $\mathrm{Mor}^2_{\mathscr{C}^{\#}} \subset \mathrm{Mor}^1_{\mathscr{C}^{\#}} \times \mathrm{Mor}^1_{\mathscr{C}^{\#}}$ given by pairs of 1-morphisms that are related via a sequence of Cerf moves.

For this to define a 2-category, one should allow for empty chains as identity 1-morphisms. Vertical composition of 2-morphisms is well defined, associative and unital since relation via Cerf moves is an equivalence relation. Horizontal composition of 2-morphisms $(\underline{L}_{12}, \underline{L}'_{12}) \circ^2_{\mathrm{h}} (\underline{L}_{23}, \underline{L}'_{23}) := (\underline{L}_{12}\#\underline{L}_{23}, \underline{L}'_{12}\#\underline{L}'_{23})$ is compatible with vertical composition because the equivalence via Cerf moves is designed to be compatible with concatenation. Now—although the composition in $\mathscr{C}$ is encoded in $\mathscr{C}^{\#}$ only via the Cerf moves—the original category $\mathscr{C}$ can be reconstructed as the quotient $\mathscr{C} \cong \mathscr{C}^{\#}/\!\sim\, = |\mathscr{C}^{\#}|$ defined in Remark 3.1.7 below.

This is exactly how the symplectic category was constructed in Definition 2.2.2. Indeed, for $\mathscr{C} = \mathrm{Symp}$ the above construction reproduces Definition 2.2.1 of the extended symplectic category $\mathscr{C}^{\#} = \mathrm{Symp}^{\#}$ and extends it to a 2-category by adding the geometric composition moves as 2-morphisms, whose vertical and horizontal compositions are well defined due to the equivalence relation $\sim$ being compatible with the horizontal 1-composition $\circ^1_{\mathrm{h}} = \#$ by concatenation in $\mathrm{Symp}^{\#}$. If one wishes to avoid empty chains, it can be viewed as a bicategory in which horizontal composition is strictly associative, but the diagonals $\Delta_M \subset M^- \times M$ only provide identities up to 2-isomorphism given by the embedded compositions $L_{01} \circ \Delta_{M_1} = L_{01}$ and $\Delta_{M_1} \circ L_{12} = L_{12}$.

Conversely, any bicategory (not just those arising from Cerf decompositions as in Example 3.1.6) gives rise to a 1-category by the following quotient construction.

Remark 3.1.7 (Quotient of a bicategory by 2-morphisms). Let $\mathscr{D}$ be a bicategory. Two 1-morphisms $f, g \in \mathrm{Mor}^1_{\mathscr{D}}(x, y)$ are called **isomorphic f $\sim$ g** if there exist 2-morphisms $\alpha, \beta \in \mathrm{Mor}^2_{\mathscr{D}}(f, g)$ whose vertical compositions $\alpha \circ_{\mathrm{v}} \beta = \mathrm{id}_f$ and $\beta \circ_{\mathrm{v}} \alpha = \mathrm{id}_g$ are the identites. This defines an evidently symmetric relation, which is moreover transitive and reflexive because the vertical composition $\circ_{\mathrm{v}}$ is associative and unital.

Now the bicategory $\mathscr{D}$ induces a quotient 1-category $|\mathscr{D}| := \mathscr{D}/\sim$ with the same objects $\mathrm{Obj}_{|\mathscr{D}|} := \mathrm{Obj}_{\mathscr{D}}$ and morphisms $\mathrm{Mor}_{|\mathscr{D}|} := \mathrm{Mor}^1_{\mathscr{D}}/\sim$ given as 1-morphisms modulo the equivalence relation $\sim$. Here, the horizontal 1-composition in $\mathscr{D}$ descends to a well-defined composition on the quotient $|\mathscr{D}|$ due to its compatibility with the equivalence relation—which is the content of the assumption that the horizontal bifunctor in $\mathscr{D}$ is compatible with identities and vertical composition.

3.2 Higher Bordism Categories

We begin with a more rigorous construction of the bordism category Bor_{d+1} in Example 2.1.2 as the quotient (as in Remark 3.1.7) of a bicategory $\mathrm{Bor}_{d+1+\varepsilon}$ comprising d-manifolds, $(d+1)$-cobordisms, and diffeomorphisms of $(d+1)$-cobordisms. After that, we restrict to the case $d=2$ and extend $\mathrm{Bor}_{2+1+\varepsilon}$ to a rigorous construction of a bordism bicategory Bor_{2+1+1} comprising 2-manifolds, 3-cobordisms, and four-dimensional manifolds with boundary and corners.

Example 3.2.1 The **$\mathbf{d+1+\varepsilon}$ bordism bicategory $\mathbf{Bor_{d+1+\varepsilon}}$ of d-manifolds, cobordisms, and diffeomorphisms** is constructed as follows, with illustrations in Figs. 8 and 9.

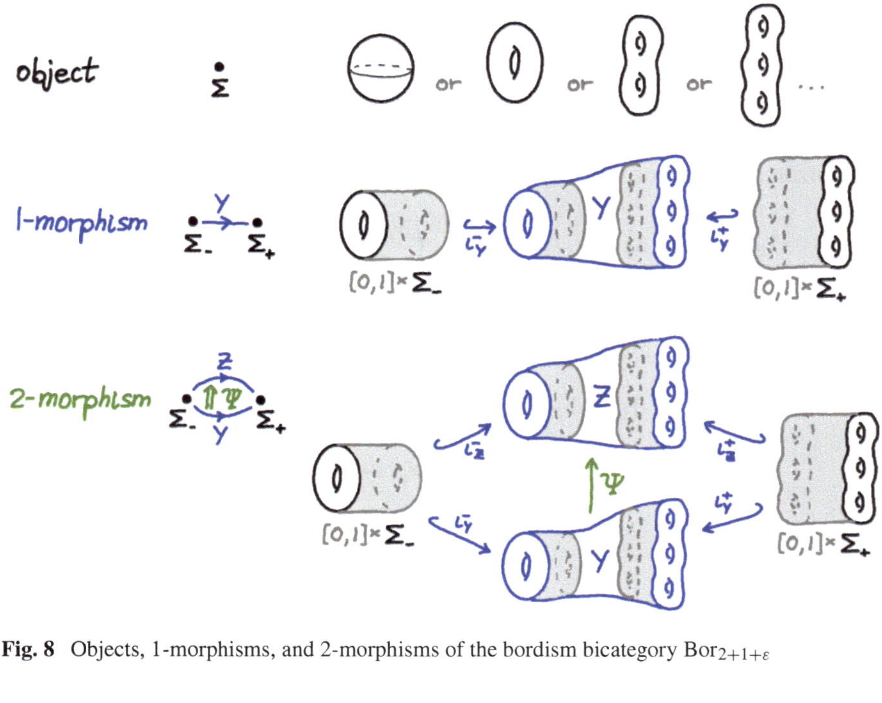

Fig. 8 Objects, 1-morphisms, and 2-morphisms of the bordism bicategory $\mathrm{Bor}_{2+1+\varepsilon}$

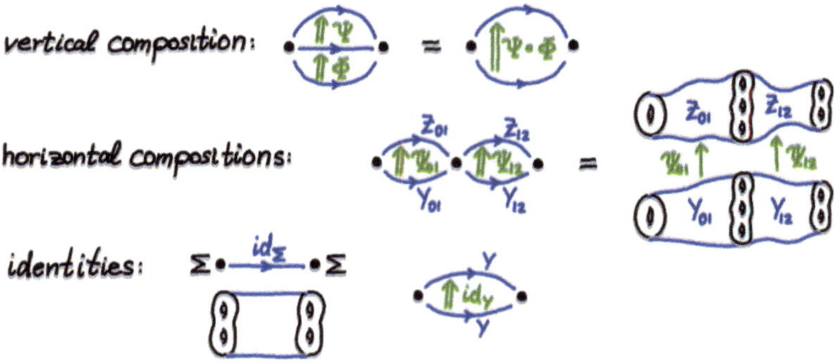

Fig. 9 Composition and identities in $\mathrm{Bor}_{2+1+\varepsilon}$

- Objects in $\mathrm{Obj}_{\mathrm{Bor}_{d+1+\varepsilon}}$ are closed, oriented, d-dimensional manifolds.
- 1-morphisms in $\mathrm{Mor}^1_{\mathrm{Bor}_{d+1+\varepsilon}}(\Sigma_-, \Sigma_+)$ are the representatives of morphisms in Bor_{d+1}, that is triples (Y, ι^-, ι^+) consisting of a compact, oriented, $(d+1)$-dimensional manifold Y with boundary and orientation perserving embeddings $\iota^\pm : [0,1] \times \Sigma_\pm \to Y$ to tubular neighborhoods of the boundary components $\partial Y = \iota^-(0, \Sigma_-) \sqcup \iota^+(1, \Sigma_+)$. For reasons that we will explain in item ∘ below, we also require the images of $\iota^\pm$ to be disjoint in Y.
- 2-morphisms in $\mathrm{Mor}^2_{\mathrm{Bor}_{d+1+\varepsilon}}((Y, \iota^\pm_Y), (Z, \iota^\pm_Z))$ between a pair of 1-morphisms $(Y, \iota^\pm_Y), (Z, \iota^\pm_Z) \in \mathrm{Mor}^1_{\mathrm{Bor}_{d+1+\varepsilon}}(\Sigma_-, \Sigma_+)$ are orientation preserving diffeomorphisms $\Psi : Y \to Z$ that intertwine the tubular neighborhood embeddings, $\Psi \circ \iota^\pm_Y = \iota^\pm_Z$.
- Vertical composition is by composition of diffeomorphisms $\Phi \circ_{\mathrm{v}} \Psi = \Psi \circ \Phi$. This is evidently associative and has units $\mathrm{id}_{(Y,\iota^\pm)} = \mathrm{id}_Y$, so the 1-morphisms $\mathrm{Mor}^1_{\mathrm{Bor}_{d+1+\varepsilon}}(\Sigma_-, \Sigma_+)$ between fixed objects $\Sigma_\pm$ form a category.
- Horizontal 1-composition is given by the gluing operation[27]

$$(Y_{01}, \iota^-_{01}, \iota^+_{01}) \circ^1_{\mathrm{h}} (Y_{12}, \iota^-_{12}, \iota^+_{12}) := \left({Y_{01} \sqcup Y_{12}} \big/ {\iota^+_{01}(s,x) \sim \iota^-_{12}(s,x)}, \iota^-_{01}, \iota^+_{12} \right).$$

- Horizontal 2-composition of $\Psi_{ij} \in \mathrm{Mor}^2_{\mathrm{Bor}_{d+1+\varepsilon}}((Y_{ij}, \iota^\pm_{Y_{ij}}), (Z_{ij}, \iota^\pm_{Z_{ij}}))$ for $ij = 01$ and $ij = 12$ is given by gluing of the diffeomorphisms,

$$\begin{aligned} \Psi_{01} \circ^2_{\mathrm{h}} \Psi_{12} \; : \; {Y_{01} \sqcup Y_{12}} \big/ {\iota^+_{Y_{01}}(s,x) \sim \iota^-_{Y_{12}}(s,x)} \; &\longrightarrow \; {Z_{01} \sqcup Z_{12}} \big/ {\iota^+_{Z_{01}}(s,x) \sim \iota^-_{Z_{12}}(s,x)} \\ y \in Y_{ij} \; &\longmapsto \; \Psi_{ij}(y), \end{aligned}$$

[27] Here, we could allow overlapping tubular neighborhoods, since $\iota^-_{01}, \iota^+_{12}$ induce well-defined tubular neighborhoods in the glued cobordism even if their images are not disjoint from the gluing region $\mathrm{im}\, \iota^+_{01} \simeq \mathrm{im}\, \iota^+_{12}$.

which is well defined since for $\iota^+_{Y_{01}}(s,x) \sim \iota^-_{Y_{12}}(s,x)$ we have $\Psi_{01}(\iota^+_{Y_{01}}(s,x)) = \iota^+_{Z_{01}}(s,x) \sim \iota^-_{Z_{12}}(s,x) = \Psi_{12}(\iota^-_{Y_{12}}(s,x))$.

- Horizontal composition is compatible with identities since for composable cobordisms Y_{01}, Y_{12} both 2-morphisms $\mathrm{id}_{Y_{01}} \circ^2_{\mathrm{h}} \mathrm{id}_{Y_{12}}$ and $\mathrm{id}_{Y_{01}\circ^1_{\mathrm{h}} Y_{12}}$ are the identity map on $(Y_{01} \sqcup Y_{12})/\!\sim$.
- Horizontal 2-composition $\circ^2_{\mathrm{h}}$ is compatible with vertical composition since, when given diffeomorphisms $\Phi_{ij} \in \mathrm{Mor}^2_{\mathrm{Bor}_{d+1+\varepsilon}}((Y_{ij}, \iota^\pm_{Y_{ij}}), (Z_{ij}, \iota^\pm_{Z_{ij}}))$ for $ij = 01, 12$, both $(\Phi_{01} \circ^2_{\mathrm{h}} \Phi_{12}) \circ_{\mathrm{v}} (\Psi_{01} \circ^2_{\mathrm{h}} \Psi_{12})$ and $(\Phi_{01} \circ_{\mathrm{v}} \Psi_{01}) \circ^2_{\mathrm{h}} (\Phi_{12} \circ_{\mathrm{v}} \Psi_{12})$ are given by the compositions $\Psi_{ij} \circ \Phi_{ij}$ on each part $Y_{ij} \subset {}^{Y_{01}\sqcup Y_{12}}/_{\iota^+\sim\iota^-}$.
- Horizontal 1-composition is strictly associative since for composable $(d+1)$-dimensional cobordisms Y_{01}, Y_{12}, Y_{23} both $Y_{01} \circ^1_{\mathrm{h}} \big(Y_{12} \circ^1_{\mathrm{h}} Y_{23}\big)$ and $\big(Y_{01} \circ^1_{\mathrm{h}} Y_{12}\big) \circ^1_{\mathrm{h}} Y_{23}$ are given by the disjoint union $Y_{01} \sqcup Y_{12} \sqcup Y_{23}$ modulo the equivalence relation[28] given by $\iota^+_{01}(s,x) \sim \iota^-_{12}(s,x)$, $\iota^+_{12}(s,x) \sim \iota^-_{23}(s,x)$.
- Horizontal 2-composition is associative since for composable Y_{ij} and Z_{ij} as above and $\Phi_{ij} \in \mathrm{Mor}^2_{\mathrm{Bor}_{d+1+\varepsilon}}(Y_{ij}, Z_{ij})$ both $\Phi_{12} \circ^2_{\mathrm{h}} (\Phi_{23} \circ^2_{\mathrm{h}} \Phi_{34})$ and $(\Phi_{12} \circ^2_{\mathrm{h}} \Phi_{23}) \circ^2_{\mathrm{h}} \Phi_{34}$ are given by Φ_{ij} on each part $Y_{ij} \subset {}^{Y_{01}\sqcup Y_{12}\sqcup Y_{23}}/_{\iota^+_{01}\sim\iota^-_{12},\iota^+_{12}\sim\iota^-_{23}}$.
- Horizontal 1-composition $\circ^1_{\mathrm{h}}$ can also be made strictly unital, thus giving rise to a 2-category, if we allow the tubular neighborhood embeddings to have overlapping image. Then $1_{\Sigma,1} := ([0,1]\times\Sigma, \iota^\pm)$ with the canonical embeddings $\iota^\pm = \mathrm{id}_{[0,1]\times\Sigma}$ would be a strict unit. However, other cylindrical cobordisms

$$1_{\Sigma,\delta} := ([0,1]\times\Sigma, \iota^\pm_\delta) \quad \text{with} \quad \iota^-_\delta(s,x) := (\delta s, x),\ \iota^+_\delta(s,x) := (1-\delta+\delta s, x)$$

for $0 < \delta < 1$ would have no 2-morphisms to the unit since such a diffeomorphism on $[0,1]\times\Sigma$ would be required to map $\mathrm{im}\,\iota^- = [0,1]\times\Sigma$ to $\mathrm{im}\,\iota^-_\delta = [0,\delta]\times\Sigma$ and $\mathrm{im}\,\iota^+ = [0,1]\times\Sigma$ to $\mathrm{im}\,\iota^+_\delta = [1-\delta,1]\times\Sigma$. So this bordism 2-category $\mathrm{Bor}'_{d+1+\varepsilon}$ would have too few 2-morphisms to achieve our topological vision of having just one morphism in $|\mathrm{Bor}'_{d+1+\varepsilon}| = \mathrm{Bor}_{d+1}$ that is represented by $[0,1]\times\Sigma$ with the canonical boundary identifications.

By requiring the tubular neighborhood embeddings to have disjoint images, we disallow $1_{\Sigma,\delta}$ for $\delta \geq \frac{1}{2}$ as 2-morphism. On the other hand, the cylindrical cobordisms for $0 < \delta \neq \delta' < \frac{1}{2}$ are all equivalent,

$$1_{\Sigma,\delta} \sim 1_{\Sigma,\delta'} \quad \text{since} \quad \mathrm{id}_{[0,1]\times\Sigma} = \Psi \circ \Psi^{-1},\ \mathrm{id}_{[0,1]\times\Sigma} = \Psi^{-1} \circ \Psi,$$

for any diffeomorphism Ψ of $[0,1]\times\Sigma$ which extends

$$(\iota^\pm_{\delta'})^{-1} \circ \iota^\pm_\delta : \big([0,\delta] \cup [1-\delta,1]\big) \times \Sigma \to \big([0,\delta'] \cup [1-\delta',1]\big) \times \Sigma.$$

- Horizontal 1-composition $\circ^1_{\mathrm{h}}$ in $\mathrm{Bor}_{d+1+\varepsilon}$ is unital up to 2-isomorphism, with weak units for any manifold $\Sigma \in \mathrm{Obj}_{\mathrm{Bor}_{d+1+\varepsilon}}$ given by the cylindrical cobor-

---

[28] If the embeddings $\iota^\pm_{12}$ had overlapping images, we could make the same construction by completing the equivalence relation with compositions $\iota^+_{01}(s,x) \sim \iota^-_{12}(s,x) = \iota^+_{12}(s',x') \sim \iota^-_{23}(s',x')$.

disms $1_{\Sigma,\delta} \in \mathrm{Mor}^1_{\mathrm{Bor}_{d+1+\varepsilon}}(\Sigma, \Sigma)$ for any $0 < \delta < \frac{1}{2}$. Indeed, for any appropriate $(Y, \iota_Y^\pm), (Z, \iota_Z^\pm) \in \mathrm{Mor}^1_{\mathrm{Bor}_{d+1+\varepsilon}}$ we have

$$(Y, \iota_Y^\pm) \circ_{\mathrm{h}}^1 1_{\Sigma,\delta} \sim (Y, \iota_Y^\pm), \qquad 1_{\Sigma,\delta} \circ_{\mathrm{h}}^1 (Z, \iota_Z^\pm) \sim (Z, \iota_Z^\pm)$$

via appropriate diffeomorphisms

$$\Psi : {Y \sqcup ([0,1] \times \Sigma)}\big/{\iota_Y^+ \sim \iota_\delta^-} \to Y, \qquad \Phi : {([0,1] \times \Sigma) \sqcup Z}\big/{\iota_\delta^+ \sim \iota_Z^-} \to Z.$$

Here Ψ (and Φ similarly) is constructed as follows: Extend ι_Y^+ to an embedding $\widetilde{\iota_Y^+} : [-1, 1] \times \Sigma \to Y \setminus \mathrm{im}\, \iota_Y^-$. Then we have a natural diffeomorphism

$$Y \sqcup ([0,1] \times \Sigma)\big/_{\iota_Y^+(s,\cdot) \sim \iota_\delta^-(s,\cdot)\ \forall s \in [0,1]} \simeq Y \sqcup ([-\delta, 1] \times \Sigma)\big/_{\widetilde{\iota_Y^+}(s,\cdot) \sim \widetilde{\iota_\delta^-}(s,\cdot)\ \forall s \in [-1,1]}$$

with $\widetilde{\iota_\delta^-} : [-1,1] \times \Sigma \to [-\delta, 1] \times \Sigma$, $(s, x) \mapsto (\delta s, x)$. Now we can construct Ψ by the identity on $Y \setminus \mathrm{im}\, \widetilde{\iota_Y^+}$ and a diffeomorphism $[-\delta, 1] \times \Sigma \to \widetilde{\iota_Y^+}([-1,1] \times \Sigma)$ given by $\widetilde{\iota_Y^+} \circ (\widetilde{\iota_\delta^-})^{-1}$ near $\{-\delta\} \times \Sigma \to \widetilde{\iota_Y^+}(-1, \Sigma)$ and $\iota_Y^+ \circ (\iota_\delta^+)^{-1}$ on $[1-\delta, 1] \times \Sigma \to \iota_Y^+([0,1] \times \Sigma)$. It intertwines the boundary embeddings $\Psi \circ \iota_\delta^+ = \iota_Y^+$ by construction, as required for a 2-morphism from $(Y, \iota_Y^\pm) \circ_{\mathrm{h}}^1 1_{\Sigma,\delta}$ to $(Y, \iota_Y^\pm)$.

This finishes the construction of the bordism bicategory $\mathrm{Bor}_{d+1+\varepsilon}$. It particularly contains representatives of the cylindrical cobordism $Z_\phi \in \mathrm{Mor}_{\mathrm{Bor}_{d+1}}(\Sigma_0, \Sigma_1)$ associated to a diffeomorphism $\phi : \Sigma_0 \to \Sigma_1$ in (2.1.1), for any $0 < \delta < \frac{1}{2}$ given by

$$\widehat{Z}_{\phi,\delta} := \left(\begin{array}{r} \times \Sigma_1\, ,\ \iota_\delta^- : (s, x) \mapsto (\delta s, \phi(x))\, , \\ \iota_\delta^+ : (s, x) \mapsto (1 - \delta + \delta s, x) \end{array} \right) \in \mathrm{Mor}^1_{\mathrm{Bor}_{d+1}}(\Sigma_0, \Sigma_1).$$

These also reproduce the identity 1-morphisms $\widehat{Z}_{\mathrm{id},\delta} = 1_{\Sigma,\delta} \in \mathrm{Mor}^1_{\mathrm{Bor}_{d+1}}(\Sigma, \Sigma)$.

The **connected $\mathbf{d+1+\varepsilon}$ bordism bicategory $\mathbf{Bor^{conn}}_{\mathbf{d+1+\varepsilon}}$ of connected d-manifolds, connected cobordisms, and diffeomorphisms** is constructed analogously, using the objects and representatives of morphisms of $\mathrm{Bor}^{\mathrm{conn}}_{d+1}$.

Remark 3.2.2 (Quotient construction of the d+1 bordism category). Taking the quotient of the bordism bicategories $\mathrm{Bor}_{d+1+\varepsilon}$ and $\mathrm{Bor}^{\mathrm{conn}}_{d+1+\varepsilon}$ by their 2-morphisms (i.e. diffeomorphisms of $d+1$-cobordisms) as in Remark 3.1.7 now yields rigorous definitions of the (connected) bordism categories $\mathrm{Bor}_{d+1} := \mathrm{Bor}_{d+1+\varepsilon}/\!\sim$ and $\mathrm{Bor}^{\mathrm{conn}}_{d+1} := \mathrm{Bor}^{\mathrm{conn}}_{d+1+\varepsilon}/\!\sim$ outlined in Examples 2.1.2 and 2.3.3.

In particular, the cylindrical cobordism Z_ϕ associated in (2.1.1) to a diffeomorphism $\phi : \Sigma_0 \to \Sigma_1$ is more rigorously defined as the equivalence class $Z_\phi := [\widehat{Z}_{\phi,\delta}] \in \mathrm{Mor}_{\mathrm{Bor}_{d+1}}(\Sigma_0, \Sigma_1)$, which is independent of $0 < \delta < \frac{1}{2}$. This also reproduces the identity morphisms $Z_{\mathrm{id}} = [1_{\Sigma,\delta}] \in \mathrm{Mor}_{\mathrm{Bor}_{d+1}}(\Sigma, \Sigma)$ in Bor_{d+1}.

For our applications we will restrict to dimension $d = 2$ and extend the above constructions to a bordism bicategory which includes “cobordisms of cobordisms”

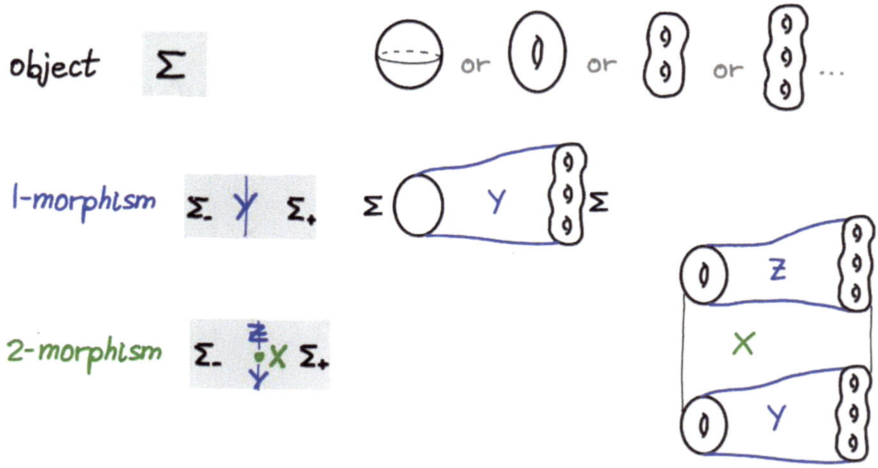

Fig. 10 Objects, 1-morphisms, and 2-morphisms of Bor_{2+1+1} and their string diagram notation. These basic diagrams represent 4-manifolds given by squares times surfaces, intervals times 3-cobordisms, and 4-cobordisms with corners

given by 4-manifolds with boundaries and corners. Again, we give general constructions for $d \geq 0$ and restrict to $d \geq 2$ to obtain a connected theory. We also use the illustrations in case $d = 2$ to give a preview of the string diagram notation in Sect. 3.4.

Example 3.2.3 The **bordism bicategory $\mathbf{Bor_{d+1+1}}$** for $d \geq 0$ consists of the following, with representation by string diagrams illustrated in Figs. 10 and 13.

- Objects in $\mathrm{Obj}_{\mathrm{Bor}_{d+1+1}}$ are closed oriented d-manifolds Σ as in $\mathrm{Bor}_{d+1+\varepsilon}$.
- 1-morphisms in $\mathrm{Mor}^1_{\mathrm{Bor}_{d+1+1}}(\Sigma_-, \Sigma_+)$ are triples (Y, ι^-, ι^+) of a compact, oriented (d+1)-cobordism Y with disjoint embeddings $\iota^\pm : [0,1] \times \Sigma_\pm \to Y$ to neighborhoods of the boundary parts $\partial Y = \iota^-(0, \Sigma_-) \sqcup \iota^+(1, \Sigma_+)$ as in $\mathrm{Bor}_{d+1+\varepsilon}$.
- 2-morphisms, i.e., morphisms in the category $\mathrm{Mor}^1_{\mathrm{Bor}_{d+1+1}}(\Sigma_-, \Sigma_+)$ are equivalence classes of tuples $\big[(X, \iota^+_X, \iota^-_X, \kappa^-, \kappa^+)\big] \in \mathrm{Mor}^2_{\mathrm{Bor}_{d+1+1}}((Y, \iota^\pm_Y), (Z, \iota^\pm_Z))$ consisting of a compact, oriented (d+2)-manifold X with boundary and corners and four orientation preserving embeddings as indicated in Fig. 11,

$$\iota^\pm_X : [0,1] \times [0,1] \times \Sigma_\pm \to X, \quad \kappa^- : [0,1] \times Y \to X, \quad \kappa^+ : [0,1] \times Z \to X.$$

The embeddings $\iota^\pm$ and $\kappa^\pm$ are required to cover the boundary

$$\partial X = \kappa^-(0, Y) \sqcup \kappa^+(1, Z) \sqcup \iota^-((0,1), 0, \Sigma_-) \sqcup \iota^+((0,1), 1, \Sigma_+)$$

in such a way that both pairs $\kappa^\pm$ and $\iota^\pm$ have disjoint images, but we have mixed overlaps on which $\kappa^\pm$ intertwines $\iota^\pm$ with the boundary identifications $\iota^\pm_Y, \iota^\pm_Z$ in the sense that for some $0 < \delta^\pm_Y, \delta^\pm_Z < \frac{1}{2}$ and all $s, t \in [0,1]$, $x \in \Sigma_\pm$ we have

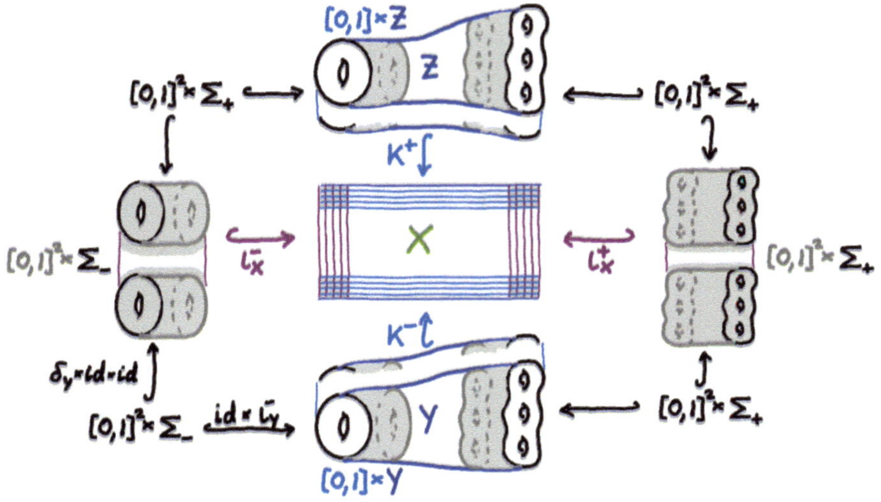

Fig. 11 A 2-morphism in Bor_{2+1+1} consists of a 4-manifold X with a number of embeddings to collar neighborhoods of its boundary strata, which are compatible near the corners

$$\kappa^-\big(s, \iota_Y^{\pm}(t,x)\big) = \iota_X^{\pm}(\delta_Y^{\pm}s, t, x), \quad \kappa^+\big(s, \iota_Z^{\pm}(t,x)\big) = \iota_X^{\pm}(1-\delta_Z^{\pm}+\delta_Z^{\pm}s, t, x). \tag{3.2.1}$$

Two such tuples are equivalent, $\big(X_0, \iota_0^+, \iota_0^-, \kappa_0^-, \kappa_0^+\big) \sim \big(X_1, \iota_1^+, \iota_1^-, \kappa_1^-, \kappa_1^+\big)$, if there exists a diffeomorphism $F : X_0 \to X_1$ that intertwines the embeddings, i.e. $F \circ \iota_0^{\pm} = \iota_1^{\pm}$ and $F \circ \kappa_0^{\pm} = \kappa_1^{\pm}$.

- The 2-morphisms $\Psi : Y \to Z$ in $\mathrm{Bor}_{d+1+\varepsilon}$ appear in Bor_{d+1+1} as the cylindrical cobordisms of cobordisms $I_\Psi := \big[([0,1]\times Z, \iota^{\pm}, \kappa_\delta^{\pm})\big] \in \mathrm{Mor}^2_{\mathrm{Bor}_{d+1+1}}$ with

$$\iota^{\pm}(s,t,x) := (s, \iota_Z^{\pm}(t,x)), \quad \kappa_\delta^-(s,y) := \big(\delta s, \Psi(y)\big), \quad \kappa_\delta^+(s,z) := \big(1-\delta+\delta s, z\big).$$

This is illustrated in Fig. 12 and may help with understanding the compatibility conditions (3.2.1) for the embeddings, since these are naturally satisfied due to

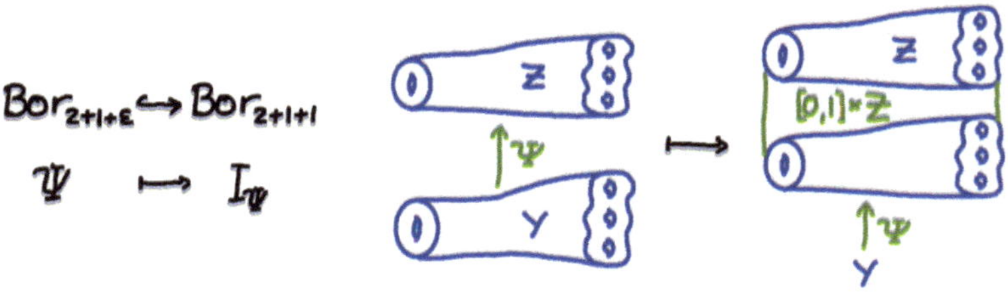

Fig. 12 Inclusion of $\mathrm{Bor}_{d+1+\varepsilon}$ in Bor_{d+1+1}

$$\Psi \circ \iota_Y^\pm = \iota_Z^\pm,$$

$$\kappa_\delta^-\big(s, \iota_Y^\pm(t,x)\big) = \big(\delta s, \Psi(\iota_Y^\pm(t,x))\big) = \big(\delta s, \iota_Z^\pm(t,x)\big) = \iota^\pm(\delta s, t, z),$$
$$\kappa_\delta^+\big(s, \iota_Z^\pm(t,x)\big) = \big(1-\delta+\delta s, \iota_Z^\pm(t,z)\big) = \iota_X^\pm(1-\delta+\delta s, t, x).$$

- Vertical composition of $\big[(X^{ij}, \iota_{ij}^\pm, \kappa_{ij}^\pm)\big] \in \mathrm{Mor}^2_{\mathrm{Bor}_{d+1+1}}\big((Y^i, \iota_{Y^i}^\pm), (Y^j, \iota_{Y^j}^\pm)\big)$ labeled by $ij = 01$ and $ij = 12$ between $(Y^i, \iota_{Y^i}^\pm) \in \mathrm{Mor}^1_{\mathrm{Bor}_{d+1+1}}(\Sigma_-, \Sigma_+)$ for $i = 0, 1, 2$ is given by gluing the $(d+2)$-manifolds and embeddings as illustrated in Fig. 13

$$\big(X^{01}, \iota_{01}^\pm, \kappa_{01}^\pm\big) \circ_{\mathrm{v}} \big(X^{12}, \iota_{12}^\pm, \kappa_{12}^\pm\big) := \left(X^{01} \sqcup X^{12} \big/_{\kappa_{01}^+(s,y) \sim \kappa_{12}^-(s,y)}, \iota_{01}^\pm \circ_{\mathrm{h}} \iota_{12}^\pm, \kappa_{01}^-, \kappa_{12}^+ \right).$$

Here the tubular neighborhoods of the common boundaries $\partial Y^i \simeq \Sigma_-^- \sqcup \Sigma_+$ are glued in the same way as the horizontal 2-composition $\circ_{\mathrm{h}}^2$ in $\mathrm{Bor}_{d+1+\varepsilon}$, that is

$$\iota_{01}^\pm \circ_{\mathrm{h}} \iota_{12}^\pm : \; Q_\pm^{01} \sqcup Q_\pm^{12} \big/_{\iota^+_{Q_\pm^{01}} \sim \iota^-_{Q_\pm^{12}}} \longrightarrow X^{01} \sqcup X^{12} \big/_{\kappa_{01}^+ \sim \kappa_{12}^-}$$
$$(s,t,x) \in Q_\pm^{ij} \longmapsto \iota_{ij}^\pm(s,t,x)$$

where $Q_\pm^{ij} := [0,1]^2 \times \Sigma_\pm$ are representatives of 2-morphisms with the embeddings $\iota^\pm_{Q_\pm^{ij}} : [0,1] \times [0,1] \times \Sigma \to Q_\pm^{ij}$ chosen so as to make the glued map well defined due to the compatibility conditions in (3.2.1) for some $\delta_{ij}^\pm > 0$,

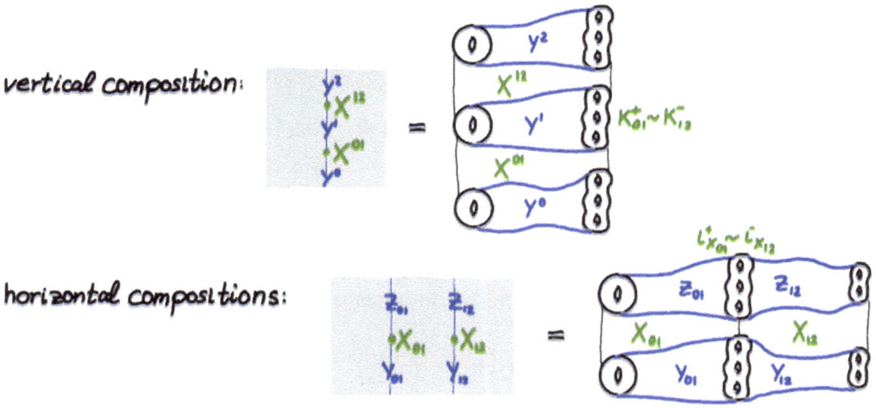

Fig. 13 Compositions of 2-morphisms in Bor_{2+1+1} and their string diagram notation. These more complicted diagrams represent constructions of 4-manifolds by gluing squares times surfaces, intervals times 3-cobordisms, and 4-cobordisms with corners along common boundary strata

$$\kappa_{01}^{+}\big(s, \iota_{Y^1}^{\pm}(t,x)\big) = \iota_{01}^{\pm}(1-\delta_{01}^{\pm}+\delta_{01}^{\pm}s, t, x) =: \iota_{01}^{\pm}\big(\iota_{Q_{\pm}^{01}}^{+}(s,t,x)\big)$$
$$\sim \kappa_{12}^{-}\big(s, \iota_{Y^1}^{\pm}(t,x)\big) = \iota_{12}^{\pm}(\delta_{12}^{\pm}s, t, x) =: \iota_{12}^{\pm}\big(\iota_{Q_{\pm}^{12}}^{-}(s,t,x)\big).$$

That is, we set $\iota_{Q_{\pm}^{01}}^{+}(s,t,x) := (1-\delta_{01}^{\pm}+\delta_{01}^{\pm}s, t, x)$ and $\iota_{Q_{\pm}^{12}}^{-}(s,t,x) := (\delta_{12}^{\pm}s, t, x)$. This gluing construction on the level of representatives yields a well defined vertical composition of equivalence classes because the equivalences are given by diffeomorphisms which intertwine the embeddings that are used to glue. Associativity follows from direct associativity of gluing, and units are provided by the cylindrical cobordisms of cobordisms $\mathrm{id}_{(Y,\iota^{\pm})} := I_{\mathrm{id}_Y}$ associated above to the identity map $\Psi = \mathrm{id}_Y : Y \to Y$ just as for $\circ_{\mathrm{h}}^{1}$ in Example 3.2.1. Thus $\mathrm{Mor}^1_{\mathrm{Bor}_{d+1+1}}(\Sigma_-, \Sigma_+)$ for fixed objects $\Sigma_\pm$ forms a category.

- Note that the vertical composition of 2-morphisms arising from diffeomorphisms in $\mathrm{Mor}^2_{\mathrm{Bor}_{d+1+\varepsilon}}$ is compatible with the vertical composition in $\mathrm{Bor}_{d+1+\varepsilon}$, that is

$$I_{\Phi_{01}} \circ_{\mathrm{v}} I_{\Phi_{12}} = I_{\Phi_{12}\circ\Phi_{01}} \qquad \forall \Phi_{ij} \in \mathrm{Mor}^2_{\mathrm{Bor}_{d+1+\varepsilon}}((Y_i, \iota_{Y_i}^{\pm}), (Y_j, \iota_{Y_j}^{\pm})). \tag{3.2.2}$$

Indeed, we have $I_{\Phi_{01}} \circ_{\mathrm{v}} I_{\Phi_{12}} = \big[\big([0,1]\times Y_1, \iota_{01}^{\pm}, \kappa_{\delta,\Phi_{01}}^{\pm}\big)\big] \circ_{\mathrm{v}} \big[\big([0,1]\times Y_2, \iota_{12}^{\pm}, \kappa_{\delta,\Phi_{12}}^{\pm}\big)\big]$ with $\iota_{01}^{\pm} = \mathrm{id}_{[0,1]}\times\iota_{Y_1}^{\pm}$, $\iota_{12}^{\pm} = \mathrm{id}_{[0,1]}\times\iota_{Y_2}^{\pm}$, $\kappa_{\delta,\Psi}^{-}(s,y) = \big(\delta s, \Psi(y)\big)$, and $\kappa_{\delta,\Psi}^{+}(s,z) = \big(1-\delta+\delta s, z\big)$ is represented by the (d+2)-manifold

$$\big([0,1]\times Y_1\big) \sqcup \big([0,1]\times Y_2\big)\Big/_{(1-\delta+\delta s,y)\sim(\delta s,\Phi_{12}(y))} \;\simeq\; [0,2-\delta]\times Y_2,$$

via the diffeomorphism induced by $\mathrm{id}_{[0,1]}\times\Phi_{12} : [0,1]\times Y_1 \to [0,1]\times Y_2$ and $(r\mapsto r+1-\delta)\times\mathrm{id}_{Y_2} : [0,1]\times Y_2 \to [1-\delta, 2-\delta]\times Y_2$. The corresponding embeddings are

$$\begin{aligned}
\iota_{01}^{\pm} \circ_{\mathrm{h}} \iota_{12}^{\pm} &= \big(\mathrm{id}_{[0,1]}\times\iota_{Y_1}^{\pm}\big) \circ_{\mathrm{h}} \big(\mathrm{id}_{[0,1]}\times\iota_{Y_2}^{\pm}\big) \;\simeq\; \mathrm{id}_{[0,2-\delta]}\times\iota_{Y_2}^{\pm},\\
\kappa_{\delta,\Phi_{01}}^{-}(s,y) &= \big(\delta s, \Phi_{01}(y)\big) \;\simeq\; \big(\delta s, \Phi_{12}\big(\Phi_{01}(y)\big)\big),\\
\kappa_{\delta,\Phi_{12}}^{+}(s,y) &= \big(1-\delta+\delta s, z\big) \;\simeq\; \big(2-2\delta+\delta s, z\big).
\end{aligned}$$

This is equivalent to the representative of $I_{\Phi_{12}\circ\Phi_{01}}$ with constant $0 < \frac{\delta}{2-\delta} < \frac{1}{2}$ via linear rescaling in the first factor $[0,2-\delta]\times Y_2 \simeq [0,1]\times Y_2$.

- Horizontal 1-composition is given by gluing as in $\mathrm{Bor}_{d+1+\varepsilon}$,

$$(Y_{01}, \iota_{01}^{-}, \iota_{01}^{+}) \circ_{\mathrm{h}}^{1} (Y_{12}, \iota_{12}^{-}, \iota_{12}^{+}) := \left(Y_{01} \sqcup Y_{12}\Big/_{\iota_{01}^{+}(s,x)\sim\iota_{12}^{-}(s,x)}, \iota_{01}^{-}, \iota_{12}^{+} \right).$$

- Horizontal 2-composition of $\big[\big(X_{ij}, \iota_{ij}^{\pm}, \kappa_{ij}^{\pm}\big)\big] \in \mathrm{Mor}^2_{\mathrm{Bor}_{d+1+1}}((Y_{ij}, \iota_{Y_{ij}}^{\pm}), (Z_{ij}, \iota_{Z_{ij}}^{\pm}))$ between $(Y_{ij}, \iota_{Y_{ij}}^{\pm}), (Z_{ij}, \iota_{Z_{ij}}^{\pm}) \in \mathrm{Mor}^1_{\mathrm{Bor}_{d+1+1}}(\Sigma_i, \Sigma_j)$ for $ij = 01$ and $ij = 12$ is given by gluing the (d+2)-manifolds and embeddings as illustrated in Fig. 13

$$\begin{aligned}
\left(X_{01}, \iota_{01}^{\pm}, \kappa_{01}^{\pm}\right) \circ_{\mathrm{h}}^{2} \left(X_{12}, \iota_{12}^{\pm}, \kappa_{12}^{\pm}\right) &:= \left({X_{01} \sqcup X_{12}}\big/_{\iota_{01}^{+} \sim \iota_{12}^{-}}, \iota_{01}^{-}, \iota_{12}^{+}, \kappa_{01}^{\pm} \circ_{\mathrm{h}} \kappa_{12}^{\pm} \right), \\
\kappa_{01}^{-} \circ_{\mathrm{h}} \kappa_{12}^{-} : [0,1] \times {Y_{01} \sqcup Y_{12}}\big/_{\iota_{Y_{01}}^{+} \sim \iota_{Y_{12}}^{-}} &\to X_{01} \circ_{\mathrm{h}}^{2} X_{12}, \quad \left(s, y \in Y_{ij}\right) \mapsto \kappa_{ij}^{-}(s,y), \\
\kappa_{01}^{+} \circ_{\mathrm{h}} \kappa_{12}^{+} : [0,1] \times {Z_{01} \sqcup Z_{12}}\big/_{\iota_{Z_{01}}^{+} \sim \iota_{Z_{12}}^{-}} &\to X_{01} \circ_{\mathrm{h}}^{2} X_{12}, \quad \left(s, z \in Z_{ij}\right) \mapsto \kappa_{ij}^{+}(s,z).
\end{aligned}$$

For the boundary embeddings $\kappa_{01}^{\pm} \circ_{\mathrm{h}} \kappa_{12}^{\pm}$ to be well defined, we need to take account of the scaling factors in (3.2.1) for the two cobordisms X_{ij} to achieve

$$\begin{aligned}
\kappa_{01}^{-}(s, \iota_{Y_{01}}^{+}(t,x)) = \iota_{01}^{+}(\delta_{Y_{01}}^{+} s, t, x) \sim \iota_{12}^{-}(\delta_{Y_{12}}^{-} s, t, x) = \kappa_{12}^{-}(s, \iota_{Y_{12}}^{-}(t,x)), \\
\kappa_{01}^{+}(s, \iota_{Z_{01}}^{+}) = \iota_{01}^{+}(1 - \delta_{Z_{01}}^{+} + \delta_{Z_{01}}^{+} s, ..) \sim \iota_{12}^{-}(1 - \delta_{Z_{12}}^{-} + \delta_{Z_{12}}^{-} s, ..) = \kappa_{12}^{+}(s, \iota_{Z_{12}}^{-}).
\end{aligned}$$

That is, we define the relation $\sim$ in the construction of the glued (d+2)-manifold $X_{01} \circ_{\mathrm{h}}^{2} X_{12} := (X_{01} \sqcup X_{12})/\sim$ by $\iota_{01}^{+}(s,t,z) \sim \iota_{12}^{-}(\phi(s),t,z)$ for some diffeomorphism $\phi : [0,1] \to [0,1]$ with $\phi(r) = \delta_{Y_{12}}^{-} r/\delta_{Y_{01}}^{+}$ for $0 \le r \le \delta_{Y_{01}}^{+}$ and $\phi(1-r) = 1 - \delta_{Z_{12}}^{-} r/\delta_{Z_{01}}^{+}$ for $0 \le r \le \delta_{Z_{01}}^{+}$. Such ϕ exists since all δ-factors in (3.2.1) are less than $\frac{1}{2}$. Finally, one needs to check that different choices of ϕ yield equivalent tuples of (d+2)-manifolds and embeddings, and thus the same 2-morphism.

- Horizontal composition is compatible with identities since for composable cobordisms Y_{12}, Y_{23} both $\mathrm{id}_{Y_{12}} \circ_{\mathrm{h}}^{2} \mathrm{id}_{Y_{23}} = I_{\mathrm{id}_{Y_{12}}} \circ_{\mathrm{h}}^{2} I_{\mathrm{id}_{Y_{23}}}$ and $\mathrm{id}_{Y_{12} \circ_{\mathrm{h}}^{1} Y_{23}} = I_{\mathrm{id}_{Y_{12} \circ_{\mathrm{h}}^{1} Y_{23}}}$ are represented by the (d+2)-manifold

$$([0,1] \times Y_{12}) \sqcup ([0,1] \times Y_{23})\big/_{\mathrm{id}_{[0,1]} \times \iota_{12}^{+} \sim \mathrm{id}_{[0,1]} \times \iota_{23}^{-}} \simeq [0,1] \times {Y_{12} \sqcup Y_{23}}\big/_{\iota_{12}^{+} \sim \iota_{23}^{-}}$$

with embeddings—arising from a universal choice of δ—given by

$$\begin{aligned}
\iota^{-} &: (s,t,x) \mapsto \left(s, \iota_{Y_{12}}^{-}(t,x)\right) = \left(s, \iota_{Y_{12} \circ_{\mathrm{h}}^{1} Y_{23}}^{-}(t,x)\right), \\
\iota^{+} &: (s,t,x) \mapsto (s, \iota_{Y_{23}}^{+}(t,x)) = \left(s, \iota_{Y_{12} \circ_{\mathrm{h}}^{1} Y_{23}}^{+}(t,x)\right), \\
\kappa_{\delta}^{-} &: (s,y) \mapsto \left(\delta \mathrm{id}_{[0,1]} \times \mathrm{id}_{Y_{12}}\right) \circ_{\mathrm{h}} \left(\delta \mathrm{id}_{[0,1]} \times \mathrm{id}_{Y_{23}}\right)(s,y) = \left(\delta s, \mathrm{id}_{Y_{12} \circ_{\mathrm{h}}^{1} Y_{23}}(y)\right), \\
\kappa_{\delta}^{+} &: (s,z) \mapsto \left(1 - \delta + \delta s, \left(\mathrm{id}_{Y_{12}} \circ_{\mathrm{h}} \mathrm{id}_{Y_{23}}\right)(z)\right) = \left(1 - \delta + \delta s, \mathrm{id}_{Y_{12} \circ_{\mathrm{h}}^{1} Y_{23}}(z)\right).
\end{aligned}$$

- Compatibility of horizontal 2-composition with vertical composition requires

$$\left([W_{01}] \circ_{\mathrm{h}}^{2} [W_{12}]\right) \circ_{\mathrm{v}} \left([X_{01}] \circ_{\mathrm{h}}^{2} [X_{12}]\right) = \left([W_{01}] \circ_{\mathrm{v}} [X_{01}]\right) \circ_{\mathrm{h}}^{2} \left([W_{12}] \circ_{\mathrm{v}} [X_{12}]\right)$$

for any pair of pairs $\left[\left(W_{ij}, \iota_{W_{ij}}^{\pm}, \kappa_{W_{ij}}^{\pm}\right)\right] \in \mathrm{Mor}^{2}_{\mathrm{Bor}_{d+1+1}}((V_{ij}, \iota_{V_{ij}}^{\pm}), (Y_{ij}, \iota_{Y_{ij}}^{\pm}))$ and $\left[\left(X_{ij}, \iota_{X_{ij}}^{\pm}, \kappa_{X_{ij}}^{\pm}\right)\right] \in \mathrm{Mor}^{2}_{\mathrm{Bor}_{d+1+1}}((Y_{ij}, \iota_{Y_{ij}}^{\pm}), (Z_{ij}, \iota_{Z_{ij}}^{\pm}))$ of equivalence classes (d+2)-cobordisms of cobordisms for $ij = 01, 12$ between (d+1)-cobordisms $(V_{ij}, \iota_{V_{ij}}^{\pm})$, $(Y_{ij}, \iota_{Y_{ij}}^{\pm})$, $(Z_{ij}, \iota_{Z_{ij}}^{\pm}) \in \mathrm{Mor}^{1}_{\mathrm{Bor}_{d+1+1}}(\Sigma_i, \Sigma_j)$ for fixed surfaces $\Sigma_0, \Sigma_1, \Sigma_2 \in \mathrm{Obj}_{\mathrm{Bor}_{d+1+1}}$.
Here both (d+2)-manifolds are of the form $\left(W_{01} \sqcup W_{12} \sqcup X_{01} \sqcup X_{12}\right)/\sim$, where in

the first gluing, the equivalence relation $\sim$ is generated by

$$\iota^+_{W_{01}} \sim \iota^-_{W_{12}} \circ \phi_W, \qquad \iota^+_{X_{01}} \sim \iota^-_{X_{12}} \circ \phi_X, \qquad \kappa^+_{W_{01}} \circ_{\mathrm{h}} \kappa^+_{W_{12}} \sim \kappa^-_{X_{01}} \circ_{\mathrm{h}} \kappa^-_{X_{12}},$$

whereas in the second gluing, the equivalence relation $\sim$ is generated by

$$\kappa^+_{W_{01}} \sim \kappa^-_{X_{01}}, \qquad \kappa^+_{W_{12}} \sim \kappa^-_{X_{12}}, \qquad \iota^+_{W_{01}} \circ_{\mathrm{h}} \iota^+_{X_{01}} \sim \left(\iota^-_{W_{12}} \circ_{\mathrm{h}} \iota^-_{X_{12}}\right) \circ \phi_{WX}.$$

This amounts to the same relation, if we choose the diffeomorphism ϕ_{WX} of $[0,1] = [0,1] \circ_{\mathrm{h}} [0,1]$ as the gluing of ϕ_W and ϕ_X. The various embeddings are identified analogously.

- Horizontal 1-composition is strictly associative as in $\mathrm{Bor}_{d+1+\varepsilon}$.
- Horizontal 2-composition is associative since for composable Y_{ij} and Z_{ij} and $\left[(X_{ij}, \ldots\right] \in \mathrm{Mor}^2_{\mathrm{Bor}_{d+1+1}}(Y_{ij}, Z_{ij})$ both $X_{12} \circ^2_{\mathrm{h}} (X_{23} \circ^2_{\mathrm{h}} X_{34})$ and $(X_{12} \circ^2_{\mathrm{h}} X_{23}) \circ^2_{\mathrm{h}} X_{34}$ are given by the same gluing of $(d+2)$-manifolds $\left(X_{01} \sqcup X_{12} \sqcup X_{34}\right)/(\iota^+_{X_{01}} \sim \iota^-_{X_{12}}, \iota^+_{X_{12}} \sim \iota^-_{X_{23}})$ and the corresponding tubular neighborhood embeddings.
- Horizontal 1-composition is unital up to 2-isomorphism as in $\mathrm{Bor}_{d+1+\varepsilon}$, that is for any surface $\Sigma \in \mathrm{Obj}_{\mathrm{Bor}_{d+1+1}}$ and $0 < \delta < \frac{1}{2}$ the cylindrical cobordism $1_{\Sigma,\delta} \in \mathrm{Mor}^1_{\mathrm{Bor}_{d+1+1}}(\Sigma, \Sigma)$ is a weak unit. To prove the latter we start by proving equivalence $(Y, \iota^\pm_Y) \circ^1_{\mathrm{h}} 1_{\Sigma,\delta} \sim (Y, \iota^\pm_Y)$ in $\mathrm{Mor}^1_{\mathrm{Bor}_{d+1+1}}(\Sigma_0, \Sigma)$ for any $(Y, \iota^\pm_Y) \in \mathrm{Mor}^1_{\mathrm{Bor}_{d+1+\varepsilon}}(\Sigma_0, \Sigma)$. For that purpose, we can use the diffeomorphism Ψ constructed in Example 3.2.1 to obtain 2-morphisms $I_\Psi, I_{\Psi^{-1}} \in \mathrm{Mor}^2_{\mathrm{Bor}_{d+1+1}}$ (represented by arrows below) whose vertical $\circ_{\mathrm{v}}$ compositions are the identities

$$(Y, \iota^\pm_Y) \xrightarrow{I_{\Psi^{-1}}} (Y, \iota^\pm_Y) \circ^1_{\mathrm{h}} 1_{\Sigma,\delta} \xrightarrow{I_\Psi} (Y, \iota^\pm_Y) \quad = \quad (Y, \iota^\pm_Y) \xrightarrow{\mathrm{id}_{(Y,\iota^\pm)}} (Y, \iota^\pm_Y),$$

$$(Y, \iota^\pm_Y) \circ^1_{\mathrm{h}} 1_{\Sigma,\delta} \xrightarrow{I_\Psi} (Y, \iota^\pm_Y) \xrightarrow{I_{\Psi^{-1}}} (Y, \iota^\pm_Y) \circ^1_{\mathrm{h}} 1_{\Sigma,\delta} \quad = \quad \mathrm{id}_{(Y,\iota^\pm_Y)\circ^1_{\mathrm{h}} 1_{\Sigma,\delta}}.$$

Indeed, we have $I_{\Psi^{-1}} \circ_{\mathrm{v}} I_\Psi = I_{\Psi \circ \Psi^{-1}} = I_{\mathrm{id}_Y} = \mathrm{id}_{(Y,\iota^\pm)}$ due to (3.2.2), and similarly $I_\Psi \circ_{\mathrm{v}} I_{\Psi^{-1}} = I_{\Psi^{-1} \circ \Psi} = I_{\mathrm{id}_{(Y \cup [0,1] \times \Sigma)/\sim}} = \mathrm{id}_{(Y,\iota^\pm_Y)\circ^1_{\mathrm{h}} 1_{\Sigma,\delta}}$. This proves the claimed equivalence for any $Y \in \mathrm{Mor}^1_{\mathrm{Bor}_{d+1+1}}(\Sigma_0, \Sigma)$, and the other required equivalences $1_{\Sigma,\delta} \circ^1_{\mathrm{h}} (Z, \iota^\pm_Z) \sim (Z, \iota^\pm_Z)$ for $(Z, \iota^\pm_Z) \in \mathrm{Mor}^1_{\mathrm{Bor}_{d+1+1}}(\Sigma, \Sigma_1)$ arise in the same way from the diffeomorphisms $\Phi : ([0,1] \times \Sigma \cup Z)/\sim \to Z$ constructed in Example 3.2.1.

This finishes the construction of the bordism bicategory Bor_{d+1+1}. Moreover, the **connected d + 1 + 1 bordism bicategory $\mathbf{Bor^{conn}}_{\mathbf{d+1+1}}$** for $d \geq 2$ is constructed analogously, using the objects and representatives of morphisms of $\mathrm{Bor}^{\mathrm{conn}}_{d+1}$.

3.3 Functors Between Bi- and 2-Categories

The purpose of this section is to make sense of a notion of extending $2+1$ Floer field theory to dimension $2+1+1=4$, which is the case $d=2$ of the following notion.

Definition 3.3.1 A **(connected) d+1+1 Floer field theory** is a 2-functor $\mathrm{Bor}_{d+1+1} \to \mathrm{Cat}$ (resp. $\mathrm{Bor}^{\mathrm{conn}}_{d+1+1} \to \mathrm{Cat}$) that factorizes through a symplectic 2-category and preserves adjunctions.

Here one should use the connected bordism bicategory $\mathrm{Bor}^{\mathrm{conn}}_{d+1+1}$ in order to fit the gauge theoretic examples from Sect. 2.5. An appropriate symplectic 2-category is constructed in [76] and will be outlined in Sect. 3.5. So it remains to spell out the functoriality requirements. We begin with 2-functors between 2-categories, and will develop the relevant notion for bicategories in Definition 3.3.5.

Definition 3.3.2 A **2-functor** $\mathscr{F} : \mathscr{C} \to \mathscr{D}$ between two 2-categories $\mathscr{C}$, $\mathscr{D}$ consists of

- a map $\mathscr{F} : \mathrm{Obj}_{\mathscr{C}} \to \mathrm{Obj}_{\mathscr{D}}$ between the sets of objects,
- functors $\mathscr{F}_{x_1,x_2} : \mathrm{Mor}_{\mathscr{C}}(x_1,x_2) \to \mathrm{Mor}_{\mathscr{D}}(\mathscr{F}(x_1),\mathscr{F}(x_2))$ for each $x_1,x_2 \in \mathrm{Obj}_{\mathscr{C}}$, i.e.,
  - maps $\mathscr{F}^1_{x_1,x_2} : \mathrm{Mor}^1_{\mathscr{C}}(x_1,x_2) \to \mathrm{Mor}^1_{\mathscr{D}}(\mathscr{F}(x_1),\mathscr{F}(x_2))$,
  - maps $\mathscr{F}^2_{x_1,x_2} : \mathrm{Mor}^2_{\mathscr{C}}(f_{12},g_{12}) \to \mathrm{Mor}^2_{\mathscr{D}}(\mathscr{F}^1_{x_1,x_2}(f_{12}),\mathscr{F}^1_{x_1,x_2}(g_{12}))$ for each pair $f_{12},g_{12} \in \mathrm{Mor}^1_{\mathscr{C}}(x_1,x_2)$,
  - compatibility with identities $\mathscr{F}^2_{x_1,x_2}(\mathrm{id}_{f_{12}}) = \mathrm{id}_{\mathscr{F}^1_{x_1,x_2}(f_{12})}$,
  - compatibility with vertical composition,

$$\mathscr{F}^2_{x_1,x_2}(f_{12} \circ_{\mathrm{v}} g_{12}) = \mathscr{F}^2_{x_1,x_2}(f_{12}) \circ_{\mathrm{v}} \mathscr{F}^2_{x_1,x_2}(g_{12}).$$

These are required to intertwine the horizontal compositions in $\mathscr{C}$ and $\mathscr{D}$ as follows:

- $\mathscr{F}$ is compatible with identities, $1_{\mathscr{F}(x)} = \mathscr{F}^1_{x,x}(1_x)$.
- $\mathscr{F}$ is compatible with composition of 1-morphisms, i.e., for each $f_{ij} \in \mathrm{Mor}^1_{\mathscr{C}}(x_i,x_j)$

$$\mathscr{F}^1_{x_1,x_3}(f_{12} \circ_{\mathrm{h}} f_{23}) = \mathscr{F}^1_{x_1,x_2}(f_{12}) \circ_{\mathrm{h}} \mathscr{F}^1_{x_2,x_3}(f_{23}).$$

- $\mathscr{F}$ is compatible with horizontal composition of 2-morphisms, i.e., for each tuple $f_{ij},g_{ij} \in \mathrm{Mor}^1_{\mathscr{C}}(x_i,x_j)$ and $\alpha_{ij} \in \mathrm{Mor}^2_{\mathscr{C}}(f_{ij},g_{ij})$

$$\mathscr{F}^2_{x_1,x_3}(\alpha_{12} \circ_{\mathrm{h}} \alpha_{23}) = \mathscr{F}^2_{x_1,x_2}(\alpha_{12}) \circ_{\mathrm{h}} \mathscr{F}^2_{x_2,x_3}(\alpha_{23}).$$

Before discussing the appropriate generalization of this notion to a 2-functor from a bicategory such as Bor_{2+1+1} to a 2-category such as Cat or Symp, let us note that 2-categories such as Symp (with canonical base objects such as the symplectic manifold consisting of a point) come with natural 2-functors to Cat. This reduces the construction of a $2+1+1$ Floer field theory to the construction of a 2-functor $\mathrm{Bor}_{2+1+1} \to \mathrm{Symp}$, which can then be composed with the Yoneda functor $\mathrm{Symp} \to \mathrm{Cat}$ that is defined below and further discussed in Lemma 3.5.6.

Lemma 3.3.3 *Let $\mathscr{C}$ be a 2-category. Then any choice of distinguished object $x_0 \in \mathrm{Obj}_{\mathscr{C}}$ induces a* **Yoneda 2-functor** $\mathscr{Y}_{x_0} : \mathscr{C} \to \mathrm{Cat}$ *as follows.*

- *To an object $x \in \mathrm{Obj}_{\mathscr{C}}$ we associate the category $\mathscr{Y}_{x_0}(x) := \mathrm{Mor}_{\mathscr{C}}(x_0, x)$.*
- *To $f \in \mathrm{Mor}^1_{\mathscr{C}}(x_1, x_2)$ we associate the functor $\mathscr{Y}_{x_0}(f) : \mathscr{Y}_{x_0}(x_1) \to \mathscr{Y}_{x_0}(x_2)$ given by horizontal composition with f and its identity 2-morphism $\mathrm{id}_f \in \mathrm{Mor}^2_{\mathscr{C}}(f,f)$,*

$$\begin{aligned}
\mathrm{Obj}_{\mathscr{Y}_{x_0}(x_1)} = \mathrm{Mor}^1_{\mathscr{C}}(x_0, x_1) &\longrightarrow \mathrm{Mor}^1_{\mathscr{C}}(x_0, x_2) = \mathrm{Obj}_{\mathscr{Y}_{x_0}(x_2)}, \\
f_{01} &\longmapsto f_{01} \circ_{\mathrm{h}} f; \\
\mathrm{Mor}_{\mathscr{Y}_{x_0}(x_1)} \supset \mathrm{Mor}^2_{\mathscr{C}}(f_{01}, g_{01}) &\longrightarrow \mathrm{Mor}^2_{\mathscr{C}}(f_{01} \circ_{\mathrm{h}} f, g_{01} \circ_{\mathrm{h}} f) \subset \mathrm{Mor}_{\mathscr{Y}_{x_0}(x_2)}, \\
\alpha &\longmapsto \alpha \circ_{\mathrm{h}} \mathrm{id}_f.
\end{aligned}$$

- *To a 2-morphism $\beta \in \mathrm{Mor}^2_{\mathscr{C}}(g_{12}, h_{12})$ between $g_{12}, h_{12} \in \mathrm{Mor}^1_{\mathscr{C}}(x_1, x_2)$ we associate the natural transformation $\mathscr{Y}_{x_0}(\beta) : \mathscr{Y}_{x_0}(g_{12}) \Rightarrow \mathscr{Y}_{x_0}(h_{12})$ which takes each $f_{01} \in \mathrm{Obj}_{\mathscr{Y}_{x_0}(x_1)} = \mathrm{Mor}^1_{\mathscr{C}}(x_0, x_1)$ to $\mathrm{id}_{f_{01}} \circ_{\mathrm{h}} \beta \in \mathrm{Mor}^2_{\mathscr{C}}(f_{01} \circ_{\mathrm{h}} g_{12}, f_{01} \circ_{\mathrm{h}} h_{12}) \subset \mathrm{Mor}_{\mathscr{Y}_{x_0}(x_2)}$.*

Proof $\mathscr{Y}_{x_0}(x)$ is a category and $\mathscr{Y}_{x_0}(f)$ is a functor by Definition 3.1.1 of a 2-category. $\mathscr{Y}_{x_0}(\beta)$ is a natural transformation since the required diagram for $\alpha \in \mathrm{Mor}_{\mathscr{C}}(f_{01}, f'_{01})$ commutes by compatibility of horizontal and vertical composition,

$$\begin{aligned}
\big(\alpha \circ_{\mathrm{h}} \mathrm{id}_{g_{12}}\big) \circ_{\mathrm{v}} \big(\mathrm{id}_{f'_{01}} \circ_{\mathrm{h}} \beta\big) &= \big(\alpha \circ_{\mathrm{v}} \mathrm{id}_{f'_{01}}\big) \circ_{\mathrm{h}} \big(\mathrm{id}_{g_{12}} \circ_{\mathrm{v}} \beta\big) = \alpha \circ_{\mathrm{h}} \beta \\
&= \big(\mathrm{id}_{f_{01}} \circ_{\mathrm{v}} \alpha\big) \circ_{\mathrm{h}} \big(\beta \circ_{\mathrm{v}} \mathrm{id}_{h_{12}}\big) = \big(\mathrm{id}_{f_{01}} \circ_{\mathrm{h}} \beta\big) \circ_{\mathrm{v}} \big(\alpha \circ_{\mathrm{h}} \mathrm{id}_{h_{12}}\big).
\end{aligned}$$

Next, we need to check that $\mathscr{F} := \mathscr{Y}_{x_0} : \mathrm{Mor}_{\mathscr{C}}(x_1, x_2) \to \mathrm{Mor}_{\mathrm{Cat}}(\mathscr{F}(x_1), \mathscr{F}(x_2))$ is a functor for each $x_1, x_2 \in \mathrm{Obj}_{\mathscr{C}}$. It is compatible with identities since both $\mathscr{F}(\mathrm{id}_{f_{12}})$ and $\mathrm{id}_{\mathscr{F}(f_{12})}$ are the natural transformation $\mathscr{F}(f_{12}) \Rightarrow \mathscr{F}(f_{12})$ which takes $f_{01} \in \mathrm{Mor}^1_{\mathscr{C}}(x_0, x_1)$ to $\mathrm{id}_{f_{01}} \circ_{\mathrm{h}} \mathrm{id}_{f_{12}} = \mathrm{id}_{f_{01} \circ_{\mathrm{h}} f_{12}}$. It is compatible with vertical composition since for $\alpha_{12}, \beta_{12} \in \mathrm{Mor}^2_{\mathscr{C}}(g_{12}, h_{12})$ both $\mathscr{F}(\alpha_{12} \circ_{\mathrm{v}} \beta_{12})$ and $\mathscr{F}(\alpha_{12}) \circ_{\mathrm{v}} \mathscr{F}(\beta_{12})$ are the natural transformation $\mathscr{F}(g_{12}) \Rightarrow \mathscr{F}(h_{12})$ which takes $f_{01} \in \mathrm{Mor}^1_{\mathscr{C}}(x_0, x_1)$ to

$$\mathrm{id}_{f_{01}} \circ_{\mathrm{h}} (\alpha_{12} \circ_{\mathrm{v}} \beta_{12}) = (\mathrm{id}_{f_{01}} \circ_{\mathrm{h}} \alpha_{12}) \circ_{\mathrm{v}} (\mathrm{id}_{f_{01}} \circ_{\mathrm{h}} \beta_{12}).$$

Finally, we check compatibility with the horizontal composition.

- Both $1_{\mathscr{F}(x)}$ and $\mathscr{F}(1_x)$ are the functor $\mathrm{Mor}_{\mathscr{C}}(x_0, x) \to \mathrm{Mor}_{\mathscr{C}}(x_0, x)$ given by $f_{01} \mapsto f_{01} = f_{01} \circ_{\mathrm{h}} 1_x$ and $\alpha \mapsto \alpha = \alpha \circ_{\mathrm{h}} \mathrm{id}_{1_x}$.
- For $f_{ij} \in \mathrm{Mor}^1_{\mathscr{C}}(x_i, x_j)$ both $\mathscr{F}(f_{12} \circ_{\mathrm{h}} f_{23})$ and $\mathscr{F}^1_{x_1,x_2}(f_{12}) \circ_{\mathrm{h}} \mathscr{F}^1_{x_2,x_3}(f_{23})$ are the functor $\mathrm{Mor}_{\mathscr{C}}(x_0, x_1) \to \mathrm{Mor}_{\mathscr{C}}(x_0, x_2)$ given by $f_{01} \mapsto f_{01} \circ_{\mathrm{h}} (f_{12} \circ_{\mathrm{h}} f_{23}) = (f_{01} \circ_{\mathrm{h}} f_{12}) \circ_{\mathrm{h}} f_{23}$ and $\alpha \mapsto \alpha \circ_{\mathrm{h}} \mathrm{id}_{f_{12} \circ_{\mathrm{h}} f_{23}} = (\alpha \circ_{\mathrm{h}} \mathrm{id}_{f_{12}}) \circ_{\mathrm{h}} \mathrm{id}_{f_{23}}$.

- For each tuple $g_{ij}, h_{ij} \in \mathrm{Mor}^1_{\mathscr{C}}(x_i, x_j)$ and $\alpha_{ij} \in \mathrm{Mor}^2_{\mathscr{C}}(g_{ij}, h_{ij})$, both $\mathscr{F}(\alpha_{12}) \circ_h \mathscr{F}(\alpha_{23})$ and $\mathscr{F}(\alpha_{12} \circ_h \alpha_{23})$ are the natural transformation $\mathscr{G} := \mathscr{F}(g_{12} \circ_h g_{23}) \Rightarrow \mathscr{H} := \mathscr{F}(h_{12} \circ_h h_{23})$ which takes $f_{01} \in \mathrm{Mor}^1_{\mathscr{C}}(x_0, x_1)$ to $\mathrm{id}_{f_{01}} \circ_h (\alpha_{12} \circ_h \alpha_{23})$. □

Remark 3.3.4 (Yoneda 2-functor for bicategories). If $\mathscr{C}$ is a bicategory, then the Yoneda construction in Lemma 3.3.3 still yields categories $\mathscr{Y}_{x_0}(x) = \mathrm{Mor}_{\mathscr{C}}(x_0, x)$, functors $\mathscr{Y}_{x_0}(f)$ given by horizontal composition with f and id_f, and natural transformations $\mathscr{Y}_{x_0}(\beta)$ given by $f_{01} \mapsto \mathrm{id}_{f_{01}} \circ_h \beta$, in such a way that $\mathscr{Y}_{x_0} : \mathrm{Mor}_{\mathscr{C}}(x_1, x_2) \to \mathrm{Mor}_{\mathrm{Cat}}(\mathscr{F}(x_1), \mathscr{F}(x_2))$ is a functor. However, $\mathscr{Y}_{x_0}$ is compatible with horizontal composition only up to isomorphisms in Cat since unitality $f_{01} \circ_h 1_x \sim f_{01}$ and associativity $f_{01} \circ_h (f_{12} \circ_h f_{23}) \sim (f_{01} \circ_h f_{12}) \circ_h f_{23}$ only hold up to 2-isomorphism in $\mathscr{C}$. Thus $\mathscr{Y}_{x_0} : \mathscr{C} \to \mathrm{Cat}$ can still be viewed as a 2-functor between bicategories in the sense of Definition 3.3.5 below.

To make the Yoneda construction for bicategories as well as our notion of 2+1+1 Floer field theory in Definition 3.3.1 precise, we define the notion of a 2-functor between bicategories $\mathscr{C}, \mathscr{D}$ by weakening Definition 3.3.2 to allow compatibility with the horizontal composition up to isomorphisms.

Definition 3.3.5 A **2-functor** $\mathscr{F} : \mathscr{C} \to \mathscr{D}$ between two bicategories $\mathscr{C}, \mathscr{D}$ consists of

- a map $\mathscr{F} : \mathrm{Obj}_{\mathscr{C}} \to \mathrm{Obj}_{\mathscr{D}}$ between the sets of objects,
- functors $\mathscr{F}_{x_1,x_2} : \mathrm{Mor}_{\mathscr{C}}(x_1, x_2) \to \mathrm{Mor}_{\mathscr{D}}(\mathscr{F}(x_1), \mathscr{F}(x_2))$ for each $x_1, x_2 \in \mathrm{Obj}_{\mathscr{C}}$,

which are compatible with the horizontal composition in the following sense:

- $1_{\mathscr{F}(x)} \sim \mathscr{F}^1_{x,x}(1_x)$ are equivalent 1-morphisms in $\mathscr{D}$ for any choice of weak units associated to $x \in \mathrm{Obj}_{\mathscr{C}}$ and $\mathscr{F}(x) \in \mathrm{Obj}_{\mathscr{D}}$.
- $\mathscr{F}^1_{x_1,x_3}(f_{12} \circ_h f_{23}) \sim \mathscr{F}^1_{x_1,x_2}(f_{12}) \circ_h \mathscr{F}^1_{x_2,x_3}(f_{23})$ are equivalent 1-morphisms in $\mathscr{D}$.
- For each tuple $f_{ij}, g_{ij} \in \mathrm{Mor}^1_{\mathscr{C}}(x_i, x_j)$ and $\alpha_{ij} \in \mathrm{Mor}^2_{\mathscr{C}}(f_{ij}, g_{ij})$ we have

$$\mathscr{F}^2_{x_1,x_3}(\alpha_{12} \circ_h \alpha_{23}) = \mathscr{F}^2_{x_1,x_2}(\alpha_{12}) \circ_h \mathscr{F}^2_{x_2,x_3}(\alpha_{23}).$$

3.4 Adjunctions, Quilt Diagrams, and Quilted Bicategories

This section will generalize the notion of string diagrams, which are graphical representations of the structure and axioms of 2-categories, as surveyed in e.g., [10, Sect. 1.1], [73, 83] in the example of topological and symplectic 2-categories. Then, we introduce a notion of quilted bicategory, in which not only string diagrams but the more general quilt diagrams define 2-morphisms, and show how bordism bicategories naturally fit into this notion.

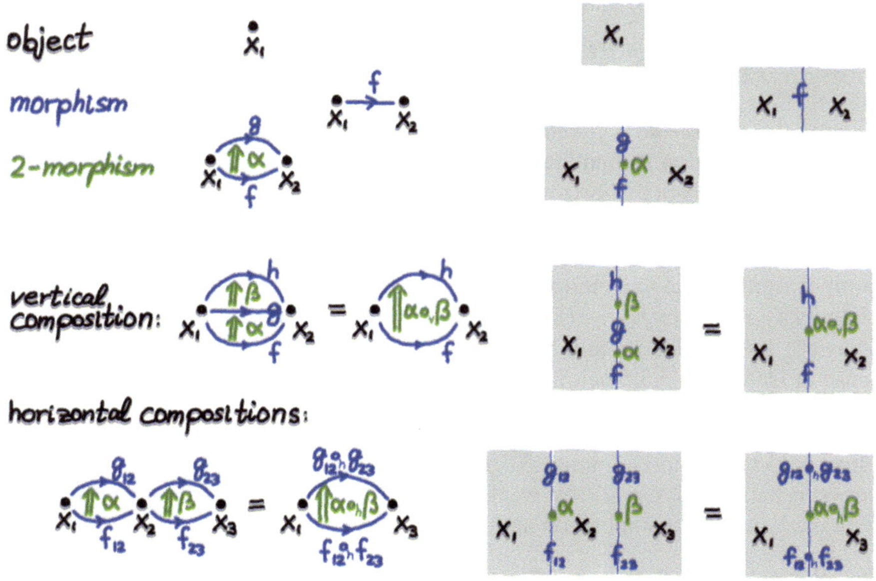

Fig. 14 The structures of a 2-category or bicategory in string diagram notation

Remark 3.4.1 (String diagrams). Roughly speaking, a string diagram in a bicategory $\mathscr{C}$ consists of vertical lines drawn in the plane, punctures on the line, and labels in $\mathscr{C}$. These in turn represent the structure of the bicategory as indicated in Fig. 14. More precisely, the lines separate the plane into connected components, called "patches," and the punctures separate the lines into connected components, called "seams," each of which lies in the intersection of the closures of exactly two patches; see the left side of Fig. 15 for illustration. The patches/seams/punctures of the diagram are labeled with objects/1-morphisms/2-morphisms in $\mathscr{C}$ in a coherent manner: a seam is labeled by a 1-morphism between the objects associated to the two adjacent patches, and a puncture is labeled by a 2-morphism between the 1-morphisms associated to the two adjacent seams. Now any such string diagram can be translated into horizontal and vertical compositions of the involved 2-morphisms, and defines a new 2-morphism between the 1-morphisms obtained from composing the labels of the seams running to $+\infty$ resp. $-\infty$. Here, we read from left to right and from bottom to top, with different choices of order of composition yielding the same result due to associativity and compatibility of horizontal and vertical composition; see the right of Fig. 15 for examples.

After compactifying the plane to a sphere, we may interpret the punctures in the plane as incoming ends—at which the 2-morphisms are prescribed—and the puncture at infinity as the outgoing end—at which the resulting 2-morphism is read off. The axioms of a 2-category or bicategory can then also be represented by string diagrams: identities between different diagrams, or the fact that diagrams have invariant

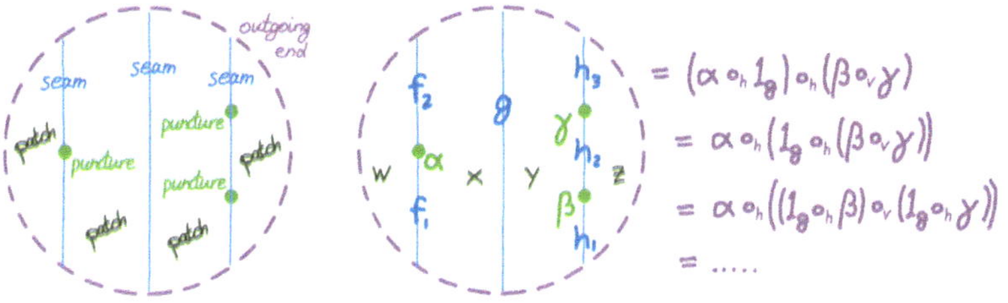

Fig. 15 String diagrams represent well-defined 2-morphisms given by iterated horizontal and vertical compositions applied to 2-morphisms given by labels and identity 2-morphisms

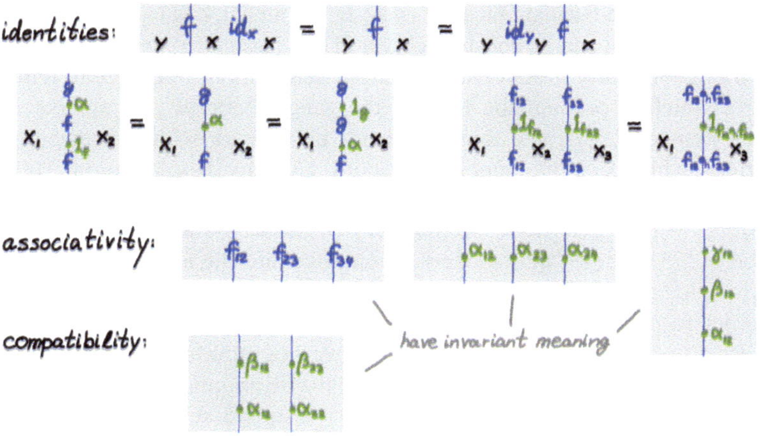

Fig. 16 2-category axioms in string diagram notation. Also see Fig. 17 for compatibility

meaning—independent of the order in which composition is being read off. See Fig. 16 for a list of the 2-category axioms as string diagrams,

The symplectic 2-category will have string diagrams—represented by pseudo-holomorphic quilts [79] described in Remark 3.5.3—which lie on more general surfaces (not just the sphere), can have any number of seams running into the punctures, do not require a left/right or top/bottom orientation, but still have exactly one outgoing end and the same meaning as a string diagram: If we prescribe Floer homology classes (the 2-morphisms) at each incoming end, then the diagram defines a Floer homology class at the outgoing end. These relative quilt invariants are applied to the basic string diagrams in [76] to construct the symplectic 2-category, but they are defined in higher generality and satisfy algebraic identities arising from forgetting the vertical/horizontal structure of string diagrams.

The purpose of this section is to cast this additional structure on the symplectic 2-category into abstract terms—giving rise to a notion akin to that of spherical 2-categories developed in [43], but expressing the algebraic properties in a graphical language rather than via monoidal structure. This is useful for a variety of

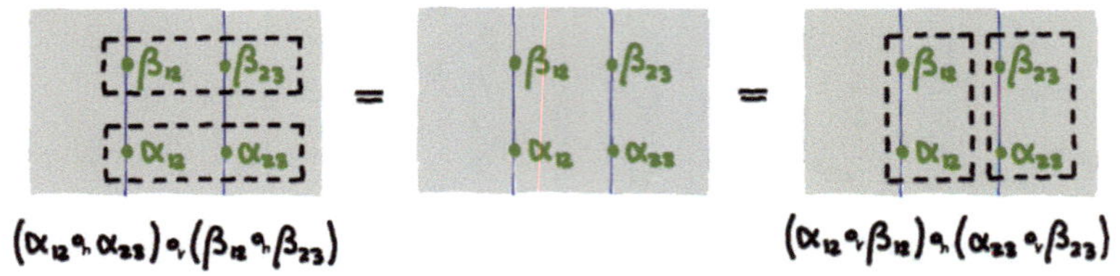

Fig. 17 Compatibility between horizontal and vertical composition in string diagram notation

reasons: First, this structure simply exists naturally, not just for the symplectic 2-category but also the bordism bicategories (see Lemma 3.4.11) and other gauge theoretic categories that can be constructed via PDE's associated to quilt diagrams (see Sect. 3.6). Second, this structure can be expressed without reference to a monoidal structure, which is problematic both in the gauge theoretic and symplectic context (see Remark 2.5.7), and thus also leads us to work with connected bordism categories—which lack the monoidal structure given by disjoint union. Third, quilt diagrams naturally appear in a generalization of Cerf decompositions from Bor_{d+1} to Bor_{d+1+1} which arise from the diagrams of Morse 2-functions in e.g., [26], as sketched in [73]. These "quilted Cerf decompositions" lie at the core of the extension principle for Floer field theories [74], as outlined in Conjecture 3.4.12.

In order to make sense of the labeling in a quilt diagram we will need some symmetry properties of the bicategory, which we will introduce before going into the actual notion of quilt diagram. First, dropping the distinguished horizontal direction in string diagrams loses the "from left to right" designation which determines that a seam is to be labeled by a 1-morphism from the object associated to the left adjacent patch to the object associated to the right adjacent patch. Instead, we will define left/right based on a choice of orientation of each seam and label the two orientations of each seam with adjoint pairs of 1-morphisms. For that purpose, the following makes the adjunction notion from Remark 2.4.3 rigorous.

Definition 3.4.2 A **2-category with adjoints** is a 2-category $\mathscr{C}$ as in Definition 3.1.1 together with an adjunction map $\mathrm{Mor}^1_{\mathscr{C}} \to \mathrm{Mor}^1_{\mathscr{C}}$, $Y \mapsto Y^T$ that associates to each $Y \in \mathrm{Mor}^1_{\mathscr{C}}(\Sigma_0, \Sigma_1)$ its **adjoint** $Y^T \in \mathrm{Mor}^1_{\mathscr{C}}(\Sigma_1, \Sigma_0)$ and satisfies:

- Adjunction is reflexive, i.e., $(Y^T)^T = Y$.
- Adjoint morphisms are dual to each other in the sense that for $Y \in \mathrm{Mor}^1_{\mathscr{C}}(\Sigma_0, \Sigma_1)$ there exist $X_Y \in \mathrm{Mor}^2_{\mathscr{C}}(1_{\Sigma_0}, Y \circ^1_{\mathrm{h}} Y^T)$ and $X_Y^T \in \mathrm{Mor}^2_{\mathscr{C}}(Y^T \circ^1_{\mathrm{h}} Y, 1_{\Sigma_1})$ satisfying identities that are illustrated in Fig. 18,

$$\left(X_Y \circ^2_{\mathrm{h}} \mathrm{id}_Y\right) \circ_{\mathrm{v}} \left(\mathrm{id}_Y \circ^2_{\mathrm{h}} X_Y^T\right) = \mathrm{id}_Y, \quad \left(\mathrm{id}_{Y^T} \circ^2_{\mathrm{h}} X_Y\right) \circ_{\mathrm{v}} \left(X_Y^T \circ^2_{\mathrm{h}} \mathrm{id}_{Y^T}\right) = \mathrm{id}_{Y^T}. \tag{3.4.1}$$

A **bicategory with adjoints** is a bicategory $\mathscr{C}$ as in Definition 3.1.2 together with an adjunction map as above, whose duality property holds for all choices of weak identity morphisms $1_{\Sigma_0}, 1_{\Sigma_1}$.

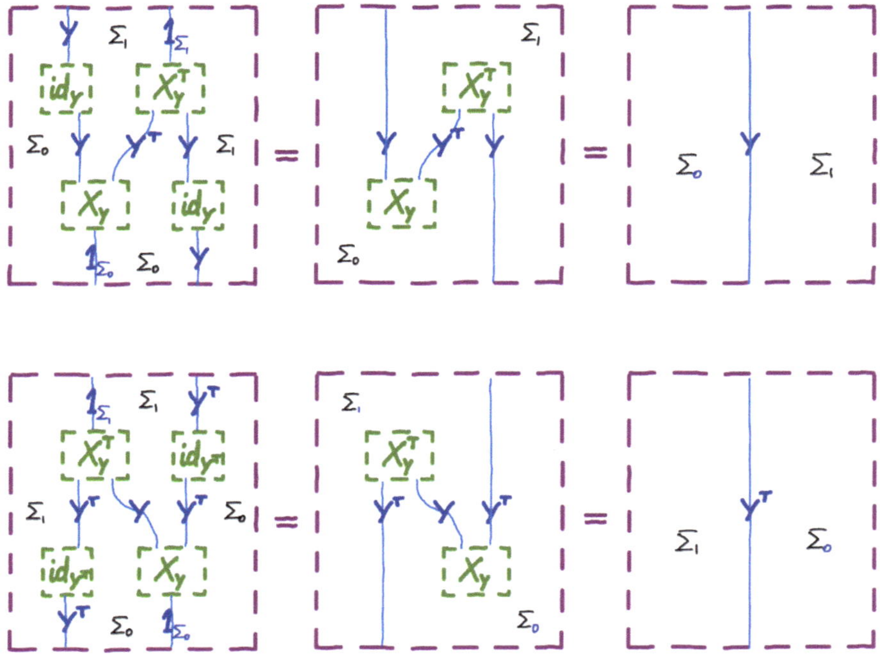

Fig. 18 The duality identities (3.4.1) can be represented by slightly generalized string diagrams

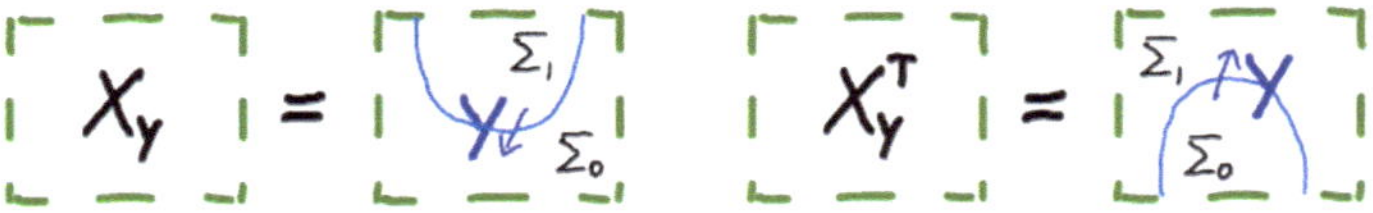

Fig. 19 In a quilted bicategory (see Sect. 3.4 below) the adjunction 2-morphisms arise from quilted structure maps that are represented by the above quilt (i.e. generalized string) diagrams

Remark 3.4.3 (a) Adjoints in Bor_{d+1+1} are obtained by orientation reversal of the 1-morphisms, as sketched for Bor_{d+1} in Remark 2.4.3. More precisely, the adjoint of a $(d+1)$-cobordism $(Y, \iota_Y^-, \iota_Y^+) \in \mathrm{Mor}^1_{\mathrm{Bor}_{d+1+1}}(\Sigma_0, \Sigma_1)$ is the cobordism $(Y, \iota_Y^-, \iota_Y^+)^T := (Y^-, \iota_Y^+ \circ \rho_1, \iota_Y^- \circ \rho_0) \in \mathrm{Mor}^1_{\mathrm{Bor}_{d+1+1}}(\Sigma_1, \Sigma_0)$ obtained by reversing the orientation on Y, switching the tubular neighborhood embeddings, and precomposing each with the orientation reversing diffeomorphism $\rho_i(t, z) = (1-t, z)$ of $[0, 1] \times \Sigma_i$.

With this reflexive operation established, the adjunction 2-morphisms X_Y and X_Y^T that are required for the duality in Definition 3.4.2 can be constructed from the further generalized string diagrams indicated in Fig. 19. For example, X_Y is obtained from a half disk times Σ_1, a square minus a half disk times Σ_0, and an interval times Y, glued along matching boundary components. This and the analogous construction for X_Y^T yields the required 2-morphisms, which satisfy

(3.4.1) since gluing them into the string diagrams in Fig. 18 yields 4-manifolds with boundary and corners that are diffeomorphic relative to the boundary.

(b) In the symplectic category Symp, the adjoint of a Lagrangian $L \subset M_0^- \times M_1$ is $L^T := \tau(L) \subset M_1^- \times M_0$ obtained by transposition $\tau(p_0, p_1) := (p_1, p_0)$, and the adjoint of a general 1-morphism $\underline{L} = (L_{01}, \ldots, L_{(k-1)k})$ is $\underline{L}^T = (L^T_{(k-1)k}, \ldots, L^T_{01})$. Again, the adjunction 2-morphisms X_Y and X_Y^T can be obtained from the fact that the generalized string diagrams in Fig. 19 have invariant meaning; see Remark 3.5.2.

A second symmetry property of a bicategory that is required to formalize quilt diagrams comes from the fact that dropping the distinguished vertical direction in string diagrams loses the "from bottom to top" designation which determines that a puncture is to be labeled by a 2-morphism from the 1-morphism associated to the bottom adjacent seam to the 1-morphism associated to the top adjacent patch. Instead, we are allowing any number of seams to intersect in a puncture of a quilt diagram, and will associate to these seams—with counterclockwise order induced from an overall orientation of the diagram—a cyclic 2-morphism space, from which the label for this puncture will be chosen. This is based on the following cyclic symmetry of the 2-morphisms in a bicategory with adjoints. Here and in the following, we will use $\mathbb{Z}_N := \mathbb{Z}/_{N\mathbb{Z}}$ to index cyclically ordered sets of N elements with no distinguished first element.

Remark 3.4.4 A **cyclic 1-morphism** in a bicategory $\mathscr{C}$ is a cyclic sequence of 1-morphisms $\underline{f} = (f_i)_{i\in\mathbb{Z}_N} : \mathbb{Z}_N \to \mathrm{Mor}^1_{\mathscr{C}}$ that is composable in the sense that we have $f_i \in \mathrm{Mor}^1_{\mathscr{C}}(x_i, x_{i+1})$ for a cyclic sequence of objects $\underline{x} = (x_i)_{i\in\mathbb{Z}_N} : \mathbb{Z}_N \to \mathrm{Obj}_{\mathscr{C}}$. This implies that the compositions $f_i \circ f_{i+1} \circ \ldots \circ f_{i+k} \in \mathrm{Mor}^1_{\mathscr{C}}(x_i, x_{i+k})$ are well defined for every $i \in \mathbb{Z}_N$, $k \in \mathbb{N}$, and in particular $f_i \circ f_{i+1} \circ \ldots \circ f_{i+N-1} \in \mathrm{Mor}^1_{\mathscr{C}}(x_i, x_{i+N} = x_i)$.

If the bicategory $\mathscr{C}$ moreover has adjoints in the sense of Definition 3.4.2, then we can associate to every cyclic 1-morphism $\underline{f} = (f_i)_{i\in\mathbb{Z}_N}$ a **cyclic 2-morphism space**

$$\mathrm{Mor}^2_{\mathscr{C}}(\underline{f}) := \mathrm{Mor}^2_{\mathscr{C}}(f_i^T, f_{i+1} \circ \ldots \circ f_{i+N-1}),$$

which is independent of the choice of $i \in \mathbb{Z}_N$ and can also be identified with the 2-morphism space $\mathrm{Mor}^2_{\mathscr{C}}\big((f_i \circ \ldots \circ f_j)^T, f_{j+1} \circ \ldots \circ f_{i-1}\big)$ for other partitions of the cyclic 1-morphism.

As a tangential note—useful for identifying different field theories as in the Atiyah–Floer type conjectures—the following remark explains an algebraic method for localizing proofs of isomorphisms between cyclic 1-morphisms or their associated cyclic 2-morphism spaces.

Remark 3.4.5 (A "local to global" principle for cyclic 1-morphisms). In a 2-category with adjoints, any "local" isomorphism between $f, g \in \mathrm{Mor}^1_{\mathscr{C}}(x_i, x_{i+1})$ implies "global isomorphisms" between any cyclic 1-morphisms that differ by replacing f with g,

$$f \sim g \quad \Longrightarrow \quad (\ldots f_{i-1}, f_i = f, f_{i+1} \ldots) \sim (\ldots f_{i-1}, f_i = g, f_{i+1} \ldots).$$

Here the local isomorphism is given by an invertible 2-morphism $\alpha \in \mathrm{Mor}^2_{\mathscr{C}}(f, g)$, i.e., $\alpha \circ_v \alpha^{-1} = 1_f$ and $\alpha^{-1} \circ_v \alpha = 1_g$ for some $\alpha^{-1} \in \mathrm{Mor}^2_{\mathscr{C}}(g, f)$. It induces global isomorphisms $\underline{f}_j := (f_j, \ldots, f_i = f, \ldots, f_{j-1}) \sim (f_j, \ldots, f_i = g, \ldots, f_{j-1}) =: \underline{g}_j$ in $\mathrm{Mor}^1(x_j, x_j)$ for any $j \in \mathbb{Z}_N$, in the sense that there exist 2-morphisms given by

$$\underline{\alpha} := \mathrm{id}_{f_j} \circ_h \ldots \mathrm{id}_{f_{i-1}} \circ_h \alpha \circ_h \mathrm{id}_{f_{i+1}} \ldots \circ_h \mathrm{id}_{f_{j-1}} \in \mathrm{Mor}^2_{\mathscr{C}}(\underline{f}_j, \underline{g}_j),$$
$$\underline{\alpha}^{-1} := \mathrm{id}_{f_j} \circ_h \ldots \mathrm{id}_{f_{i-1}} \circ_h \alpha^{-1} \circ_h \mathrm{id}_{f_{i+1}} \ldots \circ_h \mathrm{id}_{f_{j-1}} \in \mathrm{Mor}^2_{\mathscr{C}}(\underline{g}_j, \underline{f}_j),$$

which satisfy $\underline{\alpha} \circ_v \underline{\alpha}^{-1} = \mathrm{id}_{\underline{f}_j}$ and $\underline{\alpha}^{-1} \circ_v \underline{\alpha} = \mathrm{id}_{\underline{g}_j}$. Indeed, the first (and similarly the second) follows from compatibility of horizontal and vertical composition with each other as well as identities,

$$\begin{aligned}
\underline{\alpha} \circ_v \underline{\alpha}^{-1} &= (\mathrm{id}_{f_j} \circ_v \mathrm{id}_{f_j}) \circ_h \ldots (\mathrm{id}_{f_{i-1}} \circ_v \mathrm{id}_{f_{i-1}}) \circ_h (\alpha \circ_v \alpha^{-1}) \\
&\qquad \circ_h (\mathrm{id}_{f_{i+1}} \circ_v \mathrm{id}_{f_{i+1}}) \ldots \circ_h (\mathrm{id}_{f_{j-1}} \circ_v \mathrm{id}_{f_{j-1}}) \\
&= \mathrm{id}_{f_j} \circ_h \ldots \mathrm{id}_{f_{i-1}} \circ_h \mathrm{id}_f \circ_h \mathrm{id}_{f_{i+1}} \ldots \circ_h \mathrm{id}_{f_{j-1}} \\
&= \mathrm{id}_{f_j \circ_h \ldots f_{i-1} \circ_h f \circ_h f_{i+1} \ldots \circ_h f_{j-1}} = \mathrm{id}_{\underline{f}_j}.
\end{aligned}$$

Moreover the local isomorphism also implies an identification between the cyclic 2-morphism spaces,

$$f \sim g \quad \Longrightarrow \quad \mathrm{Mor}^2_{\mathscr{C}}(\ldots f_{i-1}, f_i = f, f_{i+1} \ldots) \simeq \mathrm{Mor}^2_{\mathscr{C}}(\ldots f_{i-1}, f_i = g, f_{i+1} \ldots).$$

Finally, we introduce quilt diagrams by phrasing the notions of "quilted surface" and "Lagrangian boundary conditions" from [79, Sect. 3] in abstract terms.

Definition 3.4.6 A **quilt** is a tuple $\mathcal{Q} := (q_0, Q_0, Q_1, Q_2)$ consisting of a closed oriented surface Q_2, a finite subset of points $Q_0 \subset Q_2$, a one-dimensional submanifold $Q_1 \subset Q_2 \setminus Q_0$, and one distinguished point $q_0 \in Q_0$. We moreover require $Q_1 \subset Q_2 \setminus Q_0$ to be a closed subset with finitely many connected components, as illustrated in Fig. 20.

- The **patches** $P \in \mathscr{P}_{\mathcal{Q}} \cong \pi_0(Q_2 \setminus (Q_0 \cup Q_1))$ of $\mathcal{Q}$ are the connected components $P \subset Q \setminus (E \cup S)$.
- The **seams** $S \in \mathscr{S}_{\mathcal{Q}} \cong \pi_0(Q_1)$ of $\mathcal{Q}$ are the connected components $S \subset Q_1$. The **oriented seams** $S \in \mathscr{S}^{\mathrm{or}}_{\mathcal{Q}} \simeq \mathscr{S}_{\mathcal{Q}} \times \mathbb{Z}_2$ are pairs of seams with orientations.
- For $S \in \mathscr{S}^{\mathrm{or}}_{\mathcal{Q}}$ we denote by $P^-_S, P^+_S \in \mathscr{P}_{\mathcal{Q}}$ the **adjacent patches** whose oriented boundary contains S^- resp. S (i.e., which lie to the right resp. left of S), as illustrated in Fig. 21.
- The **outgoing end** of $\mathcal{Q}$ is $\mathscr{E}^+_{\mathcal{Q}} := \{e^+\} := \{q_0\}$, and the **incoming ends** of $\mathcal{Q}$ are the points $e \in \mathscr{E}^-_{\mathcal{Q}} := Q_0 \setminus \{q_0\}$.

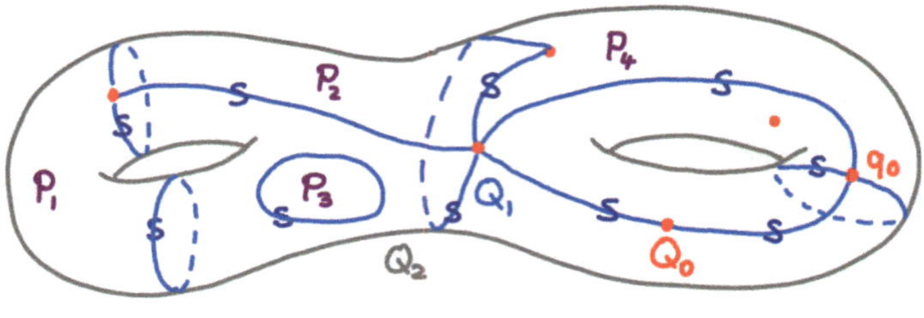

Fig. 20 A quilt (or quilted surface) is given by a closed surface Q_2 and submanifolds Q_0, Q_1, q_0. These specify patches (with arbitrary enumeration P_i above) and seams (all labeled by S above)

Remark 3.4.7 Each seam is either a circle or an open interval embedded in $Q_2 \setminus Q_0$, and cannot intersect itself or other seams by the submanifold property of Q_1. Moreover, the closure of $Q_1 \subset Q_2 \setminus Q_0$ implies that the boundary of an interval seam lies in Q_0, i.e., the seam is a closed interval immersed in Q_2 with endpoints mapping to ends in Q_0 which may or may not coincide. We had to add the finiteness condition to avoid "Hawaiian earrings"—sequences of interval seams converging to a puncture.

The finiteness condition for the sets of seams $\mathscr{S}_{\mathscr{Q}} \cong \pi_0(Q_1)$ and ends $\mathscr{E}^+_{\mathscr{Q}} \cup \mathscr{E}^-_{\mathscr{Q}} = Q_0$ also implies finiteness for the set of patches $\mathscr{P}_{\mathscr{Q}} \cong \pi_0(Q_2 \setminus (Q_0 \cup Q_1))$. Moreover, each patch is an open subset $P \subset Q_2 \setminus Q_0$, whose boundary $\overline{P} \setminus P$ is given by a union of seams. The embedding $Q_0 \setminus Q_2$ gives $\overline{P}$ the structure of an oriented 2-manifold with boundary, though some seams may lie in its interior—namely the seams which have P adjacent on both sides, i.e., the oriented seams $S \in \mathscr{S}^{\rm or}_{\mathscr{Q}}$ with $P^-_S = P^+_S = P$ (which is equivalent to $P^+_{S^-} = P^+_S = P$). By cutting along these seams and adding two copies of the seam we obtain another oriented 2-manifold $\widehat{P}$ with boundary given by the union of all oriented seams $S \in \mathscr{S}^{\rm or}_{\mathscr{Q}}$ with $P^+_S = P$; see Fig. 21 for examples. This "refinement of the closure in $Q_2 \setminus Q_0$ of each patch" comes with a natural immersion $\widehat{P} \to Q_0 \setminus Q_2$ with image $\overline{P}$ and self-intersections on the seams in the interior of $\overline{P}$.

Definition 3.4.8 A **quilt diagram** $\mathscr{Q}\mathscr{D} = \big(\mathscr{Q}, (\Sigma_P)_{P \in \mathscr{P}_{\mathscr{Q}}}, (Y_S)_{S \in \mathscr{S}_{\mathscr{Q}}}\big)$ in a bicategory $\mathscr{C}$ with adjoints consists of a quilt $\mathscr{Q}$ with labels in $\mathscr{C}$ as follows, and illustrated in Fig. 22.

- Each patch $P \in \mathscr{P}_{\mathscr{Q}}$ is labeled by an object $\Sigma_P \in {\rm Obj}_{\mathscr{C}}$.
- Each oriented seam $S \in \mathscr{S}^{\rm or}_{\mathscr{Q}}$ of $\mathscr{Q}$ is labeled by a morphism $Y_S \in {\rm Mor}^1_{\mathscr{C}}(P^-_S, P^+_S)$ such that seams of opposite orientation are labeled with adjoint morphisms, i.e., $Y_{S^-} = (Y_S)^T \in {\rm Mor}^1_{\mathscr{C}}(P^+_S, P^-_S)$, since $P^{\pm}_{S^-} = P^{\mp}_S$.

To turn a quilt diagram into a generalized string diagram, we should in addition label each incoming end $e \in \mathscr{E}^-_{\mathscr{Q}}$ by a 2-morphism $X_e \in {\rm Mor}^2_{\mathscr{C}}(\underline{Y}_e)$, and at the outgoing end $e^+ = q_0$ have the quilt diagram define a 2-morphism $X_{e^+} \in {\rm Mor}^2_{\mathscr{C}}(\underline{Y}_{e^+})$ in the **cyclic 2-morphism spaces associated to each end** as follows:

- For the outgoing end, we define a cyclic sequence of oriented seams $\underline{S}_{e^+} = (S_i)_{i \in \mathbb{Z}_{N_{e^+}}} : \mathbb{Z}_{N_{e^+}} \to \mathscr{S}^{\rm or}_{\mathscr{Q}}$ given by the oriented seams $S_i \simeq \mathbb{R}$ with $+\infty$-limit e^+,

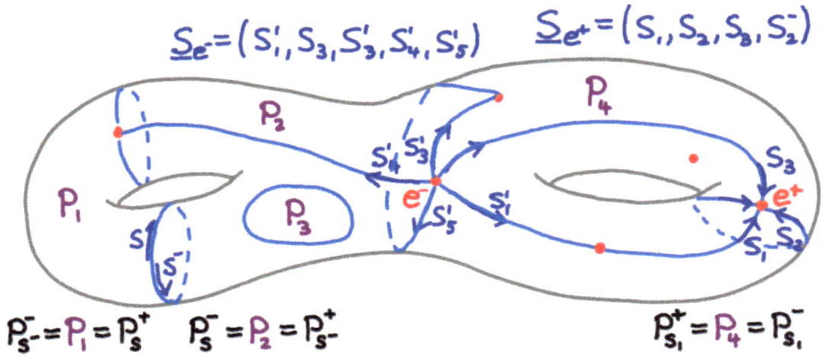

Fig. 21 A quilt with some examples of oriented seams and their adjacent patches, and cyclic sequences of oriented seams $\underline{S}_{e^\pm}$ associated to ends $e^\pm \in \mathscr{E}^\pm_{\mathscr{Q}}$. While the patch P_3 is an open disk, its refined closure $\widehat{P}_3$ simply is a closed disk. The case of $\widehat{P}_2$—a closed annulus minus one boundary puncture—shows that these refined closures are usually not compact. Finally, $\widehat{P}_4$ is an example in which the immersion to $Q_2 \setminus Q_0$ is not injective. Here P_4 is an open disk with one interior puncture, and its closure $\overline{P}_4 \subset Q_2 \setminus Q_0$ is the complement of a disk in a torus, minus 2 punctures on the boundary and 3 punctures in the interior. However, $\widehat{P}_4$ is a closed 10-gon minus the corners and one interior puncture, with oriented boundary components $S_2, S_1^-, S_1'^-, S_5', S_3'^-, S_3, S_2^-, S_3^-, S_1', S_1$

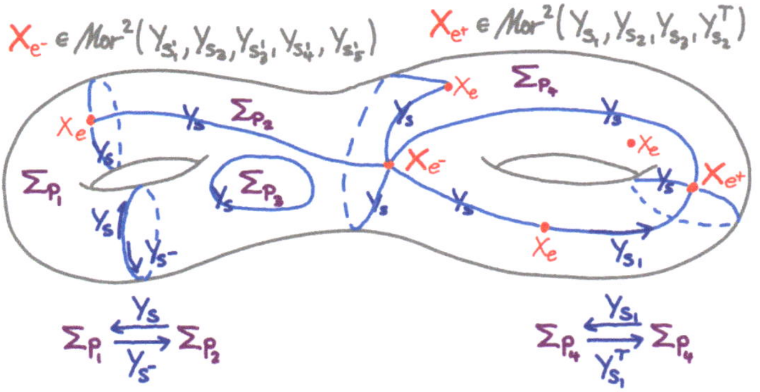

Fig. 22 In a quilt diagram, each patch P_i is labeled by an object Σ_{P_i} and each seam S is labeled by a pair of adjoint 1-morphisms Y_S, Y_S^T (corresponding to the different orientations of the seam). One could in addition label each end e by a 2-morphism X_e in the corresponding cyclic 2-morphism space, however, these will instead be viewed as inputs or outputs of a quilted composition map induced by the quilt diagram

ordered by their intersection with a counterclockwise circle around $e^+ = q_0 \in Q_2$; see Fig. 21 for an example. Then $\underline{Y}_{e^+} := \left(Y_{S_i}\right)_{i\in\mathbb{Z}_{N_{e^+}}} : \mathbb{Z}_{N_{e^+}} \to \mathrm{Mor}^1_{\mathscr{C}}$ is a cyclic 1-morphism of $\mathscr{C}$ in the sense of Remark 3.4.4, with a well defined cyclic 2-morphism space $\mathrm{Mor}^2_{\mathscr{C}}(\underline{Y}_{e^+})$.

- For each incoming end e, we obtain the cyclic 1-morphism $\underline{Y}_e$ analogously from the oriented seams $S_i \simeq \mathbb{R}$ with $-\infty$-limit e; again see Fig. 21 for an example.
- If an incoming or outgoing end e lies in the interior of a patch P, i.e., has no adjacent seams, then we associate to it the cyclic 1-morphism $\underline{Y}_e := 1_{\Sigma_P}$. (In

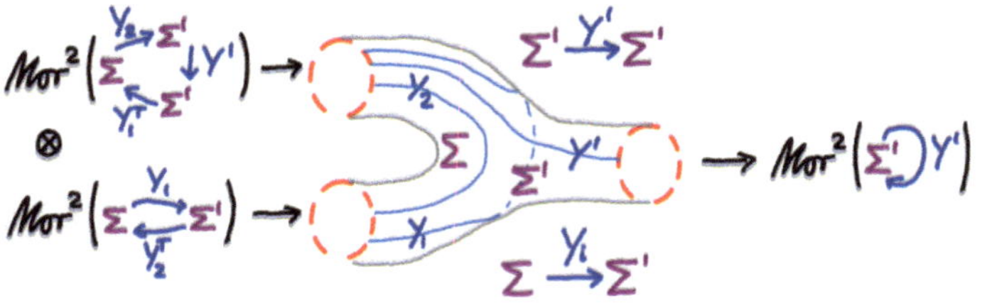

Fig. 23 Quilt diagrams induce quilted composition maps which—except for simple quilt diagrams corresponding to string diagrams—cannot be expressed in terms of the horizontal and vertical composition of 2-morphisms

the case of a bicategory $\mathscr{C}$ one should either disallow ends without seams or ensure identifications between the cyclic 2-morphism spaces associated to different choices of weak identity 1-morphisms.)

However, instead of fixing these labels, we will view the quilt diagrams as inducing maps between the cyclic 2-morphism spaces associated to the ends, as indicated in Fig. 23. Another example of a quilt map is given in Fig. 28. In particular, string diagrams already induce such maps via horizontal and vertical composition. Now, we define a quilted 2-category to be a 2-category in which not only the string diagrams but general quilt diagrams define maps on 2-morphism spaces. The analogous definition is made for bicategories. Here one could make various further specifications such as fixing the genus of the quilt diagram. (For example, spherical 2-categories as in [43] could be conjectured to correspond to 2-categories in which quilt diagrams of genus 0 yield well defined maps.)

Definition 3.4.9 A **quilted bicategory/2-category** is a bicategory/2-category $\mathscr{C}$ with adjoints in the sense of Definition 3.4.2 and with quilted composition maps[29]

$$\Phi_{\mathscr{Q}\mathscr{D}} : \otimes_{e\in\mathscr{E}^-_{\mathscr{Q}}} \mathrm{Mor}^2_{\mathscr{C}}(\underline{Y}_e) \to \mathrm{Mor}^2_{\mathscr{C}}(\underline{Y}_{e^+})$$

for each quilt diagram $\mathscr{Q}\mathscr{D} = \big(\mathscr{Q}, (\Sigma_P)_{P\in\mathscr{P}_{\mathscr{Q}}}, (Y_S)_{S\in\mathscr{S}_{\mathscr{Q}}}\big)$ that satisfy the following:

Deformation Axiom: Isomorphic quilt diagrams $\mathscr{Q}\mathscr{D} \simeq \mathscr{Q}\mathscr{D}'$ as in Fig. 24 give rise to the same quilted composition maps $\Phi_{\mathscr{Q}\mathscr{D}} = \Phi_{\mathscr{Q}\mathscr{D}'}$. Here an isomorphism

$$\big(\mathscr{Q}, (\Sigma_P)_{P\in\mathscr{P}_{\mathscr{Q}}}, (Y_S)_{S\in\mathscr{S}_{\mathscr{Q}}}\big) \simeq \big(\mathscr{Q}', (\Sigma'_{P'})_{P'\in\mathscr{P}_{\mathscr{Q}'}}, (Y'_{S'})_{S'\in\mathscr{S}_{\mathscr{Q}'}}\big)$$

is a homeomorphism $Q_2 \simeq Q_2'$ that restricts to an orientation preserving diffeomorphism $Q_2\smallsetminus Q_0 \simeq Q_2'\smallsetminus Q_0'$ and identifies the ends $Q_0 \simeq Q_0'$, in particular $q_0 \simeq q_0'$, and seams $Q_1 \simeq Q_1'$ in such a way that the labels $\Sigma_P = \Sigma'_{P'}$ and $Y_S = Y'_{S'}$ coincide under the induced identification of patches $\mathscr{P}_{\mathscr{Q}} \simeq \mathscr{P}_{\mathscr{Q}'}$ and oriented seams $\mathscr{S}^{\mathrm{or}}_{\mathscr{Q}} \simeq \mathscr{S}^{\mathrm{or}}_{\mathscr{Q}'}$.

[29]Here and below we write a tensor product $\otimes$ to indicate a Cartesian product of sets which can be replace by a tensor product in the case of 2-morphism spaces given by Floer homology groups.

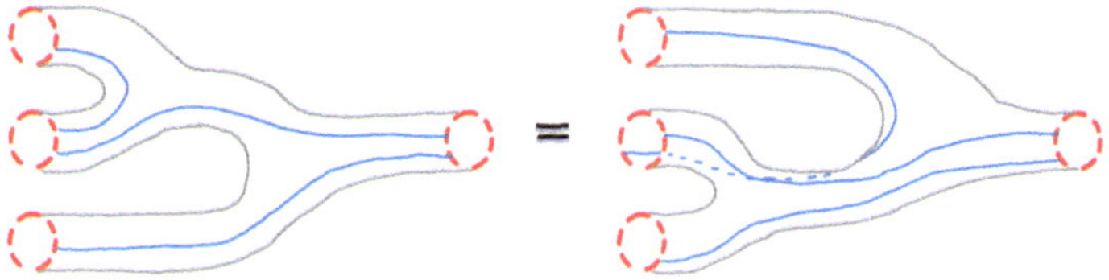

Fig. 24 Isomorphic quilt diagrams yield the same quilted composition maps

The identity $\Phi_{\mathscr{Q}\mathscr{D}} = \Phi_{\mathscr{Q}\mathscr{D}'}$ is with respect to the identification of ends $\mathscr{E}^{\pm}_{\mathscr{Q}} \simeq \mathscr{E}^{\pm}_{\mathscr{Q}'}$ induced by the bijection $Q_0 \simeq Q'_0$.

Cylinder Axiom: The invariant associated to a quilted cylinder as in Fig. 25—that is $Q_2 \setminus Q_0 \simeq \mathbb{R} \times S^1$ with parallel seams $Q_1 \simeq \mathbb{R} \times \{s_1, \dots, s_N\}$—is the identity map on the associated cyclic morphism space.

Gluing Axiom: Gluing of quilt diagrams as in Fig. 26—identifying the outgoing end of one diagram with an incoming end of another diagram—corresponds to composition of the associated quilted composition maps.

Strip shrinking Axiom: Strip or annulus shrinking as in Fig. 27—removing a patch $P \simeq \mathbb{R} \times S^1$ or $P \simeq [0, 1] \times S^1$ and replacing the its two adjacent seams S, S' by a single seam labeled with the composed 1-morphism $Y_S \circ Y_{S'}$ (and its adjoint)—corresponds to an equality of quilted composition maps.

An example of using the axioms to make graphical calculations for quilt maps is given in Fig. 29.

Remark 3.4.10 The adjunction 2-morphisms between adjoint 1-morphisms (see Definition 3.4.2) are in practice often constructed from quilted composition maps corresponding to Fig. 19 with no incoming ends; e.g., as in Remarks 3.4.3 and 3.5.2.

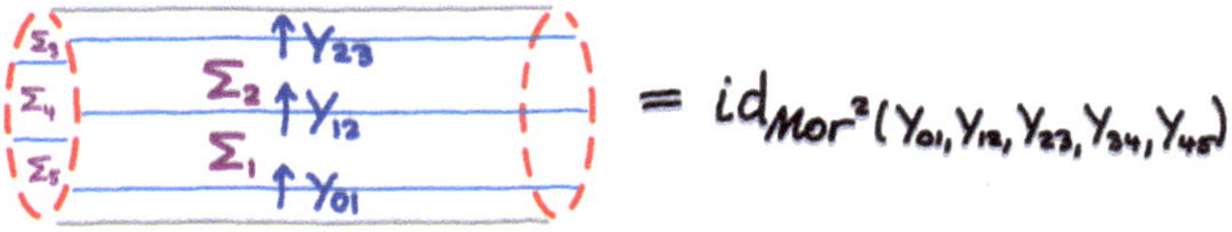

Fig. 25 Cylindrical quilt diagrams yield the identity on the corresponding cyclic morphism space

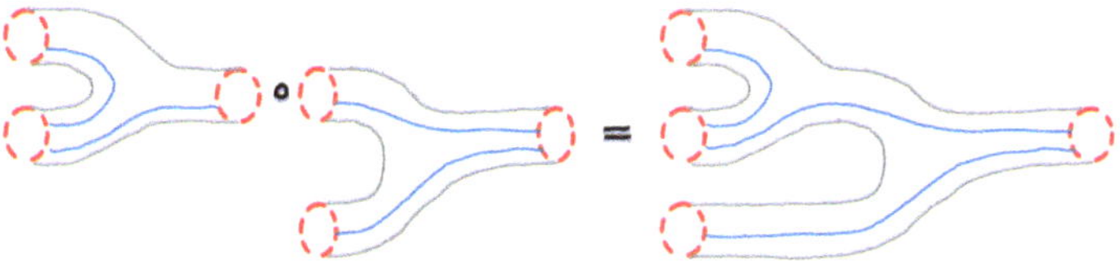

Fig. 26 Composition of quilted composition maps corresponds to gluing of the quilt diagrams

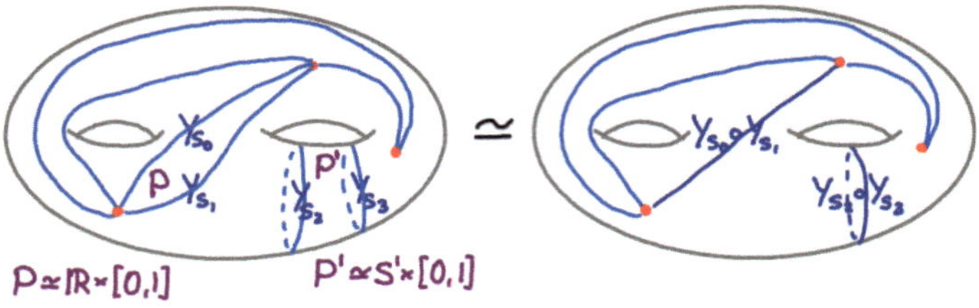

Fig. 27 Quilt diagrams related by annulus shrinking yield the same quilted composition maps; for strip shrinking they are intertwined via isomorphisms between the cyclic morphism spaces

A more fitting notion of quilted bicategory might thus be to require only the reflexive operation on 1-morphisms in Definition 3.4.2 together with well defined cyclic 2-morphism spaces as in Remark 3.4.4 and quilted composition maps satisfying the same axioms as in Definition 3.4.9.

In Sects. 3.5 and 3.6 we will see (sketches of) examples in which the quilted composition maps arise from counting solutions to a nonlinear PDE. In those settings, the strip and annulus shrinking is highly nontrivial—requiring the identification of solution spaces under a degeneration of the PDE as in [78]. On the other hand, bordism bicategories have natural quilted composition maps given by appropriate gluing of manifolds with boundaries and corners, in which also strip and annulus shrinking is naturally satisfied. We give a rough explanation here in dimension $d = 2$, though more care would be required to construct the cyclic 2-morphism spaces and smooth structures coherently and check the axioms.

Lemma 3.4.11 Bor_{2+1+1} *is a quilted bicategory with adjoints as in Remark 3.4.3.*

Proof (Sketch). Since adjunction 2-morphisms X_Y, X_Y^T are constructed from quilt diagrams in Remark 3.4.3, it remains to associate to any given quilt diagram $\mathcal{QD} = \big(\mathcal{Q}, (\Sigma_P)_{P\in\mathcal{P}_\mathcal{Q}}, (Y_S)_{S\in\mathcal{S}_\mathcal{Q}}\big)$ and labels $\big(X_e \in \mathrm{Mor}^2_{\mathrm{Bor}_{2+1+1}}(\underline{Y}_e)\big)_{e\in\mathcal{E}^+_\mathcal{Q}}$ of the incoming ends a 4-cobordism in the cyclic 2-morphism space $\mathrm{Mor}^2_{\mathrm{Bor}_{2+1+1}}(\underline{Y}_{e^+})$ associated to the outgoing end. We do so by gluing 4-manifolds as shown in Figs. 10 and 13:

- A patch $P \in \mathcal{P}_\mathcal{Q}$ that is labeled by a surface Σ is represented by the oriented 4-manifold $X_P := \widehat{P} \times \Sigma$ with boundary $\partial X_P = \bigcup_{S\in\mathcal{S}^{\mathrm{or}}_\mathcal{Q}, P_S^+ = P} S \times \Sigma$.
- A seam $S \in \mathcal{S}_\mathcal{Q}$, i.e., a pair of oriented seams $\{S, S^-\} \in \mathcal{S}^{\mathrm{or}}_\mathcal{Q}$, that is labeled by a 3-cobordism $Y_S \in \mathrm{Bor}_{2+1}(\Sigma_{P_S^-}, \Sigma_{P_S^+})$ and its adjoint $Y_{S^-} = Y_S^-$ is represented by the oriented 4-manifold $X_S := S \times Y_S$ with boundary

$$\partial X_S = \big(S \times \Sigma_{P_S^-}\big) \cup \big(S \times \Sigma^-_{P_S^+}\big) = \big(S \times \Sigma_{P_S^-}\big) \cup \big(S^- \times \Sigma_{P_{S^-}^-}\big).$$

  (Note that this is independent of the choice of orientation on $S \in \mathcal{S}_\mathcal{Q}$.)
- We now glue the 4-manifold $\bigsqcup_{P\in\mathcal{P}_\mathcal{Q}} X_P$ with boundary $\bigcup_{S\in\mathcal{S}^{\mathrm{or}}_\mathcal{Q}} S \times \Sigma_{P_S^+}$ to the 4-manifold $\bigsqcup_{S\in\mathcal{S}_\mathcal{Q}} X_S$ with boundary $\bigsqcup_{S\in\mathcal{S}^{\mathrm{or}}_\mathcal{Q}} S \times \Sigma_{P_S^-}$ via the orientation reversing diffeomorphisms

$$\partial X_S \supset S \times \Sigma_{P_S^-} \longrightarrow S^- \times \Sigma_{P_S^-} = S^- \times \Sigma_{P_{S^-}^+} \subset \partial X_{P_{S^-}^+} = \partial X_{P_S^-}.$$

If we extend the smooth structure by gluing with appropriate collar neighborhoods, then this yields an oriented 4-manifold $X_{\mathcal{Q}\mathcal{D}}$ without boundary, which is compact up to cylindrical ends $\mathbb{R}^{\pm} \times Y_e \subset X_{\mathcal{Q}\mathcal{D}}$ for each end $e \in \mathcal{E}_{\mathcal{Q}}^{\pm}$. If we now delete a little neighborhood of $e \in Q_0 \subset Q_2$ from all patches and seams, then each cylindrical end is replaced by boundary and corners as follows:

The boundary strata near e are the 3-cobordisms $Y_{S_i} \in \mathrm{Mor}^1_{\mathrm{Bor}_{2+1+1}}(\Sigma_{P_i^-}, \Sigma_{P_i^+})$ in the cyclic 1-morphism $\underline{Y}_e = \big(Y_{S_i}\big)_{i \in \mathbb{Z}_{N_e}}$ and identity cobordisms $1_{\Sigma_{P_i^{\pm}}} = [0, \delta_i] \times \Sigma_{P_i^{\pm}}$. The corners are formed by identifications $\mathrm{im}\, \iota^-_{Y_{S_i}}(0, \cdot) \sim \{1\} \times \Sigma_{P_i^-}$ and $\mathrm{im}\, \iota^+_{Y_{S_i}}(1, \cdot) \sim \{0\} \times \Sigma_{P_i^+}$. This boundary&corner structure corresponds (with reversed orientations) to the boundary&corners of $X_e \in \mathrm{Mor}^2_{\mathrm{Bor}_{2+1+1}}(\underline{Y}_e)$, so that we can glue in these 4-manifolds at each incoming end to obtain a 4-manifold with boundary and corners arising from the outgoing end. This defines the result of the quilted composition map $\Phi_{\mathcal{Q}\mathcal{D}}\big(\otimes_{e \in \mathcal{E}_{\mathcal{Q}}^-} X_e\big) \in \mathrm{Mor}^2_{\mathcal{C}}(\underline{Y}_{e^+})$.

This construction is fairly evidently compatible with isomorphisms of quilt diagrams, gluing, and strip shrinking, thus satisfies the axioms required in Definition 3.4.9 of a quilted bicategory. □

The notion of quilted 2-categories now allows us to formulate the following extension principle which we will further discuss in [74]. Its proof is outlined in [73] and makes crucial use of the fact that, as a result of the theory of Morse 2-functions (see, e.g., [26]), Bor_{2+1+1} is not just a quilted bicategory but in an appropriate sense is quilt-generated by the Cerf decompositions of Bor_{2+1} of Theorem 2.3.4.

Conjecture 3.4.12 (Extension principle for Floer field theories). Let $\mathcal{C}$ be a quilted 2-category as in Definition 3.4.9 whose underlying 1-category has Cerf decompositions as in Definition 2.3.2. Then any Cerf-compatible partial functor $\mathcal{F}$: $(\mathrm{Obj}_{\mathrm{Bor}_{2+1}}, \mathrm{SMor}_{\mathrm{Bor}_{2+1}}) \to (\mathrm{Obj}_{\mathcal{C}}, \mathrm{SMor}_{\mathcal{C}})$ as in Lemma 2.4.1, which preserves adjunctions as in Remark 2.5.1 and satisfies a quilted naturality axiom (see [73]), has a natural extension to a 2-functor $\mathrm{Bor}_{2+1+1} \to \mathcal{C}$.

An analogous extension principle can be formulated for bordism bicategories Bor_{d+1+1} in any dimension $d \geq 0$ and connected bordism bicategories $\mathrm{Bor}^{\mathrm{conn}}_{d+1+1}$ in dimension $d \geq 0$. We propose to apply this principle to the Floer field theories outlined in Sect. 2.5, where $\mathcal{C}$ the symplectic 2-category oulined in Sect. 3.5 or other gauge theoretic 2-categories outlined in Sect. 3.6. It should yield "2+1+1 Floer field theories" $\mathrm{Bor}_{2+1+1} \to \mathrm{Cat}$ by composition with the Yoneda 2-functor $\mathcal{C} \to \mathrm{Cat}$ from Lemma 3.3.3. We moreover expect equivalences between these field theories, as phrased in the quilted Atiyah–Floer Conjecture 3.6.4.

3.5 The Symplectic 2-Category

This section gives a brief overview of the construction of a symplectic 2-category in [76]. Conceptually, it can be thought of as starting with the construction of a 2-category in Example 3.1.6 from the Cerf decompositions in the extended symplectic category $\mathrm{Symp}^{\#}$ of Definition 2.2.1, and then replacing the 2-morphisms that were defined from the abstract Cerf moves by a geometrically more meaningful notion, while preserving the isomorphisms $(L_{12}, L_{23}) \sim L_{12} \circ L_{23}$ in the sense of Remark 3.1.7, as mentioned at the end of Sect. 2.2.

In the following, sketch of the symplectic 2-category we use the same horizontal 1-composition $\circ_{\mathrm{h}}^1$ as in $\mathrm{Symp}^{\#}$, the 2-morphisms are given by (quilted) Floer homology groups as defined in [75]. These arise from a complex whose differential is constructed from moduli spaces of solutions of an elliptic PDE that is closely connected to the PDE that we associate to quilt diagrams in Remark 3.5.3. We also use these moduli spaces to construct the vertical and horizontal 2-composition $\circ_{\mathrm{v}}$, $\circ_{\mathrm{h}}^2$ from their respective string diagrams. In order to obtain well-defined structures, we have to make further restrictions on the allowable symplectic objects and morphisms as in Remark 3.5.4, or generalize the notion of 2-category, as discussed in Remark 3.5.5.

Example 3.5.1 The **symplectic 2-category Symp** roughly consists of the following.

- Objects are symplectic manifolds M.
- For each pair $M, N \in \mathrm{Obj}_{\mathrm{Symp}}$ the category of 1-morphisms is the following Donaldson–Fukaya category of generalized Lagrangians $\mathrm{Mor}_{\mathrm{Symp}}(M, N)$.
  - 1-morphisms $\underline{L} = (L_{01}, \ldots, L_{(k-1)k}) \in \mathrm{Mor}^1_{\mathrm{Symp}}(M, N)$ are the composable chains of simple Lagrangians $L_{ij} \subset M_i^- \times M_j$ between symplectic manifolds $M = M_0, M_1, \ldots, M_k = N$.
  - 2-morphisms between $\underline{L}, \underline{L}' \in \mathrm{Mor}^1_{\mathrm{Symp}}(M, N)$ are the elements of the quilted Floer homology group $\mathrm{Mor}^2_{\mathrm{Symp}}(\underline{L}, \underline{L}') = HF(\underline{L}, \underline{L}')$; see Remark 3.5.3.
  - Vertical composition $\circ_{\mathrm{v}} : HF(\underline{L}, \underline{L}') \otimes HF(\underline{L}', \underline{L}'') \to HF(\underline{L}, \underline{L}'')$ for $\underline{L}, \underline{L}', \underline{L}'' \in \mathrm{Mor}^1_{\mathrm{Symp}}(M, N)$ arises from counts of pseudoholomorphic quilts representing the associated string diagram. It is associative by a gluing theorem as in [45].
  - The identity $\mathrm{id}_{\underline{L}} \in \mathrm{Mor}^2_{\mathrm{Symp}}(\underline{L}, \underline{L})$ for $\underline{L} \in \mathrm{Mor}^1_{\mathrm{Symp}}(M, N)$ arises from counts of pseudoholomorphic quilts representing the associated string diagram.
- The composition functor $\mathrm{Mor}_{\mathrm{Symp}}(M, N) \times \mathrm{Mor}_{\mathrm{Symp}}(N, P) \to \mathrm{Mor}_{\mathrm{Symp}}(M, P)$ is defined as follows.
  - Horizontal composition of 1-morphisms

    $$\circ_{\mathrm{h}} : \mathrm{Mor}^1_{\mathrm{Symp}}(M, N) \times \mathrm{Mor}^1_{\mathrm{Symp}}(N, P) \to \mathrm{Mor}^1_{\mathrm{Symp}}(M, P), \quad (\underline{L}, \underline{L}') \mapsto \underline{L}\#\underline{L}'$$

    is given by the evidently associative concatenation

    $$(L_{01}, \ldots, L_{(k-1)k})\#(L'_{01}, \ldots, L'_{(k'-1)k'}) := (L_{01}, \ldots, L_{(k-1)k}, L'_{01}, \ldots, L'_{(k'-1)k'}).$$

- The identities $1_M = (\,) \in \mathrm{Mor}^1_{\mathrm{Symp}}(M, M)$ for $\circ_{\mathrm{h}}$ are given by the trivial chains.
- Horizontal composition of 2-morphisms arises from counts of pseudoholomorphic quilts representing the associated string diagram,

$$\circ_{\mathrm{h}} : HF(\underline{L}_{12}, \underline{L}'_{12}) \times HF(\underline{L}_{23}, \underline{L}'_{23}) \longrightarrow HF(\underline{L}_{12}\#\underline{L}_{23}, \underline{L}'_{12}\#\underline{L}'_{23}).$$

Compatibility with identities and vertical composition follows from gluing theorems as in [45].

While this gives a well-defined symplectic 2-category, we still have to relate it to the symplectic category defined in Sect. 2.2, in which horizontal composition of morphisms is given by the geometric composition of Lagrangians—if the latter is embedded. Example 3.1.6 shows how the same can be achieved up to isomorphism in a 2-categorical setting. However, the 2-morphisms in the present 2-category are quilted Floer homology classes, so the following becomes a nontrivial result—proven in [78] as isomorphism of Floer homologies, which is formulated categorically in [76].

- For any pair of Lagrangians $L_{12} \subset M_1^- \times M_2$, $L_{23} \subset M_2^- \times M_3$ with embedded geometric composition $L_{12} \circ L_{23} \subset M_1^- \times M_2$ as defined in (2.2.1), the 1-morphisms $L_{12} \circ_{\mathrm{h}} L_{23} = L_{12}\#L_{23} \sim L_{12} \circ L_{23}$ are isomorphic in $\mathrm{Mor}^1_{\mathrm{Symp}}(M_1, M_3)$ in the sense of Remark 3.1.7: We have $\alpha \circ_{\mathrm{v}} \beta = \mathrm{id}_{L_{12}\#L_{23}}$ and $\beta \circ_{\mathrm{v}} \alpha = \mathrm{id}_{L_{12}\circ L_{23}}$ for some 2-morphisms $\alpha \in \mathrm{Mor}^2_{\mathrm{Symp}}(L_{12}\#L_{23}, L_{12} \circ L_{23})$, $\beta \in \mathrm{Mor}^2_{\mathrm{Symp}}(L_{12} \circ L_{23}, L_{12}\#L_{23})$.

The last item ensures that the symplectic category $\mathrm{Symp}^1 := \mathrm{Symp}^\#/\!\sim$ of Definition 2.2.2 and Example 3.1.6 and the quotient $|\mathrm{Symp}| = \mathrm{Symp}/\!\sim$ as in Remark 3.1.7 of the symplectic 2-category by isomorphisms are related by a functor

$$\mathrm{Symp}^1 \to |\mathrm{Symp}|, \qquad M \mapsto M, \qquad {}^{\mathrm{Mor}_{\mathrm{Symp}^\#}}\!/_{\sim} \ni [\underline{L}] \mapsto [\underline{L}] \in {}^{\mathrm{Mor}^1_{\mathrm{Symp}}}\!/_{\sim}$$

since equivalence $\sim$ in $\mathrm{Mor}_{\mathrm{Symp}^\#}$ implies isomorphism $\sim$ in Symp. This functor is full, i.e., surjective on morphism spaces, but it is not faithful, i.e., injective, since generalized Lagrangians $\underline{L}$ in the symplectic 2-category may be Floer-theoretic isomorphic without being related by embedded geometric compositions. In fact, the difference can already be seen for simple Lagrangians $L, L' \subset M^- \times N$, which are equivalent in $\mathrm{Mor}_{\mathrm{Symp}^\#}(M, N)$ only if they are identical, but whenever $L' = \phi(L)$ is the image of L under a Hamiltonian symplectomorphism $\phi : M \to M$, then standard Floer theoretic arguments show that $L \sim L'$ are isomorphic as 1-morphisms in $\mathrm{Mor}^1_{\mathrm{Symp}}(M, N)$.

Remark 3.5.2 (Adjoints and quilted composition maps in Symp*).* The symplectic 2-category has adjoints as follows:

- For a Lagrangian[30] $L \subset M_0^- \times M_1$ the adjoint 1-morphism $L^T \subset M_1^- \times M_0$ is given by the image of L under transposition of factors $M_0 \times M_1 \to M_1 \times M_0$.
- For a general 1-morphism $\underline{L} = (L_{01}, \ldots, L_{(k-1)k}) \in \mathrm{Mor}^1_{\mathrm{Symp}}(M, N)$ the adjoint is given by reversal and transposition, $\underline{L}^T = (L^T_{(k-1)k}, \ldots, L^T_{01}) \in \mathrm{Mor}^1_{\mathrm{Symp}}(N, M)$.
- Duality for $\underline{L} \in \mathrm{Mor}^1_{\mathrm{Symp}}(M, N)$ is guaranteed by the identity elements

$$X_{\underline{L}} := \mathrm{id}_{\underline{L}} = \mathrm{id}_{\underline{L}^T} \in HF(\underline{L}\#\underline{L}^T), \qquad X^T_{\underline{L}} := \mathrm{id}_{\underline{L}} = \mathrm{id}_{\underline{L}^T} \in HF(\underline{L}^T\#\underline{L})$$

since the quilted Floer homology in [75] has canonical cyclic symmetries

$$\begin{aligned} HF(\underline{L}\#\underline{L}^T) &= \mathrm{Mor}^2_{\mathscr{C}}(1_M, \underline{L} \circ_{\mathrm{h}} \underline{L}^T) = \mathrm{Mor}^2_{\mathscr{C}}(\underline{L}, \underline{L}) \\ = HF(\underline{L}^T\#\underline{L}) &= \mathrm{Mor}^2_{\mathscr{C}}(\underline{L}^T \circ_{\mathrm{h}} \underline{L}, 1_N) = \mathrm{Mor}^2_{\mathscr{C}}(\underline{L}^T, \underline{L}^T) \end{aligned}$$

which identify these morphism spaces and their identity elements $\mathrm{id}_{\underline{L}}$, $\mathrm{id}_{\underline{L}^T}$ defined in [76], so that the required identities reduce to the compatibility of horizontal and vertical composition with identities,

$$\left(X_{\underline{L}} \circ_{\mathrm{h}} \mathrm{id}_{\underline{L}}\right) \circ_{\mathrm{v}} \left(\mathrm{id}_{\underline{L}} \circ_{\mathrm{h}} X^T_{\underline{L}}\right) = \mathrm{id}_{\underline{L}}, \qquad \left(\mathrm{id}_{\underline{L}^T} \circ_{\mathrm{h}} X_{\underline{L}}\right) \circ_{\mathrm{v}} \left(X^T_{\underline{L}} \circ_{\mathrm{h}} \mathrm{id}_{\underline{L}^T}\right) = \mathrm{id}_{\underline{L}^T}.$$

Moreover, Symp is a quilted 2-category whose quilted composition maps

$$\Phi_{\mathscr{Q}\mathscr{D}} : \otimes_{e \in \mathscr{E}^-_{\mathscr{Q}}} HF(\underline{L}_e) \to HF(\underline{L}_{e^+})$$

are defined in [79], which also proves the axioms for quilted cylinders and gluing of diagrams by standard Floer theoretic arguments. However, strip and annulus shrinking—as required in Definition 3.4.9 for a quilted 2-category—requires the adiabatic limit analysis in [78], which may be obstructed by a novel "codimension 0 in the boundary" singularity—figure eight bubbles.

Remark 3.5.3 (PDE associated to quilt diagrams in Symp*).* The key step in the construction [79] of quilted composition maps is to associate to every quilt diagram $\mathscr{Q}\mathscr{D} = \left(\mathscr{Q}, (M_P)_{P \in \mathscr{P}_{\mathscr{Q}}}, (\underline{L}_S)_{S \in \mathscr{S}_{\mathscr{Q}}}\right)$ an elliptic PDE as follows:

- A patch P labeled by a symplectic manifold M_P is represented by a pseudoholomorphic map $u_P : \widehat{P} \to M_P$ whose domain is the oriented 2-manifold $\widehat{P}$ that covers the closure $\overline{P} \subset Q_2 \setminus Q_0$ as in Remark 3.4.7.
- A seam S labeled by a Lagrangian submanifold $L_S \subset M^-_{P^-_S} \times M_{P^+_S}$ is represented by a Lagrangian seam condition: The map $u_{P^-_S}|_S \times u_{P^+_S}|_S : S \to M^-_{P^-_S} \times M_{P^+_S}$ induced by boundary restrictions of the pseudoholomorphic maps associated to the adjacent patches is required to take values in L_S.
- A seam $S \simeq \mathbb{R}$ labeled by a sequence of Lagrangians $L_{i(i+1)} \subset M_i^- \times M_{i+1}$ which form a general 1-morphism $\underline{L} = (L_{01}, \ldots, L_{(k-1)k} \in \mathrm{Mor}^1_{\mathrm{Symp}}(M_{P^-_S}, M_{P^+_S})$, repre-

[30]Note that an overall sign change of the symplectic form does not affect the Lagrangian property.

sents pseudoholomorphic strips $u_i : \mathbb{R} \times [0, 1] \to M_i$ for $i = 1, \ldots, k-1$ with Lagrangian seam conditions $(u_i|_{\{1\}\times\mathbb{R}} \times u_{i+1}|_{\{0\}\times\mathbb{R}})(\mathbb{R}) \subset L_{i(i+1)}$ and

$$(u_{P_S^-}|_S \times u_1|_{\{0\}\times\mathbb{R}})(\mathbb{R}) \subset L_{01}, \qquad (u_{k-1}|_{\{1\}\times\mathbb{R}} \times u_{P_S^+}|_S)(\mathbb{R}) \subset L_{(k-1)k}.$$

- A seam $S \simeq S^1$ labeled by a general 1-morphism $\underline{L}$ represents pseudoholomorphic annuli $u_i : S^1 \times [0, 1] \to M_i$ with the analogous seam conditions.

The tuple $(u_P)_{P\in\mathscr{P}_{\mathscr{Q}}}$ (together with the additional maps $(u_i)_{i=1,\ldots k-1}$ from each seam with generalized Lagrangian label) is called a **pseudoholomorphic quilt**.

This notion generalizes pseudoholomorphic maps $u : Q \to M$—which arise from quilt diagrams $\mathscr{Q}\mathscr{D} = (Q, M, \emptyset)$ that consist of a closed Riemann surface $Q_2 = Q$ without seams or punctures, labeled by a symplectic manifold M—as well as pseudoholomorphic maps with Lagrangian boundary conditions $u : (Q, \partial Q) \to (M, L)$. To build in boundary, we can for example represent the latter by a quilt diagram $\mathscr{Q}\mathscr{D} = (\mathscr{Q}, M, L)$ whose quilted surface has patches Q and $\overline{Q}$ (with reversed orientation), a seam for each boundary component of Q (identified with the corresponding boundary component of $\overline{Q}$), labels M for Q, pt for $\overline{Q}$, and L for each seam.

We can moreover build in any number of punctures on boundaries, seams, or in the interior. Note in particular that interior punctures on a patch P are associated to the cyclic 2-morphism space $\mathrm{Mor}^2_{\mathrm{Symp}}(1_{M_P}) = HF(\Delta_{M_P}) \cong HF(M_P)$, which can be identified with the Hamiltonian Floer homology of the symplectic manifold M_P. For an introduction to Floer homology see, e.g., [58, 63]. These also provide good introductions to the technique of "counting" (very specific) moduli spaces of PDEs to construct Floer chain complexes and chain maps between them—whose homology and induced map on homologies are independent of choices (most notably of perturbations that are chosen to regularize the moduli spaces).

Remark 3.5.4 (Monotonicity assumptions). Moduli spaces of pseudoholomorphic quilts—just as moduli spaces of pseudoholomorphic curves—are rarely compact and often do not carry a smooth structure which allows us to "count" or "integrate over" them to define the structure maps in the symplectic 2-category. While the "Gromov compactification" of these spaces (in terms of breaking of Floer trajectories and bubbling trees of pseudoholomorphic spheres and disks; see, e.g., [48, 63]) is well understood, the regularization of the compactified moduli spaces still remains a challenge in general settings (see [46] for a survey). In fact, bubbling gives actual obstructions to the algebraic requirements for a 2-category—beginning with disk bubbling obstructing the definition of the 2-morphism spaces (since the Floer differential may fail to square to zero), via additional algebraic terms in the structure equations arising from disk bubbles, to figure eight bubbles obstructing the desired isomorphism $L_{12}\#L_{23} \sim L_{12} \circ L_{23}$ for embedded geometric composition.

The present state of the art is that a rigorous symplectic 2-category Symp^τ is constructed in [76] by restriction to monotone or exact symplectic manifolds and oriented Lagrangian submanifolds with minimal Maslov index ≥ 3. While the latter assumption is made to ensure that the Floer differential squares to zero (so that

Floer homology is well defined), the monotonicity requires that the Maslov index $I(\underline{u})$ and symplectic area $A(\underline{u})$ of the quilted maps $\underline{u} = (u_P)_{P\in\mathscr{P}_{\mathscr{Q}}}$ are proportional $I(\underline{u}) = \tau A(\underline{u})$ via a constant $\tau \geq 0$. This helps with excluding bubbling because it relates Fredholm indices (i.e., expected dimension of moduli spaces) to the energy of the solutions, so that bubbling (i.e. loss of energy) forces loss of Fredholm index—which in the relevant moduli spaces would yield solutions of negative expected dimension. Once these are ruled out by appropriate regularization, the bubbling can be excluded without actually constructing the compactified moduli space. The same argument is used to exclude bubbling in the strip and annulus shrinking of [78] to prove the isomorphism $L_{12}\#L_{23} \sim L_{12} \circ L_{23}$ in Symp^τ when the latter geometric composition is embedded.

Remark 3.5.5 (Generalized notions of symplectic 2-categories). In order to extend the construction of a symplectic 2-category to non-monotone settings, and more generally study the relationship between the algebraic and geometric compositions $L_{12}\#L_{23}$ and $L_{12} \circ L_{23}$, a Gromov compactification for strip and annulus shrinking—involving multilevel trees of pseudoholomorphic disks, figure eights, and spheres—is constructed in [9] with the help of removable singularity results for the figure eight bubble in [6]. By analyzing the boundary strata of the resulting compactified moduli spaces, and supported by the upcoming Fredholm theory [7] for moduli spaces of figure eights, we then predict a 2-categorical structure that comprises all (compact) symplectic manifolds and Lagrangians, and in which composition of 1-morphisms is given by geometric composition of Lagrangians. It takes the form of a curved A_∞ 2-category, and in fact motivates the definition of this new algebraic notion in [8].

We end this section by disclosing the categorical ignorance in the first publications on the symplectic 2-category in [76]. While that paper painstakingly constructs a 2-functor $\mathrm{Symp}^\tau \to \mathrm{Cat}$, this directly coincides with the Yoneda construction.

Lemma 3.5.6 *The functor* $\mathrm{Symp}^\tau \to \mathrm{Cat}$ *constructed in [76] is identical to the functor* $\mathscr{F}_{\mathrm{pt}}$ *given by Lemma 3.3.3 with the distinguished object* $x_0 = \mathrm{pt}$.

Similarly, [75] proves isomorphisms between quilted Floer homology groups for cyclic 1-morphisms in Symp which are related by a geometric composition by arguing that the adiabatic limit analysis in [78] transfers directly. In the 2-categorical setup with adjoints, this can now be proven more directly by the categorical "local to global" argument of Remark 3.4.5.

Remark 3.5.7 (Isomorphisms of Floer homology under geometric composition). The "local to global" principle discussed in Sect. 2.6 and Remark 3.4.5 translates to the quilted Floer homology groups via identifications

$$HF\big(\underline{L} = (L_{i(i+1)})_{i\in\mathbb{Z}_N}\big) = \mathrm{Mor}^2_{\mathrm{Symp}}\big(\underline{L} = (L_{i(i+1)})_{i\in\mathbb{Z}_N}\big).$$

So for purely algebraic reasons (which are interpreted geometrically in Remark 3.5.8), we obtain the implications

$$L_{12}\#L_{23} \sim L_{12} \circ L_{23} \quad \Longrightarrow \quad (\ldots, L_{12}\#L_{23}, \ldots) \sim (\ldots L_{12} \circ L_{23} \ldots)$$
$$\Longrightarrow \quad HF(\ldots, L_{12}, L_{23}, \ldots) \simeq HF(\ldots L_{12} \circ L_{23} \ldots).$$

Thus to prove that quilted Floer homology is invariant (up to isomorphism) under embedded geometric composition, it suffices to prove that any embedded geometric composition $L_{12} \circ L_{23}$ as defined in (2.2.1) gives rise to an isomorphism $L_{12}\#L_{23} \sim L_{12} \circ L_{23} =: L_{13}$ between algebraic and geometric compositions. Such local isomorphisms require the construction of quilted Floer homology classes

$$\alpha \in \mathrm{Mor}^2_{\mathrm{Symp}}(L_{12}\#L_{23}, L_{12} \circ L_{23}) = HF(L_{12}, L_{23}, L_{13}^T),$$
$$\alpha^{-1} \in \mathrm{Mor}^2_{\mathrm{Symp}}(L_{12} \circ L_{23}, L_{12}\#L_{23}) = HF(L_{13}, L_{23}^T, L_{12}^T)$$

that satisfy

$$\alpha \circ_{\mathrm{v}} \alpha^{-1} = \mathrm{id}_{L_{12}\#L_{23}}, \qquad \alpha^{-1} \circ_{\mathrm{v}} \alpha = \mathrm{id}_{L_{13}}.$$

To find such classes suppose that we have isomorphisms of the "local" quilted Floer homologies under embedded geometric composition (with $L_{23}^T \circ L_{12}^T = (L_{12} \circ L_{23})^T$),

$$HF(L_{12}, L_{23}, L_{13}^T) \stackrel{\sim}{\to} HF(L_{12} \circ L_{23}, L_{13}^T) = HF(L_{13}, L_{13}) = \mathrm{Mor}^2_{\mathrm{Symp}}(L_{13}, L_{13}),$$
$$HF(L_{13}, L_{23}^T, L_{12}^T) \stackrel{\sim}{\to} HF(L_{13}, L_{23}^T \circ L_{12}^T) = HF(L_{13}, L_{13}) = \mathrm{Mor}^2_{\mathrm{Symp}}(L_{13}, L_{13}),$$
$$HF(L_{12}, L_{23}, L_{13}^T) \stackrel{\sim}{\to} HF(L_{12}, L_{23}, L_{23}^T, L_{12}^T) = \mathrm{Mor}^2_{\mathrm{Symp}}(L_{12}\#L_{23}, L_{12}\#L_{23}),$$
$$HF(L_{13}, L_{23}^T, L_{12}^T) \stackrel{\sim}{\to} HF(L_{12}, L_{23}, L_{23}^T, L_{12}^T) = \mathrm{Mor}^2_{\mathrm{Symp}}(L_{12}\#L_{23}, L_{12}\#L_{23}),$$

and suppose that these isomorphisms are compatible with identities and products. Then, we may pull back the identities $\mathrm{id}_{L_{13}} \in \mathrm{Mor}^2_{\mathrm{Symp}}(L_{13}, L_{13})$ and $\mathrm{id}_{L_{12}\#L_{23}} \in Mor^2_{\mathrm{Symp}}(L_{12}\#L_{23}, L_{12}\#L_{23})$ to obtain two well defined classes α, α^{-1} as required,

$$\alpha \circ_{\mathrm{v}} \alpha^{-1} = \mathrm{id}_{L_{12}\#L_{23}} \circ_{\mathrm{v}} \mathrm{id}_{L_{12}\#L_{23}} = \mathrm{id}_{L_{12}\#L_{23}}, \qquad \alpha^{-1} \circ_{\mathrm{v}} \alpha = \mathrm{id}_{L_{13}} \circ_{\mathrm{v}} \mathrm{id}_{L_{13}} = \mathrm{id}_{L_{13}}.$$

Finally, we will clarify some confusions regarding the generality and possible obstructions to the isomorphism of Floer homology under geometric composition.

Remark 3.5.8 (Genearlized Floer isomorphisms under geometric composition). The isomorphism $L_{12}\#L_{23} \sim L_{12} \circ L_{23}$ should generalize directly to exact noncompact settings as long as the Lagrangians have a conical structure near infinity that allows one to use maximum principles to guarantee compactness. An application to the construction of a Floer field theory that extends the link invariants [64] was proposed in [56] but unfortunately seems to be lacking this conical structure.

On the other hand, extensions of this isomorphism to negative monotone settings announced in [37] overlooked obstructions arising from Morse–Bott trajectories.[31]

---

[31] The published arguments in [37] are insufficient to exclude breaking at the Morse–Bott end in any case other than exactness, and its Theorem 3—the isomorphism in the new case of negative

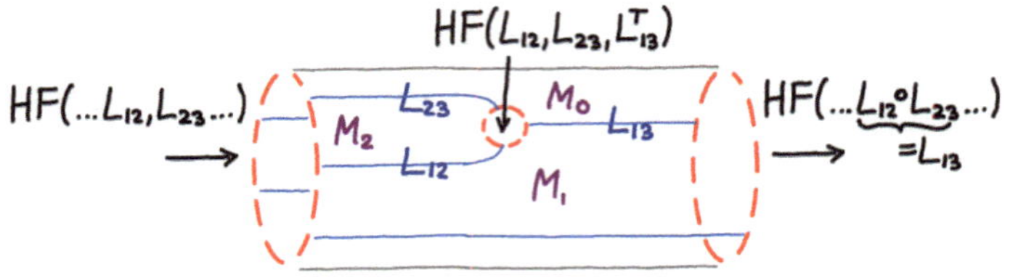

Fig. 28 The isomorphism $HF(\dots, L_{12}, L_{23}, \dots) \simeq HF(\dots L_{12} \circ L_{23} \dots)$ in Remark 3.5.7 arises from a quilted Floer homology class $\alpha \in HF(L_{12}, L_{23}, L_{13}^T)$ via the quilted composition map induced by the above quilt diagram

In fact, these obstructions are homotopically identical to the figure eight bubbles conjectured in [78] and established in [9], so that true generalizations of this isomorphism are expected only from the compactification and Fredholm theory for figure eight bubbles in [6, 7, 9], towards capturing the obstructions algebraically. It is, however, worthwhile to discuss the approach by Matthias Schwarz which [37] attempted to implement: It is a geometric version of the "local to global" approach in Remark 3.5.7, which aims for an explicit construction of a direct homomorphism

$$HF(\dots, L_{12}, L_{23}, \dots) \to HF(\dots, L_{13}, \dots) \tag{3.5.1}$$

from a relative quilt invariant with canonical asymptotics at a Morse–Bott end for the cyclic 1-morphism $(L_{12}, L_{23}, L_{13}^T)$. This corresponds to a map $\beta \mapsto \Phi_{\mathcal{Q}\mathcal{D}}(\beta, \alpha)$ given by plugging a canonical element α into a quilted composition map

$$\Phi_{\mathcal{Q}\mathcal{D}} : HF(\dots, L_{12}, L_{23}, \dots) \otimes HF(L_{12}, L_{23}, L_{13}^T) \to HF(\dots, L_{13}, \dots)$$

that arises from the quilt diagram in Fig. 28. Using the classes α, α^{-1} from Remark 3.5.7, we obtain an inverse $\gamma \mapsto \Phi_{\mathcal{Q}\mathcal{D}'}(\gamma, \alpha^{-1})$ to (3.5.1) from

$$\Phi_{\mathcal{Q}\mathcal{D}'} : HF(\dots, L_{13}, \dots) \otimes HF(L_{13}, L_{23}^T, L_{12}^T) \to HF(\dots, L_{12}, L_{23}, \dots),$$

a quilted composition map arising from another quilt diagram $\mathcal{Q}\mathcal{D}'$ that is obtained by reflecting $\mathcal{Q}\mathcal{D}$. Indeed, $\mathcal{Q}\mathcal{D}$, $\mathcal{Q}\mathcal{D}'$ glue—in two orders, one of which is shown in Fig. 29—to diagrams which also correspond to the gluing of quilt diagrams $\mathcal{Q}\mathcal{D}$,'' $\mathcal{Q}\mathcal{D}'''$ with the string diagram for $\circ_{\rm v}$. Moreover, if in the latter gluings we replace the $\circ_{\rm v}$ diagram with the string diagram for the corresponding identity, then the glued diagram is a quilted cylinder. Now the gluing and cylinder axioms imply

(Footnote 31 continued)
monotonicity—is in the corrigendum only claimed under the additional assumption of "absence of quantum contributions to the Morse–Bott differential" as discussed below. More details on the known issues in [37] are given in Sect. 4.

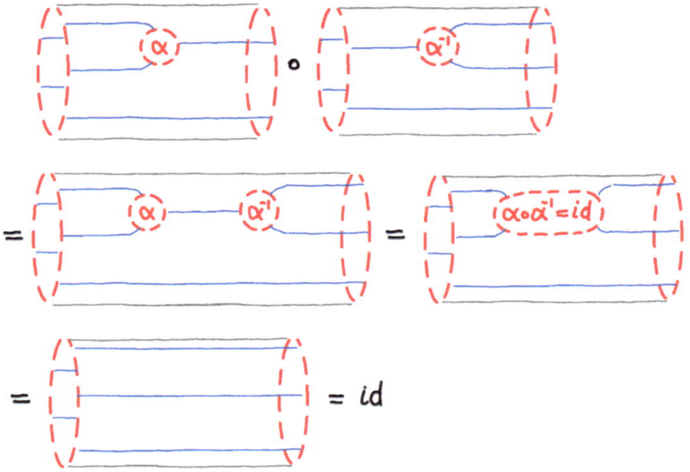

Fig. 29 The identity $\Phi_{\mathcal{Q}\mathcal{D}'}(\Phi_{\mathcal{Q}\mathcal{D}}(\beta,\alpha),\alpha^{-1}) = \beta$ follows from applying the gluing, deformation, and cylinder axioms for the quilted composition maps and the identity $\alpha \circ_{\mathrm{v}} \alpha^{-1} = \mathrm{id}_{L_{12}\#L_{23}}$

$$\Phi_{\mathcal{Q}\mathcal{D}'}\big(\Phi_{\mathcal{Q}\mathcal{D}}(\beta,\alpha),\alpha^{-1}\big) = \Phi_{\mathcal{Q}\mathcal{D}''}(\beta,\alpha \circ_{\mathrm{v}} \alpha^{-1}) = \Phi_{\mathcal{Q}\mathcal{D}''}(\beta,\mathrm{id}_{L_{12}\#L_{23}}) = \beta,$$
$$\Phi_{\mathcal{Q}\mathcal{D}}\big(\Phi_{\mathcal{Q}\mathcal{D}'}(\gamma,\alpha^{-1}),\alpha\big) = \Phi_{\mathcal{Q}\mathcal{D}'''}(\gamma,\alpha^{-1} \circ_{\mathrm{v}} \alpha) = \Phi_{\mathcal{Q}\mathcal{D}'''}(\beta,\mathrm{id}_{L_{13}}) = \gamma,$$

which proves that (3.5.1) is an isomorphism. The Morse–Bott end amounts to an implicit construction of $\alpha := [L_{13}] \in HF(L_{12}, L_{23}, L_{13}^T)$ from the fundamental class of L_{13} and an identification of the chain groups (but not the differentials) which yield the Floer homology resp. the Morse homology of the Lagrangian intersection,

$$CF(L_{12}, L_{23}, L_{13}^T) \simeq CM(\cap(L_{12}, L_{23}, L_{13}^T)) \simeq CM(L_{13}).$$

The resulting Morse homology $HM(L_{13})$ is isomorphic to singular homology of L_{13} since the Lagrangian intersection is diffeomorphic to $L_{13} = L_{12} \circ L_{23}$,

$$\begin{aligned} \cap(L_{12}, L_{23}, L_{13}^T) &= (L_{12} \times L_{23} \times L_{13}^T) \cap (\Delta_{M_1} \times \Delta_{M_2} \times \Delta_{M_3})^T \\ &\cong ((L_{12} \circ L_{23}) \times L_{13}^T) \cap (\Delta_{M_1} \times \Delta_{M_3})^T \cong L_{13}. \end{aligned}$$

Schwarz proposed to prove that (3.5.1) is an isomorphism by arguing that it has "upper triangular form," but also observed that the crucial step is to construct a chain map in the first place that induces (3.5.1), which amounts to showing that $\alpha = [L_{13}]$ lies in the kernel of the Floer differential. The only cases beyond the monotone case covered in [78] in which this is claimed to be known at this point (after correction of [37]) are those in which not just the generators but also the differentials of the chain complexes $CF(L_{13}, L_{23}^T, L_{12}^T)$, $CF(L_{12}, L_{23}, L_{23}^T, L_{12}^T)$, $CF(L_{12}, L_{23}, L_{13}^T)$, $CF(L_{13}, L_{13})$ all agree with the Morse chain complex $CM(L_{13})$. In other words, we assume absence of quantum contributions to the Floer–Bott differential.[32] On the one hand, this allows one to define $\alpha := [L_{13}] \in HF(L_{12}, L_{23}, L_{13}^T)$ and thus obtain a homomorphism (3.5.1). On the other hand, this also completes the two previous algebraic arguments for the isomorphism in simple ways that require neither [78] nor [37]. In Remark 3.5.7, an absence of quantum differentials yields the required identification (compatible with composition and identities) of Floer homologies (in a Morse–Bott setup in which only Morse trajectories contribute) $HF(L_{13}, L_{23}^T, L_{12}^T) \simeq HF(L_{12}, L_{23}, L_{23}^T, L_{12}^T) \simeq HF(L_{12}, L_{23}, L_{13}^T) \simeq HF(L_{13}, L_{13})$. In the above construction of a direct homomorphism (3.5.1), the absence of quantum differentials yields α (as above) and $\alpha^{-1} := [L_{13}] \in HF(L_{13}, L_{23}^T, L_{12}^T)$ so that $\alpha \circ_v \alpha^{-1} = [L_{13}] \cap [L_{13}] = [L_{13}] = \mathrm{id}_{L_{12}\#L_{23}}$ and $\alpha^{-1} \circ_v \alpha = [L_{13}] \cap [L_{13}] = [L_{13}] = \mathrm{id}_{L_{13}}$.

The bottom line is that we do not get around proving—implicitly or explicitly—the isomorphism $L_{12}\#L_{23} \sim L_{12} \circ L_{23}$ as 1-morphisms in the symplectic 2-category.

3.6 Gauge Theoretic 2-Categories and Quilted Atiyah Floer Conjectures

This section takes the quilt diagram approach in the construction of the symplectic 2-category and applies it to the gauge theoretic ASD Yang–Mills PDE to obtain proposals for various 2-categories which mix gauge theoretic, symplectic, and topological data. On the one hand, this categorical framework allows us to rigorously apply the abstract "local to global" approach of Remark 3.4.5 to Atiyah–Floer type conjectures, as already sketched in Remark 2.6.4; also see (2.6.1). On the other hand, it yields various approaches to constructing $2+1+1$ Floer-type field theories $\mathrm{Bor}_{2+1+1}^{\mathrm{conn}} \to \mathrm{Cat}$, which in turn leads us to formulate quilted Atiyah–Floer conjectures relating them.

Throughout, we fix a compact Lie group G and should also fix bundle types via characteristic classes. Ideally, this would avoid reducible connections as in the case of nontrivial $SO(3)$-bundles over 3-manifolds. However, this cannot generally be achieved in a coherent fashion when manifolds are decomposed to yield a field theory. Moreover, the Donaldson invariants of 4-manifolds [15, 17]—defined for $G = SU(2)$ or $G = SO(3)$—successfully deal with reducibles by encoding them as ends of the ASD moduli spaces, which yields a polynomial structure. On the other hand, instanton Floer homology for 3-manifolds [22] is currently only constructed in the absence of reducibles (using trivial $SU(2)$-bundles over homology 3-spheres, or nontrivial $SO(3)$-bundles), and thought to require an equivariant theory to deal with reducibles. For the following, we will assume that such theories can be con-

[32]In the corrigendum to [37], this assumption is misleadingly labeled "additional monotonicity." While (exact/positive/negative) monotonicity assumptions for the relevant quilted Floer cylinders also had to be added, the crucial extra assumption in the negative monotone case is that solutions of positive energy (i.e., possible quantum differentials) have sufficiently negative Fredholm index—exactly such that their occurrence in the Floer–Bott differential can be excluded by transversality.

structed from the same ASD moduli spaces. Then, the $3+1$ field theory outlined by Donaldson [16] for 4-cobordisms between appropriate 3-manifolds should have a refinement to $2+1+1$ dimensions which can be cast as the following 2-category.

Example 3.6.1 The **Donaldson 2+1 bordism 2-category DBor** should consist of:

- Objects in $\mathrm{Obj}_{\mathrm{DBor}} := \mathrm{Obj}_{\mathrm{Bor}_{2+1+1}}$ are closed oriented surfaces Σ.
- 1-morphisms in $\mathrm{Mor}^1_{\mathrm{DBor}}(\Sigma_+, \Sigma_-) := \mathrm{Mor}^1_{\mathrm{Bor}_{2+1+1}}(\Sigma_+, \Sigma_-)$ are 3-cobordism Y with boundary collars $\iota^{\pm} : [0,1] \times \Sigma_{\pm} \to Y$ as in Example 3.2.1.
- Horizontal 1-composition is gluing $Y_{01} \circ_h Y_{12} := \big(Y_{01} \sqcup Y_{12}\big)/\iota^+_{01}(s,x) \sim \iota^-_{12}(s,x)$ as in Example 3.2.1.
- 2-morphisms in $\mathrm{Mor}^2_{\mathrm{Dbor}}(Y, Y') := HF_{\mathrm{inst}}(\#(Y, Y'))$ for $Y, Y' \in \mathrm{Mor}^1_{\mathrm{DBor}}(\Sigma_+, \Sigma_-)$ are instanton Floer homology classes constructed analogous to [16, 22] on the closed 3-manifold $\#(Y, Y') := \big(Y^- \sqcup Y'\big)/\iota^{\pm}_Y \sim \iota^{\pm}_{Y'}$ obtained by reversing the orientation of Y and gluing at both incoming and outgoing boundaries.
- Vertical and horizontal composition of 2-morphisms and $\mathrm{id}_Y \in \mathrm{Mor}^2_{\mathrm{Symp}_G}(Y, Y)$ arise from the ASD moduli spaces representing the associated string diagrams; see below.

Here, we associate to every quilt diagram $\mathcal{QD} = \big(\mathcal{Q}, (\Sigma_P)_{P\in\mathcal{P}_{\mathcal{Q}}}, (Y_S)_{S\in\mathcal{S}_{\mathcal{Q}}}\big)$ an elliptic PDE as follows:

- A patch P labeled by a surface Σ is represented by a connection $\Theta_P \in \mathcal{A}(\widehat{P} \times \Sigma)$ satisfying the ASD equation $F_{\Theta_P} + *F_{\Theta_P} = 0$. Here, $\widehat{P}$ is an oriented 2-manifold that covers the closure $\overline{P} \subset Q_2 \setminus Q_0$ as in Remark 3.4.7.
- A seam S labeled by a 3-cobordism $Y_S \in \mathrm{Bor}_{2+1}(\Sigma_{P_S^-}, \Sigma_{P_S^+})$ is represented by a connection $\Theta_S \in \mathcal{A}(S \times Y)$ satisfying the ASD equation $F_{\Theta_S} + *F_{\Theta_S} = 0$ and diagonal seam condition: The restrictions of the connections $\Theta_{P_S^-}, \Theta_{P_S^+}$ for the adjacent patches $P_S^{\pm} \in \mathcal{P}_{\mathcal{Q}'}$ to the boundary slices $\{s\} \times \Sigma_{P_S^{\pm}}$ for $s \in S$ are required to coincide with $\Theta_S|_{\{s\}\times\partial Y}$ over $\partial Y = \Sigma^-_{P_S^-} \sqcup \Sigma_{P_S^+}$.

Up to challenges with reducibles, this should yield a quilted 2-category DBor with adjunction given by orientation reversal as in Lemma 3.4.11. Note in particular that the matching conditions for the connections at the seams, after applying an appropriate gauge transformation, simply become a smooth extension on the glued 4-manifolds. Thus, the moduli space constructed here is the moduli space of ASD connections on the 4-manifold constructed in Lemma 3.4.11 from the quilt diagram $\big(\mathcal{Q}, (\Sigma_P)_{P\in\mathcal{P}_{\mathcal{Q}}}, (Y_S)_{S\in\mathcal{S}_{\mathcal{Q}}}\big)$ in Bor_{2+1+1}. We may replace the boundary and corners of this 4-manifold at each end $e \in \mathcal{E}^{\pm}_{\mathcal{Q}}$ by a cylindrical end over the 3-manifold $\#(\underline{Y}_e) := \big(\bigsqcup_{i\in\mathbb{Z}_{N_e}} Y_{S_i}\big)/\iota^+_{Y_{S_i}} \sim \iota^-_{Y_{S_{i+1}}}$ that is obtained by gluing the components of the cyclic 1-morphism $\underline{Y}_e = (Y_{S_i})_{i\in\mathbb{Z}_{N_e}}$ associated to this end. Then, the quilted composition map should be given by the relative Donaldson invariant for the resulting 4-manifold with cylindrical ends,

$$\Phi_{\mathcal{QD}} : \otimes_{e\in\mathcal{E}^-_{\mathcal{Q}}} \mathrm{Mor}^2_{\mathcal{C}}(\underline{Y}_e) = HF(\#(\underline{Y}_e)) \to HF(\#(\underline{Y}_{e^+})) = \mathrm{Mor}^2_{\mathcal{C}}(\underline{Y}_{e^+}).$$

Next, we build an analogous target 2-category for the infinite dimensional Floer field theory outlined in Example 2.5.2. In order to obtain differentiable structures this requires the choice of an integrability constant $p > 2$.

Example 3.6.2 The **symplectic instanton 2-category SIn** should consist of:

- Objects are closed, oriented surfaces Σ—thought to represent the symplectic Banach space $\mathscr{A}(\Sigma) = L^p(\Sigma, \mathrm{T}^*\Sigma \otimes \mathfrak{g})$ of trivial G-connections on Σ.
- 1-morphisms $\underline{\mathscr{L}} = (\mathscr{L}_{01}, \ldots, \mathscr{L}_{(k-1)k}) \in \mathrm{Mor}^1_{\mathrm{SIn}}(\Sigma, \Sigma')$ are chains of Lagrangian submanifolds $\mathscr{L}_{ij} \subset \mathscr{A}(\Sigma_i)^- \times \mathscr{A}(\Sigma_j) = \mathscr{A}(\Sigma_i^- \sqcup \Sigma_j)$ in the symplectic spaces of connections over a chain of surfaces $\Sigma = \Sigma_0, \Sigma_1, \ldots, \Sigma_k = \Sigma'$, which are gauge invariant, $\mathscr{G}(\Sigma_i^- \sqcup \Sigma_j)^* \mathscr{L}_{ij} = \mathscr{L}_{ij}$.
- Horizontal composition of 1-morphisms $\underline{\mathscr{L}} \circ_{\mathrm{h}} \underline{\mathscr{L}}' := \underline{\mathscr{L}} \# \underline{\mathscr{L}}'$ is concatenation $(\ldots, \mathscr{L}_{(k-1)k}) \# (\mathscr{L}'_{01}, \ldots) := (\ldots, \mathscr{L}_{(k-1)k}, \mathscr{L}'_{01}, \ldots)$, with identities $1_\Sigma = (\,) \in \mathrm{Mor}^1_{\mathrm{SIn}}(\Sigma, \Sigma)$ for $\circ_{\mathrm{h}}$ given by trivial chains.
- 2-morphisms in $\mathrm{Mor}^2_{\mathrm{SIn}}(\underline{\mathscr{L}}, \underline{\mathscr{L}}') := HF_{\mathrm{inst}}(\underline{\mathscr{L}}, \underline{\mathscr{L}}')$ are the elements of quilted instanton Floer homology groups [59] outlined below.
- Vertical and horizontal composition of 2-morphisms and $\mathrm{id}_{\underline{\mathscr{L}}} \in \mathrm{Mor}^2_{\mathrm{SIn}}(\underline{\mathscr{L}}, \underline{\mathscr{L}})$ arise from ASD quilts representing the associated string diagrams; see below.

Up to challenges with reducibles, this approach should yield a quilted 2-category with adjunction given by transposition analogous to Remark 3.5.2. As in Remark 3.5.3 the key step is to associate an elliptic PDE to any quilt diagram $\mathscr{Q}\mathscr{D} = \big(\mathscr{Q}, (\Sigma_P)_{P \in \mathscr{P}_\mathscr{Q}}, (\underline{\mathscr{L}}_S)_{S \in \mathscr{S}_\mathscr{Q}}\big)$. For that purpose we first replace any seam labeled by a sequence of Lagrangians $\mathscr{L}_{i(i+1)} \subset \mathscr{A}(\Sigma_i)^- \times \mathscr{A}(\Sigma_{i+1})$ with strips resp. annuli labeled by the Σ_i and seams between them labeled by the simple Lagrangians $\mathscr{L}_{i(i+1)}$. Then the new quilt diagram $\mathscr{Q}\mathscr{D}' = (\mathscr{Q}', \ldots)$ determines a moduli space of **ASD quilts** $(\Theta_P)_{P \in \mathscr{P}_{\mathscr{Q}'}}$, which satisfy the following PDE:

- A patch P labeled by a surface Σ is represented by a connection $\Theta_P \in \mathscr{A}(\widehat{P} \times \Sigma)$ satisfying the ASD equation $F_{\Theta_P} + *F_{\Theta_P} = 0$.
- A seam S labeled by a Lagrangian submanifold $\mathscr{L}_S \subset \mathscr{A}(\Sigma_{P_S^-})^- \times \mathscr{A}(\Sigma_{P_S^+})$ is represented by a Lagrangian seam condition: The restrictions of the connections $\Theta_{P_S^-}, \Theta_{P_S^+}$ for the adjacent patches $P_S^\pm \in \mathscr{P}_{\mathscr{Q}'}$ to the boundary slices $\{s\} \times \Sigma_{P_S^\pm}$ for $s \in S$ induce connections $(\Theta_{P_S^-}, \Theta_{P_S^+})_s \in \mathscr{A}(\Sigma_{P_S^-})^- \times \mathscr{A}(\Sigma_{P_S^+})$, which are required to lie in $\mathscr{L}_S$.

These moduli spaces can be given compactifications and Fredholm descriptions by the nonlinear elliptic analysis for ASD connections with Lagrangian boundary conditions that is developed in [69, 70] for 4-manifolds with boundary space-time splitting such as $\bigsqcup_{P \in \mathscr{P}_{\mathscr{Q}'}} P \times \Sigma_P$. They should hence induce quilted composition maps

$$\Phi_{\mathscr{Q}\mathscr{D}} : \otimes_{e \in \mathscr{E}^-_\mathscr{Q}} \mathrm{Mor}^2_{\mathrm{SIn}}(\underline{\mathscr{L}}_e) \to \mathrm{Mor}^2_{\mathrm{SIn}}(\underline{\mathscr{L}}_{e^+}),$$

where the cyclic 2-morphism spaces $\mathrm{Mor}^2_{\mathrm{SIn}}(\underline{\mathscr{L}}) = HF_{\mathrm{inst}}(\underline{\mathscr{L}})$ for cyclic 1-morphisms $\underline{\mathscr{L}} = (\mathscr{L}_{i(i+1)})_{i \in \mathbb{Z}_N}$ are defined to be the quilted instanton Floer homology. The latter is the homology of a Floer complex whose differential arises from moduli spaces of ASD quilts on a quilted cylinder with seam conditions in the $\mathscr{L}_{i(i+1)}$ (modulo an overall $\mathbb{R}$-shift). This also defines the usual 2-morphisms of pairs of 1-morphisms $\underline{\mathscr{L}} = (\mathscr{L}_{(i-1)i})_{i=1\dots k}, \underline{\mathscr{L}}' = (\mathscr{L}'_{(i-1)i})_{i=1\dots k'} \in \mathrm{Mor}^1_{\mathrm{SIn}}(\Sigma, \Sigma')$,

$$\mathrm{Mor}^2_{\mathrm{SIn}}(\underline{\mathscr{L}}, \underline{\mathscr{L}}') := HF_{\mathrm{inst}}(\underline{\mathscr{L}}, \underline{\mathscr{L}}') := HF_{\mathrm{inst}}(\#(\underline{\mathscr{L}}, \underline{\mathscr{L}}')),$$

by concatenation to a cyclic 1-morphism indexed by $\mathbb{Z}_{k+k'}$

$$\#(\underline{\mathscr{L}}, \underline{\mathscr{L}}') := \left(\mathscr{L}^T_{(k-1)k}, \dots, \mathscr{L}^T_{01}, \mathscr{L}'_{01}, \dots, \mathscr{L}'_{(k'-1)k'}\right).$$

A first case of instanton Floer theory with Lagrangian boundary conditions is developed in [59] to construct a Floer homology $HF_{\mathrm{inst}}(Y, \mathscr{L})$ for pairs of a 3-manifold Y with boundary and a Lagrangian $\mathscr{L} \subset \mathscr{A}(\partial Y)$, using ASD connections $\Theta \in \mathscr{A}(\mathbb{R} \times Y)$ with boundary conditions $\Theta|_{\{s\}\times\partial Y} \in \mathscr{L}\ \forall s \in \mathbb{R}$. It requires an exclusion of nontrivial reducible connections, as is guaranteed for pairs $(Y, \mathscr{L}_H)$ when $Y \cup_\Sigma H$ is a homology 3-sphere (i.e. has the same homology with integer coefficients). For more general pairs, an equivariant Floer homology would be required to deal with the reducible flat connections on Y with boundary restriction in $\mathscr{L}$. Apart from dealing with the reducibles, the quilted setup above with $\mathscr{L}_{i(i+1)} \subset \mathscr{A}(\Sigma_i)^- \times \mathscr{A}(\Sigma_{i+1})$ can be reformulated as the instanton Floer theory with Lagrangian boundary conditions $HF_{\mathrm{inst}}(\underline{\mathscr{L}}) = HF_{\mathrm{inst}}\big(\sqcup_{i=0}^{N-1}[0,1] \times \Sigma_i, (\mathscr{L}_{i(i+1)})_{i\in\mathbb{Z}_N}\big)$.

For the quilted Atiyah–Floer Conjectures 3.6.4 we will restrict to the topologically generated part of this 2-category: The **symplectic instanton 2+1 bordism 2-category $\mathbf{SIn_{2+1}}$** is given as above with the restriction that the gauge invariant Lagrangian submanifolds $\mathscr{L}_{ij} = \mathscr{L}(Y_{ij}) \subset \mathscr{A}(\Sigma_i)^- \times \mathscr{A}(\Sigma_j) = \mathscr{A}(\Sigma_i^- \sqcup \Sigma_j)$ must be those associated in Example 2.5.2 to handle attachments $Y_{ij} \in \mathrm{Mor}^1_{\mathrm{Bor}_{2+1+1}}(\Sigma_i, \Sigma_j)$.

The analogous 2-category associated to the finite dimensional Floer field theory arising from G-representation spaces as outlined in Example 2.5.4 is a restriction of the symplectic 2-category as follows.

Example 3.6.3 The **symplectic 2+1 bordism 2-category of G-representations**, $\mathbf{Symp^G_{2+1}}$, is given by the topologically generated part of Symp in Example 3.5.1, as follows.

- Objects are the symplectic representation spaces M_Σ of Example 2.5.4.
- 1-morphisms $\underline{L} = (L_{01}, \dots, L_{(k-1)k}) \in \mathrm{Mor}^1_{\mathrm{Symp}^G_{2+1}}(M_\Sigma, M_{\Sigma'})$ are chains of Lagrangian submanifolds $L_{ij} = L_{Y_{ij}} \subset M^-_{\Sigma_i} \times M_{\Sigma_j}$ which were associated in Example 2.5.4 to handle attachments $Y_{ij} \in \mathrm{Mor}^1_{\mathrm{Bor}_{2+1+1}}(\Sigma_i, \Sigma_j)$.
- Horizontal composition of 1-morphisms $\underline{L} \circ_{\mathrm{h}} \underline{L}' := \underline{L}\#\underline{L}'$ is concatenation as in Symp, with identities $1_{M_\Sigma} = (\,) \in \mathrm{Mor}^1_{\mathrm{Symp}^G_{2+1}}(M_\Sigma, M_\Sigma)$ given by trivial chains.
- 2-morphism spaces $\mathrm{Mor}^2_{\mathrm{Symp}^G_{2+1}}(\underline{L}, \underline{L}') := HF(\underline{L}, \underline{L}')$ are quilted Floer homology.

- Vertical and horizontal composition of 2-morphisms and $\mathrm{id}_{\underline{L}} \in \mathrm{Mor}^2_{\mathrm{Symp}^G_{2+1}}(\underline{L}, \underline{L})$ arise from pseudoholomorphic quilts representing the associated string diagrams as in Remark 3.5.3.

Here the challenge of reducibles is more severe since it leads to singular symplectic spaces M_Σ. However, working for example with nontrivial bundles as in Remark 2.5.5 yields a quilted 2-category with adjunction given by transposition as in Remark 3.5.2—as a subcategory of the monotone symplectic 2-category Symp^τ.

Along with these three gauge theoretic 2-categories $\mathscr{C} = \mathrm{DBor}, \mathrm{SIn}_{2+1}, \mathrm{Symp}^{\mathrm{G}}_{2+1}$ we have three proposals for Floer field theories via functors $\mathrm{Bor}_{2+1} \to \mathscr{C}$, to the 1-category level of one of these 2-categories:

- $\mathrm{Bor}_{2+1} \to \mathrm{Symp}^G_{2+1}$ is determined by the G-representation spaces $\Sigma \mapsto M_\Sigma$ and $Y \mapsto L_Y$ in Example 2.5.4.
- $\mathrm{Bor}_{2+1} \to \mathrm{SIn}_{2+1}$ is determined by the G-connection spaces $\Sigma \mapsto \mathscr{A}(\Sigma)$ and $Y \mapsto \mathscr{L}_Y$ in Example 2.5.2.
- $\mathrm{Bor}_{2+1} \to \mathrm{DBor}_{2+1}$ is determined by $\Sigma \mapsto \Sigma$ and $Y \mapsto Y$.

The Floer field theory extension principle outlined in Conjecture 3.4.12 should yield natural extensions $\mathrm{Bor}_{2+1+1} \to \mathscr{C} \to \mathrm{Cat}$ for each of these, after making coherent choices of bundles as in Remark 2.5.5 or otherwise resolving the challenge of reducibles. (This may require a restriction to the connected bordism category.) Now the natural extension of the Atiyah–Floer Conjectures 2.6.1 and 2.6.2 for Heegaard splittings and cyclic Cerf decompositions is the following for Donaldson theory.

Conjecture 3.6.4 (Quilted Atiyah–Floer conjecture). The three extended Floer field theories $\mathrm{Bor}_{2+1+1} \to \mathrm{Symp}^G_{2+1}$, $\mathrm{Bor}_{2+1+1} \to \mathrm{SIn}_{2+1}$, $\mathrm{Bor}_{2+1+1} \to \mathrm{DBor}_{2+1}$ arising from appropriate G-bundles induce isomorphic 2-functors $\mathrm{Bor}_{2+1+1} \to \mathrm{Cat}$.

Note here that the last extended Floer field theory $\mathrm{Bor}_{2+1+1} \to \mathrm{DBor}_{2+1} \to \mathrm{Cat}$ should comprise the Donaldson invariants of 4-manifolds, as is visible not from the essentially trivial generating Floer field theory $\mathrm{Bor}_{2+1} \to \mathrm{DBor}_{2+1}$, but from the 2-morphism level of the target 2-category DBor_{2+1}. Analogous conjectures can be made for Seiberg–Witten theory and have been partially proven in [35].

Remark 3.6.5 (Quilted Atiyah–Floer conjecture for Seiberg–Witten–Heegaard–Floer theory). Example 2.5.6 should induce a Floer field theory $\mathrm{Bor}_{2+1} \to \mathrm{Symp}^{\mathrm{symm}}_{2+1}$ to a subcategory of Symp^τ; both being generated by symmetric products $\Sigma \mapsto \mathrm{Sym}^{g+n}\Sigma$ and $Y_\alpha \mapsto L_\alpha$. Another Floer feld theory $\mathrm{Bor}_{2+1} \to \mathrm{SWBor}_{2+1}$ should arise from the partial functor $\Sigma \mapsto \Sigma$ and $Y \mapsto Y$ to a 2-category defined as in Example 3.6.1, with instanton Floer theory resp. Donaldson invariants replaced by monopole Floer theory resp. Seiberg–Witten invariants.

Now the two resulting extended Floer field theories $\mathrm{Bor}_{2+1+1} \to \mathrm{Symp}^{\mathrm{symm}}_{2+1}$, $\mathrm{Bor}_{2+1+1} \to \mathrm{SWBor}_{2+1}$ should induce isomorphic 2-functors $\mathrm{Bor}_{2+1+1} \to \mathrm{Cat}$.

Finally, we will outline two further 2-categories which will serve to compare the above 2-categories and related Floer field theories, by embedding both into a "convex span." The first will serve to compare Don with SIn.

Example 3.6.6 The **instanton Atiyah–Floer 2-category InAF** should consist of:

- Objects are closed, oriented surfaces Σ.
- $\mathrm{Mor}_{\mathrm{InAF}}(\Sigma, \Sigma')$ combines $\mathrm{Mor}_{\mathrm{SIn}}(\Sigma, \Sigma')$ and $\mathrm{Mor}_{\mathrm{Don}}(\Sigma, \Sigma')$ as follows:
  - 1-morphisms are chains $\underline{f} = (f_{i(i+1)})_{i=0,\dots k-1}$ of morphisms between surfaces $\Sigma = \Sigma_0, \Sigma_1, \dots, \Sigma_k = \Sigma'$, where for each $i = 0, \dots k-1$ we either have $f_{i(i+1)} = \mathscr{L}_{i(i+1)} \in \mathrm{Mor}^1_{\mathrm{SIn}}(\Sigma_i, \Sigma_{i+1})$ a gauge invariant Lagrangian submanifold of $\mathscr{A}(\Sigma_i)^- \times \mathscr{A}(\Sigma_{i+1})$, or $f_{i(i+1)} = Y_{i(i+1)} \in \mathrm{Mor}^1_{\mathrm{Don}}(\Sigma_i, \Sigma_{i+1})$ a 3-cobordism.
  - $\mathrm{Mor}^2_{\mathrm{InAF}}(\underline{f}, \underline{g}) := HF_{\mathrm{inst}}(\underline{f}, \underline{g})$ is the quilted instanton Floer homology group arising from quilted cylinders with seams labeled by the entries of $\underline{f}, \underline{g}$.
  - Vertical composition $\circ_{\mathrm{v}}$ and its identities $\mathrm{id}_{\underline{f}} \in \mathrm{Mor}^2_{\mathrm{InAF}}(\underline{f}, \underline{f})$ arise from moduli spaces of ASD quilts representing the associated string diagrams.
- The composition functor $\mathrm{Mor}_{\mathrm{InAF}}(\Sigma, \Sigma') \times \mathrm{Mor}_{\mathrm{InAF}}(\Sigma', \Sigma'') \to \mathrm{Mor}_{\mathrm{InAF}}(\Sigma, \Sigma'')$ is defined by concatenation $\underline{f} \circ_{\mathrm{h}} \underline{f}' := \underline{f} \# \underline{f}'$ on 1-morphisms, with identities given by trivial chains $1_\Sigma = (\,) \in \mathrm{Mor}^1_{\mathrm{InAF}}(\Sigma, \Sigma)$, and horizontal 2-composition arises from moduli spaces of ASD quilts representing the associated string diagram.

The quilt diagrams here are represented by the same moduli spaces of ASD quilts as in Example 3.6.2, where as in Example 3.6.1 a seam S labeled by a 3-cobordism Y_S represents an ASD connection $\Theta_S \in \mathscr{A}(S \times Y)$ that matches (slice-wise, or completely after gauge) with the restrictions of the connections $\Theta_{P_S^-}, \Theta_{P_S^+}$ for the adjacent patches. Up to challenges with reducibles, this should yield a quilted 2-category with adjunction given by transposition as in Remark 3.5.2.

The final outline of a 2-category is the "convex span" of Don with Symp^G_{2+1}, after which one can easily imagine a combination of SIn with Symp^G_{2+1}, or a 2-category comprising all three of the basic gauge theoretic 2-categories Don, SIn, Symp^G_{2+1}.

Example 3.6.7 The **Atiyah–Floer 2-category AtFl** roughly consists of the following.

- Objects in $\mathrm{Obj}_{\mathrm{AtFl}} = \mathrm{Obj}_{\mathrm{Don}} \cup \mathrm{Obj}_{\mathrm{Symp}^G_{2+1}}$ are either closed, oriented surfaces Σ or symplectic representation spaces M_Σ associated to a surface as in Example 2.5.4.
- $\mathrm{Mor}_{\mathrm{AtFl}}(\Sigma, \Sigma')$ extends $\mathrm{Mor}_{\mathrm{Symp}^G_{2+1}}(M_\Sigma, M_{\Sigma'})$ and $\mathrm{Mor}_{\mathrm{Don}}(\Sigma, \Sigma')$ as follows:
  - Simple morphisms all arise from 3-cobordisms $Y \in \mathrm{Mor}^1_{\mathrm{Bor}_{2+1+1}}(\Sigma, \Sigma')$, but depending on the type of objects they relate, they appear as
    3-cobordisms $Y \in \mathrm{SMor}_{\mathrm{AtFl}}(\Sigma, \Sigma') := \mathrm{Mor}^1_{\mathrm{Don}}(\Sigma, \Sigma')$,
    Lagrangians $L_Y \in \mathrm{SMor}_{\mathrm{AtFl}}(M_\Sigma, M_{\Sigma'}) := \mathrm{Mor}^1_{\mathrm{Symp}^G_{2+1}}(M_\Sigma, M_{\Sigma'})$,
    Lagrangians $\mathscr{L}_Y/\mathscr{G}(\Sigma') \subset \mathscr{A}(\Sigma)^- \times M_{\Sigma'}$ in $\mathrm{SMor}_{\mathrm{AtFl}}(\Sigma, M_{\Sigma'})$,
    Lagrangians $\mathscr{L}_Y/\mathscr{G}(\Sigma) \subset M_\Sigma^- \times \mathscr{A}(\Sigma')$ in $\mathrm{SMor}_{\mathrm{AtFl}}(M_\Sigma, \Sigma')$.
    Here, the three types of Lagrangians are only associated to handle attachments Y, and the last two types are projections of the gauge invariant Lagrangian $\mathscr{L}_Y \subset \mathscr{A}(\Sigma) \times \mathscr{A}(\Sigma')$ from Example 2.5.2, which by construction lies in the flat connections, $\mathscr{L} \subset \mathscr{A}_{\mathrm{flat}}(\Sigma) \times \mathscr{A}_{\mathrm{flat}}(\Sigma')$, so that quotienting by the

gauge group on the first factor yields a projection to the representation space $M_\Sigma = \mathscr{A}_{\text{flat}}(\Sigma)/\mathscr{G}(\Sigma)$. The same goes for projection in the second factor, and projecting in both factors yields $L_Y = \mathscr{L}_Y/(\mathscr{G}(\Sigma) \times \mathscr{G}(\Sigma'))$.

- 1-morphisms are chains $\underline{f} = (f_{i(i+1)})_{i=0,\dots k-1}$ of simple morphisms $f_{i(i+1)} \in \text{SMor}_{\text{AtFl}}(x_i, x_{i+1})$ for a sequence $x_0, \dots, x_{k-1} \in \text{Obj}_{\text{AtFl}}$ of objects, i.e. a sequence in which each entry is of one of the types $Y_{i(i+1)}, L_{Y_{i(i+1)}}, \mathscr{L}_{Y_{i(i+1)}}/\mathscr{G}(\Sigma_i)$, $\mathscr{L}_{Y_{i(i+1)}}/\mathscr{G}(\Sigma_{i+1})$ for a chain of 1-morphisms $Y_{i(i+1)} \in \text{Mor}^1_{\text{Bor}_{2+1+1}}(\Sigma_i, \Sigma_{i+1})$.
- $\text{Mor}^2_{\text{AtFl}}(\underline{f}, \underline{g}) := HF_{\text{inst}}(\underline{f}, \underline{g})$ is the quilted instanton Floer homology group arising from quilted cylinders with seams labeled by the entries of $\underline{f}, \underline{g}$.
- Vertical composition $\circ_{\text{v}}$ and its identities $\text{id}_{\underline{f}} \in \text{Mor}^2_{\text{AtFl}}(\underline{f}, \underline{f})$ arise from moduli spaces of ASD quilts representing the associated string diagrams.

- The composition functor $\text{Mor}_{\text{AtFl}}(x, x') \times \text{Mor}_{\text{AtFl}}(x', x'') \to \text{Mor}_{\text{AtFl}}(x, x'')$ is defined by concatenation $\underline{f} \circ_{\text{h}} \underline{f}' := \underline{f}\#\underline{f}'$ on 1-morphisms, with identities given by trivial chains $1_\Sigma = (\,) \in \text{Mor}^1_{\text{AtFl}}(x, x)$, and horizontal 2-composition arises from moduli spaces of ASD quilts representing the associated string diagram.

The quilt diagrams here are represented by a coupling of the pseudoholomorphic and ASD moduli spaces in Examples 3.5.1 and 3.6.1 via Lagrangian seam conditions similar to those in Example 3.6.2. Combining the PDE representations from those constructions, it remains to give PDE meaning to seams labeled with simple morphisms in $\text{SMor}_{\text{AtFl}}(M_\Sigma, \Sigma')$ or $\text{SMor}_{\text{AtFl}}(\Sigma, M_{\Sigma'})$. Since these are related by transposition, it suffices to consider the first:

- A seam S with adjacent patches $P^\pm_S \in \mathscr{P}_{\mathscr{Q}'}$ labeled by a Lagrangian submanifold $\mathscr{L}_{Y_S}/\mathscr{G}(\Sigma_{P^-_S}) \subset M^-_{\Sigma_{P^-_S}} \times \mathscr{A}(\Sigma_{P^+_S})$ is represented by a seam condition between the pseudoholomorphic map $u_{P^-_S} : \widehat{P^-_S} \to M_{\Sigma_{P^-_S}}$ and the ASD connection $\Theta_{P^+_S} \in \mathscr{A}(\widehat{P^+_S} \times \Sigma_{P^+_S})$. Their restrictions to the seam induce a map
$$S \to M^-_{\Sigma_{P^-_S}} \times \mathscr{A}(\Sigma_{P^+_S}), \qquad s \mapsto (u_{P^-_S}(s), \Theta_{P^+_S}|_{\{s\}\times\Sigma_{P^+_S}}),$$
which is required to take values in $\mathscr{L}_{Y_S}/\mathscr{G}(\Sigma_{P^-_S})$.

For this to rigorously define a quilted 2-category with adjunction given by transposition as in Remark 3.5.2, one again has to resolve the challenge of reducibles by, e.g., working with nontrivial bundles as in Remark 2.5.5. Once the symplectic spaces M_Σ are all smooth, the analytic setup for ASD connections with Lagrangian boundary conditions in Example 3.6.2 directly transfers to prove the basic Fredholm and compactness properties for these moduli spaces. An exposition of the compactness results in an explicitly quilted setting can be found in [41]; the following paragraph provides a brief summary for experts.

The only missing analytic pieces are regularity and estimates for the pair (u, A) near a seam S labeled by a Lagrangian $\mathscr{L}/\mathscr{G}(\Sigma) \subset M^-_\Sigma \times \mathscr{A}(\Sigma')$. This could be achieved by the analytic setup developed in [68–70], which proceeds by splitting $\Theta_P = \Phi \mathrm{d}s + \Psi \mathrm{d}t + A$ into functions $\Phi, \Psi : P \to \mathfrak{g}$ in a neighborhood $U = \{(s,t) | t \geq 0\}$ of the seam $S = \{t = 0\}$, and a map $A : U \to \mathscr{A}(\Sigma')$. Coulomb gauge fixing conditions include a Dirichlet boundary condition $\Psi|_{t=0} = 0$ and—via the flatness part of the Lagrangian boundary condition $F_A|_{t=0} = 0$—induce a Neumann condition $\partial_t \Phi|_{t=0} = 0$. Thus estimates for Φ, Ψ are obtained from Dirichlet resp. Neumann problems with lower order contributions from A. Once these are established, the map $A : U \to \mathscr{A}(\Sigma')$ satisfies a Cauchy–Riemann equation with respect to the complex structure given by the Hodge $*$ on Σ and lower order or controlled inhomogeneous terms. Moreover, it is coupled with the pseudoholomorphic map $u^- : U \to M_\Sigma^-$ (obtained from u by reflection on the seam) by a Lagrangian seam condition. Thus estimates on (u, A) in a neighborhood of the seam would follow from the general theory for the Cauchy–Riemann equation in Banach spaces [68].

4 Known Issues in the Bibliography

The following is a brief guide to known issues and associated errata of the literature presented in the Bibliography:
[18]: Known issues resolved in [19].
[36]: Issues arising from the correction to [37] are presently unresolved.
[37]: Assumptions added in a 2015 corrigendum at arxiv:1003.4493v7 resolve known issues listed below, but reduce the result to known resp. essentially trivial cases; see Remark 3.5.8.

The published version of [37] claims in Lemmas 11, 14 that "bubbling at the Morse–Bott end" is captured topologically in terms of a pair of disks with boundary on L_{01} and L_{12}. This is generally false. First, a resolution of the $L_{02} = L_{01} \circ L_{12}$ seam yields a quilt with seam conditions in the order $(L_{01}, L_{12}, L_{12}^T, L_{01}^T)$ instead of $(L_{01}, L_{12}, L_{01}, L_{12})$. The latter in fact is not well defined unless $M_0 = M_2$. Second, one should note that folding of the quilt indeed yields a strip in "$\underline{M} = M_0 \times M_1 \times M_1 \times M_2$" with both boundary conditions given by Lagrangian embeddings of $L_{01} \times L_{12}$, but the correct symplectic structure on $\underline{M}$ is $(\omega_0, -\omega_1, -\omega_1, \omega_2)$, and the two embeddings differ by a permutation of the two M_1 factors.

Finally, even if torsion assumptions would allow to deform a Morse–Bott trajectory of positive symplectic area into a sum of disk classes, these are generally no longer pseudoholomorphic or of nonnegative area. Thus, an argument implicitly used in an earlier corrigendum (along the lines of "$a + b > 0 \Rightarrow a, b > 0$") does not suffice to exclude bubbling in the torsion case.
[39]: Known issues resolved in [40].
[56]: Known issues (e.g., conical structure at infinity) not resolved at present.
[75]: Known issues resolved in [77].

Acknowledgments I would like to credit and thank Denis Auroux, Chris Douglas, Dan Freed, David Gay, Robert Lipshitz, Tim Perutz, Dietmar Salamon, Chris Schommer-Priess, Peter Teichner, and Chris Woodward for illuminations of various aspects of the ideas presented here. Moreover, thanks to the organizers and participants of the 2015 AWM Symposium for sharing an amazing breadth of quality math and inspiring me to try and make the Floer field theory ideas rigorous. Finally, thanks to the diligent and speedy referees for help with cleaning up various details and encouragement to make known issues known.

References

1. M. Aschenbrenner, S. Friedl, H. Wilton, 3-manifold groups. arXiv:1205.0202
2. M.F. Atiyah, Topological quantum field theories. Publications Mathematiques de l'IHES **68**, 175–186 (1988)
3. M.F. Atiyah, New invariants of three and four dimensional manifolds. Proc. Symp. Pure Math. **48**, (1988)
4. M.F. Atiyah, R. Bott, The Yang Mills equations over Riemann surfaces. Phil. Trans. R. Soc. Lond. A **308**, 523–615 (1982)
5. D. Auroux, Fukaya categories of symmetric products and bordered Heegaard-Floer homology. J. Gökova Geom. Topol. **4**, 1–54 (2010)
6. N. Bottman, Pseudoholomorphic quilts with figure eight singularity. arXiv:1410.3834
7. N. Bottman, Fredholm property for figure eight quilts. In preparation
8. N. Bottman, H. Tanaka, An A_∞ 2-nerve. In preparation
9. N. Bottman, K. Wehrheim, Gromov compactness for squiggly strip shrinking in pseudoholomorphic quilts. arXiv:1503.03486
10. A. Caldararu, S. Willerton, The Mukai pairing. I: a categorical approach. N. Y. J. Math. **16**, 61–98 (2010)
11. A. Cannas da Silva, *Lectures on Symplectic Geometry* (Springer, Berlin, 2001)
12. A. Cannas da Silva, A Chiang-type Lagrangian in $\mathbb{CP}^2$. arXiv:1511.02041
13. J. Cerf, La stratification naturelle des espaces de fonctions différentiables réelles et le théorème de la pseudo-isotopie. Inst. Hautes Études Sci. Publ. Math. (39): 5D173 (1970)
14. S.K. Donaldson, An application of gauge theory to four-dimensional topology. J. Differential Geom. **18**(2), 279–315 (1983)
15. S.K. Donaldson, Polynomial invariants of smooth four-manifolds. Topology **29**, 257–315 (1990)
16. S.K. Donaldson, *Floer Homology Groups in Yang-Mills Theory* (Cambridge University Press, Cambridge, 2002)
17. S.K. Donaldson, P. Kronheimer, *The Geometry of Four-Manifolds* (Oxford University Press, Oxford, 1990)
18. S. Dostoglou, D.A. Salamon, Self-dual instantons and holomorphic curves. Ann. Math. **139**, 581–640 (1994). [see Sect. 4]
19. S. Dostoglou, D.A. Salamon, Corrigendum: self-dual instantons and holomorphic curves. Ann. Math. **165**, 665–673 (2007)
20. D.L. Duncan, Compactness results for the quilted Atiyah-Floer conjecture. arXiv:1212.1547
21. Y. Eliashberg, A.B. Givental, H. Hofer, Introduction to symplectic field theory. Geom. Funct. Anal. **10**, 560–673 (2000)
22. A. Floer, Instanton invariant for 3-manifolds. Comm. Math. Phys. **118**, 215–240 (1988)
23. A. Floer, Morse theory for Lagrangian intersections. J. Diff. Geom. **28**, 513–547 (1988)
24. K. Fukaya, SO(3)-Floer homology of 3-manifolds with boundary 1. arXiv:1506.01435
25. O. Garcia-Prada, A direct existence proof for the vortex equations over a compact Riemann surface. Bull. Lond. Math. Soc. **26**(1), 88–96 (1994)
26. D. Gay, R. Kirby, Indefinite Morse 2-functions; broken fibrations and generalizations. Geom. Topol. **19**, 2465–2534 (2015)
27. D. Gay, K. Wehrheim, C.T. Woodward, Connected Cerf theory. Link to preprint
28. M. Gromov, Pseudo holomorphic curves in symplectic manifolds. Invent. Math. **82**, 307–347 (1985)
29. V. Guillemin, S. Sternberg, The moment map revisited. J. Differe. Geom. **69**(1), 137–162 (2005)
30. A. Hatcher, The Classification of 3-Manifolds—A Brief Overview, talk at the 2004 Cornell Topology Festival. www.math.cornell.edu/~hatcher
31. M. Hedden, C. Herald, P. Kirk, The pillowcase and perturbations of traceless representations of knot groups. Geom. Topol. **18**, 211–287 (2014)

32. D.L. Johnson, *Presentations of Groups* (Cambridge University Press, Cambridge, 1990)
33. P.B. Kronheimer, T.S. Mrowka, *Floer Homology for Seiberg-Witten monopoles* (Cambridge University Press, Cambridge, 2007)
34. P.B. Kronheimer, T.S. Mrowka, Knot homology groups from instantons. J. Topol. **4**(4), 835–918 (2011)
35. C. Kutluhan, Y.-J. Lee, C.H. Taubes, HF=HM I–V: Heegaard Floer homology and Seiberg–Witten Floer homology, arXiv:1007.1979; Reeb orbits and holomorphic curves for the ech/Heegaard-Floer correspondence, arXiv:1008.1595; Holomorphic curves and the differential for the ech/Heegaard Floer correspondence, arXiv:1010.3456; The Seiberg-Witten Floer homology and ech correspondence, arXiv:1107.2297; Seiberg-Witten-Floer homology and handle addition, arXiv:1204.0115
36. Y. Lekili, Heegaard Floer homology of broken fibrations over the circle. Adv. Math. **244**, 268–302 (2013). [see Sect. 4]
37. Y. Lekili, M. Lipyanskiy, Geometric composition in quilted Floer theory. Adv. Math. **236**, 1–23 (2013). [see Sect. 4]
38. Y. Lekili, T. Perutz, Lagrangian correspondences and invariants of three-manifolds with boundary. In preparation
39. R. Lipshitz, A cylindrical reformulation of Heegaard-Floer homology. Geom. Topol. **10**, 955–1096 (2006). [see Sect. 4]
40. R. Lipshitz, Errata to 'A cylindrical reformulation of Heegaard Floer homology'. Geom. Topol. **18**, 17–30 (2014)
41. M. Lipyanskiy, Gromov-Uhlenbeck Compactness. arXiv:1409.1129
42. J. Lurie, On the classification of topological field theories. Curr. Dev. Math. **2008**, 129–280 (2009)
43. M. Mackaay, Spherical 2-categories and 4-manifold invariants. Adv. Math. **143**, 288–348 (1999)
44. C. Manolescu, C.T. Woodward, Floer homology on the extended moduli space. Progr. Math. **296**, 283–329 (2012)
45. S. Ma'u, Gluing pseudoholomorphic quilted disks. arXiv:0909.3339
46. D. McDuff, K. Wehrheim, Kuranishi atlases with trivial isotropy. arXiv:1208.1340
47. D. McDuff, D.A. Salamon, *Introduction to Symplectic Topology* (Oxford University Press, Oxford, 1995)
48. D. McDuff, D.A. Salamon, J-holomorphic curves and symplectic topology. AMS (2012)
49. J. Milnor, A procedure for killing homotopy groups of differentiable manifolds. Proc. Sympos. Pure Math. **III**, 39–55 (1961)
50. J. Milnor, *Lectures on the H-Cobordism Theorem* (Princeton University Press, Princeton, 1965). Notes by L. Siebenmann and J. Sondow
51. J.W. Morgan, *The Seiberg-Witten Equations and Applications to the Topology of Smooth Four-Manifolds* (Princeton University Press, Princeton, 1995)
52. P. Ozsváth, Z. Szabó, Holomorphic disks and topological invariants for closed 3-manifolds. Ann. Math. **159**(3), 1027–1158 (2004)
53. T. Perutz, Lagrangian matching invariants for fibred four-manifolds I. Geom. Topol. **11**, 759–828 (2007)
54. T. Perutz, Hamiltonian handleslides for Heegaard Floer homology. J. Gökova Geom. Topol. **2007**, 15–35 (2007)
55. T. Perutz, Lagrangian matching invariants for fibred four-manifolds II. Geom. Topol. **12**, 1461–1542 (2008)
56. R. Rezazadegan, Seidel-Smith cohomology for tangles. Selecta Math. **15:3**, 487–518 (2009). [see Sect. 4]
57. D.A. Salamon, Lagrangian intersections, 3-manifolds with boundary, and the Atiyah-Floer conjecture. Proc. ICM, Zürich **1**, 526–536 (1994)
58. D.A. Salamon, Lectures on floer homology. Park City Math. Ser. **7**, 145–229 (1999)
59. D.A. Salamon, K. Wehrheim, Instanton Floer homology with Lagrangian boundary conditions. Geom.Topol. **12**, 745–918 (2008)

60. D.A. Salamon, K. Wehrheim, An open-closed isomorphism for instanton Floer homology. In preparation
61. A. Scorpan, The Wild World of 4-Manifolds. AMS (2005). www.maths.ed.ac.uk/~aar/papers
62. G. Segal, Topological structures in string theory. R. Soc. **359**, 1389–1398 (2001)
63. P. Seidel, *Fukaya categories and Picard-Lefschetz theory*, Zürich Lectures in Advanced Mathematics (European Mathematical Society (EMS), Zürich, 2008)
64. P. Seidel, I. Smith, A link invariant from the symplectic geometry of nilpotent slices. Duke Math. J. **134**(3), 453–514 (2006)
65. C.H. Taubes, Casson's invariant and gauge theory. J. Diff. Geom. **31**, 547–599 (1990)
66. M. Usher, Vortices and a TQFT for Lefschetz fibrations on 4-manifolds. Algebr. Geom. Topol. **6**(4), 1677–1743 (2006)
67. K. Wehrheim, *Uhlenbeck Compactness*, EMS Series of Lectures in Mathematics (2004)
68. K. Wehrheim, Banach space valued Cauchy-Riemann equations with totally real boundary conditions. Comm. Cont. Math. **6**(4), 601–635 (2004)
69. K. Wehrheim, Anti-self-dual instantons with Lagrangian boundary conditions I: elliptic theory. Comm. Math. Phys. **254**(1), 45–89 (2005)
70. K. Wehrheim, Anti-self-dual instantons with Lagrangian boundary conditions II: bubbling. Comm. Math. Phys. **258**(2), 275–315 (2005)
71. K. Wehrheim, Lagrangian boundary conditions for anti-self-dual connections and the Atiyah-Floer conjecture. J. Symp. Geom. **3**(4), 703–747 (2005)
72. K. Wehrheim, Bubble zoology for large structure limit from anti-self-dual connections to pseudoholomorphic curves. Talks at Columbia, UW Madison, fall (2005)
73. K. Wehrheim, String diagrams in Topology, Geometry, and Analysis. In: Slides from AWM symposium (2015). https://math.berkeley.edu/~katrin/slides/AWMslides.pdf
74. K. Wehrheim, Extensions of Floer field theories. Work in progress
75. K. Wehrheim, C.T. Woodward, Quilted Floer cohomology. Geom. Topol. **14**, 833–902 (2010). [see Sect. 4]
76. K. Wehrheim, C.T. Woodward, Functoriality for Lagrangian correspondences in Floer theory. Quantum Topol. **1**(2), 129–170 (2010)
77. K. Wehrheim, C.T. Woodward, Quilted Floer trajectories with constant components. Geom. Topol. **16**, 127–154 (2012)
78. K. Wehrheim, C.T. Woodward, Floer cohomology and geometric composition of Lagrangian correspondences. Adv. Math. **230**(1), 177–228 (2012)
79. K. Wehrheim, C.T. Woodward. Pseudoholomorphic quilts. J. Symp. Geom.. arXiv:0905.1369
80. K. Wehrheim, C.T. Woodward, Floer field theory. Link to preprint
81. K. Wehrheim, C.T. Woodward, Floer field theory for tangles. arXiv:1503.07615
82. A. Weinstein, The symplectic "category", *Differential Geometric Methods in Mathematical Physics*, vol. 905, Lecture Notes in Mathematics (1982), pp. 45–51
83. S. Willerton, String diagrams. Youtube playlist (2007)
84. E. Witten, Supersymmetry and Morse theory. J. Differ. Geom. **17**(4), 661–692 (1982)
85. E. Witten, Topological quantum field theory. Commun. Math. Phys. **117**(3), 353–386 (1988)

Part II
Low Dimensional Topology

An Elementary Fact About Unlinked Braid Closures

J. Elisenda Grigsby and Stephan M. Wehrli

Abstract Let $n \in \mathbb{Z}^+$. We provide two short proofs of the following classical fact, one using Khovanov homology and one using Heegaard–Floer homology: if the closure of an n-strand braid σ is the n-component unlink, then σ is the trivial braid.

Keywords Braids · Khovanov homology · Heegaard–floer homology

Mathematics Subject Classification 20F36 (primary) · 57M27 · 57R58 · 81R50 (secondary)

Let $\mathcal{B}_n$ denote the n-strand braid group, $\mathbb{1}_n \in \mathcal{B}_n$ the n-strand trivial braid, and U_n the n-component unlink in S^3. Denote by $\widehat{\sigma}$ the closure of $\sigma \in \mathcal{B}_n$, considered as a link in S^3. The following fact first appears in the literature in [4, Thm. 4.1]. It can also be obtained as an immediate corollary of [3, Thm.1]:

Proposition 1 *Let $\sigma \in B_n$. If $\widehat{\sigma} = U_n$, then $\sigma = \mathbb{1}_n$.*

The primary purpose of this note is to provide two short proofs of Proposition 1, one using Khovanov homology and one using Heegaard–Floer homology. Although the classical proof contained in [4] is straightforward, we hope these new proofs will also be of interest, since they suggest ways in which algebraic properties of link homology theories can give information about braid dynamics.

The key geometric idea underlying both proofs is the following pair of simple but powerful observations, from [8]:

J. Elisenda Grigsby—Partially supported by NSF grant DMS-0905848 and NSF CAREER award DMS-1151671.
Stephan M. Wehrli—Partially supported by NSF grant DMS-1111680.

J.E. Grigsby (✉)
Mathematics Department, Boston College, 522 Maloney Hall, Chestnut Hill, MA 02467, USA
e-mail: grigsbyj@bc.edu

S.M. Wehrli
Department of Mathematics, Syracuse University, 215 Carnegie, Syracuse, NY 13244, USA
e-mail: smwehrli@syr.edu

G. Letzter et al. (eds.), *Advances in the Mathematical Sciences*, Association for Women in Mathematics Series 6, DOI 10.1007/978-3-319-34139-2_2

- A self-diffeomorphism of a surface with non-empty boundary that fixes the boundary pointwise is isotopic to the identity rel boundary iff it is both right- and left-veering (cf. [9] for a definition);
- If the Heegaard–Floer contact invariant (resp., Plamenevskaya's transverse invariant) is nonzero, then every open book supporting the contact structure (resp., every braid closure representing the transverse link) is right-veering [9, 16] (resp., [1]).

The two proofs of Proposition 1 we present are formally analogous. The proof involving Khovanov homology (Sect. 1) requires little more than the two facts above, while the Heegaard–Floer homology proof (Sect. 2) involves applying the facts above to fibered links in connected sums of copies of $S^1 \times S^2$, using ideas of Birman–Hilden [2]. Recall that the preimage of the braid axis in the double branched cover of a braid closure in S^3 is a fibered link. When the closure of the braid is the unlink, one obtains a fibered link in a connected sum of copies of $S^1 \times S^2$. The existence of a nontrivial n-strand braid whose closure is U_n would imply the existence of a non-trivial fibered link of minimal complexity (i.e., maximal Euler characteristic) in $\#^{n-1}(S^1 \times S^2)$.

More precisely, let Y_n denote $\#^n(S^1 \times S^2)$. For L a fibered link with fiber F, we will abuse terminology and refer to $\chi(F)$ as the *Euler characteristic of L*.

Define
$$\mathcal{L}_n := \{\ell \in \mathbb{Z}^+ \mid \ell \leq (n+1) \text{ and } \ell \equiv (n+1)\text{mod}2\}.$$

Note that for each $\ell \in \mathcal{L}_n$, it is straightforward to construct a fibered ℓ-component link, $\mathbf{L}_\ell \subset Y_n$, of Euler characteristic $1-n$. See Fig. 1. The monodromy of $\mathbf{L}_\ell$ is trivial, and the pair $(Y_n, \mathbf{L}_\ell)$ is well-defined up to diffeomorphism.

The following result appears in [14].

Proposition 2 *[14, Prf. of Thm. 1.3] Let $L_\ell \subset Y_n$ be a fibered, ℓ-component link with $\ell \in \mathcal{L}_n$ and Euler characteristic $1-n$. Then the pair (Y_n, L_ℓ) is diffeomorphic to the pair $(Y_n, \mathbf{L}_\ell)$.*

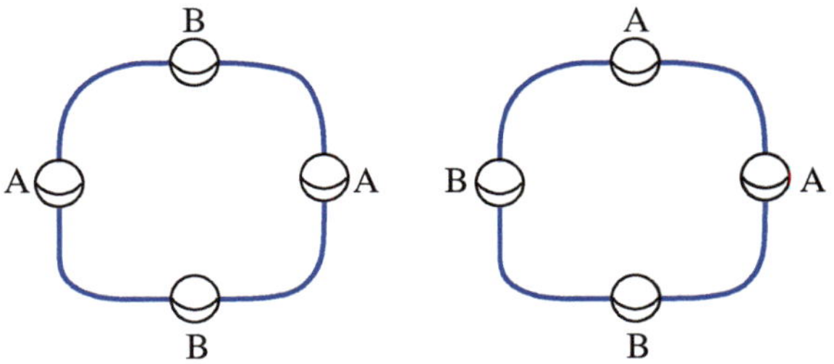

Fig. 1 Kirby diagrams of the links $\mathbf{L}_1$ (*left*) and $\mathbf{L}_3$ (*right*) in $Y_2 := \#^2 S^1 \times S^2$. The S^2's (boundaries of the feet of 4-dimensional 1-handles) are identified as labeled, via a reflection in the plane perpendicular to the *straight line* joining their centers. The fibered link in each case is drawn in *blue*. To construct $\mathbf{L}_\ell \in Y_n$ in general, arrange n pairs of S^2's along an unknot in S^3 so that attaching 2-dimensional one-handles to the disk bounded by the unknot, via the chosen configuration, forms an oriented surface with ℓ boundary components

It is clear (cf. Lemma 1) that if $\ell \notin \mathcal{L}_n$, then an ℓ-component link cannot have Euler characteristic $1-n$. It is also clear (cf. Lemma 2) that $1-n$ is the maximal possible Euler characteristic among all fibered links in Y_n. Informally, Proposition 2 therefore says that for allowable ℓ, maximal Euler characteristic fibered ℓ-component links in $\#^n(S^1 \times S^2)$ are unique up to diffeomorphism. After the first version of this note appeared, it was pointed out in [5, Cor. 1.3] that Proposition 2 implies Proposition 1, by the main result of [2].

In Sect. 2.1 we present an alternative proof of Proposition 2 using Heegaard–Floer homology. We thank John Baldwin for pointing out that this proof of Proposition 2 implies:

Corollary 1 *If $Y \not\cong Y_n$ is a closed, oriented 3-manifold with the same Heegaard–Floer module structure as Y_n, then Y contains no fibered links of Euler characteristic $1-n$.*

There is a unique maximal Euler characteristic fibered link in S^3 (namely, the unknot) whose corresponding open book supports the standard tight contact structure. Ken Baker (cf. [11]) asked the following interesting question:

Question 1 Fix a contact structure, ξ, on a 3-manifold, Y, and let

$$\overline{\chi}_\xi := \max\{\chi(L) \mid L \text{ is a fibered link whose open book supports } \xi\}.$$

Up to diffeomorphism, are there finitely many fibered links L supporting ξ with $\chi(L) = \overline{\chi}_\xi$?

Proposition 2 tells us that for the standard tight contact structure on Y_n the answer is yes.

1 Khovanov Homology Proof of Proposition 1

Proof (Proposition 1) Choose a diagram, $D(\widehat{\sigma})$, for $\widehat{\sigma}$ obtained as the closure of a diagram for σ, and mark the n points on the diagram corresponding to the intersection with the closure arc. Recall that the ($\mathbb{F} = \mathbb{Z}/2\mathbb{Z}$) Khovanov homology, $\mathrm{Kh}(\widehat{\sigma})$, of $\widehat{\sigma}$ is an invariant of the isotopy class of $\widehat{\sigma} \subset S^3$ that takes the form of a bigraded vector space over $\mathbb{F}$. Since we have also chosen a basepoint on each of the n link components, [7, Prop.1] tells us that $\mathrm{Kh}(\widehat{\sigma})$ inherits the structure of a module over the ring

$$\mathcal{A}_n := \mathbb{F}[x_1, \ldots, x_n]/(x_1^2, \ldots, x_n^2)$$

as follows.

Associated to the diagram of $\widehat{\sigma}$ is a cube of resolutions whose vertices are in one-to-one correspondence with complete resolutions (i.e., Kauffman states) of the diagram. The basis elements (generators) of the underlying vector space of the Khovanov chain complex, $\mathrm{CKh}(D(\widehat{\sigma}))$, are, in turn, in one-to-one correspondence with markings of the components of each resolution with either a 1 or an x (i.e., *enhanced* Kauffman states).

Let $\mathcal{I}_{braid}$ be the unique "braid-like" complete resolution of $D(\widehat{\sigma})$, and denote by Ψ^+ (resp., Ψ^-) the basis element $1 \otimes \ldots \otimes 1$ (resp., $x \otimes \ldots \otimes x$) in the vector space associated to $\mathcal{I}_{braid}$. Ψ^- is a cycle, hence represents an element in $\mathrm{Kh}(\widehat{\sigma})$. Indeed, $[\Psi^-] \in \mathrm{Kh}(\widehat{\sigma})$ is precisely Plamenevskaya's invariant [17] of the transverse isotopy class of the transverse link represented by $\widehat{\sigma}$.

We are now ready to understand the $\mathcal{A}_n$ structure induced by the n points $p_1, \ldots, p_n$. For each complete resolution, $\mathcal{I}$, choose a numbering of its $\ell_{\mathcal{I}}$ connected components, and let $v_1 \otimes \ldots \otimes v_{\ell_{\mathcal{I}}}$ represent the Khovanov generator whose jth component in $\mathcal{I}$ is marked with $v_j \in \{1, x\}$. Suppose p_i lies on the kth component of $\mathcal{I}$. Then the action of $x_i \in \mathcal{A}_n$ is the $\mathbb{F}$-linear extension of the assignment:

$$x_i \cdot (v_1 \otimes \ldots \otimes v_k \otimes \ldots \otimes v_{\ell_{\mathcal{I}}}) := v_1 \otimes \ldots \otimes x \otimes \ldots \otimes v_{\ell_{\mathcal{I}}}$$

if $v_k = 1$ and 0 otherwise.

It is straightforward to check that the Khovanov differential commutes with the action of $\mathcal{A}_n$, and it is shown in [7] (see also [12, 13]) that the homotopy equivalences associated to Reidemeister moves respect the $\mathcal{A}_n$-module structure, and moving a basepoint past a crossing yields a homotopic map. The homology, $\mathrm{Kh}(\widehat{\sigma})$, therefore inherits the structure of an $\mathcal{A}_n$-module, and this $\mathcal{A}_n$-module structure is an invariant of the link.

With these preliminaries in place, assume that $\widehat{\sigma} = U_n$. A quick calculation using the standard diagram of U_n tells us that $\mathrm{Kh}(U_n) \cong \mathcal{A}_n$ as an $\mathcal{A}_n$-module. Let $\theta \in \mathrm{CKh}(D(\widehat{\sigma}))$ be a cycle representing the homology class $1 \in \mathrm{Kh}(U_n) \cong \mathcal{A}_n$.

We now claim that when θ is expressed as a linear combination of the standard Khovanov generators, the coefficient of Ψ^+ must be 1. To see this, note that $x_1 \cdots x_n(\theta)$ represents the non-zero homology class $x_1 \cdots x_n \in \mathrm{Kh}(\widehat{\sigma})$, but if v is any basis element not equal to Ψ^+, then $x_1 \cdots x_n(v) = 0$. We see this immediately for $v \neq \Psi^+ \in \mathcal{I}_{braid}$, and any complete resolution $\mathcal{I} \neq \mathcal{I}_{braid}$ contains at least one connected component intersecting the closure arc more than once, hence containing at least two basepoints p_i, p_j, $i \neq j$. We conclude that any basis element v associated to $\mathcal{I} \neq \mathcal{I}_{braid}$ satisfies $x_i x_j(v) = 0$, hence also satisfies $x_1 \cdots x_n(v) = 0$.

The arguments in the previous paragraph imply that $x_1 \cdots x_n(\theta) = x_1 \cdots x_n(\Psi^+) = \Psi^-$, so $[\Psi^-] = x_1 \cdots x_n \in Kh(\widehat{\sigma})$. In particular, $[\Psi^-] \neq 0$.

But [1, Prop. 3.1] then implies that σ is *right-veering*.

Repeat the argument above on $m(\sigma)$, the mirror of σ, to conclude that σ is also *left-veering*. Since the only braid which is both left- and right-veering is the identity braid (cf. [1, Lem. 3.1]), $\sigma = \mathbb{1}_n$, as desired.

2 Fibred Links in $\#^n(S^1 \times S^2)$

Recall that $\mathcal{L}_n := \{\ell \in \mathbb{Z}^+ \mid \ell \leq (n+1) \text{ and } \ell \equiv (n+1) \bmod 2\}$.

Lemma 1 *If an* ℓ*-component link* L *has Euler characteristic* $1-n$*, then* $\ell \in \mathcal{L}_n$.

Proof Let S denote the fiber surface of L, $\chi(S)$ its Euler characteristic, and $g(S)$ its genus. Then $\chi(S) = 1 - n = (2 - 2g(S)) - \ell$. Since $g(S) \in \mathbb{Z}^{\geq 0}$, we obtain $\ell \equiv (n+1) \bmod 2$ and $\ell \leq n+1$.

Lemma 2 *If* $L \subset Y_n$ *is a fibered link, then* $\chi(L) \leq 1 - n$.

Proof Suppose L has ℓ components, and let S denote the fiber surface of L, and h its monodromy. $H_1(S)$ is free of rank $1 - \chi(S) = 2g(S) + (\ell - 1)$. Viewing $Y_n - L$ as the mapping torus of h (cf. Sect. 2.1), we obtain a corresponding presentation of $H_1(Y_n) \cong \mathbb{Z}^n$ with $1 - \chi(L)$ generators, hence $1 - \chi(L) \geq n$.

2.1 *Heegaard–Floer Homology Proof of Proposition 2*

We begin with some background on Heegaard–Floer homology.

2.1.1 Heegaard–Floer Module

Recall that in [15], Ozsváth-Szabó associate to a closed, oriented 3-manifold Y a graded vector space (for simplicity we work over $\mathbb{F} = \mathbb{Z}/2\mathbb{Z}$), $\widehat{HF}(Y)$, which splits over $\text{Spin}^c(Y)$, the set of spin^c structures on Y:

$$\widehat{HF}(Y) = \bigoplus_{\mathfrak{s} \in \text{Spin}^c(Y)} \widehat{HF}(Y, \mathfrak{s})$$

For appropriate choices of symplectic and almost complex structures, $\widehat{HF}(Y)$ is the Lagrangian Floer homology of a natural pair of Lagrangian tori, $\mathbb{T}_\alpha$ and $\mathbb{T}_\beta$, in the g-fold symmetric product of a pointed Heegaard surface, (Σ, w), for Y.

$\widehat{HF}(Y)$ can be given the structure of a module over $\Lambda^*(H_1(Y; \mathbb{F}))$, as described in [15, Sect. 4.2.5]. Explicitly, let

$$(\Sigma, \alpha = \{\alpha_1, \ldots, \alpha_g\}, \beta = \{\beta_1, \ldots, \beta_g\}, z)$$

be a pointed, genus g Heegaard splitting of Y, and consider $\zeta \in H_1(Y; \mathbb{F})$. Ozsváth-Szabó define an associated chain map,

$$A_\zeta : \widehat{CF}(\Sigma, \alpha, \beta, z) \to \widehat{CF}(\Sigma, \alpha, \beta, z),$$

on the Heegaard–Floer chain complex as follows ([15, Rmk. 4.20]). Let $\mathbf{x}, \mathbf{y} \in \mathbb{T}_\alpha \cap \mathbb{T}_\beta$ be generators of the chain complex. Recall that $\pi_2(\mathbf{x}, \mathbf{y})$ denotes the set of domains in Σ representing topological Whitney disks connecting $\mathbf{x}$ to $\mathbf{y}$, in the sense of [15, Sect. 2.4]. If $\phi \in \pi_2(\mathbf{x}, \mathbf{y})$, we follow the notation in [14, Sect. 2.1], letting $\partial_\alpha \phi := (\partial \phi) \cap \mathbb{T}_\alpha$, regarded as a 1-chain with boundary $\mathbf{y} - \mathbf{x}$.

Choose an immersed curve,

$$\gamma_\zeta \subset \Sigma - \{\alpha_i \cap \beta_j\}_{i,j \in \{1,\dots,g\}},$$

representing $\zeta \in H_1(Y; \mathbb{F})$ and define

$$a(\gamma_\zeta, \phi) := \#\widehat{\mathcal{M}}(\phi)(\gamma_\zeta \cdot \partial_\alpha \phi),$$

where $\gamma_\zeta \cdot \partial_\alpha \phi$ is the algebraic intersection number of γ_ζ and $\partial_\alpha \phi$. Then the chain map associated to ζ is given by:

$$A_\zeta(\mathbf{x}) = \sum_{\mathbf{y} \in \mathbb{T}_\alpha \cap \mathbb{T}_\beta} \sum_{\left\{\phi \in \pi_2(\mathbf{x},\mathbf{y}) \,\middle|\, \mu(\phi)=1, n_w(\phi)=0\right\}} a(\gamma_\zeta, \phi) \cdot \mathbf{y}.$$

The map A_ζ is well-defined (independent of the choice of γ) up to chain homotopy (cf. [14, Lem. 2.4]).

2.1.2 Heegaard–Floer Contact Invariant

We now recall the definition of the Heegaard–Floer contact invariant [16], following the alternative construction given in [10]. Let ξ be a contact structure on a closed, connected, oriented 3-manifold Y. Then Giroux tells us [6] that there exists some fibered link L whose corresponding open book supports ξ. One can then build a Heegaard diagram for $-Y$ (Y with the opposite orientation) using

- a choice of *basis*, $\{a_1, \dots, a_n\}$, for a page S (of Euler characteristic $1 - n$) of the open book [10, Sect. 3.1], and
- the data of the monodromy, h, of the open book.

Honda-Kazez-Matić then identify a distinguished cycle in the corresponding chain complex, $\widehat{CF}(-Y)$, and prove both that the class it represents in $\widehat{HF}(-Y)$ is invariant of the choices used in its construction and that it agrees with the contact invariant defined in [16].

We will need the following, which appears in [8] and follows immediately from [16, Thm. 1.4] and [9, Thm. 1.1]:

Lemma 3 *If $L \subset Y$ is a fibered link whose monodromy, h, is not right-veering, then the Heegaard–Floer contact invariant associated to the contact structure supported by L vanishes.*

We now proceed to the proof.

Proof (Proposition 2) Let $L_\ell \subset Y_n$ be an ℓ-component fibered link of Euler characteristic $1-n$. Construct a corresponding Heegaard diagram for $-Y_n$ as in [10, Sect. 3].

The module structure on $\widehat{HF}(-Y_n)$ has been computed in [15, Lemma. 9.1]. Explicitly, $\widehat{HF}(-Y_n) \cong \mathcal{A}_n$ as a module over

$$\Lambda^*(H_1(-Y_n;\mathbb{F})) \cong \mathcal{A}_n := \mathbb{F}[\zeta_1, \dots, \zeta_n]/(\zeta_1^2, \dots, \zeta_n^2).$$

In particular, $\zeta_1 \cdots \zeta_n \neq 0 \in \widehat{HF}(-Y_n)$.

We can understand the module action explicitly in our setting as follows. All of our notation matches [10]. Examine the Honda-Kazez-Matić pointed Heegaard diagram

$$(\Sigma = S_{1/2} \cup -S_0, \{\beta_1, \dots, \beta_n\}, \{\alpha_1, \dots, \alpha_n\}, z)$$

associated to the fibered link, L_ℓ. In particular, choose a small perturbation, $b_i \subset S$, of each arc $a_i \subset S$, as described in [10, Sect. 3.1]. Subject to the identifications $S \cong S_{1/2} \cong S_0$, form:

$$\alpha_i := (a_i \subset S_{1/2}) \cup (a_i \subset -S_0)$$
$$\beta_i := (b_i \subset S_{1/2}) \cup (h(b_i) \subset -S_0).$$

By construction, $|S_{1/2} \cap (\alpha_i \cap \beta_j)| = \delta_{ij}$. Let x_i denote the unique intersection point in $(S_{1/2} \cap \alpha_i \cap \beta_i)$, and let $\mathbf{x} = (x_1, \dots, x_n) \in \mathbb{T}_\alpha \cap \mathbb{T}_\beta \subset Sym^n(\Sigma)$. Honda-Kazez-Matić prove that $\mathbf{x}$ is a cycle in the Heegaard–Floer chain complex and that it represents the Heegaard–Floer contact class $c(\xi_{(S,h)}) \in \widehat{HF}(-Y_n)$ associated to the contact structure $\xi_{(S,h)}$ compatible with the open book (S, h).

Now choose a *dual basis*, $\{\gamma_1, \dots, \gamma_n\}$, of simple closed curves on $S_{1/2}$ satisfying $|a_i \cap \gamma_j| = \delta_{ij}$. The set of homology classes, $\{[\gamma_1], \dots, [\gamma_n]\}$, obtained by viewing the γ_i as 1-cycles in $-Y_n$, forms a basis for $H_1(-Y_n;\mathbb{F})$. Hence, for each $i \in \{1, \dots, n\}$, the corresponding map on homology induced by the chain map $A_{[\gamma_i]}$ can be identified with $\zeta_i \in \mathcal{A}_n$.

Let $\theta \in \widehat{CF}(-Y_n)$ be any cycle representing $1 \in \widehat{HF}(-Y_n)$. Since $\zeta_1 \cdots \zeta_n \neq 0 \in \widehat{HF}(-Y_n)$, we know that there exists at least one generator $\mathbf{y} \in \mathbb{T}_\alpha \cap \mathbb{T}_\beta$ satisfying

$$\langle A_{[\gamma_1]} \cdots A_{[\gamma_n]} \cdot \theta, \mathbf{y}\rangle \equiv 1 \bmod 2.$$

Associated to such a generator $\mathbf{y}$ is an odd number of corresponding Maslov index n domains in $\pi_2(\theta, \mathbf{y})$, each of which can be realized as the sum of n of the Maslov index 1 domains contributing to the chain maps $A_{[\gamma_1]}, \dots, A_{[\gamma_n]}$. Consider the local multiplicity of such a Maslov index n domain, ψ, in the *four* regions adjacent to one of the constituent intersection points, x_i, of the distinguished cycle $\mathbf{x} = (x_1, \dots, x_n)$ representing the contact class. We know (see Fig. 2) that the local multiplicity of ψ in the two regions adjacent to x_i that contain the basepoint, z_0, must be 0 and also that

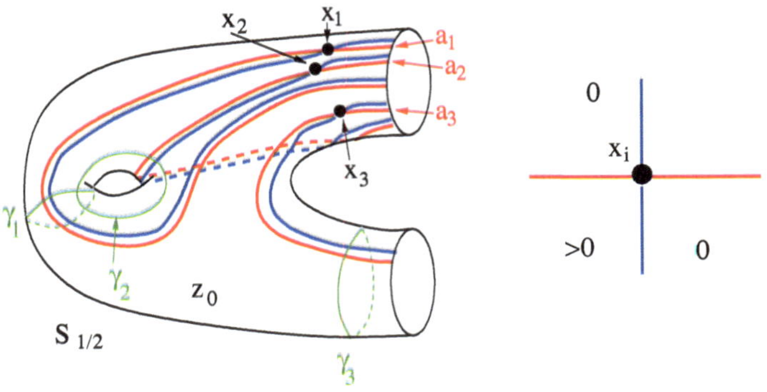

Fig. 2 The contact class on (half of) the Honda-Kazez-Matić Heegaard diagram associated to a fibered link $L_2 \subset Y_3$. The *right-hand picture* is a close-up of one of the constituent *intersection points* of the contact class and restrictions on the local multiplicities of the Maslov index n domain ψ. The NW, SE domains must have multiplicity 0 since they contain the basepoint z_0. One of the other two domains must have positive multiplicity, since it is the unique domain intersecting γ_i

the local multiplicity in the region adjacent to the unique intersection point between γ_i and a_i must be nonzero (hence positive, since ψ is a sum of domains representing holomorphic disks). Since the fourth region must have non-negative multiplicity, we conclude that x_i must be a corner, of multiplicity at least one, in the boundary of ψ, implying that x_i must be a constituent intersection point of the generator $\mathbf{y}$.

Since the above argument holds for each of the x_i, we conclude that, in fact, $\mathbf{y}$ is actually the distinguished contact class, $\mathbf{x}$, and it follows that (working mod 2) $A_{[\gamma_1]} \cdots A_{[\gamma_n]} \cdot \theta = \mathbf{x}$. Therefore,

$$[A_{[\gamma_1]} \cdots A_{[\gamma_n]} \cdot \theta] = [\mathbf{x}] = \zeta_1 \cdots \zeta_n \neq 0 \in \widehat{HF}(-Y_n),$$

so the Heegaard–Floer contact invariant associated to the contact structure supported by L_ℓ is nonzero. By Lemma 3, the monodromy, h, of L_ℓ is right-veering.

Now consider the *mirror* of L, i.e., the fibered link $L \subset -Y_n$ with monodromy h^{-1}. By running the same argument above, we conclude that the contact invariant associated to the contact structure supported by the mirror of L is also nonzero. Hence, h^{-1} is right-veering, implying that h is left-veering.

But if h is both right- and left-veering, then h is isotopic to the identity mapping class, and hence (Y_n, L_ℓ) is diffeomorphic as a pair to $(Y_n, \mathbf{L}_\ell)$.

Acknowledgments We thank Ken Baker, John Baldwin, Rob Kirby, Tony Licata, and Danny Ruberman for interesting conversations and Joan Birman and Bill Menasco for a useful e-mail correspondence. We are especially grateful to Ian Biringer for telling us about Hopfian groups, to Matt Hedden for pointing out that Proposition 2 appears in [14], and to Tim Cochran for making us aware that historical references to Proposition 1 in the literature appear under the slogan, "Milnor's invariants detect the trivial braid." We would also like to thank the referee for making a number of useful suggestions that greatly improved the exposition.

References

1. J.A. Baldwin, J.E. Grigsby, Categorified invariants and the braid group. Proc. Am. Math. Soc. **143**(7), 2801–2814 (2015)
2. J.S. Birman, H.M. Hilden, On isotopies of homeomorphisms of Riemann surfaces. Annl. Math. **2**(97), 424–439 (1973)
3. J.S. Birman, W.W. Menasco, Studying links via closed braids. V. The unlink. Trans. Am. Math. Soc. **329**(2), 585–606 (1992)
4. T.D. Cochran, Non-trivial links and plats with trivial Gassner matrices. Math. Proc. Cambr. Philos. Soc. **119**(1), 43–53 (1996)
5. P. Ghiggini, P. Lisca, Open book decompositions versus prime factorizations of closed, oriented 3–manifolds. math.GT/1407.2148, (2014)
6. E. Giroux. Géométrie de contact: de la dimension trois vers les dimensions supérieures. In *Proceedings of the International Congress of Mathematicians*, vol. 2 (Higher Education Press, Beijing, 2002), pp. 405–414
7. M. Hedden, Y. Ni, Khovanov module and the detection of unlinks. math.GT/1204.0960, (2012)
8. M. Hedden, L. Watson, On the geography and botany of knot Floer homology. math.GT/1404.6913, (2014)
9. K. Honda, W.H. Kazez, G. Matić, Right-veering diffeomorphisms of compact surfaces with boundary. Inven. Math. **169**(2), 427–449 (2007)
10. K. Honda, W.H. Kazez, G. Matić, On the contact class in Heegaard Floer homology. J. Differ. Geom. **83**(2), 289–311 (2009)
11. J. Johnson, Heegaard splittings and open books. math.GT/1110.2142, (2011)
12. M. Khovanov, A categorification of the Jones polynomial. Duke Math. J. **101**(3), 359–426 (2000)
13. M. Khovanov, Patterns in knot cohomology. I. Exp. Math. **12**(3), 365–374 (2003)
14. Y. Ni, Homological actions on sutured Floer homology. math.GT/1010.2808, (2010)
15. P. Ozsváth, Z. Szabó, Holomorphic disks and topological invariants for closed three-manifolds. Annl. Math. **159**(3), 1027–1158 (2004)
16. P. Ozsváth, Z. Szabó, Heegaard Floer homology and contact structures. Duke Math. J. **129**(1), 39–61 (2005)
17. O. Plamenevskaya, Transverse knots and Khovanov homology. Math. Res. Lett. **13**(4), 571–586 (2006)

Symmetric Unions Without Cosmetic Crossing Changes

Allison H. Moore

Abstract A symmetric union of two knots is a classical construction in knot theory which generalizes connected sum, introduced by Kinoshita and Terasaka in the 1950s. We study this construction for the purpose of finding an infinite family of hyperbolic non-fibered three-bridge knots of constant determinant which satisfy the well-known cosmetic crossing conjecture. This conjecture asserts that the only crossing changes which preserve the isotopy type of a knot are nugatory.

Keywords Cosmetic crossings · Nugatory crossing conjecture · Symmetric union · Khovanov homology · Branched double cover

Mathematics Subject Classification 57M25 · 57M27

1 Introduction

In the 1950s, Kinoshita and Terasaka defined the *union* of two knots as a generalization of a connected sum [12]. An aesthetically appealing variation of this construction is a *symmetric union*, in which the connected sum of a knot and its mirror image is modified by a certain tangle replacement, and the resulting diagram admits an axis of mirror symmetry. In this note, we use symmetric unions to construct a new family of knots satisfying a well-known conjecture.

Theorem 1 *There exists an infinite family of hyperbolic non-fibered three-bridge knots of fixed determinant which satisfy the cosmetic crossing conjecture.*

An embedded disk D in S^3 intersecting K twice with zero algebraic intersection number is called a *crossing disk*. If ∂D bounds an embedded disk in the complement

A.H. Moore
Department of Mathematics, Rice University, MS-136 Box 1892,
Houston, TX 77251-1892, USA
e-mail: allison.h.moore@rice.edu

G. Letzter et al. (eds.), *Advances in the Mathematical Sciences*, Association for Women in Mathematics Series 6, DOI 10.1007/978-3-319-34139-2_3

of K, then the corresponding crossing c is called *nugatory* and a crossing change at c preserves the isotopy type of K. *Cosmetic* crossing changes are non-nugatory crossing changes which preserve the oriented isotopy type of the knot. The cosmetic crossing conjecture asserts that no such crossings exist.

Conjecture 1 (X.S. Lin) *If K admits a crossing change at crossing c which preserves the oriented isotopy class of the knot, then c is nugatory.*

The cosmetic crossing conjecture also appears in the literature as the "nugatory" crossing conjecture; see Problem 1.58 in Kirby's List [13]. To prove Theorem 1 we will apply an obstruction of the author and Lidman.

Theorem 2 ([16]) *Let K be a knot in S^3 whose branched double cover $\Sigma(K)$ is an L-space. If each summand of the first singular homology of $\Sigma(K)$ has square-free order, then K admits no cosmetic crossing changes.*

Recall that *L-spaces* are the rational homology spheres with the simplest possible Heegaard Floer homology, meaning that rank $\widehat{HF}(Y) = |H_1(Y;\mathbb{Z})|$. By work of Ozsváth and Szabó [22], knots that are reduced Khovanov homology thin have branched double covers that L-spaces. Thus Khovanov homology will be one of the tools we use to prove that the knots of Theorem 1 satisfy the conditions of Theorem 2.

Prior to Theorem 2, the main classes of knots known to satisfy Conjecture 1 were fibered knots, two-bridge knots, and Whitehead doubles of prime, non-cabled knots [2, 9, 30], and it was shown in [1] that any genus one knot which might admit a cosmetic crossing change must be algebraically slice. The infinite family of knots we construct here are shown in Sect. 3 to be non-alternating, non-fibered, hyperbolic, of genus two, and bridge number three. In a different direction, Theorem 2 was applied in [16] to settle the status of Conjecture 1 for all knots with up to nine crossings, families of pretzel knots of arbitrarily high genus, and certain knots arising as the branched sets of surgeries on strongly invertible L-space knots. In particular, the examples constructed in [16] were of nonconstant determinant. The present knots have fixed determinant and branched double covers with noncyclic first homology. These properties differentiate them from all other knots known to satisfy Conjecture 1, adding further variety to the landscape of knots for which this fundamental conjecture has been settled.

2 Symmetric Unions

Let K denote an oriented knot in S^3. The mirror of K is denoted $m(K)$. We will abuse notation and let K refer to both the knot and its planar diagram. We will use J to denote an oriented knot as well. Elementary rational tangles will be denoted by T_n for $n \in \{\mathbb{Z}, \infty\}$, as indicated in Fig. 1.

Definition 1 A *symmetric union of J* is an (unoriented) knot diagram obtained by replacing an elementary 0-tangle T_0 with an elementary n-tangle T_n, with $n \neq 0, \infty$,

Fig. 1: Examples of elementary rational tangles

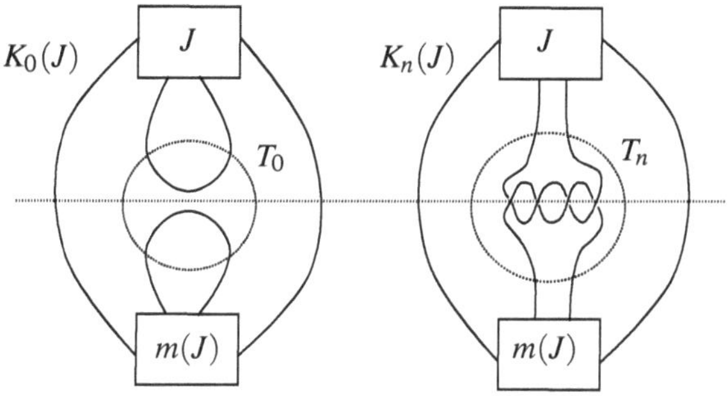

Fig. 2: A symmetric tangle replacement of a 0-tangle T_0 with an elementary n-tangle T_n. (Here, $n = 4$.) The diagrams of J and $m(J)$ in this schematic are assumed to be mirror symmetric with respect to the *horizontal axis*

along an axis of mirror symmetry in a diagram of $J\#m(J)$ as in Fig. 2. A knot which admits a symmetric union diagram is called a *symmetric union*, and we denote a symmetric union of J by $K_n(J)$. The (unoriented) knot J is called the *partial knot* of $K_n(J)$, and $K_0(J)$ is $J\#m(J)$.

The definition is due to Kinoshita and Teraksa [12]. Note that when J is oriented and n is even, $K_n(J)$ inherits an orientation from the connected sum of J with its reverse mirror image, but when n is odd, the orientation of $K_n(J)$ is not well defined. To construct an oriented symmetric union, we will adopt the convention that the northeast strand of $T_n \subset K_n(J)$ in Fig. 2 is oriented so that it agrees with the orientation of the northeast strand in $T_0 \subset K_0(J)$.[1] With $K_n(J)$ oriented, the crossings in the tangle T_n are positive whenever $n > 0$.

Elsewhere in the literature, a symmetric union may refer the generalization of this construction in which multiple symmetric tangle replacements are made, but we will call these *generalized symmetric unions*. The reader is warned that the symmetric union construction is not unique; the isotopy type of $K_n(J)$ depends on both the diagram of $J\#m(J)$ and the location of the tangle replacement. For example, two distinct symmetric unions of the unknot are pictured in Fig. 3. Despite this dependence on the diagram, a classical fact about symmetric unions is that when n is even, the Alexander polynomial of $K_n(J)$ depends on neither n nor the choice of diagram.

Theorem 3 ([12]) *If $K_n(J)$ is any symmetric union of the knot J and n is even, then*

$$\Delta_{K_n(J)}(t) = (\Delta_J(t))^2.$$

[1]This orientation convention is somewhat artificial; however, our choice of orientation ultimately will not matter because the knot invariants which we study in Sects. 2 and 3 are not sensitive to orientation reversal.

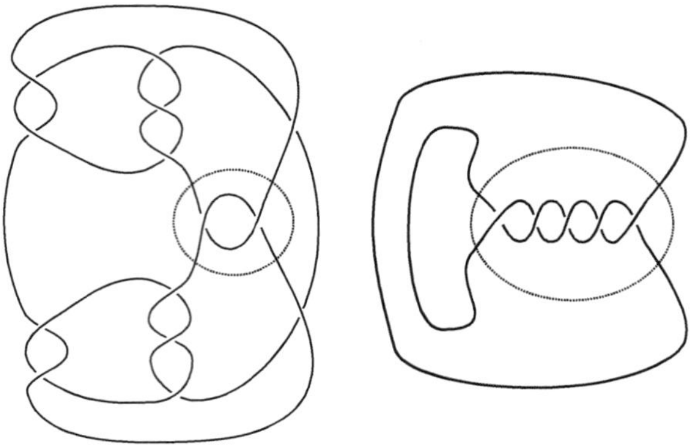

Fig. 3: The knot on the *left* is the Kinoshita–Terasaka knot $11n42$, and the knot on the *right* is an unknot. Both have partial knot the unknot

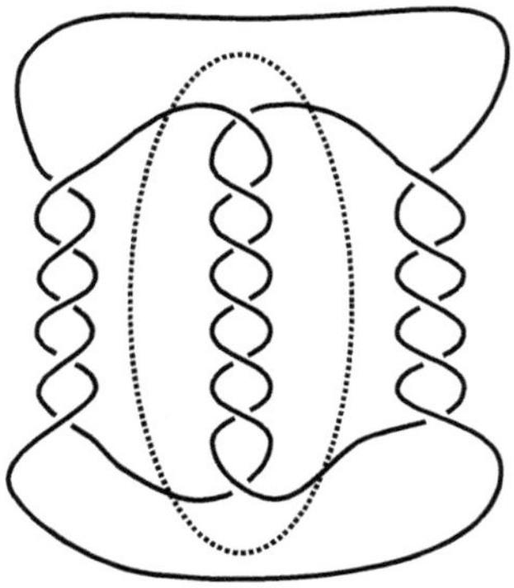

Fig. 4: Pretzel knots of the form $(p, q, -p)$, for $p, q \in \mathbb{Z}$ with p odd, have symmetric fusion number one. The axis of mirror symmetry is *vertical* in this example

Moreover, $\det(K_n(J)) = \det(J)^2$ for any n (cf. [14, Theorem 2.6]).

A symmetric union is always ribbon, which is evidenced by the existence of a symmetric ribbon disk in its symmetric diagram, similar to the one that occurs in any symmetric diagram of $J\#m(J)$. The Ozsváth and Szabó τ-invariant [20, Corollary 1.3] gives a lower bound on the smooth four-ball genus, $|\tau(K)| \leq g_4(K)$, as does Rasmussen's s-invariant [26, Theorem 1]. Hence these invariants will vanish for any symmetric union, a feature we will utilize in Sect. 3.1.

Definition 2 If replacing the elementary n-tangle T_n of a symmetric union diagram $K_n(J)$ with the ∞-tangle T_∞ results in a two-component unlink, we say that the diagram $K_n(J)$ has *symmetric fusion number one*. A knot which admits a symmetric union diagram of symmetric fusion number one is also said to have symmetric fusion number one.

For example, the pretzel knots of the form $(p, q, -p)$, for $p, q \in \mathbb{Z}$ with p odd, have symmetric fusion number one. See Fig. 4. Note that if $K_n(J)$ has symmetric fusion number one, then $K_m(J)$ has symmetric fusion number one for any $m \neq 0, \infty$. A knot of symmetric fusion number one is necessarily the band sum of a two-component unlink.

Generalized symmetric unions of the figure eight knot were used by Kanenobu to construct infinite families of knots with different Alexander modules and the same Jones polynomials [10]. More recently, Kanenobu's knots have become popular in the study of knot polynomials and knot homology theories (for instance [6, 17, 31]). The proof of Theorem 1 will make use of some of the same techniques as Kanenobu [10] and Greene and Watson [6].

2.1 Knot Floer Homology

Let $\widehat{HFK}_m(K, s)$ refer to the knot Floer homology of $K \subset S^3$ with $\mathbb{Z}/2\mathbb{Z}$ coefficients, due to Ozsváth and Szabó [21] and Rasmussen [24]. This knot invariant is a bigraded vector space with Maslov grading m and Alexander grading s. Because knot Floer homology categorifies the symmetrized Alexander polynomial, one may wonder if a statement generalizing Theorem 3 holds for the knot Floer groups, and in particular whether a Künneth formula like the one satisfied by connected sums,

$$\widehat{HFK}(K_1 \# K_2) \cong \widehat{HFK}(K_1) \otimes \widehat{HFK}(K_2),$$

holds. Unfortunately no such property can hold for symmetric unions in general. Knot Floer homology detects the unknot [21], therefore any nontrivial symmetric union of an unknot (e.g., the one in Fig. 3) will have $\widehat{HFK}(K)$ nontrivial, contradicting any general analogy. However, when $K_n(J)$ has symmetric fusion number one, Kinoshita and Terasaka's characterization of the Alexander polynomial of a symmetric union does indeed generalize.

Theorem 4 *Let $K_n(J)$ be a symmetric union of a knot J such that $K_n(J)$ has symmetric fusion number one. When n is even, there is a graded isomorphism*

$$\widehat{HFK}(K_n(J)) \cong \widehat{HFK}(J) \otimes \widehat{HFK}(m(J)),$$

and when n is odd, we have that

$$\widehat{HFK}(K_n(J)) \cong \widehat{HFK}(K_1(J)).$$

This follows as a special case of [7, Theorem 1] (alternatively [18, Theorem 3.3]) whose proof we will not repeat here. The key observation is that after perhaps mirroring, the knots $K_n(J)$ and $K_{n-2}(J)$ form an oriented skein triple with the two-component unlink and their knot Floer groups fit into a long exact sequence [23, Theorem 1.1]. Using that symmetric unions are ribbon, hence slice, the concordance invariant $\tau(K_n(J))$ vanishes for all n. This fact, taken together with the skein triple and the observation that $K_0(J)$ is $J \# m(J)$, gives the statement of the theorem.

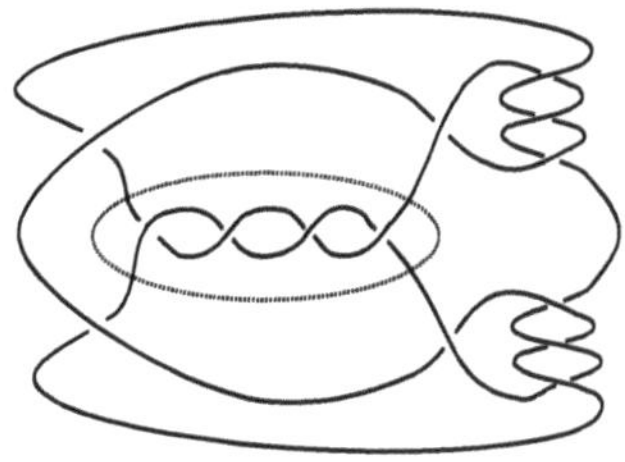

Fig. 5: The symmetric unions $K_n(5_2)$ of the knot 5_2. For the knot pictured here, $n = 4$

Because knot Floer homology detects genus [21] and fiberedness [19], and satisfies a Künneth formula under connected sum, the following corollaries are immediate.

Corollary 1 *Let $K_n(J)$ be a symmetric union of a knot J such that $K_n(J)$ has symmetric fusion number one. If n is even, then $K_n(J)$ is fibered if and only if J is fibered, and $g(K_n(J)) = 2g(J)$. If n is odd, then $g(K_n(J)) = g(K_1(J))$.*

Corollary 2 *Let $K_n(J)$ be a symmetric union of symmetric fusion number one with n is even. Then $K_n(J)$ is nontrivial if and only if the partial knot J is nontrivial.*

2.2 *Main Examples*

We now define the main examples of interest in this note. Denote by $\mathcal{K}$ the subset of symmetric unions

$$\{K_n \mid n \equiv 0 \pmod{14},\ n \neq 0\} \subset \{K_n := K_n(5_2) \mid n \in \mathbb{Z}\} \tag{1}$$

where the symmetric unions $K_n := K_n(5_2)$ are constructed from the knot 5_2 as shown in Fig. 5. The knot $K_n(5_2)$ has symmetric fusion number one for all $n \neq 0, \infty$. For the remainder of this note we will assume that K_n denotes the specific symmetric union $K_n(5_2)$ for $n \in \mathbb{Z}$.

3 Proof of Main Theorem

Before addressing the main theorem, we need to prove a lemma about the Khovanov homology of K_n. This will allow us to deduce that the branched double cover of K_n is an L-space for all n.

3.1 Khovanov Homology

Let $Kh^{q,u}(L)$ refer to the Khovanov homology of a link $L \in S^3$ with quantum grading q and (co)homological grading u, and coefficients in $\mathbb{Q}$. The Khovanov homology groups are a link invariant which categorify a normalized Jones polynomial [11]. Our grading and notational conventions follow Rasmussen [25]. For example, the Khovanov homology of the knot 5_2 with all positive crossings is described by the Poincaré polynomial

$$P_{Kh(5_2)}(q,u) = q + q^3 + q^3u + q^5u^2 + q^7u^2 + q^9u^3 + q^9u^4 + q^{13}u^5. \quad (2)$$

The Khovanov thin knots are those with homology supported in two diagonals $\delta = q - 2u$ of the gradings.[2] Khovanov homology satisfies an unoriented skein exact sequence (cf [25, Lemma 4.2]). With our conventions this is

$$\xrightarrow{\cdot u} q^{2+3\varepsilon}u^{1+\varepsilon}Kh(\asymp) \longrightarrow Kh(\times) \longrightarrow qKh()() \xrightarrow{\cdot u} \quad (3)$$

where ε is the difference between the number of negative crossings in the unoriented resolution $(\asymp)$ and the original diagram $(\times)$. As in [25], the notation $qKh()()$ means the complex $Kh()()$ is shifted in such a way as to multiply its Poincaré polynomial by q. The arrow marked with $\cdot u$ is the boundary map and it raises the homological grading by 1. Though computations similar to Lemma 1 can be found in [18, 28], for concreteness we provide a proof.

Lemma 1 *The knot K_n is Khovanov homology thin with $\mathbb{Q}$–coefficients for all n. Moreover, $Kh(K_n)$ for $n \geq 0$ is given by the closed formula*

$$\begin{aligned} Kh(K_n;\mathbb{Q}) = \mathbf{1}^0_{-1} + \mathbf{1}^0_1 + \mathbf{1}^{n-5}_{2(n-5)-1} &\cdot (\mathbf{1}^0_0 + \mathbf{1}^1_2 + \mathbf{3}^2_4 + \mathbf{3}^3_6 + \mathbf{4}^4_8 + \mathbf{4}^5_{10} + \mathbf{3}^6_{12} + \mathbf{3}^7_{14} + \mathbf{1}^8_{16} + \mathbf{1}^9_{18}) \\ + \quad \mathbf{1}^{n-4}_{2(n-4)+1} &\cdot (\mathbf{1}^0_0 + \mathbf{1}^1_2 + \mathbf{3}^2_4 + \mathbf{3}^3_6 + \mathbf{4}^4_8 + \mathbf{4}^5_{10} + \mathbf{3}^6_{12} + \mathbf{3}^7_{14} + \mathbf{1}^8_{16} + \mathbf{1}^9_{18}) \end{aligned} \quad (4)$$

where for brevity, $\mathbf{d}^u_q$ denotes $\mathbb{Q}^{\mathbf{d}}$ in bigrading (q,u).

Proof Without loss of generality, we assume $n \geq 0$; a similar proof holds in the case that $n \leq 0$ with minor changes in bigradings. Alternatively, the result for $n \leq 0$ follows from the isotopy $K_{-n}(J) \simeq m(K_n(J))$ obtained by rotating about the axis of symmetry and the identity $Kh^{q,u}(m(K)) \cong Kh^{-q,-u}(K)$ for all q, u and any knot K.

We proceed by induction on n. The cases $Kh(K_n)$ for $0 \leq n \leq 7$ have been verified computationally using the *KnotTheory`* package for Mathematica [29]. Assume that $n > 7$. For the inductive hypothesis, $Kh(K_{n-1})$ is thin and described by (4).

[2]When working with $\mathbb{Z}$–coefficients, thin knots must also have homology that is free over $\mathbb{Z}$.

Any crossing in the tangle T_n gives rise to an unoriented triple $\{(\times), (\asymp), ()(\}$, where $K_n = (\times)$ and $K_{n-1} = (\asymp)$. Since K_n has symmetric fusion number one for all $n \neq 0, \infty$, the resolution $)($ corresponds with the two-component unlink. Because $n \geq 0$, the crossings in T_n are positive, and so the number of negative crossings in the diagram K_n for any n is equal to the total number of crossings in the diagram of the partial knot. Therefore the difference ε in negative crossings between the resolutions $(\asymp)$ and $(\times)$ is zero and the skein triple becomes

$$\xrightarrow{\cdot u} q^2 u Kh(\asymp) \longrightarrow Kh(\times) \longrightarrow qKh()() \xrightarrow{\cdot u} .$$

The two-component unlink has Khovanov homology $\mathbb{Q}_{(-2)} \oplus \mathbb{Q}^2_{(0)} \oplus \mathbb{Q}_{(2)}$ supported in homological grading zero. Using this and the inductive hypothesis, whenever $u \neq 0, 1$ or $q \neq 1, 3$ the sequence splits as

$$0 \longrightarrow q^2 u Kh(\asymp) \longrightarrow Kh(\times) \longrightarrow 0, \tag{5}$$

implying the isomorphism $Kh^{q,u}(\times) \cong Kh^{q-2,u-1}(\asymp)$ for all $u \neq 0, 1$ or $q \neq 1, 3$. For $(q, u) = (1, 0)$ and $(1, 1)$, the sequence splits as

$$0 \longrightarrow Kh(\times) \longrightarrow q\mathbb{Q}^2 \xrightarrow{\cdot u} q^2 u \mathbb{Q} \longrightarrow Kh(\times) \to 0, \tag{6}$$

and for $(q, u) = (3, 0)$ and $(3, 1)$, the sequence splits as

$$0 \longrightarrow Kh(\times) \longrightarrow q\mathbb{Q} \xrightarrow{\cdot u} q^2 u \mathbb{Q} \longrightarrow Kh(\times) \to 0. \tag{7}$$

Exactness yields two solutions for each of (6) and (7),

$$\begin{aligned} Kh^{1,0}(\times) \oplus Kh^{1,1} &= \mathbb{Q} \oplus 0 \text{ or } \mathbb{Q}^2 \oplus \mathbb{Q} \\ Kh^{3,0}(\times) \oplus Kh^{3,1} &= 0 \oplus 0 \text{ or } \mathbb{Q} \oplus \mathbb{Q}. \end{aligned} \tag{8}$$

We aim to show that the first choice in each line of (8) is the correct one, so let us assume for the contrary that the second outcome of (6) holds.

Because symmetric unions are ribbon, and therefore slice, the concordance invariant $s(K_n(J))$ vanishes for all n. In particular, the Lee spectral sequence [15] must converges to two copies of $\mathbb{Q}$ in quantum gradings that average to zero, hence these surviving elements live in $q = \pm 1$. Suppose the two survivors are in gradings $(-1, 0)$ and $(1, 1)$. See Table 1. With our current conventions, the induced differential on the rth page of the Lee spectral sequence increases the homological grading by 1 and the quantum grading by $2r$. By assumption, there is a $\mathbb{Q}^2$ summand of the E^1 page in bigrading $(1, 0)$, and it must cancel via d^r, for some $r \geq 1$, with a term of rank two. However, by (5) and (7) there is at most one copy of $\mathbb{Q}$ in the bigradings $(q, 2)$ for $q \geq 3$ and $Kh(\times)$ vanishes in $u = -1$, so no such term exists. Hence it must be the case that the surviving generators live in bigradings $(\pm 1, 0)$.

Again by assumption to the contrary, there is a copy of $\mathbb{Q}$ in (1, 1) which must now die in the spectral sequence. Since it cannot cancel with the surviving generator in bigrading $(-1, 0)$, it must cancel with a generator in $(q, 2)$ for $q \geq 3$. Yet (5) implies that $Kh(\times)$ vanishes in $u = 2$, so no such generator exists. It must be the case that $Kh^{1,0}(\times) \cong \mathbb{Q}$ and $Kh^{1,1} = 0$.

Let us now assume that the second outcome of (7) holds. The two $\mathbb{Q}$ summands in gradings (3, 0) and (3, 1) must die in the spectral sequence. There are no incoming d^r differentials from gradings $u = 0$ or $u = -1$ otherwise a surviving generator is killed. And again by (5) there are no terms in the bigradings $(q, 2)$ or $(q, 1)$ for $q > 3$ with which they may cancel. It must be the case that $Kh^{3,0}(\times) = 0$ and $Kh^{3,1} = 0$, and we conclude that $Kh(\times)$ is thin. The closed formula (4) follows immediately from the discussion above. □

Remark 1 Important to our application is the fact that the branched double cover of a reduced Khovanov thin knot with $\mathbb{Z}/2\mathbb{Z}$–coefficients is an L-space, which follows from the symmetry of Heegaard Floer homology under orientation reversal and the spectral sequence from reduced Khovanov homology of a link to the Heegaard Floer homology of the branched double cover of the mirror of the link [22]. Notice that in the argument of Lemma 1, there is a single location not contained on the diagonals $\delta = \pm 1$, and this is bigrading (3, 0). Had we used $\mathbb{Z}$-coefficients to write down the skein exact sequence, we would have seen

$$0 \longrightarrow Kh^{3,0}(\times) \longrightarrow q\mathbb{Z}^{(2,0)} \xrightarrow{\cdot u} \cdots . \tag{9}$$

Since $Kh^{(3,0)}(\times)$ injects, it is torsion-free, and the argument of Lemma 1 shows there are no free summands in bigrading (3, 0). Thus $Kh(K_n; \mathbb{Z}/2\mathbb{Z})$ is also thin, and therefore K_n is reduced Khovavnov thin with $\mathbb{Z}/2\mathbb{Z}$ coefficients as well. We deduce that $\Sigma(K_n)$ is an L-space for all n.

Table 1: A portion of the E^1 page of the spectral sequence with u-grading vertically and q-grading horizontally

$(n-5)+1$						$\mathbb{Q}$	$\mathbb{Q}$
$(n-5)$					$\mathbb{Q}$		
⋮							
2							
1		$\mathbb{Q}$ or 0	$\mathbb{Q}$ or 0				
0	$\mathbb{Q}$	$\mathbb{Q}^2$ or $\mathbb{Q}$	$\mathbb{Q}$ or 0				
	-1	1	3	…	$2(n-5)-1$	$2(n-5)+1$	…

The induced differentials d^r for $r \geq 1$ map the regions in question to the gray regions, whereas the incoming differentials come from the yellow regions

3.2 Proof of Main Theorem

We now set about to prove that the infinite family of knots $\mathcal{K}$ satisfies the cosmetic crossing conjecture, among several other properties. Our main obstruction for a knot to admit a cosmetic crossing change is

Theorem 5 ([16, Theorem 2]) *Let K be a knot in S^3 whose branched double cover $\Sigma(K)$ is an L-space. If each summand of the first singular homology of $\Sigma(K)$ has square-free order, then K admits no cosmetic crossing changes.*

With this obstruction in hand, we prove

Theorem 6 *The set $\mathcal{K}$ describes an infinite family of knots which have determinant 49 and noncyclic $H_1(\Sigma(K); \mathbb{Z})$. These knots are non-alternating, non-fibered, hyperbolic, of genus two, bridge number three, and satisfy the cosmetic crossing conjecture.*

Proof By Theorem 4, for all n even, $\widehat{HFK}(K_n) \cong \widehat{HFK}(5_2) \otimes \widehat{HFK}(m(5_2))$ and for all n odd, $\widehat{HFK}(K_n) \cong \widehat{HFK}(K_1)$. The knot Floer groups for $\widehat{HFK}(K_n)$ for n odd can be found after identifying K_1 as the alternating knot 10_{22}, whose knot Floer homology is determined by its Alexander polynomial and signature. Represented as a Poincaré polynomial, the knot Floer homology groups are thus

$$P_{\widehat{HFK}(K_n)}(s,m) = 4s^{-2}m^{-2} + 12s^{-1}m^{-1} + 17 + 12sm + 4s^2m^2 \qquad (10)$$

for n even, and

$$P_{\widehat{HFK}(K_n)}(s,m) = 2s^{-3}m^{-3} + 6s^{-2}m^{-2} + 10s^{-1}m^{-1} + 13 + 10sm + 6s^2m^2 + 2s^3m^3 \qquad (11)$$

for n odd. By Corollary 1, K_n is non-fibered for all n and of genus two when n is even and genus three when n is odd.

Equations 10 and 11 imply that $\det(K_n) = 49$ for all n. This is also implied by [14, Theorem 2.6] as well as Lemma 2 below. Because there are only a finite number of alternating knots with any fixed determinant (see for instance [18, Lemma 14]), the knots K_n, for $n \in \mathbb{Z}$, are generically non-alternating. Lemma 1 and Remark 1 imply K_n is reduced Khovanov homology thin over $\mathbb{Z}/2\mathbb{Z}$ for all n, ensuring that $\Sigma(K_n)$ is an L-space for all n.

The rest of the proof will follow after we verify Lemmas 2, 3, and 4.

Lemma 2 *For each n, the knot K_n has $H_1(\Sigma(K_n); \mathbb{Z}) \cong \mathbb{Z}/7\mathbb{Z} \oplus \mathbb{Z}/7\mathbb{Z}$ if n is a multiple of 7 and $H_1(\Sigma(K_n); \mathbb{Z}) \cong \mathbb{Z}/49\mathbb{Z}$ otherwise.*

Proof Recall that the Goeritz matrix associated to a checkerboard coloring of a knot diagram gives a presentation matrix for $H_1(\Sigma(K_n); \mathbb{Z})$ [4]. Indeed, to compute the Goeritz matrix of a knot diagram K, enumerate the white regions of a checkerboard

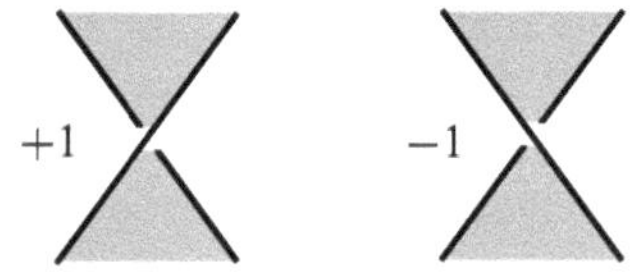

Fig. 6: Incidence numbers $\eta(c)$ assigned to each crossing in a checkerboard coloring

coloring of K by $X_1, \ldots, X_m$, and define the symmetric $m \times m$ integral matrix $G'(K) = (g_{ij})$ by

$$g_{ij} = \begin{cases} -\sum_{c \in X_{ij}} \eta(c) & i \neq j \\ -\sum_{\ell \neq i} g_{i\ell} & i = j, \end{cases}$$

where the incidence numbers $\eta(c)$ are assigned as in Fig. 6 and $X_{ij} = \overline{X}_i \cap \overline{X}_j$. The Goeritz matrix $G := G(K)$ is then obtained by deleting the first row and column of $G'(K)$. It provides a presentation for $H_1(\Sigma(K); \mathbb{Z})$ and $\det(K) = |\det G|$.

From the diagram in Fig. 5, we obtain a Goeritz matrix presentation for $H_1(\Sigma(K_n); \mathbb{Z})$,

$$\begin{pmatrix} 4 & 0 & -1 & 0 \\ 0 & -4 & 0 & 1 \\ -1 & 0 & 2-n & n \\ 0 & 1 & n & -n-2 \end{pmatrix}.$$

It is straightforward to verify that this is equivalent to the presentation matrix $\begin{pmatrix} 7 & 4n \\ 0 & 7 \end{pmatrix}$. This presents $\mathbb{Z}/7\mathbb{Z} \oplus \mathbb{Z}/7\mathbb{Z}$ if and only if 7 divides $4n$, which is equivalent to n being a multiple of 7. Otherwise, the matrix presents $\mathbb{Z}/49\mathbb{Z}$. □

As in [6], we adopt the strategy of [10, Lemmas 4 and 5] for the following two arguments.

Lemma 3 *All K_n with $n \equiv 0 \pmod 7$ have bridge number three.*

Proof From the diagram in Fig. 5, the $b(K_n)$ is bounded above by three for all n. Lemma 2 implies that whenever $n \equiv 0 \pmod 7$, the branched double cover of K_n cannot be a lens space because its first homology is noncyclic, thus K_n cannot be a two-bridge knot by Hodgson-Rubinstein [8]. Alternatively, recall that two-bridge knots are alternating, so at most finitely many K_n are two-bridge anyway. □

Lemma 4 *All K_n with $n \equiv 0 \pmod{14}$ and $n \neq 0$ are hyperbolic.*

Proof By Lemma 3, $b(K_n) = 3$ and since n is even, $g(K_n) = 2$. By Riley [27], a three-bridge knot is either hyperbolic, a torus knot, or a connected sum. The only torus knot of genus two is the (5, 2)–torus knot, and its Alexander polynomial distinguishes it from K_n for all n. Suppose now K_n is composite. Then $K_n = K\#K'$ for some knots K and K' each of genus one. Since

$$br(K\#K') = br(K) + br(K') - 1,$$

this implies that K and K' are both two-bridge knots. The branched double cover $\Sigma(K\#K')$ is a nontrivial connected sum of lens spaces with $|H_1(\Sigma(K\#K'; \mathbb{Z})| = 49$, so each summand must have order 7. By [3, Proposition 12.26], a genus one, two-bridge knot or its mirror is of the form $\mathfrak{b}(\alpha, \beta)$, where

$$\beta = 2c, \qquad \alpha = 4bc \pm 1, \qquad b, c, \in \mathbb{Z}.$$

The branched double cover of $\mathfrak{b}(\alpha, \beta)$ is $L(\alpha, \beta)$, therefore $7 = |H_1(L(\alpha, \beta); \mathbb{Z})| = 4bc \pm 1$. The only integral solutions are when $b, c = 2, 1$ or $b, c = 1, 2$, both of which correspond with the knot 5_2 or its mirror. In this case the Jones polynomial distinguishes K_n from connected sums of 5_2 with itself or its mirror. There can be no such K and K', hence K_n is prime. Since K_n is neither a torus knot nor a connected sum, it must be hyperbolic. □

Excluding K_0 and the finitely many knots K_n which may be alternating, the properties in the statement of Theorem 6 are simultaneously satisfied whenever $n \equiv 0 \pmod{14}$. This completes the proof of the theorem. □

3.3 Observations

We close with several observations.

Remark 2 For the sake of concreteness, we chose the knot 5_2 with which to construct the set in (1). However, one can carry out similar constructions using other base knots. For example, the partial knot of the pretzel knot $P = (p, q, -p)$, where p is odd (see Fig. 4), is the $(2, p)$–torus knot, which is reduced Khovanov homology thin. The general strategy of Lemma 1 applies and was carried out by Starkston to investigate their Khovanov homology in [28]. A computation similar to that of Lemma 2 would show that $H_1(\Sigma(P); \mathbb{Z}) \cong \mathbb{Z}_p \oplus \mathbb{Z}_p$ if and only if $p \mid q$. In this case, Theorem 2 applies when p is square-free, and we similarly obtain that such knots satisfy the cosmetic crossing conjecture. However, when q is odd, P is genus one, and when q is even, P is fibered. So no new information is gained with these pretzel knots, unlike the knots $K_n \in \mathcal{K}$.

Remark 3 The symmetric unions K_n, as well as the symmetric pretzel knots and Kanenobu knots have constant determinant and are Khovanov homology thin. Greene conjectured that there exist only finitely many quasi-alternating links with a given determinant [5, Conjecture 3.1]. We suspect that the present examples, like the Kanenobu knots and pretzel knots, also fail to be quasi-alternating, and that an argument similar to that made by Greene and Watson in [6] for the case of the Kanenobu knots can be made.

Remark 4 Recall that an *L-space knot* is a knot which admits a positive Dehn surgery to an L-space. Because the knots K_n are obtained by rational tangle replacement in

$K_0 = 5_2\#m(5_2)$, there exists a knot $\tilde{\gamma}$ in $\Sigma(K_0)$ which admits surgeries to the L-space $\Sigma(K_n)$ for all n. In particular, this knot $\tilde{\gamma}$ is the lift of a crossing arc γ in the trivial 0–tangle $T_0 \subset K_0$. Since $\Sigma(K_0)$ is the connected sum of lens spaces $L(7,2)\#L(7,3)$, we therefore observe that the lift $\widetilde{\gamma} \in L(7,2)\#L(7,3)$ is an example of an L-space knot in a reducible L-space. An alternate proof to Lemma 2 could be obtained by studying presentation matrices for $H_1(\Sigma(K_n);\mathbb{Z})$ where $\Sigma(K_n)$ is obtained by Dehn surgery along the primitive curve $\widetilde{\gamma} \in L(7,2)\#L(7,3)$.

Acknowledgments The author is especially grateful to Tye Lidman for his interest. She also thanks Laura Starkston and Liam Watson for helpful correspondence. This work is partially supported by NSF grant DMS-1148609.

References

1. C. Balm, S. Friedl, E. Kalfagianni, M. Powell, Cosmetic crossings and Seifert matrices. Comm. Anal. Geom. **20**(2), 235–253 (2012)
2. C.J. Balm, E. Kalfagianni, Knots without cosmetic crossings. Preprint (2014). arXiv:1406.1755 [math.GT]
3. G. Burde, H. Zieschang, M. Heusener, Knots, in *De Gruyter Studies in Mathematics*, vol. 5, extended ed. (De Gruyter, Berlin, 2014)
4. C.M. Gordon, R.A. Litherland, On the signature of a link. Inven. Math. **47**(1), 53–69 (1978)
5. J.E. Greene, Homologically thin, non-quasi-alternating links. Math. Res. Lett. **17**(1), 39–49 (2010)
6. J.E. Greene, L. Watson. Turaev torsion, definite 4-manifolds, and quasialternating knots. Bull. Lond. Math. Soc. **45**(5), 962–972 (2013)
7. M. Hedden, L. Watson, On the geography and botany of knot Floer homology. Preprint (2014). arXiv:1404.6913v2 [math.GT]
8. C. Hodgson, J.H. Rubinstein, Involutions and isotopies of lens spaces, in *Knot theory and manifolds (Vancouver, B.C., 1983)*, Lecture Notes in Mathematics, vol. 1144 (Springer, Berlin, 1985), pp. 60–96
9. E. Kalfagianni, Cosmetic crossing changes of fibered knots. J. Reine Angew. Math. **669**, 151–164 (2012)
10. T. Kanenobu, Infinitely many knots with the same polynomial invariant. Proc. Am. Math. Soc. **97**(1), 158–162 (1986)
11. M. Khovanov, A categorification of the Jones polynomial. Duke Math. J. **101**(3), 359–426 (2000)
12. S. Kinoshita, H. Terasaka, On unions of knots. Osaka Math. J. **9**, 131–153 (1957)
13. R. Kirby, Problems in low dimensional manifold theory, in *Algebraic and geometric topology (Proc. Sympos. Pure Math., Stanford Univ., Stanford, Calif., 1976), Part 2*, Proceedings of Symposia in Pure Mathematics, XXXII (American Mathematical Society, Providence, RI, 1978), pp. 273–312
14. C. Lamm, Symmetric unions and ribbon knots. Osaka J. Math. **37**(3), 537–550 (2000)
15. E.S. Lee, An endomorphism of the Khovanov invariant. Adv. Math. **197**(2), 554–586 (2005)
16. T. Lidman, A.H. Moore, Cosmetic surgery in L-spaces and nugatory crossings. Preprint (2015)
17. A. Lobb, The Kanenobu knots and Khovanov-Rozansky homology. Proc. Am. Math. Soc. **142**(4), 1447–1455 (2014)
18. A.H. Moore, L. Starkston, Genus-two mutant knots with the same dimension in knot Floer and Khovanov homologies. Algebr. Geom. Topol. **15**(1), 43–63 (2015)
19. Y. Ni, Knot Floer homology detects fibred knots. Invent. Math. **170**(3), 577–608 (2007)

20. P. Ozsváth, Z. Szabó, Knot Floer homology and the four-ball genus. Geom. Topol. **7**, 615–639 (2003)
21. P. Ozsváth, Z. Szabó, Holomorphic disks and knot invariants. Adv. Math. **186**(1), 58–116 (2004)
22. P. Ozsváth, Z. Szabó, On the Heegaard Floer homology of branched double-covers. Adv. Math. **194**(1), 1–33 (2005)
23. P. Ozsváth, Z. Szabó, On the skein exact sequence for knot Floer homology. Preprint (2007). arXiv:0707.1165v1 [math.GT]
24. J. Rasmussen, Floer homology and knot complements. Ph.D. thesis, Harvard University (2003)
25. J. Rasmussen, Knot polynomials and knot homologies in *Geometry and topology of manifolds*. Fields Institute Communications vol. 47 (American Mathematical Society, Providence, RI, 2005), pp. 261–280
26. J. Rasmussen, Khovanov homology and the slice genus. Inven. Math. **182**(2), 419–447 (2010)
27. R. Riley, An elliptical path from parabolic representations to hyperbolic structures, in *Topology of low-dimensional manifolds (Proc. Second Sussex Conf., Chelwood Gate, 1977)*. Lecture Notes in Mathematical, vol. 722 (Springer, Berlin, 1979), pp. 99–133
28. L. Starkston. The Khovanov homology of $(p, -p, q)$ pretzel knots. J. Knot Theor. Ramif. **21**(5), 1250056 (2012)
29. The Knot Atlas. The knot theory' package (2015). http://katlas.org/
30. I. Torisu, On nugatory crossings for knots. Topol. Appl. **92**(2), 119–129 (1999)
31. L. Watson, Knots with identical Khovanov homology. Algebr. Geom. Topol. **7**, 1389–1407 (2007)

The Total Thurston–Bennequin Number of Complete and Complete Bipartite Legendrian Graphs

Danielle O'Donnol and Elena Pavelescu

Abstract We study the Thurston–Bennequin number of complete and complete bipartite Legendrian graphs. We define a new invariant called the total Thurston–Bennequin number of the graph. We show that this invariant is determined by the Thurston–Bennequin numbers of 3-cycles for complete graphs and by the Thurston–Bennequin number of 4-cycles for complete bipartite graphs. We discuss the consequences of these results for K_4, K_5 and $K_{3,3}$.

Keywords Legendrian graph · Thurston–Bennequin number · Minimal embedding · Complete graph · Complete bipartite graph

Mathematics Subject Classification 57M15 · 57M50

1 Introduction

Motivated by their appearance in important results, the authors began a systematic study of Legendrian graphs [6]. Two nice examples of such results are Giroux's proof of existence of open book decompositions compatible with a given contact structure [5], and Eliashberg and Fraser's proof of the Legendrian simplicity of the unknot [3]. We anticipate that with a better understanding of Legendrian graphs, they will become an even more robust tool.

Danielle O'Donnol—Partially supported by the National Science Foundation grant DMS-1406481.

D. O'Donnol (✉)
Department of Mathematics, Indiana University, Rawels Hall, 831 E. 3rd St, Bloomington, IN 47405, USA
e-mail: odonnol@indiana.edu

E. Pavelescu
Department of Mathematics and Statistics, University of South Alabama, 411 University Boulevard North, Mobile, AL 36688, USA
e-mail: elenapavelescu@southalabama.edu

G. Letzter et al. (eds.), *Advances in the Mathematical Sciences*, Association for Women in Mathematics Series 6, DOI 10.1007/978-3-319-34139-2_4

Throughout this article, we work in $\mathbb{R}^3$ with the standard contact structure, ξ_{std}. For a given graph, we will denote the *abstract graph* with G, the *Legendrian embedding* of G with G, and a *front projection* of G with $\widetilde{G}$. In [6], the authors extended the classical invariants Thurston–Bennequin number, *tb*, and rotation number, *rot*, from Legendrian knots to Legendrian graphs. The Thurston–Bennequin number measures the number of times the contact planes twist around the knot as the knot is traversed once, and can be computed from the front projection as $tb(K) = w(\widetilde{K}) - \frac{1}{2}c(\widetilde{K})$, where $w(\widetilde{K})$ is the writhe and $c(\widetilde{K})$ is the number of cusps. For a Legendrian graph G with a fixed order on its cycles, the *Thurston–Bennequin number*, $tb(G)$, is the ordered list of the Thurston–Bennequin numbers of its cycles. The cycles of a Legendrian graph are piecewise smooth Legendrian knots. However, the computations from the front projection are done in the same way. For a thorough treatment of $tb(G)$ see [6]. Similarly, the *rotation number*, $rot(G)$, is the ordered list of the rotation numbers of its cycles. It is known [2] that if K is a Legendrian knot in $(\mathbb{R}^3, \xi_{std})$ and Σ is a Seifert surface for K, then

$$tb(K) + |rot(K)| \leq -\chi(\Sigma).$$

This inequality puts an upper bound on the $tb(K)$. In particular, if K is the unknot, then $tb(K) \leq -1$. There is a unique Legendrian unknot with maximal $tb(K) = -1$, called the *trivial unknot*. For a topological knot type $\mathcal{K}$, we denote by $tb_{max}(\mathcal{K})$ the maximum *tb* among all Legendrian knots of topological type $\mathcal{K}$.

In this paper, we introduce a new invariant of Legendrian graphs, called the total Thurston–Bennequin number. The *total Thurston–Bennequin number*, $TB(G)$, is the sum of *tb*s over all cycles of G. We derive a simplified diagrammatic means of computing $TB(G)$ for complete graphs and complete bipartite graphs, and show that it depends only on *tb*s of the smallest cycles. In particular cases, we show that *tb* of a graph (given an ordering on the cycles) is determined by *TB*.

The main theorems relate the total Thurston–Bennequin number of complete and complete bipartite graphs with the sum of *tb*s of their smallest cycles. For a Legendrian embedding of the complete graph on n vertices, K_n, in $(\mathbb{R}^3, \xi_{std})$, we show:

$$\begin{aligned} TB(K_n) &= TB_3(K_n) \cdot \sum_{r=3}^{n} \frac{(n-3)!}{(n-r)!} \\ &= \left[(n-2)w_e(\widetilde{K}_n) + w_{ae}(\widetilde{K}_n) - \tfrac{1}{2}[(n-2)c_e(\widetilde{K}_n) + c_v(\widetilde{K}_n)]\right] \cdot \sum_{r=3}^{n} \frac{(n-3)!}{(n-r)!}, \end{aligned}$$

where $TB_3(K_n)$ is the sum of *tb*s over all 3-cycles of K_n, and w_* and c_* indicate different writhes and cusp counts described in Sect. 2. For a Legendrian embedding of the complete bipartite graph, $K_{n,m}$ with $m \leq n$ in $(\mathbb{R}^3, \xi_{std})$, we show

$$TB(K_{n,m}) = TB_4(K_{n,m}) \cdot \sum_{r=2}^{m} \frac{(m-2)!(n-2)!}{(m-r)!(n-r)!},$$

where $TB_4(K_{n,m})$ is the sum of *tb*s over all 4-cycles of $K_{n,m}$, and

$$TB_4(K_{n,m}) =$$

$$(n-1)(m-1)w_e(\widetilde{K}_{n,m}) + (n-1)w_{ae[P]}(\widetilde{K}_{n,m}) + (m-1)w_{ae[Q]}(\widetilde{K}_{n,m}) + w_{ne}(\widetilde{K}_{n,m})$$

$$-\tfrac{1}{2}[(n-1)(m-1)c_e(\widetilde{K}_{n,m}) + (n-1)c_{v[P]}(\widetilde{K}_{n,m}) + (m-1)c_{v[Q]}(\widetilde{K}_{n,m})].$$

Here w_* and c_* indicate different writhes and cusp counts described in Sect. 3.

Let a *minimal embedding* of G be one where all minimal length cycles are unknots with $tb = -1$, also called a *minimal Legendrian embedding*. Any abstract graph could have a minimal embedding. An embedded graph is said to be *unknotted* if all of its cycles are unknots. There is a particular subset of minimal embeddings which we call unknotted minimal embeddings. An *unknotted minimal embedding* of G is a minimal embedding where all the cycles are unknots. In this article, we predominately look at unknotted minimal embeddings, however not all abstract graphs will have such an embedding. There are graphs which are intrinsically knotted (every embedding of the graph contains a nontrivial knot among its cycles). In particular, the smallest complete graph which is intrinsically knotted is K_7, and $K_{5,5}$ is the smallest intrinsically knotted complete bipartite graph. So only those graphs smaller than K_7 and $K_{5,5}$ can have unknotted minimal embeddings.

Here, we explain ways to get minimal embeddings for complete graphs and complete bipartite graphs. For the front project of a minimal Legendrian K_n, place all vertices on the same horizontal line in the front projection. Then place the edges adjacent to a given vertex as nested arcs, with no intersections between adjacent edges. See the left image of Fig. 1 for such a K_6. All 3-cycles in this embedding have a front projection like that of the unknot on the right in the picture and therefore are trivial unknots.

Similarly, for the complete bipartite graph $K_{n,m}$, place all vertices of $K_{n,m}$ on the same horizontal line in the front projection, with the vertices in one partition first. Then place the edges adjacent to a given vertex as nested arcs, with no intersections between adjacent edges. See the left image of Fig. 2 for such a $K_{5,3}$. All 4-cycles in this embedding have a front projection like that of the unknot on the right in the picture and therefore are trivial unknots.

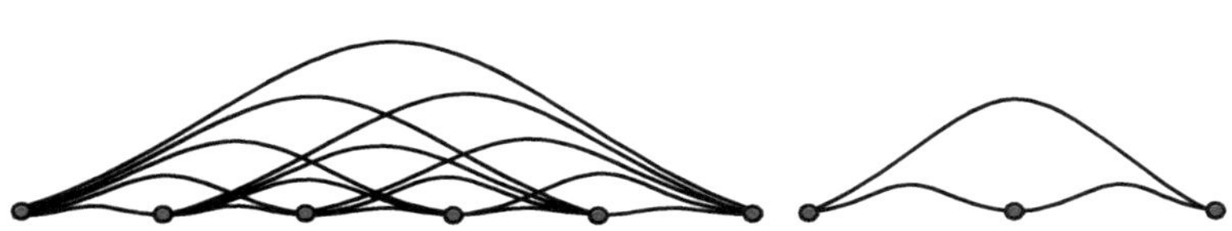

Fig. 1 A minimal embedding of K_6 and a 3-cycle in this embedding

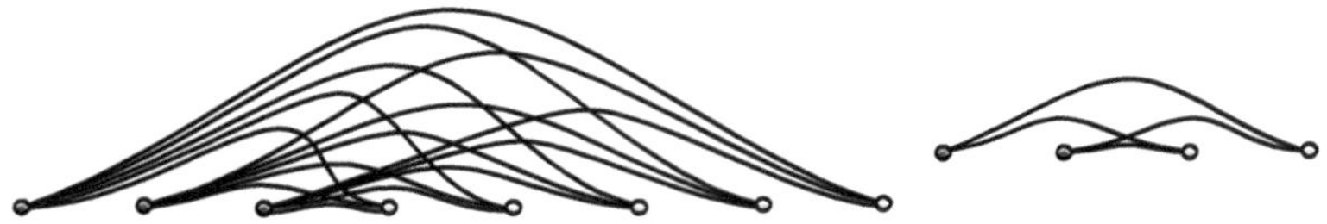

Fig. 2 Minimal embedding of $K_{5,3}$ and a 4-cycle in this embedding

We give some examples of minimal K_4, K_5 and $K_{3,3}$. For unknotted minimal K_4s and $K_{3,3}$s, using our understanding of *TB* and the graphs structure, we show they have unique *tb* and $|rot|$ up to relabelling of the cycles. We give a lower bound for *tb* of an unknotted *r*-cycle in a Legendrian K_n with all 3-cycles trivial unknots.

2 The Total Thurston–Bennequin Number for Complete Graphs

In this section, we introduce much of the notation and definitions that are used throughout this article. This section is focused on the total Thurston–Bennequin number of a complete graph. Here, we prove the main theorem for the *TB* of complete graphs and look at the consequences for K_4 and K_5.

We first introduce notation and definitions. A *cycle* is a path whose first vertex is also the last, up to a choice of starting vertex. So in an embedded graph, a *cycle* is a subgraph which is a simple closed curve. An *r-cycle* is a cycle with *r* edges. No orientation or base point is assumed for a cycle.

Definition 1 For a Legendrian graph G, we define the *total Thurston–Bennequin number of* G, $TB(G)$, as the sum of Thurston–Bennequin numbers over all cycles of G. For a Legendrian graph G, we define $TB_r(G)$ as the sum of Thurston–Bennequin numbers over all r-cycles of G.

Notice that both $TB(G)$ and $TB_r(G)$ are invariants of the Legendrian type of the graph G. We denote by $G[\mathcal{K}]$ an embedding of the graph G for which all cycles have the knot type $\mathcal{K}$. We denote the unknot with $\mathcal{U}$. The set of all edges of G is $E(G)$ and the set of vertices is $V(G)$. The set of all r-cycles of a graph G is denoted by $\Gamma_r(G)$, or simply Γ_r when G is understood. Fix a choice of front diagram $\widetilde{G}$. For a cycle γ of G, we denote by $w(\widetilde{\gamma})$ and $c(\widetilde{\gamma})$ the *writhe of* $\widetilde{\gamma}$ (signed sum of crossings of $\widetilde{\gamma}$) and the *number of cusps of* $\widetilde{\gamma}$, respectively. For adjacent edges e, f, and vertex v, we denote by $w(\widetilde{e})$ the writhe of edge $\widetilde{e}$ with itself, by $w(\widetilde{e},\widetilde{f})$ the signed sum of crossings between adjacent edges $\widetilde{e}$ and $\widetilde{f}$, by $c(\widetilde{e})$ the number of cusps along edge $\widetilde{e}$, and by $c(\widetilde{v})$ the number of cusps at vertex $\widetilde{v}$, looking at each pair of edges going through v. For a front diagram of a Legendrian graph G, we define

- the *edge writhe of* $\widetilde{G}$ as the sum of writhes over all edges of G, $w_e(\widetilde{G}) = \sum_{f \in E(G)} w(\widetilde{f})$.

- the *adjacent edge writhe of* $\widetilde{G}$ as the sum of writhes over all pairs of adjacent edges of G, $w_{ae}(\widetilde{G}) = \sum_{e,f \in E(G)\text{adj}} w(\widetilde{e}, \widetilde{f})$.
- the *edge cusps of* $\widetilde{G}$ as the number of cusps along all edges of G, $c_e(\widetilde{G}) = \sum_{f \in E(G)} c(\widetilde{f})$.
- the *vertex cusps of* $\widetilde{G}$ as the count of cusps at all vertices of G, $c_v(\widetilde{G}) = \sum_{v \in V(G)} c(\widetilde{v})$.

None of the writhes nor the cusp counts are invariants, they all depend on the front projection.

Theorem 1 *Let K_n be a Legendrian embedding of the complete graph on n vertices in $(\mathbb{R}^3, \xi_{std})$. Then*

$$TB_r(K_n) = \frac{(n-3)!}{(n-r)!} TB_3(K_n).$$

As a consequence,

$$TB(K_n) = TB_3(K_n) \cdot \sum_{r=3}^{n} \frac{(n-3)!}{(n-r)!}.$$

The quantities $TB_i(K_n)$, $3 \le i \le n$, can be computed from writhe and cusp counts of vertices and edges in the front projection rather than summing the tbs of the cycles.

Proof When computing the sum of writhes over the 3-cycles or the r-cycles of K_n, we consider crossings of an edge with itself, crossings between adjacent edges, and crossings between non-adjacent edges.

1. Each edge of K_n appears in $n-2$ of the 3-cycles and in $\binom{n-2}{r-2}(r-2)! = \frac{(n-2)!}{(n-r)!}$ r-cycles.
2. Each pair of adjacent edges of K_n appears in one 3-cycle and in $\frac{(n-3)!}{(n-r)!}$ r-cycles.
3. Non-adjacent edges do not contribute to the total writhe of the graph. The cycles containing both edges *ab* and *cd* can be paired as *...ab...cd...* and *...ab...dc....* The signed intersection of the two edges in one of the cycles is the negative of their signed intersection in the other cycle. This means that the crossings between the two edges do not contribute to the sum of the writhes.

   Items (1)–(3) above give

$$\begin{aligned}\textstyle\sum_{\gamma \in \Gamma_r} w(\widetilde{\gamma}) &= \tfrac{(n-2)!}{(n-r)!} w_e(\widetilde{K}_n) + \tfrac{(n-3)!}{(n-r)!} w_{ae}(\widetilde{K}_n) \\ &= \tfrac{(n-3)!}{(n-r)!}[(n-2)w_e(\widetilde{K}_n) + w_{ae}(\widetilde{K}_n)] = \tfrac{(n-3)!}{(n-r)!} \textstyle\sum_{\gamma \in \Gamma_3} w(\widetilde{\gamma}).\end{aligned}$$

Cusps occur either at a vertex, that is, at each pair of adjacent edges, or along one edge. Using (1) and (2) above, we have

$$\sum_{\gamma\in\Gamma_r} c(\widetilde{\gamma}) = \tfrac{(n-2)!}{(n-r)!} c_e(\widetilde{K}_n) + \tfrac{(n-3)!}{(n-r)!} c_v(\widetilde{K}_n)$$

$$= \tfrac{(n-3)!}{(n-r)!}[(n-2)c_e(\widetilde{K}_n) + c_v(\widetilde{K}_n)] = \tfrac{(n-3)!}{(n-r)!} \sum_{\gamma\in\Gamma_3} c(\widetilde{\gamma}).$$

For every cycle γ, $tb(\gamma) = w(\widetilde{\gamma}) - \frac{1}{2}c(\widetilde{\gamma})$. Then the two identities above give

$$TB_r(K_n) = \tfrac{(n-3)!}{(n-r)!} \left[(n-2)w_e(\widetilde{K}_n) + w_{ae}(\widetilde{K}_n) - \tfrac{1}{2}[(n-2)c_e(\widetilde{K}_n) + c_v(\widetilde{K}_n)]\right]$$

$$= \tfrac{(n-3)!}{(n-r)!} TB_3(K_n).$$

Adding over r gives

$$TB(K_n) = \left[(n-2)w_e(\widetilde{K}_n) + w_{ae}(\widetilde{K}_n) - \tfrac{1}{2}[(n-2)c_e(\widetilde{K}_n) + c_v(\widetilde{K}_n)]\right] \cdot \sum_{r=3}^{n} \frac{(n-3)!}{(n-r)!}$$

$$= TB_3(K_n) \cdot \sum_{r=3}^{n} \frac{(n-3)!}{(n-r)!}. \qquad \square$$

While Theorem 1 has many consequences, we focus on minimal embeddings.

2.1 Remarks About K_4

In [6], the authors showed there does not exist a Legendrian embedding $K_4[\mathcal{U}]$ where all cycles are trivial unknots. Moreover, $TB(K_4[\mathcal{U}])$ is shown to be at most -8. Theorem 1 gives a refinement of this result. For $n = 4$, we have

$$TB_4(K_4) = TB_3(K_4).$$

The graph K_4 has seven cycles, four cycles of length 3 and three of length 4. This means that for a Legendrian embedding $K_4[\mathcal{U}]$ with all trivial 3-cycles we have

$$TB_4(K_4) = TB_3(K_4) = -4.$$

For such an embedding, exactly one of its three 4-cycles has $tb = -2$ and the other two have $tb = -1$. This is the only way to have $TB(K_4[\mathcal{U}]) = -8$. If all the 4-cycles have $tb = -1$, then $TB(K_4[\mathcal{U}]) = -6$, which is a contradiction.

The 4-cycles of K_4 are equivalent under graph automorphism, so up to a relabelling of the cycles there is a unique $tb(K_4)$ for an unknotted minimal embedding K_4. Since unknots with $tb = -2$ and $tb = -1$ are unique, there is only one possible $tb(K_4)$ and $|rot(K_4)|$ for an unknotted minimal embedding K_4. In Fig. 3, we show two diagrams

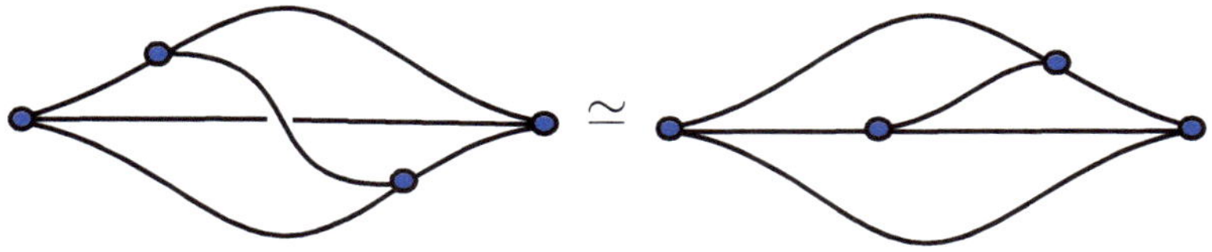

Fig. 3 Two diagrams of an unknotted minimal K_4

of an unknotted minimal K_4. (See Figs. 16 and 17 for the equivalence.) While there is a unique *tb* and $|rot|$, this does not mean there is only one unknotted minimal embedding K_4.

Remark 1 In more generality, for K_n with all 3-cycles trivial unknots and 4-cycles unknots, one third of 4-cycles have $tb = -2$ and two thirds of 4-cycles have $tb = -1$. This is because for every K_4 subgraph of K_n, exactly one of three 4-cycles has $tb = -2$ and the other two have $tb = -1$.

Remark 2 The graph K_4 is adaptable [8], that is, given any set of seven knot types, there exists an embedding of K_4 with its seven cycles realizing the seven knot types. Each topological embedding of a graph has a Legendrian realization [6]. Since $TB_4(K_4) = TB_3(K_4)$, if all cycles of K_4 are of the same knot type L with $tb_{max}(L) > 0$, then at least one of the 3-cycles has non-maximal *tb*. On the other hand, if all cycles of K_4 are of the same knot type L with $tb_{max}(L) < 0$, then at least one of the 4-cycles has non-maximal *tb*.

In Sect. 4, questions about existence and uniqueness of minimal embeddings of K_4 will be discussed.

2.2 Remarks About K_5

We look at unknotted minimal Legendrian embeddings of K_5. The graph K_5 has fifteen 4-cycles. By Remark 1, five of the 4-cycles have $tb = -2$ and ten have $tb = -1$. For $n = 5$ and $r = 5$, Theorem 1 says

$$TB_5(K_5) = 2TB_3(K_5) = -20.$$

In Proposition 1, we show that the minimum *tb* for an unknotted 5-cycle is -4. There are ten possible ways to write -20 as a sum of twelve integers in the set $\{-4, -3, -2, -1\}$. These ten sequences are candidates for *tb*s of the 5-cycles of K_5:

$$s_1 = (-1, -1, -1, -1, -2, -2, -2, -2, -2, -2, -2, -2)$$

$$s_2 = (-1, -1, -1, -1, -1, -2, -2, -2, -2, -2, -2, -3)$$

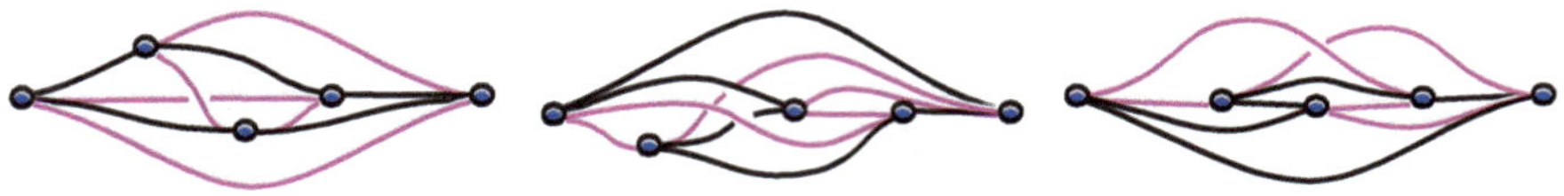

Fig. 4 Unknotted minimal embeddings of K_5 realizing s_2 (*left* and *middle*) and s_3 (*right*). The highlighted cycles have $tb = -3$

$$s_3 = (-1, -1, -1, -1, -1, -1, -2, -2, -2, -2, -3, -3)$$

$$s_4 = (-1, -1, -1, -1, -1, -1, -1, -2, -2, -3, -3, -3)$$

$$s_5 = (-1, -1, -1, -1, -1, -1, -1, -1, -3, -3, -3, -3)$$

$$s_6 = (-1, -1, -1, -1, -1, -1, -2, -2, -2, -2, -2, -4)$$

$$s_7 = (-1, -1, -1, -1, -1, -1, -1, -2, -2, -2, -3, -4)$$

$$s_8 = (-1, -1, -1, -1, -1, -1, -1, -1, -2, -3, -3, -4)$$

$$s_9 = (-1, -1, -1, -1, -1, -1, -1, -1, -2, -2, -4, -4)$$

$$s_{10} = (-1, -1, -1, -1, -1, -1, -1, -1, -1, -3, -4, -4)$$

Unlike unknotted minimal embeddings of K_4, an unknotted minimal embedding of K_5 can have cycles with $tb < -2$. So it can have cycles where the Bennequin inequality given by $tb(\mathcal{U}) + |rot(\mathcal{U})| \leq -1$ (here $\mathcal{U}$ denotes an unknotted cycle) is strict. We give examples realizing sequences s_2 and s_3. See Fig. 4. The leftmost K_5 in Fig. 4, has one 5-cycle with $tb = -3$. This cycle has rotation number 0. The middle K_5 in Fig. 4, has one 5-cycle with $tb = -3$. This cycle has rotation number ± 2, depending on the chosen orientation. This middle K_5 is the only embedding we have found where all its cycles are unknots U such that $tb(U) + |rot(U)| = -1$. Thus the Bennequin bound is also sharp for $K_5(\mathcal{U})$. The rightmost K_5 in Fig. 4, has two 5-cycles with $tb = -3$ (the highlighted cycle and its reflection about the middle vertical) and both of these cycles have rotation number 0.

In Sect. 4, questions about existence and uniqueness of minimal embeddings of K_5 will be discussed.

In view of Proposition 1 we introduce a few more definitions. The graph G' is a *subdivision* of G if G' can be obtained from G by adding any number of vertices to any of its edges. If vertices of a graph are labeled say $v_1, \ldots, v_n$, then the edge between v_i and v_j is denoted v_iv_j and a list of more than two vertices indicates the cycle given by those vertices in the indicated order, so $v_iv_jv_kv_l$ is a 4-cycle with the vertices in the order shown.

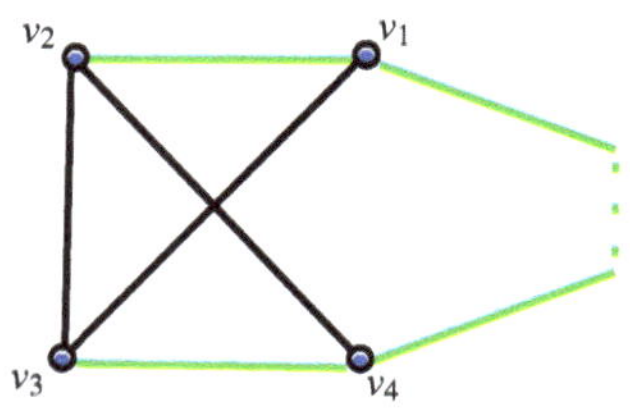

Fig. 5 Subdivision of K_4 containing the cycle γ_r. The edges of γ_r contained only in unknotted cycles are highlighted in *green*

Proposition 1 *Let γ_r be an unknotted r-cycle in K_n with all 3-cycles trivial unknots. Assume that $r-1$ of the edges of γ_r are only in unknotted cycles. Then*

$$tb(\gamma_r) \geq b_r := -2^{r-3}.$$

Proof By hypothesis, $tb(\gamma_3) = -1$. Note that $b_3 = -1 = -2^0$. Consider the K_4 subgraph determined by the four vertices in γ_4. Since three of the edges of γ_4 are assumed to be in only unknotted cycles, all the cycles of this K_4 are unknots, i.e., it is unknotted. From our discussion in Sect. 2.1, we know that $tb(\gamma_4) \geq -2$. Note that $b_4 = -2$. Assume the statement holds for $(r-1)$-cycles and let γ_r be an r-cycle as in the hypothesis. We label the vertices of γ_r as $v_1, v_2, \ldots, v_r$, with v_2v_3 the edge not necessarily in unknotted cycles only. We look at the subdivision of K_4 obtained by adding the edges v_1v_3 and v_2v_4 to γ_r. See Fig. 5. Notice that this K_4 subdivision is unknotted, as all cycles contain some of the edges of γ_r that are only in unknotted cycles. For this subdivision of K_4, the identity $TB_3(K_4) = TB_4(K_4)$ gives

$$tb(v_1v_2v_3) + tb(v_2v_3v_4) + tb(\gamma_{r-1}) + tb(\gamma'_{r-1}) = tb(v_1v_3v_4v_2) + tb(\gamma_r) + tb(\gamma'_r),$$

where $\gamma_{r-1}, \gamma'_{r-1}$ are $(r-1)$-cycles and γ'_r is an r-cycle. The two $(r-1)$-cycles $\gamma_{r-1}, \gamma'_{r-1}$ both have all but one of their edges only in unknotted cycles. By hypothesis, $tb(v_1v_2v_3) = tb(v_2v_3v_4) = -1$, $tb(\gamma_{r-1}) \geq -2^{(r-4)}$, and $tb(\gamma'_{r-1}) \geq -2^{(r-4)}$. So the left-hand side of the equality is at least $-2 - 2\left(2^{(r-4)}\right)$. The right-hand side of the equality is at most $-2 + tb(\gamma_r)$, when $v_1v_3v_4v_2$ and γ'_r have their maximal possible tb of -1. Together this gives $-2 - 2\left(2^{(r-4)}\right) \leq -2 + tb(\gamma_r)$, so $tb(\gamma_r) \geq -2^{(r-3)}$. □

Notice in the proof of Proposition 1, we only needed a certain set of K_4 subdivisions to be unknotted. The assumption that $r-1$ edges of γ_r are only in unknotted cycles is stronger than needed. While the following gives the same bound for $r \in \{3, 4, 5\}$, it gives a stronger bound when $r \geq 6$.

Proposition 2 *For $r \leq 14$, let γ_r be an unknotted r-cycle in K_n with all 3-cycles trivial unknots. Assume all edges of γ_r are only contained in unknotted cycles. Then*

$$tb(\gamma_r) \geq c_r := \frac{2(s_r^2 - 1)}{3} - (r-2)s_r,$$

where $s_r = 2^{\lfloor \log_2(r-2) \rfloor}$ *and* $\lfloor a \rfloor$ *represents the largest integer not greater than a.*

Proof By hypothesis, $tb(\gamma_3) = -1$. Also $c_3 = -1 = \frac{2(s_3^2-1)}{3} - (3-2)s_3$, since $s_3 = 1$. Consider the K_4 subgraph determined by the four vertices in γ_4. Since all edges of γ_4 are only in unknotted cycles, this K_4 is unknotted. From our discussion in Sect. 2.1, we know that $tb(\gamma_4) \geq -2$. Note $c_4 = -2 = \frac{2(s_4^2-1)}{3} - 2s_4$.

We will now outline the general case. Think of a k-cycle γ_k as obtained by adding $k-4$ vertices to the edges of a 4-cycle of K_4. There are many possible choices, and in each case we get a subdivision of K_4 where we use the identity $TB_3(K_4) = TB_4(K_4)$. Since all of the edges of γ_k are only in unknotted cycles, this K_4 subdivision is unknotted. On the right-hand side of the equality, we always have the k-cycle γ_k and two other cycles which have at most $tb = -1$. So the right-hand side of the identity is at most $tb(\gamma_k) - 2$. Since we added $k-4$ vertices to various edges of K_4, the total length of the four cycles on the left-hand side of the identity is $4 \cdot 3 + 2(k-4) = 2k+4$.

The left-hand side of the identity can take on various forms

$$S_{n_1,n_2,n_3,n_4} = tb(\gamma_{n_1}) + tb(\gamma_{n_2}) + tb(\gamma_{n_3}) + tb(\gamma_{n_4}),$$

with $n_1 + n_2 + n_3 + n_4 = 2k+4$, $n_i \geq 3$.

The assumption needed for induction is again weaker than that stated in the proposition. In order to obtain the bound, we need an unknotted even K_4 subdivision for γ_k, and for some smaller cycles that we will describe. An *even* K_4 *subdivision for* γ_k is a subdivided K_4 where γ_k appears as one of the subdivided 4-cycles, and is constructed by adding vertices to two non-adjacent edges of a 4-cycle in a K_4, either an equal number $\frac{k}{2} - 2$ to each edge if k is even, or $\frac{k-1}{2} - 1$ and $\frac{k-1}{2} - 2$ to the edges if k is odd. See Fig. 6. This can also be thought of as adding two edges to the γ_k in a particular way. For the following argument the bound must also hold for the four cycles appearing on the left-hand side of our identity. (These cycles would be 3-cycles in the K_4 if it were not subdivided.)

To have the bound hold for them, they must be unknots, and there must exist an even K_4 subdivision for each of them. In Fig. 6 we show even K_4 subdivisions for γ_k with $k \in \{8, 10, 12, 14\}$ and all of the smaller subdivisions needed. In the case of k odd the small subdivisions needed are those shown for $k-1$ and $k+1$. The green (grey) highlighted edges indicate those that are always in unknotted cycles. The black edges may be in knotted cycles. In the subdivisions pictured all cycles must be unknots because the only cycles that are in all black are 3-cycles. The following shows that under these conditions the c_k bound holds. For $k = 15$ it is no longer possible to obtain the needed subdivisions without using cycles that may be knotted.

In Fig. 7, we show the cycles and subdivisions which appear in the analysis for γ_7. For our induction process to work, we need that the bound also holds for the subdivided 3-cycles on the left-hand side of the identity. Figure 7 presents the subdivided 3-cycles of the even K_4 subdivision for γ_7, together with subsequent subdivided K_4s where these cycles appear. All the cycles are unknots, either because they share an edge with γ_7 or because they are 3-cycles in the original K_n.

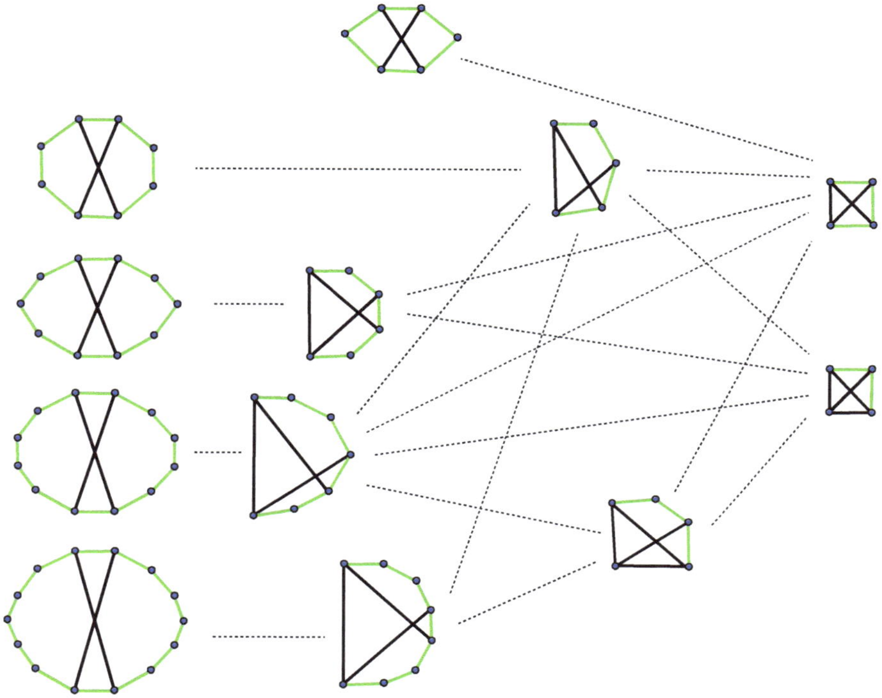

Fig. 6 Subdivisions of K_4 used to obtain the stronger bound. The edges that are only in unknotted cycles are highlighted in *green*

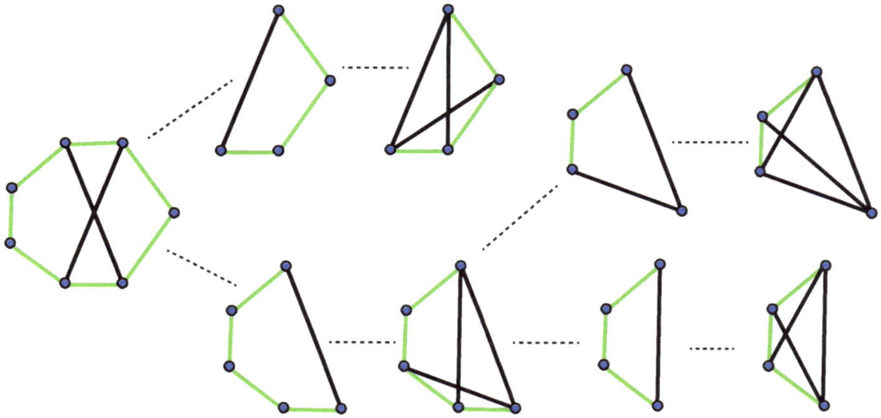

Fig. 7 An even K_4 subdivision of γ_7, together with subsequent 3-cycles of this K_4 each subdivided evenly. All 3-cycles are either triangles or share an edge (in *green*) with γ_7

For k even: Let $k = 2k_1$. Consider the even K_4 subdivision for γ_k where $k_1 - 2$ vertices are placed on two non-adjacent edges of the 4-cycle, in this case $n_1 = n_2 = n_3 = n_4 = k_1 + 1$. Then $tb(\gamma_k) \geq 2 + 4tb(\gamma_{k_1+1})$. We use mathematical induction to prove that $tb(\gamma_k) \geq \frac{2(s_k^2-1)}{3} - (k-2)s_k$. Assume

$$tb(\gamma_{k_1+1}) \geq \frac{2(s_{k_1+1}^2 - 1)}{3} - (k_1 - 1)s_{k_1+1}.$$

Then

$$tb(\gamma_k) \geq 2 + 4tb(\gamma_{k_1+1}) \geq 2 + \frac{8}{3}(s_{k_1+1}^2 - 1) - 4(k_1 - 1)s_{k_1+1}.$$

We show this last quantity is equal to $\frac{2}{3}(s_k^2 - 1) - (k-2)s_k$. We have

$$\begin{aligned}
\tfrac{2}{3}(s_k^2 - 1) - (k-2)s_k &= \tfrac{2}{3}(2^{2\lfloor \log_2(2k_1-2)\rfloor} - 1) - (2k_1 - 2)2^{\lfloor \log_2(2k_1-2)\rfloor} \\
&= \tfrac{2}{3}(2^{2+2\lfloor \log_2(k_1-1)\rfloor} - 1) - 2(k_1 - 1)2^{1+\lfloor \log_2(k_1-1)\rfloor} \\
&= \tfrac{8}{3}2^{2\lfloor \log_2(k_1-1)\rfloor} - \tfrac{2}{3} - 4(k_1 - 1)2^{\lfloor \log_2(k_1-1)\rfloor} \\
&= \tfrac{8}{3}s_{k_1+1}^2 - \tfrac{2}{3} - 4(k_1 - 1)s_{k_1+1} \\
&= 2 + \tfrac{8}{3}(s_{k_1+1}^2 - 1) - 4(k_1 - 1)s_{k_1+1}.
\end{aligned}$$

Thus

$$tb(\gamma_k) \geq \frac{2}{3}(s_k^2 - 1) - (k-2)s_k.$$

For k odd: Let $k = 2k_1 + 1$. Consider the even K_4 subdivision for γ_k where $k_1 - 2$ vertices and $k_1 - 1$ vertices are placed on two non-adjacent edges of the 4-cycle, in this case $n_1 = n_2 = k_1 + 1$ and $n_3 = n_4 = k_1 + 2$. Then $tb(\gamma_k) \geq 2 + 2tb(\gamma_{k_1+1}) + 2tb(\gamma_{k_1+2})$. Assume

$$tb(\gamma_{k_1+1}) \geq \frac{2(s_{k_1+1}^2 - 1)}{3} - (k_1 + 1 - 2)s_{k_1+1} \text{ and}$$

$$tb(\gamma_{k_1+2}) \geq \frac{2(s_{k_1+2}^2 - 1)}{3} - (k_1 + 2 - 2)s_{k_1+2}.$$

Then

$$tb(\gamma_k) \geq 2 + \frac{4(s_{k_1+1}^2 - 1)}{3} - 2(k_1 - 1)s_{k_1+1} + \frac{4(s_{k_1+2}^2 - 1)}{3} - 2k_1 s_{k_1+2}.$$

Table 1 Values of b_r and c_r, for $3 \leq r \leq 14$

r	3	4	5	6	7	8	9	10	11	12	13	14
b_r	−1	−2	−4	−8	−16	−32	−64	−128	−256	−512	−1024	−2048
c_r	−1	−2	−4	−6	−10	−14	−18	−22	−30	−38	−46	−54

We show this last quantity is equal to

$$\frac{2}{3}(s_k^2 - 1) - (k - 2)s_k.$$

- For $k_1 = 2^t$, $\lfloor \log_2(k_1 - 1) \rfloor = t - 1$ and $\lfloor \log_2(2k_1 - 1) \rfloor = t$. So we have $s_{k_1+1} = 2^{t-1}$, $s_{k_1+2} = 2^t$, and $s_k = 2^t$. One can check that the two quantities are both equal to $-\frac{2}{3} + 2^t - \frac{4}{3}2^{2t}$.
- For $2^t < k_1 < 2^{t+1}$ we have $2^{t+1} - 1 < 2k_1 - 1 < 2^{t+2} - 1$. Then $\lfloor \log_2 k_1 \rfloor = t$, $\lfloor \log_2(k_1 - 1)] \rfloor = t$ and $\lfloor \log_2(2k_1 - 1) \rfloor = t + 1$. So $s_{k_1+1} = 2^t$, $s_{k_1+2} = 2^t$, and $s_k = 2^{t+1}$. One can check that the two quantities are both equal to $-\frac{2}{3} + \frac{8}{3}2^{2t} - (4k_1 - 2)2^t$. □

The lower bounds on $tb(\gamma_r)$, given by b_r from Proposition 1 and c_r from Proposition 2 are summarized in Table 1.

Remark 3 In Proposition 1, we assume that all of the 3-cycle are trivial unknots and that all of the edges of the γ_r of interest are only in unknotted cycles. This assumption is stronger than needed for $r \leq 14$ but incomplete for $r > 14$. To be able to have the c_r bound, there needs to be unknotted even K_4 subdivisions for γ_r, the cycles that would be 3-cycles in the K_4 if it were not subdivided need to be unknots, and any of their cycles that would be 3-cycles need to be unknots, continuing this way until one gets to an unknotted K_4 graph that is not subdivided. Let such a set of even K_4 subdivisions which start with γ_r be called a *special set of* K_4 *subdivisions for* γ_r. So for $r \geq 3$, if γ_r is an unknotted r-cycle in K_n with all 3-cycles trivial unknots, and all the graphs in a special set of K_4 subdivisions for γ_r are unknotted, then $tb(\gamma_r) \geq c_r$.

3 The Total Thurston–Bennequin Number for Complete Bipartite Graphs

This section is concerned with the total Thurston–Bennequin number of Legendrian complete bipartite graphs $K_{n,m}$. For such a graph, denote by P and Q the subsets of vertices in the n-partition and m-partition, respectively. Fix a choice of front projection $\widetilde{K}_{n,m}$. Let $w_{ae[P]}(\widetilde{K}_{n,m})$ denote the total signed sum of crossings over all pairs of edges adjacent to a vertex in P. Let $c_{v[P]}(\widetilde{K}_{n,m})$ denote the total number of cusps at vertices in P, taken over all pairs of adjacent edges. Let $w_{ne}(\widetilde{K}_{n,m})$ be the sum of writhes over all pairs of non-adjacent edges of G, where the orientation on each pair of edges is given by a choice of orientation on the 4-cycle they define.

Notice that the writhe of a pair (e,f) is independent of the choice of orientation on the 4-cycle, since the orientation of both edges will be changed if the orientation of the cycle is changed. The writhes and cusp counts depend on the front projection.

Theorem 2 *Let $K_{n,m}$ be a Legendrian embedding of a complete bipartite graph in $(\mathbb{R}^3, \xi_{std})$, with $n \geq m \geq 3$. Then*

$$TB_{2r}(K_{n,m}) = TB_4(K_{n,m}) \cdot \frac{(m-2)!(n-2)!}{(m-r)!(n-r)!}.$$

As a consequence,

$$TB(K_{n,m}) = TB_4(K_{n,m}) \cdot \sum_{r=2}^{m} \frac{(m-2)!(n-2)!}{(m-r)!(n-r)!}.$$

The quantities $TB_i(K_{n,m})$, $4 \leq i \leq m$, can be computed from writhe and cusp counts of vertices and edges in the front projection rather than summing the tbs of the cycles.

Proof We consider the writhe and number of cusps in a Legendrian front projection for $K_{n,m}$. For the writhe, we consider crossings of an edge with itself, crossings between adjacent edges and crossings between non-adjacent edges. To compute the number of cusps, we look at cusps along each edge and at cusps occurring at the vertices (between a pair of adjacent edges).

1. Each edge appears in $(n-1)(m-1)$ cycles of length 4 and in $\frac{(n-1)!(m-1)!}{(n-r)!(m-r)!}$ cycles of length $2r$.
2. Each pair of adjacent edges appears in $(n-1)$ cycles of length 4 if the two edges are adjacent at a vertex in P (the n-partition). Each pair of adjacent edges appears in $(m-1)$ cycles of length 4 if the two edges are adjacent at a vertex in Q (the m-partition). Each pair of adjacent edges appears in $\frac{(m-2)!(n-1)!}{(m-r)!(n-r)!}$ cycles of length $2r$ if the two edges are adjacent at a vertex in P. Each pair of adjacent edges appears in $\frac{(m-1)!(n-2)!}{(m-r)!(n-r)!}$ cycles of length $2r$ if the two edges are adjacent at a vertex in Q.
3. Each pair of non-adjacent edges appears in one cycle of length 4 and in $(2r-3)\frac{(m-2)!(n-2)!}{(m-r)!(n-r)!}$ cycles of length $2r$. We obtain this count as follows: The $r-2$ vertices in each partition that are in the cycle and not in one of the two non-adjacent edges can be chosen in $\binom{n-2}{r-2}\binom{m-2}{r-2}$ ways. Without loss of generality choose one of the edges of interest to start constructing the cycle. There are $2r-3$ positions where the second edge of interest can be placed in a $2r$-cycle, since it cannot be adjacent. See the first row of Fig. 8. Each of the $r-2$ vertices in each partition can occupy $r-2$ remaining positions relative to the chosen two non-adjacent edges. There are $(r-2)!(r-2)!$ ways to place these remaining vertices.
   Fix an arbitrary embedding $K_{n,m}$. In $(r-1)\frac{(m-2)!(n-2)!}{(m-r)!(n-r)!}$ of these cycles, the non-adjacent edges intersect with one orientation, the same orientation as their intersection in the 4-cycle. In the other $(r-2)\frac{(m-2)!(n-2)!}{(m-r)!(n-r)!}$ cycles, they intersect with

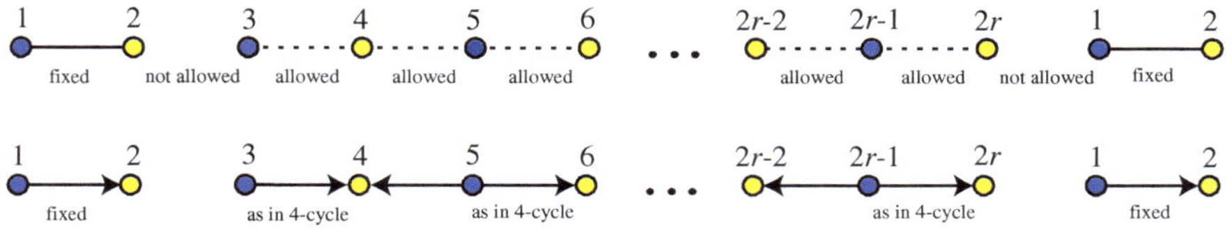

Fig. 8 Placement of two non-adjacent edges in a $2r$-cycle

the opposite orientation. See the second row of Fig. 8. This means that the contribution of the two edges to the writhe comes from $\frac{(m-2)!(n-2)!}{(m-r)!(n-r)!}$ of the $2r$-cycles. For each of these cycles, the contribution is the same as the one from the 4-cycle containing the two non-adjacent edges.

Items (1)–(3) above give

$$TB_{2r}(K_{n,m}) = TB_4(K_{n,m}) \cdot \frac{(m-2)!(n-2)!}{(m-r)!(n-r)!}.$$

Adding over all cycles gives

$$TB(K_{n,m}) = TB_4(K_{n,m}) \cdot \sum_{r=2}^{m} \frac{(m-2)!(n-2)!}{(m-r)!(n-r)!}.$$

We can also compute $TB_4(K_{n,m})$ from the writhe and cusp count for edges and vertices as follows:

$$TB_4(K_{n,m}) = (n-1)(m-1)w_e(\widetilde{K}_{n,m}) + (n-1)w_{ae[P]}(\widetilde{K}_{n,m}) + (m-1)w_{ae[Q]}(\widetilde{K}_{n,m}) + w_{ne}(\widetilde{K}_{n,m})$$
$$-\tfrac{1}{2}[(n-1)(m-1)c_e(\widetilde{K}_{n,m}) + (n-1)c_{v[P]}(\widetilde{K}_{n,m}) + (m-1)c_{v[Q]}(\widetilde{K}_{n,m})]. \qquad \square$$

Proposition 3 *For an unknotted minimal Legendrian $K_{3,3}$, three of its cycles of length six have $tb = -2$ and the other three have $tb = -1$. Additionally, any pair of 6-cycles with the same tb share three non-adjacent edges, and any pair of 6-cycles with different tbs share two sets of adjacent edges.*

Proof For $n = m = 3$ and $r = 3$, Theorem 2 says

$$TB_6(K_{3,3}) = TB_4(K_{3,3}).$$

The graph $K_{3,3}$ has nine cycles of length 4 and six cycles of length 6. If all nine 4-cycles are of maximal $tb = -1$, then the sum of *tb*s over all six cycles is -9. So there are at most three 6-cycles with $tb = -2$. In the following, we use our understanding of embeddings of K_4 to show that there are three 6-cycles with $tb = -2$. If we delete

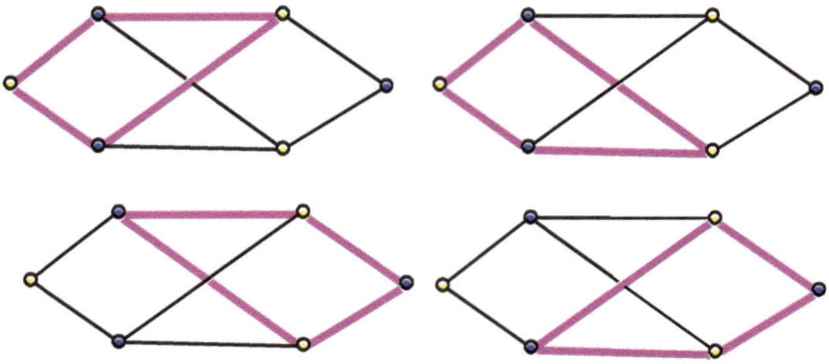

Fig. 9 The 3-cycles of the K_4 subdivision correspond to 4-cycles of $K_{3,3}$

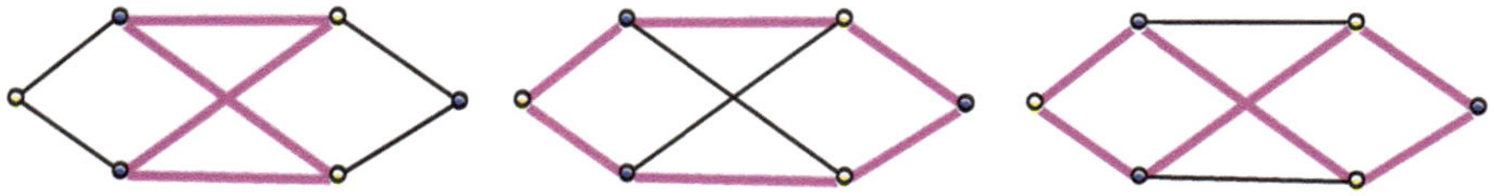

Fig. 10 The 4-cycles of the K_4 subdivision correspond to a 4-cycle of $K_{3,3}$ (*left*) and two 6-cycles of $K_{3,3}$ (*middle, right*)

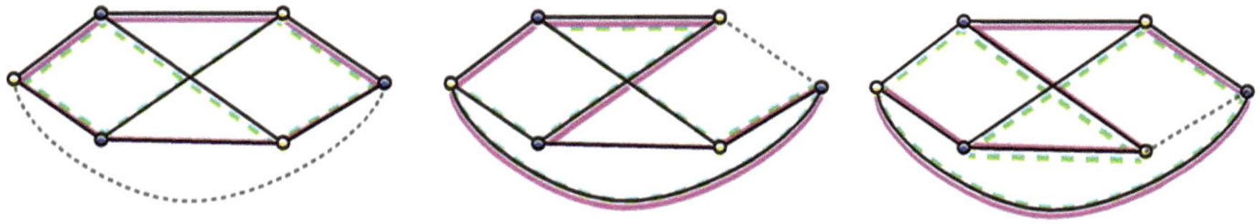

Fig. 11 By deleting the three edges adjacent to a vertex we see the six 6-cycles of $K_{3,3}$ as 4-cycles in the subdivision of K_4

one of the edges of $K_{3,3}$, we obtain a subdivision of K_4, call it K. We will describe K as a K_4 graph and ignore the valence 2 vertices, to simplify the explanation. The 3-cycles of K are 4-cycles of $K_{3,3}$. See Fig. 9.

One 4-cycle of K is a 4-cycle of $K_{3,3}$, while the other two 4-cycles of K are 6-cycles of $K_{3,3}$. See Fig. 10.

By assumption, all 4-cycles of $K_{3,3}$ have maximal $tb = -1$, so all 3-cycles of K have maximal $tb = -1$. This means exactly one of the 4-cycles of K has $tb = -2$, with the other cycles having $tb = -1$. The 4-cycles of K are coming from a 4-cycle of $K_{3,3}$ and two 6-cycles of $K_{3,3}$. Since all 4-cycles of $K_{3,3}$ have maximal $tb = -1$, one of the two 6-cycles must have $tb = -2$ and the other must have $tb = -1$.

All edges of $K_{3,3}$ are equivalent up to graph automorphism. So all K_4 subdivisions obtained by deleting a single edge of $K_{3,3}$ have the same structure. The set of three K_4 subdivisions shown in Fig. 11 contain all of the 6-cycles of $K_{3,3}$, with each cycle appearing once. So for each subdivision there is a different 6-cycle with $tb = -2$. Thus there are exactly three 6-cycles with $tb = -2$ and three 6-cycles with $tb = -1$.

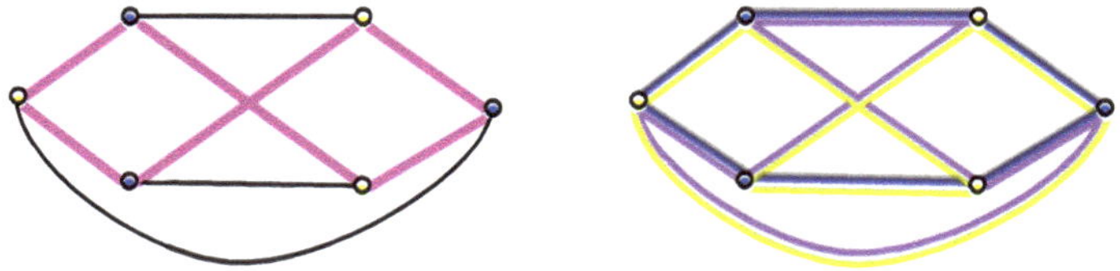

Fig. 12 The 6-cycle on the *left* shares two pairs of adjacent edges with each of three 6-cycles on the *right*

Fig. 13 An unknotted minimal embedding of $K_{3,3}$

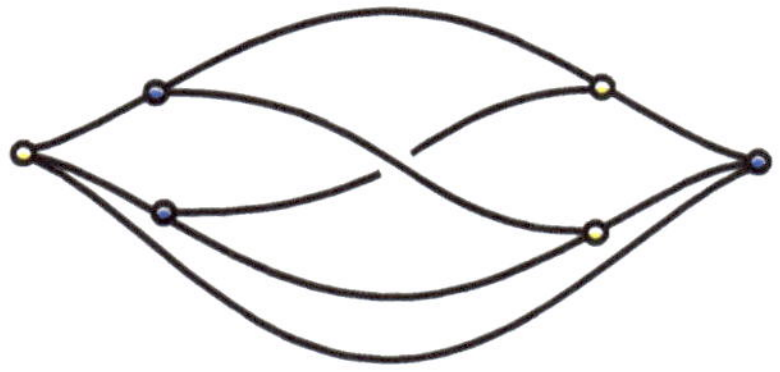

Since all of the K_4 subdivisions obtained by deleting a single edge of $K_{3,3}$ have the same structure, any pair of 6-cycles with different *tb*s will have the same structure as those in Fig. 10. Thus, any two 6-cycles with different *tb*s will share two pairs of adjacent edges. Let γ_6 be an arbitrary 6-cycle in an embedding $K_{3,3}$. Consider the set of three 6-cycles that share two pairs of adjacent edges with γ_6. See Fig. 12. The cycles in this set will all have the same *tb*s. Any pair of these 6-cycles with same *tb*s share three non-adjacent edges. □

From Proposition 3, we understand the structure of an unknotted minimal embedding $K_{3,3}$ well. Up to relabelling the cycles, this gives one possible $tb(K_{3,3})$ for an unknotted minimal embedding. Since unknots with $tb = -2$ and $tb = -1$ are unique, there is a unique *tb* and $|rot|$ for an unknotted minimal embedding $K_{3,3}$. In Fig. 13, we give an unknotted minimal embedding of $\mathrm{K}_{3,3}$. In Sect. 4, we show this embedding is equivalent to the one described after Theorem 2.

Remark 4 In a $K_{3,n}$ the vertices of a 6-cycles define a unique $K_{3,3}$ subgraph. Thus Proposition 3 implies for unknotted minimal embeddings $K_{3,n}$, half of the 6-cycles have $tb = -2$ and half of them have $tb = -1$.

Here, we take a moment to consider complete bipartite graphs $K_{n,m}$, with $n \geq m$ and $m < 3$. There are no cycles in the complete bipartite graphs $K_{n,1}$, so they are of little interest. The complete bipartite graphs with $m = 2$, i.e., $K_{n,2}$, are subdivisions of the θ_n-graphs. For the θ_n-graphs, the smallest cycles are 2-cycles. These are the only cycles. So there cannot be a nice relationship between the cycles, like that seen earlier. For completeness, we give a formula for the *TB* of an embedded θ_n-graph in terms of writhe and cusp counts of vertices and edges in the front projection. Let θ_n be a Legendrian embedding of the θ_n-graph. In a θ_n-graph there are no non-adjacent edges. Each edge appears in $(n-1)$ cycles and each pair of adjacent edges makes up one of the cycles. This gives

$$TB(\theta_n) = (n-1)w_e(\widetilde{\theta_n}) + w_{ae}(\widetilde{\theta_n}) - \frac{1}{2}[(n-1)c_e(\widetilde{\theta_n}) + c_v(\widetilde{\theta_n})].$$

4 Questions and Examples of Embeddings

In this section, we consider minimal embeddings of K_4, K_5, and $K_{3,3}$. We show the equivalence of diagrams of unknotted minimal embeddings discussed earlier. First we will recall the Reidemeister moves for Legendrian graphs. Two generic front projections of a Legendrian graph are related by Reidemeister moves I, II, and III together with three moves given by the mutual position of vertices and edges [1]. See Fig. 14.

In Fig. 3, we show two diagrams for the one unknotted minimal embedding K_4 that is known. In Fig. 15, we show four diagrams of unknotted minimal K_4s without crossings. To go between the top and bottom diagrams, in each column of Fig. 15, it takes two Reidemeister IV moves (the vertices are number to make it easier to see how this is done). In Fig. 16, we show how to go between the two diagrams in the top row of Fig. 15. Thus all diagrams in Fig. 15 are equivalent.

Finally in Fig. 17, we give the more complicated sequence showing that the left diagram from Fig. 3 is equivalent with the final diagram in Fig. 16. Thus the diagrams in Fig. 3 are equivalent.

We do not know of a different unknotted minimal Legendrian K_4, which leads us to the following question:

Question 1 Is this the unique unknotted minimal Legendrian K_4?

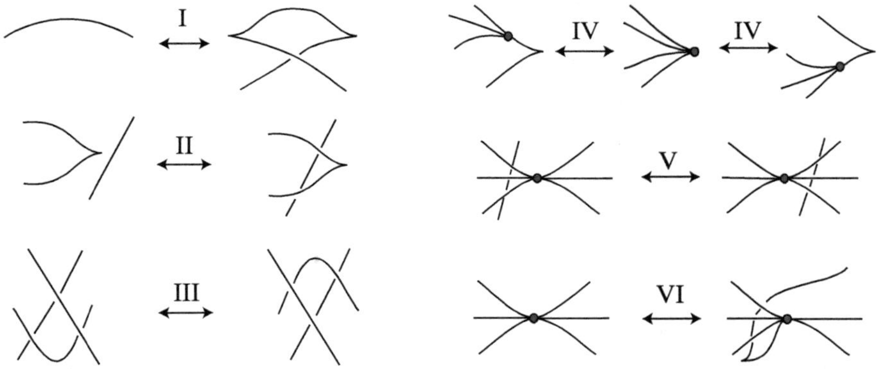

Fig. 14 Legendrian isotopy moves for graphs: Reidemeister moves I, II, and III, a vertex passing through a cusp (IV), an edge passing under or over a vertex (V), an edge adjacent to a vertex rotates to the other side of the vertex (VI). Reflections of these moves that are Legendrian front projections are also allowed

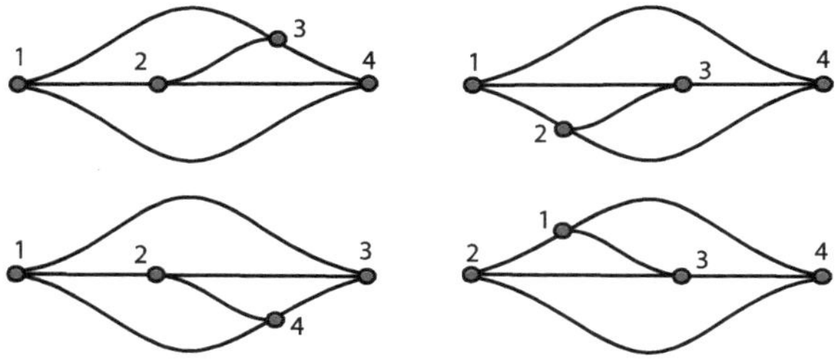

Fig. 15 Four unknotted minimal Legendrian K_4s

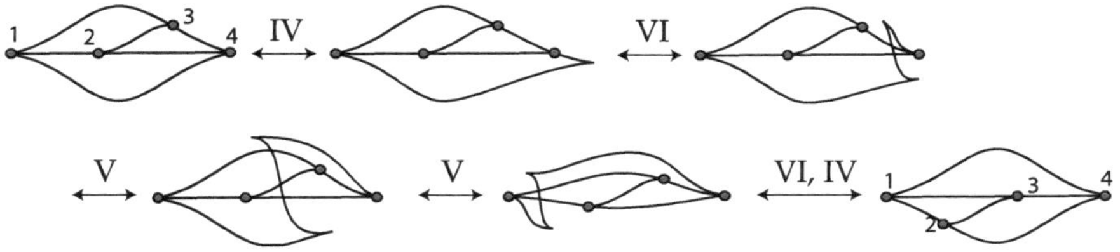

Fig. 16 A sequence of Legendrian K_4s related by Reidemeister moves

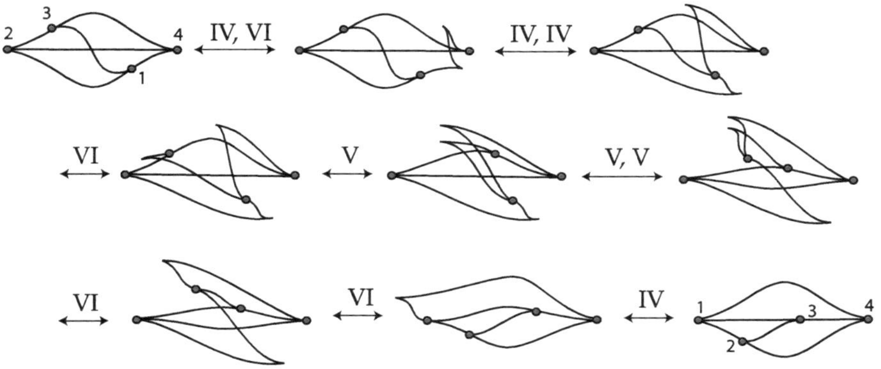

Fig. 17 A sequence of Legendrian K_4s related by Reidemeister moves

If we consider minimal embeddings rather than unknotted minimal embeddings there are a number of other possibilities. In Fig. 18, we give an infinite family of examples. For each $k \in \mathbb{Z}$ odd, this is a minimal Legendrian K_4 where:

- one 4-cycle is an unknot with $tb = -1$,
- one 4-cycle is an unknot with $tb = -k - 1$ ($rot = \pm 1$), and
- one 4-cycle is a $(2, k)$-torus knot with $tb = k - 2$ ($rot = 0$).

Question 2 Is there an embedding $K_4(L)$ where $tb_{max}(L) = 0$ and all cycles have maximal tb?

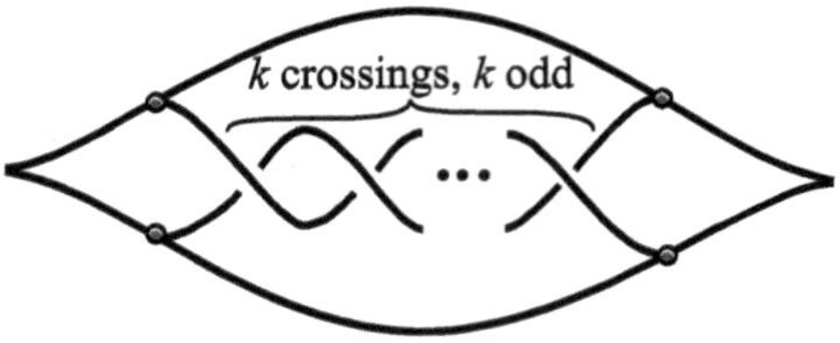

Fig. 18 Minimal Legendrian K_4 with one 4-cycle a $(2, k)$-torus knot with $tb = k - 2$

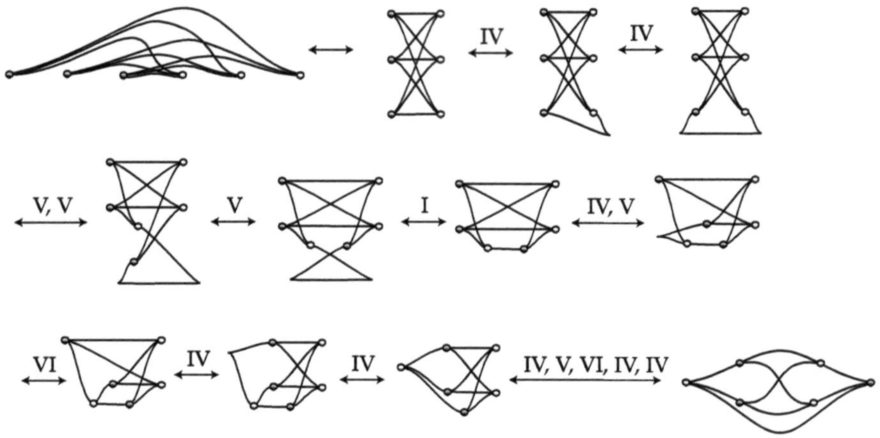

Fig. 19 A sequence of Legendrian $K_{3,3}$s related by Reidemeister moves and planar isotopy

For K_5, we have shown three different unknotted minimal embeddings, two with the sequence s_2 for the *tb*s of the 5-cycles and one with s_3 for the *tb*s of the 5-cycles. However, based on our calculations there are many other possible minimal embeddings.

Question 3 Is there an unknotted minimal Legendrian K_5 realizing the sequence s_3 different than the one in Fig. 4? (One way this could occur is if one or both 5-cycles with $tb = -3$ had rotation number ± 2.)

Question 4 Are there unknotted minimal Legendrian K_5s realizing any of the other sequences of *tb*s for the 5-cycles $(s_1, s_4, s_5, s_6, s_7, s_8, s_9, s_{10})$?

Conjecture 1. Any unknotted minimal Legendrian K_5 will contain at least one 5-cycle with $tb = -3$, and will not contain a 5-cycle with $tb = -4$.

Building on our examples of minimal Legendrian K_4s in Fig. 18, there are also infinitely many different possible minimal Legendrian K_5s which are not unknotted.

In Fig. 19, we show that the $K_{3,3}$ embedding described after Theorem 2 is the same as that shown in Fig. 13. For the graph $K_{3,3}$, having an embedding with all its smallest cycles trivial unknots seems to be a more rigid constraint than it is for complete graphs. We have not found any other unknotted minimal embeddings or minimal embedding of $K_{3,3}$.

Question 5 Is the embedding shown in Fig. 13 the unique unknotted minimal Legendrian $K_{3,3}$?

Question 6 Is the embedding shown in Fig. 13 the unique minimal Legendrian $K_{3,3}$?

Acknowledgments The authors thank Youngjin Bae, Byung Hee An and Gabriel C. Drummond-Cole for useful conversation, and Tim Cochran and John Etnyre for their interest and support. They also thank the referee for their careful reading and valuable suggestions.

References

1. S. Baader, M. Ishikawa, Legendrian graphs and quasipositive diagrams. Ann. Fac. Sci. Toulouse Math. **18**, 285–305 (2009)
2. Y. Eliashberg, Contact 3-manifolds twenty years since. J. Martinet's work. Ann. Inst. Fourier (Grenoble) **42**(1–2), 165–192 (1992)
3. Y. Eliashberg, M. Fraser, Topologically trivial legendrian knots. J. Symplectic Geom. **7**(2), 77–127 (2009)
4. H. Geiges, An introduction to contact topology, in *Cambridge Studies in Advanced Mathematics* vol. 109 (Cambridge University Press, Cambridge, 2008)
5. E. Giroux, Contact geometry: from dimension three to higher dimensions, in *Proceedings of the International Congress of Mathematicians*, vol. II, pp. 405–414 (Higher Ed. Press, Beijing 2002)
6. D. O'Donnol, E. Pavelescu, On Legendrian graphs. Algebraic Geom. Topol. **12**(3), 1273–1299 (2012)
7. D. O'Donnol, E. Pavelescu, Legendrian θ-graphs. Pac. J. Math. **270**, 191–210 (2014)
8. M. Yamamoto, Knots in spatial embeddings of the complete graph on four vertices. Topol. Appl. **36**, 291–298 (1990)

Coverings of Open Books

Tetsuya Ito and Keiko Kawamuro

Abstract We study a coverings of open books and virtually overtwisted contact manifolds using open book foliations. We show that open book coverings produces interesting examples such as transverse knots with depth greater than 1. We also demonstrate explicit examples of virtually overtwisted open books.

Keywords Open book foliation • Virtually overtwisted contact structure • Coverings

Mathematics Subject Classification Primary 57M25 · 57M27 · Secondary 57M50

1 Introduction

In the classification of contact structures on oriented 3-manifolds, there is a dichotomy between tight and overtwisted contact structures. The classification of overtwisted contact structures is reduced to homotopy theory by Eliashberg [4]. This is not the case for tight contact structures and study of tight contact structures is an active topic in contact geometry. A tight contact structure is called *universally tight* if its universal cover is tight, and *virtually overtwisted* if it has a *finite* cover that is overtwisted. As a consequence of the geometrization, the fundamental groups of 3-manifolds are residually finite, which implies that every tight contact structure is either universally tight or virtually overtwisted (cf. [13]). Namely, *universally* overtwisted is equivalent to virtually overtwisted.

The idea of coverings plays important roles in many areas of mathematics, including study of contact structures. In this note, we identify a covering map of contact

T. Ito
Department of Mathematics, Graduate School of Science, Osaka University Toyonaka, Osaka 560-0043, Japan
e-mail: tetito@math.sci.osaka-u.ac.jp

K. Kawamuro (✉)
Department of Mathematics, University of Iowa, Iowa City, IA 52242, USA
e-mail: kawamuro@iowa.uiowa.edu

G. Letzter et al. (eds.), *Advances in the Mathematical Sciences*, Association for Women in Mathematics Series 6, DOI 10.1007/978-3-319-34139-2_5

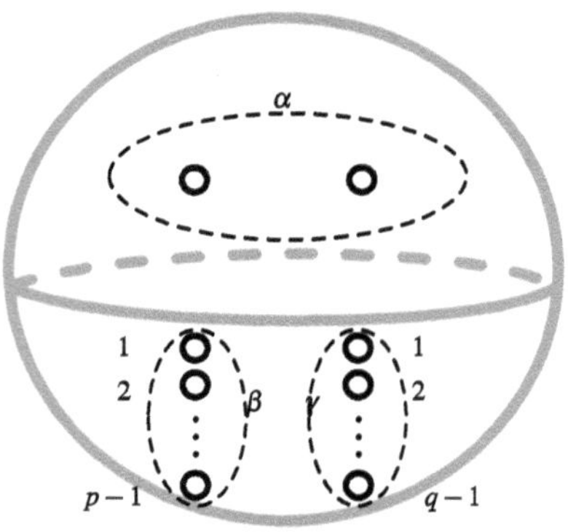

Fig. 1 The planar surface S with $p+q$ boundary components

manifolds with an open book covering map (see Sect. 2), and study virtually overtwisted contact manifolds using open book foliations. Here is one of the results.

Corollary 1. *Let B be the binding of an open book (S, ϕ). Then the* depth [1] *of the binding is* 1 *if and only if ϕ is not right-veering.*

In Sect. 4 we study examples of open books which have interesting properties. We give a family of planar open books that supports overtwisted, virtually overtwisted, and universally tight contact structures. Some non-planar examples are also discussed.

Proposition 1 *Let $S = S_{0,p+q}$ be a sphere with $p+q$ holes, where $p, q \geq 2$. Let $\alpha, \beta, \gamma \subset S$ be circles as shown in Fig. 1. Let $\phi \in \mathrm{Aut}(S, \partial S)$ be a diffeomorphism given by*

$$\phi = T \circ {T_\alpha}^n \circ T_\beta \circ T_\gamma$$

where T is the product of one positive Dehn twist along each of the $p+q$ boundary components and T_α is the positive Dehn twist along the curve α.

1. *If $n \leq -2$ then (S, ϕ) supports an overtwisted contact structure.*
2. *If $n = -1$ then (S, ϕ) supports a virtually overtwisted tight contact structure.*
3. *If $n \geq 0$ then (S, ϕ) supports a universally tight contact structure.*

2 Giroux Correspondence and Coverings

Let $S = S_{g,r}$ be an oriented genus g surface with r boundary components and $\phi \in \mathrm{Aut}(S, \partial S)$ be an orientation preserving diffeomorphism of S fixing the boundary ∂S pointwise. The pair (S, ϕ) is called an *abstract open book* (in this note the adjective "abstract" is omitted for simplicity) and $M_{(S,\phi)}$ denotes the closed oriented 3-manifold obtained by gluing the mapping torus of ϕ and solid tori. See Etnyre's lecture note [6] for basics (and more) of open books. The Giroux correspondence

[11] states that there is a one-to-one correspondence between open books (up to positive stabilization) and contact manifolds (up to isotopy). We denote by $\xi_{(S,\phi)}$ the (isotopy class of) contact structure on the manifold $M_{(S,\phi)}$ *compatible with* (or we often say *supported by*) the open book (S, ϕ) via the Giroux correspondence.

Throughout this note a covering means a *finite* covering. Suppose that $\pi : \tilde{S} \to S$ is a covering map.

Definition 1 If there exists a diffeomorphism $\tilde{\phi} \in \mathrm{Aut}(\tilde{S}, \partial \tilde{S})$ satisfying

$$\pi \circ \tilde{\phi} = \phi \circ \pi$$

then we call $(\tilde{S}, \tilde{\phi})$ a *covering* of the open book (S, ϕ). We write $\pi : (\tilde{S}, \tilde{\phi}) \to (S, \phi)$ abusing the notation and call it an *open book covering map.*

$$\begin{array}{ccc} \tilde{S} & \xrightarrow{\tilde{\phi}} & \tilde{S} \\ \pi \downarrow & & \downarrow \pi \\ S & \xrightarrow{\phi} & S \end{array}$$

Theorem 1 *Let $\pi : (\tilde{S}, \tilde{\phi}) \to (S, \phi)$ be an open book covering map. Then the compatible contact structures for the open books, via the Giroux correspondence [11], yield a covering map*

$$P : (M_{(\tilde{S},\tilde{\phi})}, \xi_{(\tilde{S},\tilde{\phi})}) \to (M_{(S,\phi)}, \xi_{(S,\phi)})$$

compatible *with π, namely the restriction of P to each page $\tilde{S}_t$ ($t \in [0, 1]$) satisfies $P|_{\tilde{S}_t} = \pi$.*

Proof For simplicity, we denote the covering space $(M_{(\tilde{S},\tilde{\phi})}, \xi_{(\tilde{S},\tilde{\phi})})$ by $(\tilde{M}, \tilde{\xi})$, and the base space $(M_{(S,\phi)}, \xi_{(S,\phi)})$ by (M, ξ). We naturally extend the projection $\pi : \tilde{S} \to S$ to a map $P : \tilde{S} \times [0, 1] \to S \times [0, 1]$ between the product manifolds such that the restriction of P to each page $\tilde{S}_t (\simeq \tilde{S})$ satisfies $P|_{\tilde{S}_t} = \pi$. By the commutativity $\pi \circ \tilde{\phi} = \phi \circ \pi$ we have $\phi(S_1) = \phi \circ \pi(\tilde{S}_1) = \pi \circ \tilde{\phi}(\tilde{S}_1) = \pi(\tilde{S}_0) = S_0$ thus the map $P : \tilde{S} \times [0, 1] \to S \times [0, 1]$ extends to the mapping tori $P : (\tilde{S} \times [0, 1])/\tilde{\phi} \to (S \times [0, 1])/\phi$ and then over to the bindings. Namely, the map P induces a covering map $P : \tilde{M} \longrightarrow M$.

Let α be a contact 1-form on M such that $\xi = \ker \alpha$. Let $\tilde{\alpha} := P^*\alpha$ be the pullback of α then $\tilde{\alpha} \wedge d\tilde{\alpha} = P^*(\alpha \wedge d\alpha) > 0$ and $\ker \tilde{\alpha}$ gives a contact structure on $\tilde{M}$ such that $P_*(\ker \tilde{\alpha}) = \ker \alpha = \xi$. This shows that $P : (\tilde{M}, \ker \tilde{\alpha}) \to (M, \xi)$ is a covering map. We also see that $(\tilde{M}, \ker \tilde{\alpha})$ is supported by the open book $(\tilde{S}, \tilde{\phi})$, that is, $\tilde{\alpha} > 0$ on the binding of the open book $(\tilde{S}, \tilde{\phi})$ and $d\tilde{\alpha} > 0$ on each page $\tilde{S}_t$. Thus the Giroux correspondence implies that $(\tilde{M}, \ker \tilde{\alpha})$ and $(\tilde{M}, \tilde{\xi})$ are isotopic, and the map $(\tilde{M}, \tilde{\xi}) \xrightarrow{P} (M, \xi)$ is a covering map. □

Conversely, we have the following.

Theorem 2 *Let $P : (\tilde{M}, \tilde{\xi}) \to (M, \xi)$ be a covering map for contact manifolds. For every open book (S, ϕ) supporting (M, ξ) there exists an open book $(\tilde{S}, \tilde{\phi})$ supporting $(\tilde{M}, \tilde{\xi})$ and giving an open book covering map $\pi : (\tilde{S}, \tilde{\phi}) \to (S, \phi)$ compatible with P.*

Proof Let S_t ($t \in [0, 1]$) denote the pages of the open book decomposition (S, ϕ) of M. Let $\tilde{S}_t := P^{-1}(S_t)$ and $\tilde{B} = P^{-1}(B)$, where $B \subset M$ is the binding for (S, ϕ). All the $\tilde{S}_t$ have the same topological type, denoted by $\tilde{S}$, and P induces a covering map $\pi : \tilde{S} \to S$. There exists $\tilde{\phi} \in \mathrm{Aut}(\tilde{S}, \partial\tilde{S})$ such that

$$\tilde{M} \backslash \tilde{B} \simeq (\tilde{S} \times [0, 1])/(x, 1) \sim (\tilde{\phi}(x), 0)$$

Since the pages S_0 and S_1 are identified under ϕ the commutativity $\pi \circ \tilde{\phi} = \phi \circ \pi$ holds. Thus we get an open book covering map $\pi : (\tilde{S}, \tilde{\phi}) \to (S, \phi)$ compatible with P.

By the same argument as in the proof of Proposition 1, we can show that $(\tilde{S}, \tilde{\phi})$ supports the contact manifold $(\tilde{M}, \tilde{\xi})$. □

Remark 1 For a covering map of contact 3-manifolds $P : (\tilde{M}, \tilde{\xi}) \to (M, \xi)$, not every open book decomposition $(\tilde{S}, \tilde{\phi})$ of $(\tilde{M}, \tilde{\xi})$ arises as an open book covering compatible with P.

To see this statement we recall the following simple fact, which easily follows from the definition of right-veering diffeomorphisms [14].

Lemma 1 *Let $\pi : (\tilde{S}, \tilde{\phi}) \to (S, \phi)$ be an open book covering map. Then ϕ is right-veering if and only if $\tilde{\phi}$ is right-veering.*

Now consider the case that (M, ξ) is tight and $(\tilde{M}, \tilde{\xi})$ is overtwisted. Then by [14, Theorem 1.1] there is an open book decomposition $(\tilde{S}, \tilde{\phi})$ of $(\tilde{M}, \tilde{\xi})$ such that $\tilde{\phi}$ is not right-veering. On the other hand, since (M, ξ) is tight every open book (S, ϕ) of (M, ξ) has right-veering ϕ. Hence Lemma 1 shows that the non-right-veering open book $(\tilde{S}, \tilde{\phi})$ cannot cover (S, ϕ).

3 The Overtwisted Complexity, Depth of Bindings and Open Book Coverings

In this section, we study properties of open book coverings using the notion of *right-veeringness* [14] and the open book foliation method [15].

Let us recall the *overtwisted complexity* $n(S, \phi)$ introduced in [17, Definition 6.4]. It is a nonnegative integer given by

$$n(S, \phi) = \min \{e_-(\mathcal{F}_{ob}(D)) \mid D \text{ is a transverse overtwisted disk in } (S, \phi)\},$$

if (S, ϕ) supports an overtwisted contact structure, and $n(S, \phi) = 0$ otherwise. Here, $e_-(\mathcal{F}_{ob}(D))$ denotes the number of negative elliptic points in the open book foliation on D. See Definition 4.1 of [15] for the definition of a *transverse overtwisted disk*, which can be understood as a transverse push-off of a usual overtwisted disk, or, the spanning disk of a transverse unknot K with $sl(K) = +1$.

The following property is proved in [17].

Proposition 2 ([17] Corollary 6.5)

1. $n(S, \phi) = 0$ *if and only if* $\xi_{(S,\phi)}$ *is tight (and hence* ϕ *is right-veering).*
2. $n(S, \phi) = 1$ *if and only if* $\xi_{(S,\phi)}$ *is overtwisted and* ϕ *is not right-veering.*
3. $n(S, \phi) \geq 2$ *if and only if* $\xi_{(S,\phi)}$ *is overtwisted and* ϕ *is right-veering.*

As a consequence we can show the following:

Proposition 3 *Let* $\pi : (\tilde{S}, \tilde{\phi}) \to (S, \phi)$ *be an open book covering such that* $n(\tilde{S}, \tilde{\phi}) = 1$ *then* (S, ϕ) *supports an overtwisted contact structure.*

Proof Suppose that (S, ϕ) supports a tight contact structure. Then ϕ is right-veering for every boundary component of S. By Lemma 1 $\tilde{\phi}$ is also right-veering for every boundary component of $\tilde{S}$. The property (3) of Proposition 3 implies that $n(\tilde{S}, \tilde{\phi}) \geq 2$, which is a contradiction. □

The overtwisted complexity is closely related to the depth of transverse knots or links introduced by Baker and Onaran in [1]: The *depth* of a transverse knot or link[1] K in an overtwisted contact 3-manifold (M, ξ) is defined by

$$d(K) = \min\{|D \cap K| \mid D \text{ is an overtwisted disk in } (M, \xi)\}$$

and K is called *non-loose* if $d(K) > 0$, that is, ξ is tight on $M \setminus K$.

Theorem 3 *Let B be the binding of an open book (S, ϕ) supporting an overtwisted contact structure. If (S, ϕ) supports an overtwisted contact structure then $d(B) = n(S, \phi)$.*

Proof Let D_{trans} be a transverse overtwisted disk realizing $n(S, \phi)$, that is, the open book foliation $\mathcal{F}_{ob}(D_{\text{trans}})$ has $n(S, \phi)$ negative elliptic points.

Let (B, π) be an open book decomposition of $M_{(S,\phi)}$ that is determined by the abstract open book (S, ϕ). By [15, Theorem 2.21] we may choose a contact structure ξ supported by (B, π) such that the characteristic foliation $\mathcal{F}_\xi(D_{\text{trans}})$ and the open book foliation $\mathcal{F}_{ob}(D_{\text{trans}})$ are topologically conjugate. Moreover we may assume that the set of positive/negative elliptic points of $\mathcal{F}_\xi(D_{\text{trans}})$ coincides exactly with the set of positive/negative elliptic points of $\mathcal{F}_{ob}(D_{\text{trans}})$.

Recall that a positive/negative elliptic point of the open book foliation on a surface F is just a positive/negative intersection point of F and the binding B. With this in mind, we denote by $B \cap^{\pm} D_{\text{trans}}$ the set of $\pm$-intersection points of D_{trans} and B.

---

[1]As mentioned in Remark 5.2.4 of [1] the depth can be defined for links though it is originally defined for knots.

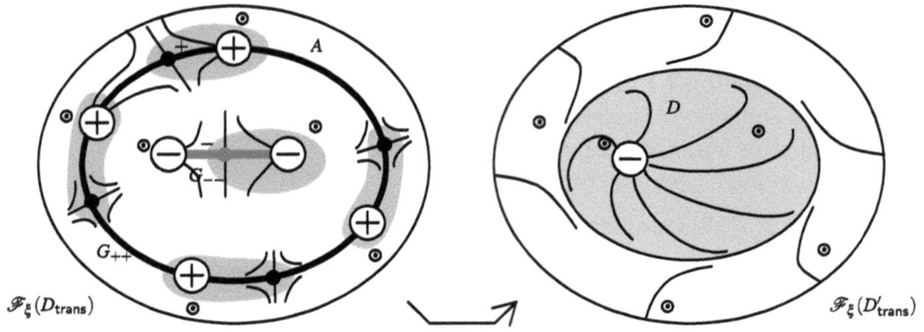

Fig. 2 Giroux elimination lemma is applied to the *gray* regions in the left disk. The *dots* ⊙ represent the intersection points $B' \cap D_{\mathrm{trans}}$ and $B' \cap D'_{\mathrm{trans}}$

Let B' be a transverse link that is obtained from B by transverse isotopy *only near* the intersection points $B \cap D_{\mathrm{trans}}$ so that

- $|B' \cap^+ D_{\mathrm{trans}}| = |B \cap^+ D_{\mathrm{trans}}|$ and $(B' \cap^+ D_{\mathrm{trans}}) \subset A$
- $|B' \cap^- D_{\mathrm{trans}}| = |B \cap^- D_{\mathrm{trans}}|$ and $B' \cap G_{--}(\mathcal{F}_\xi(D_{\mathrm{trans}})) = \emptyset$

where $A \subset D_{\mathrm{trans}}$ is the annulus bounded by the graph $G_{++}(\mathcal{F}_\xi(D_{\mathrm{trans}}))$ and the boundary $\partial D_{\mathrm{trans}}$, and $G_{++}(\mathcal{F}_\xi(D_{\mathrm{trans}}))$ (resp. $G_{--}(\mathcal{F}_\xi(D_{\mathrm{trans}}))$) is the Giroux graph in the characteristic foliation consisting of positive (resp. negative) elliptic points and stable (resp. unstable) separatrices of positive (resp. negative) hyperbolic points (see [11, page 646] and [15, Definition 2.17]). Since the two foliations $\mathcal{F}_\xi(D_{\mathrm{trans}})$ and $\mathcal{F}_{ob}(D_{\mathrm{trans}})$ are topologically conjugate, the graphs $G_{\pm\pm}(\mathcal{F}_\xi(D_{\mathrm{trans}}))$ and $G_{\pm\pm}(\mathcal{F}_{ob}(D_{\mathrm{trans}}))$ are topologically conjugate. By the definition of a transverse overtwisted disk [15, Definition 4.1] the graph G_{--} is a tree and G_{++} is a circle enclosing G_{--}. See Fig. 2, where $G_{--}(\mathcal{F}_\xi(D_{\mathrm{trans}}))$ and $G_{++}(\mathcal{F}_\xi(D_{\mathrm{trans}}))$ are depicted by the grey and the black bold arcs, respectively.

Note that B' is not used as a binding but it is just a transverse link. We also keep using the same contact structure ξ, hence the characteristic foliation $\mathcal{F}_\xi(D_{\mathrm{trans}})$ does not change.

We apply the *Giroux elimination lemma* [9, Lemma 3.3] to small 3-ball neighborhoods (gray regions in Fig. 2) of $G_{\pm\pm}(\mathcal{F}_\xi(D_{\mathrm{trans}}))$ each of which contains a pair of consecutive elliptic and hyperbolic points (of the same sign) and is disjoint from B'. We can find a disk, D'_{trans}, and a subdisk, D, of D'_{trans} with the following properties:

- D'_{trans} is C^0 close to D_{trans}.
- D is a standard overtwisted disk, i.e., its characteristic foliation contains exactly one elliptic singularity and $\mathrm{tb}(\partial D) = 0$.
- $\{B' \cap^+ D'_{\mathrm{trans}}\} = \{B' \cap^+ D_{\mathrm{trans}}\}$ and $|B' \cap^+ D| = 0$.
- $\{B' \cap^- D_{\mathrm{trans}}\} = \{B' \cap^- D'_{\mathrm{trans}}\} = \{B' \cap^- D\}$

Here the third property follows from the condition $(B' \cap^+ D_{\mathrm{trans}}) \subset A$. The fourth property follows from the condition $B' \cap G_{--}(\mathcal{F}_\xi(D_{\mathrm{trans}})) = \emptyset$. Though D'_{trans} may not admit an open book foliation this would not be a problem. We have

$$d(B) = d(B') \leq |B' \cap D| = |B' \cap^- D| = |B' \cap^- D_{\text{trans}}| = |B \cap^- D_{\text{trans}}| = n(S, \phi).$$

Thus $d(B) \leq n(S, \phi)$.

Conversely, let D be an overtwisted disk realizing $d(B)$, that is, $|B \cap D| = d(B)$. Taking the positive transverse push-off of the Legendrian boundary ∂D we find a transverse unknot, K, with $sl(K) = 1$. A spanning disk D' of K still intersects B at $d(B)$ points. By Pavalescu's proof of Alexander theorem [22, Theorem 3.2], there is an isotopy preserving each page and moving the non-braided parts of K to neighborhoods of the binding. In the neighborhoods, we can move D' so that $K = \partial D'$ becomes a closed braid without introducing negative intersection points of D' and B.

Following the discussion in the proof of [15, Theorem 4.3], from D' we can construct a transverse overtwisted disk D_{trans} whose open book foliation has no more than $d(B)$ negative elliptic points, hence $n(S, \phi) \leq d(B)$. □

As a consequence of Proposition 2 and Theorem 3 we have the following characterization of depth one bindings, which generalizes [1, Theorem 5.2.3] (except for the part regarding the tension invariant).

Corollary 1 *Let B be the binding of an open book (S, ϕ). Then $d(B) = 1$ if and only if ϕ is not right-veering.*

Corollary 1 gives a construction of Legendrian or transverse knots and links with large depth (cf. [1, Problems 6.1 and 6.4]).

Corollary 2 *Let B be the binding of an open book (S, ϕ) supporting an overtwisted contact structure. Let L be a Legendrian approximation of B. If ϕ is right-veering then $1 < d(B) \leq d(L)$.*

The inequality $d(B) \leq d(L)$ holds even without the right-veering assumption.

In fact, there are several constructions of right-veering open books supporting overtwisted contact structures as listed below:

1. [14, Proposition 6.1] Every open book can be made right-veering after a sequence of positive stabilizations.
2. By Theorem 2, for a covering map $P : (\tilde{M}, \tilde{\xi}) \to (M, \xi)$ between a tight (M, ξ) and an overtwisted $(\tilde{M}, \tilde{\xi})$ with an open book (S, ϕ) supporting (M, ξ), there exists an open book covering $\pi : (\tilde{S}, \tilde{\phi}) \to (S, \phi)$ compatible with P. By [14, Theorem 1.1] ϕ is right-veering and Lemma 1 implies that $\tilde{\phi}$ is right-veering. Such a family of examples is discussed in Proposition 1 where $d(\tilde{B}) = 2$.

If the bindings of a open book is not connected then by further positive stabilizations, which preserve the right-veering property, we can always make the binding connected. Hence it is fairly easy to construct a transverse or Legendrian knot with depth greater than 1.

We point out that if an open book $(\tilde{S}, \tilde{\phi})$ in the construction (2) is not destabilizable then it gives rise to an example of right-veering, non-destabilizable open book supporting an overtwisted contact structure. The existence (or nonexistence) of such open books is asked in [14] and many examples have been found [16, 19–21].

Presumably, under certain condition, open book coverings would provide non-destabilizable open books: In [8], it is shown that a right-veering open book (S, ϕ) is destabilizable if and only if the *translation distance* (see [8] for the definition) of ϕ is equal to one. Although the behavior of the translation distance under a covering operation is not clear, it is likely that if ϕ has a large translation distance then so does $\tilde{\phi}$, hence open book covering is non-destabilizable.

4 Illustration of Overtwisted Coverings and a Pants Pattern

In this section we study a sequence of open books that supports overtwisted, virtually overtwisted tight, and universally tight contact structures.

We begin with a proof of Proposition 1.

Proof We prove the assertion (1). Applying the proof of Theorem 4.1 in [16] we can construct a transverse overtwisted disk in the open book (S, ϕ). By the definition every transverse overtwisted disk has the self-linking number 1, that is, the Bennequin–Eliashberg inequality [5] is violated. Thus (S, ϕ) supports an overtwisted contact structure.

The assertion (3) follows from the same argument in Example 5.2 of [7].

Finally we prove the assertion (2). By the lantern relation (see for example [2, Proposition 5.1]) the mapping class ϕ can be written in the product of positive Dehn twists. Therefore, results of Giroux [11] and Eliashberg–Gromov [3] imply that (M, ξ) is tight.

Below we consider the following four cases. We find a transverse overtwisted disk in an open book covering for each case.
(Case 1) $p - 1 \equiv q - 1 \equiv 1 \pmod 2$;
(Case 2) $p - 1 \equiv q - 1 \equiv 0 \pmod 2$;
(Case 3) $p - 1 \equiv 1$ and $q - 1 \equiv 0 \pmod 2$;
(Case 4) $p - 1 \equiv 0$ and $q - 1 \equiv 1 \pmod 2$:

For each case we cut two copies of S along the thick gray arcs as shown in Fig. 3, then glue them along the cut arcs to get a connected surface $\tilde{S}$. Clearly $\tilde{S}$ is a double cover of S. We call the projection map $\Pi : \tilde{S} \to S$. (For Cases 3 and 4, the base space S is disconnected after the cut but it is easy to verify that the covering space $\tilde{S}$ is connected.)

Choose base points $x_0 \in S$ and $\tilde{x}_0 \in \Pi^{-1}(x_0)$. Let G be the index two subgroup of $\pi_1(S, x_0)$ defined by $G = \{\gamma \in \pi_1(S, x_0) \mid \langle[\gamma], [c]\rangle = 0\}$, where $\langle -, - \rangle : H_1(S) \times H_1(S, \partial S) \to \mathbb{Z}_2$ is the mod 2 algebraic intersection pairing and $[c] \in H_1(S, \partial S)$ is the relative homology class represented by the set of cutting arcs (with any choice of orientation). Note that the covering space $\Pi : \tilde{S} \to S$ has $\Pi_*(\pi_1(\tilde{S}, \tilde{x}_0)) = G$. Since $\phi_*(G) = G$ there is a homeomorphism $\tilde{\phi} : \tilde{S} \to \tilde{S}$ such that $\tilde{\phi}(\tilde{x}_0) = \tilde{x}_0$ and $\Pi \circ \tilde{\phi} = \phi \circ \Pi$. We call $\tilde{\phi}$ a lift of ϕ. For an arc $\tilde{\gamma}$ in $\tilde{S}$ the image $\tilde{\phi}(\tilde{\gamma})$ is nothing

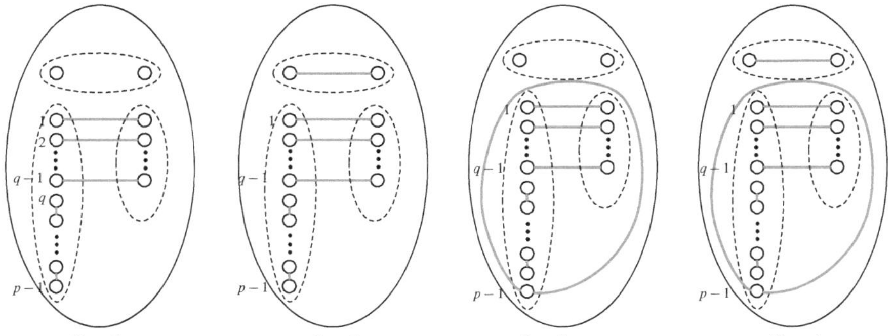

Fig. 3 Cutting arcs (*highlighted gray*) in S to construct $\tilde{S}$

but the lift of the arc $\phi(\pi(\tilde{\gamma}))$ in S. This allows us to compute $\tilde{\phi}$ and one can check that $\tilde{\phi}$ fixes the boundary $\partial\tilde{S}$ pointwise. In general it may be hard to write $\tilde{\phi}$ as the product of Dehn twists.

Figure 4 (resp. Fig. 5) gives a movie presentation of a transverse overtwisted disk for Case 1 (resp. Case 2). For Case 3 and Case 4 combining the ideas of Figs. 4 and 5 one can also find transverse overtwisted disks. We leave it to readers as an exercise. Therefore, the open book $(\tilde{S}, \tilde{\phi})$ supports an overtwisted contact structure. □

Remark 2 Let $K_{p,q} \subset (S^3, \xi_{st})$ be a Legendrian unknot in the standard contact 3-sphere with the Thurston–Bennequin number $\mathrm{tb}(K_{p,q}) = -(p+q)+1$ and the rotation number $\mathrm{rot}(K_{p,q}) = p-q$ or $q-p$, where $p, q \geq 2$. See Fig. 6.

Let (M, ξ) denote the contact structure obtained from (S^3, ξ_{st}) by the Legendrian surgery along $K_{p,q}$. When $n = -1$ in Proposition 1 we can verify that the open book (S, ϕ) supports the contact manifold (M, ξ) applying Schönenberger's algorithm [23] and the lantern relation.

With this identification of (S, ϕ) and (M, ξ) the assertion (2) of Proposition 1 can also be proved applying Gompf's criterion of virtually overtwisted contact structures [12].

Lastly, we note that if $p = 1$ or $q = 1$ then ξ is known to be universally tight due to Honda [13] and Giroux [10].

Remark 3 Under the projection $\pi : \tilde{S}_t \to S_t$ for each page we can see how the transverse overtwisted disk is 'folded' (in other words, self-intersecting) in the base tight manifold (M, ξ). For example when $t = t_1$ of Case 1 in Fig. 4 the projected image of the transverse overtwisted disk has two intersection points marked with black dots as in Fig. 9.

One can generalize the construction of overtwisted disks to a k-fold cover using the same cutting arcs of Fig. 3. For example:

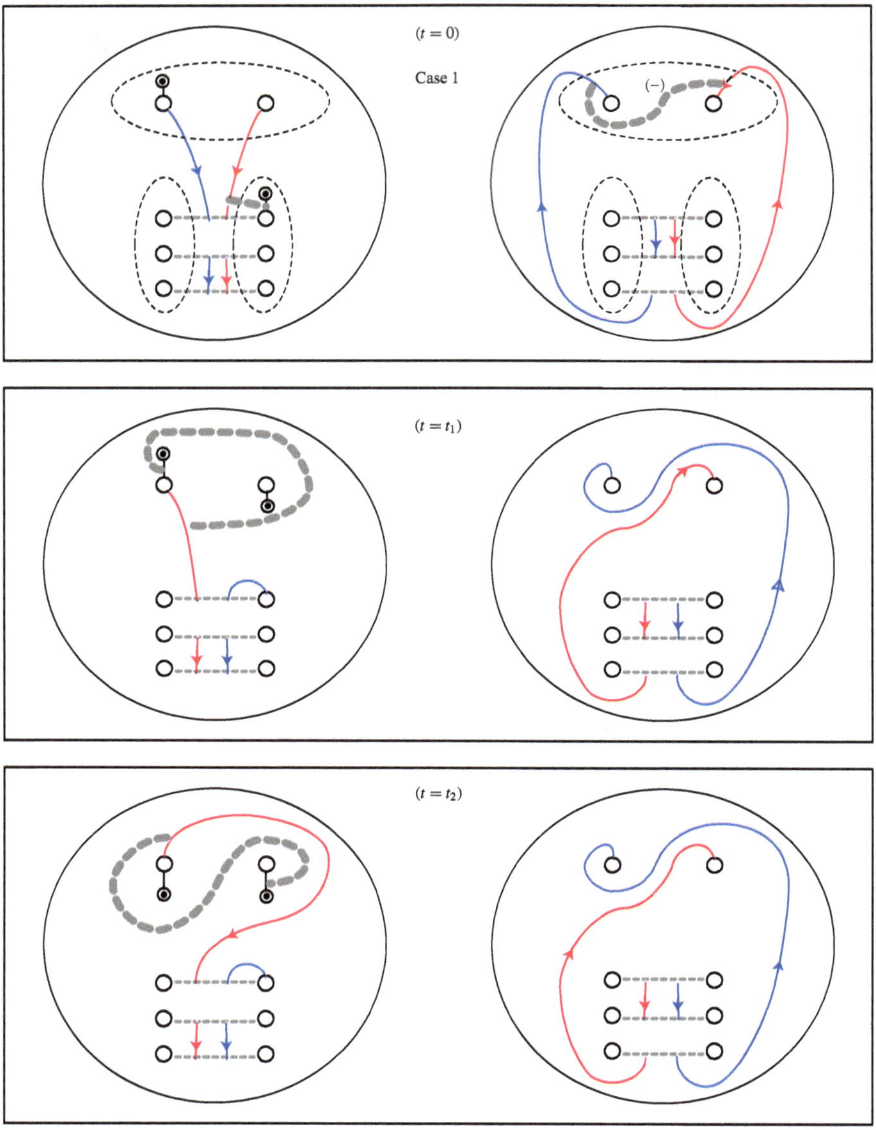

Fig. 4 (Case 1, $p-1=q-1=3$): A movie presentation of a transverse overtwisted disk. Each *arrow* indicates orientation of the b-arc. The starting (resp. ending) point of a b-arc is a positive (resp. negative) elliptic point. The end point of an a-arcs is marked with a $\odot$. This transverse overtwisted disk has two negative elliptic points and four positive elliptic points. Thick dashed arcs are describing arcs for hyperbolic points. The signs of the hyperbolic points are all positive except the one marked with (−) in the page $t=0$. One can easily generalize this to any p and q with $p-1 \equiv q-1 \equiv 1 \pmod 2$

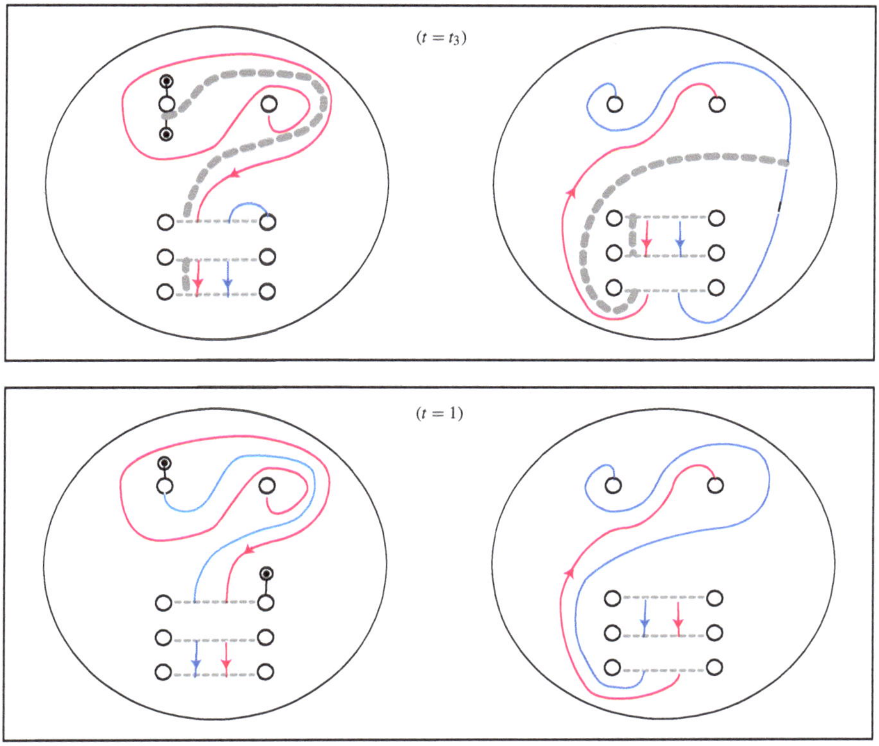

Fig. 4 (continued)

Example 1 Let S be a 2-sphere with four holes. Let $a, b, c, d, e \subset S$ be simple closed curves parallel to the boundary as shown in Fig. 10. Let

$$\Phi_{\alpha,\beta} = T_a^{\alpha+1} T_b^{\beta+1} T_c T_d T_e^{-1}.$$

Suppose that $\alpha, \beta > 0$ and there exists a number $k \geq 2$ that divides both $\alpha + 1$ and $\beta + 1$. Then there exists a k-fold cover of $(S, \Phi_{\alpha,\beta})$ that supports an overtwisted contact structure, i.e., $(S, \Phi_{\alpha,\beta})$ supports a virtually overtwisted tight contact structure.

So far we have only seen planar open books. In fact, our example can be applied to higher genus open books: Suppose that an open book (S', Ψ) supports a tight contact structure, the Nielsen–Thurston type of ϕ is reducible, and 'containing' $(S, \Phi_{\alpha,\beta})$ as a subspace, that is, $S \subset S'$ and $\Psi|_S = \Phi_{\alpha,\beta}$. Then (S', Ψ) supports a virtually overtwisted contact structure.

Example 2 Let S' be a genus 4 surface with two holes, d and c, see Fig. 7. Let $\Psi = T_a^2 T_b^2 T_c T_d T_e^{-1} T_f$ be a diffeomorphism of S'. The open book (S', Ψ) contains $(S, \Phi_{1,1})$ of Example 1 and (S', Ψ) supports a tight contact structure. Take a double

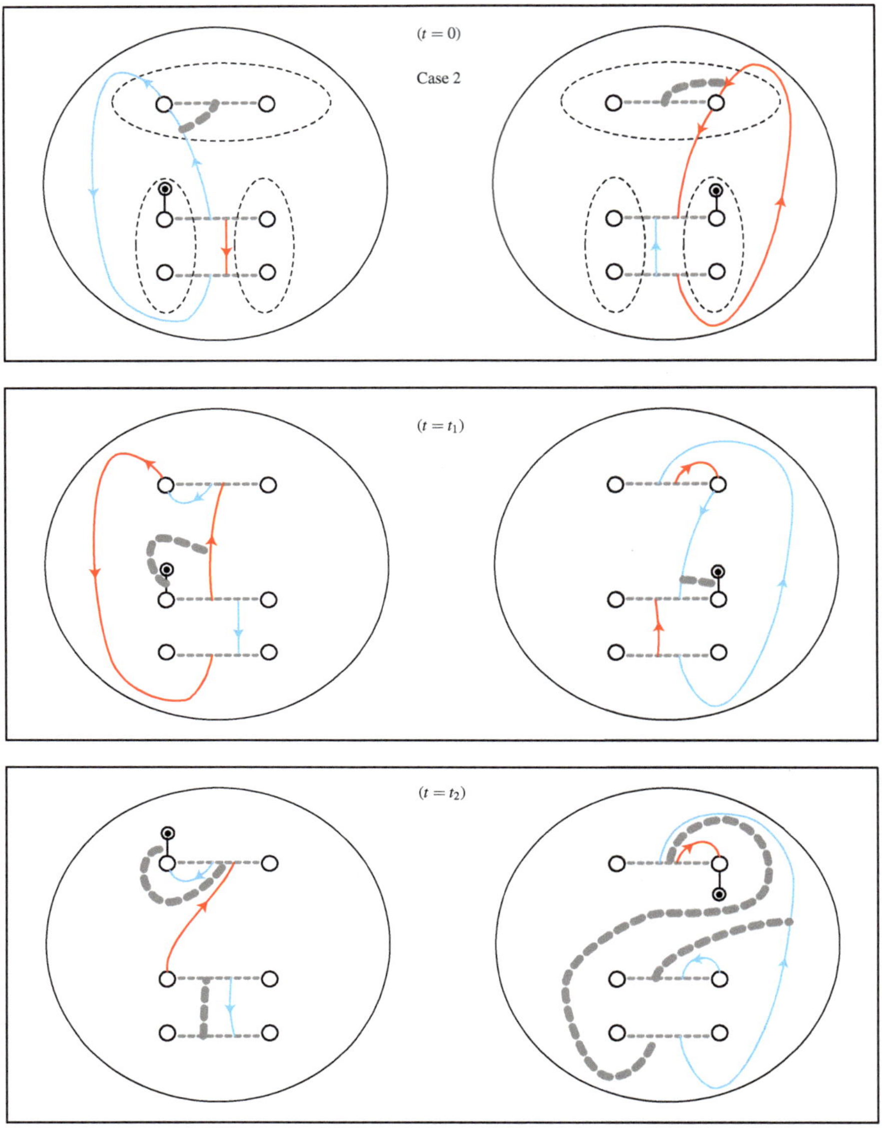

Fig. 5 (Case 2, $p-1=q-1=2$): A movie presentation of a transverse overtwisted disk. This transverse overtwisted disk has two negative elliptic points and four positive elliptic points. The signs of the hyperbolic points are all positive except the one in the page $t=0$. One can generalize this to any p and q with $p-1\equiv q-1\equiv 0 \pmod 2$

cover $\tilde{S}'$ of S' tjat os a genus 7 surface with four boundary components, $\tilde{c}, \tilde{d}, \tilde{c}', \tilde{d}'$. The monodromy Ψ lifts to a diffeomorphism $\tilde{\Psi} = T_{\tilde{a}} T_{\tilde{b}} T_{\tilde{c}} T_{\tilde{d}} T_{\tilde{e}}^{-1} T_{\tilde{f}} T_{\tilde{c}'} T_{\tilde{d}'} T_{\tilde{e}'}^{-1} T_{\tilde{f}'}$ of

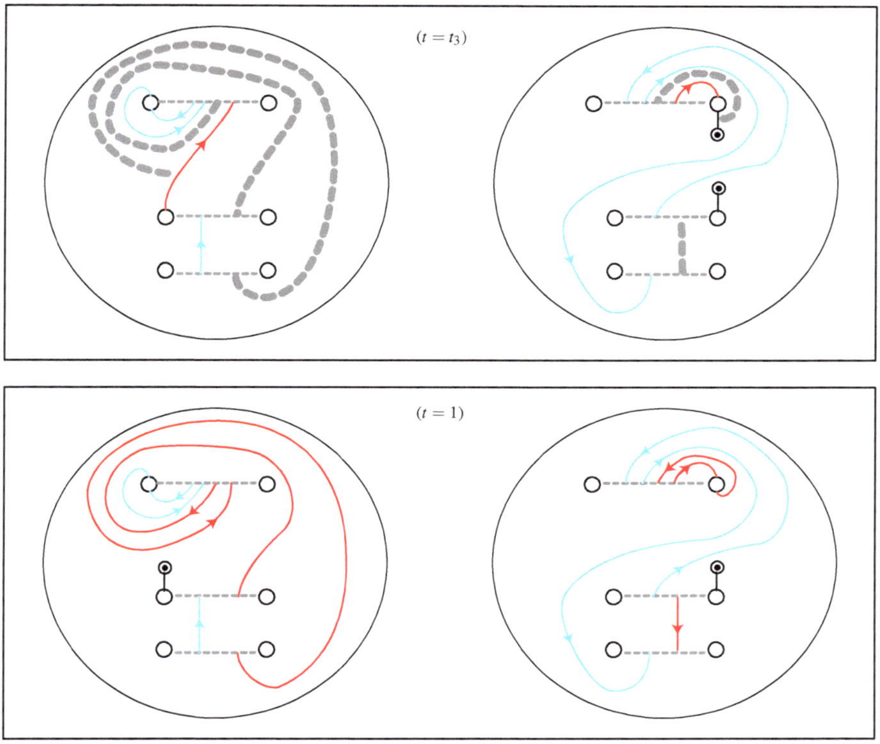

Fig. 5 (continued)

Fig. 6 The front projection of $K_{p,q}$

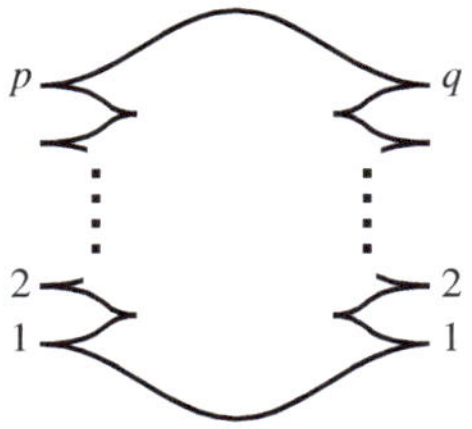

Fig. 7 Self-intersection points of the transverse overtwisted disk (of Fig. 4) under the projection $\pi : \tilde{S}_t \to S_t$

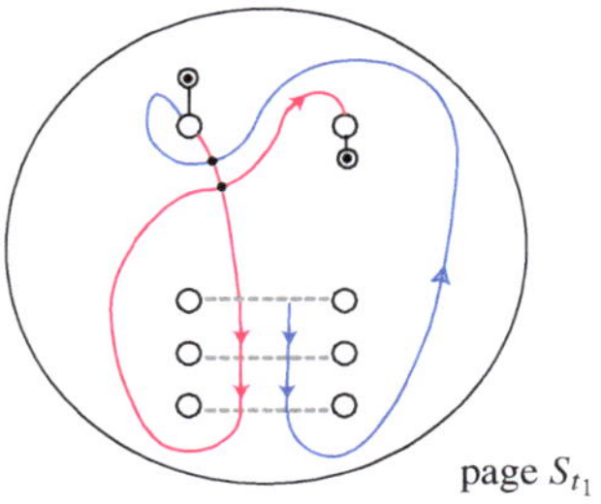

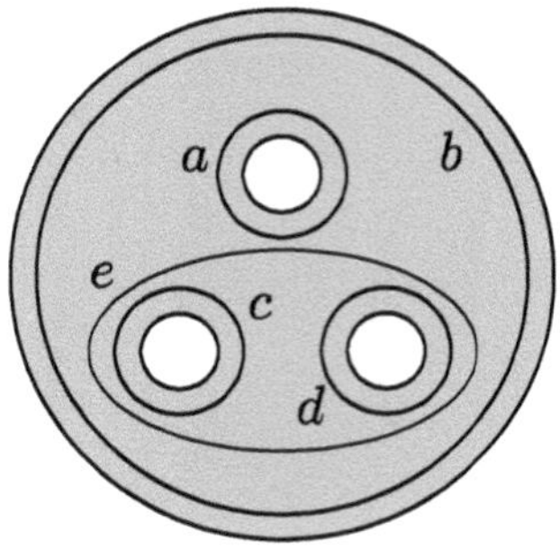

Fig. 8 Pants region P

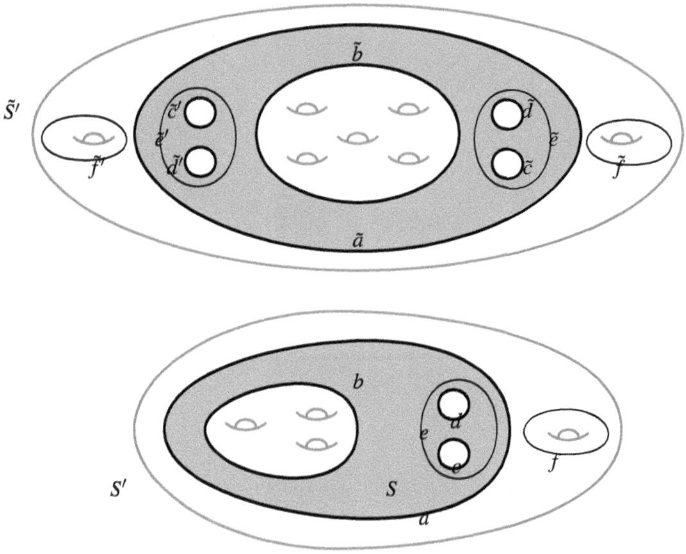

Fig. 9 (Top) A genus 7 surface $\tilde{S}'$ with four holes. (Bottom) An genus 4 surface S' with two holes

Fig. 10 The surface S

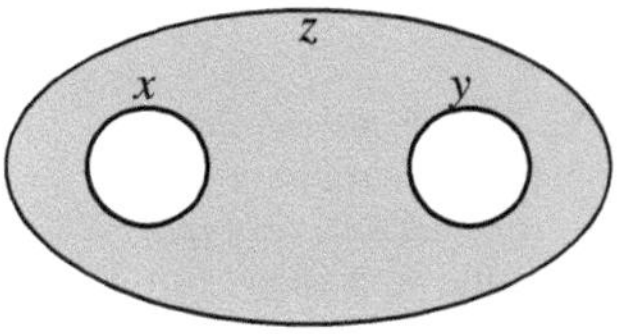

$\tilde{S}'$. Example 1 guarantees that the open book $(\tilde{S}', \tilde{\Psi})$ supports an overtwisted contact structure.

Remark 4 Lemma 1 and the discussion in Sect. 3 imply that if (S, ϕ) is a virtually overtwisted contact structure then its overtwisted cover has the overtwisted complexity (see Sect. 3) $n(\tilde{S}, \tilde{\phi}) \geq 2$. We notice that all the examples of virtually overtwisted open books (S, ϕ) we study in this note have $n(\tilde{S}, \tilde{\phi}) = 2$.

Moreover, these open books (S, ϕ) all contain a pants region $P \subset S$ (see Fig. 8) with the following properties

- P is bounded by curves x, y, z with $x, y \subset \partial S$ and $z \subset \mathrm{Int}(S)$,
- the monodromy ϕ preserves P and $\phi|_P = T_x T_y T_z{}^{-1}$.

(The curve z corresponds to α of Fig. 1 and e of Fig. 10). Such a pants region P plays a crucial role in our construction of transverse overtwisted disks because the two negative elliptic points of each transverse overtwisted disk lie on the lifts of x and y.

Question: Do there exist open book patterns, like the above pants pattern, that give virtually overtwisted contact structures?

Acknowledgments The authors would like to thank John Etnyre for useful conversation on Theorem 2, and John Etnyre, Jeremy Van Horn-Morris and Amey Kaloti for Proposition 4.1-(3). They also thank the referee for numerous comments that helped improving the paper significantly. TI was partially supported by JSPS KAKENHI Grant Numbers 25887030 and 15K17540. KK was partially supported by NSF grant DMS-1206770.

References

1. K. Baker, S. Onaran, Nonlooseness of nonloose knots. Algebr. Geom. Topol. **15**(2), 1031–1066 (2015)
2. B. Farb, D. Margalit, *A Primer on Mapping Class Groups*. Princeton Mathematical Series, 49. (Princeton University Press, Princeton, 2012)
3. Y. Eliashberg, M. Gromov, *Convex Symplectic Manifolds. Several Complex Variables and Complex Geometry, Part 2* (Santa Cruz, CA, 1989), Proc. Sympos. Pure Math., 52, Part 2, Amer. Math. Soc., Providence, RI (1991), pp. 135–162
4. Y. Eliashberg, Classification of overtwisted contact structures on 3-manifolds. Invent. Math. **98**(3), 623–637 (1989)
5. Y. Eliashberg, Contact 3-manifolds twenty years since J Martinet's work. Ann. Inst. Fourier (Grenoble) **42**, 165–192 (1992)
6. J. Etnyre, *Lectures on open book decompositions and contact structures.* In: Floer Homology, Gauge Theory, and Low-Dimensional Topology, Clay Math. Proc., vol. 5, pp. 103-141, American Mathematical Society, Providence, RI (2006)
7. J. Etnyre, J. V. Horn-Morris, Monoids in the Mapping Class Group. Geom. Topol. Monogr. **19**, 319–365 (2015)
8. J. Etnyre, Y. Li, The arc complex and contact geometry: nondestabilizable planar open book decompositions of the tight contact 3-Sphere. Int. Math. Res. Not. **2015**(5), 1401–1420 (2015)
9. E. Giroux, Convexité en topologie de contact. Comment. Math. Helv. **66**(4), 637–677 (1991)
10. E. Giroux, Structures de contact en dimension trois et bifurcations des feuilletages de surfaces Invent. Math. **141**(3), 615–689 (2000)
11. E. Giroux, *Géométrie de contact: de la dimension trois vers les dimensions supérieures*, in *Proceedings of the International Congress of Mathematicians*, Vol. II, (Higher Ed. Press, Beijing, 2002) pp. 405–414
12. R. Gompf, Handlebody construction of stein surfaces. Ann. Math. **148**(2), 619–693 (1998),
13. K. Honda, On the classification of tight contact structures. I. Geom. Topol. **4**, 309–368 (2000)

14. K. Honda, W. Kazez, G. Matić, Right-veering diffeomorphisms of compact surfaces with boundary. Invent. Math. **169**(2), 427–449 (2007)
15. T. Ito, K. Kawamuro, Open book foliation. Geom. Topol. **18**(3), 1581–1634 (2014)
16. T. Ito, K. Kawamuro, Visualizing overtwisted discs in open books. Publ. Res. Inst. Math. Sci. **50**, 169–180 (2014)
17. T. Ito, K. Kawamuro, *Essential open book foliations and fractional Dehn twist coefficient*, (Preprint)
18. T. Ito, K. Kawamuro, Operations on open book foliations. Algebr. Geom. Topol. **14**, 2983–3020 (2014)
19. W. Kazez, R. Roberts, Fractional Dehn twists in knot theory and contact topology. Algebr. Geom. Topol. **13**, 3603–3637 (2013)
20. Y. Lekili, Planar open books with four binding components. Algebr. Geom. Topol. **11**, 909–928 (2011)
21. P. Lisca, On overtwisted, right-veering open books. Pacific J. Math. **257**(1), 219–225 (2012)
22. E. Pavelescu, Braiding knots in contact 3-manifolds. Pacific J. Math. **253**, 475–487 (2011)
23. S. Schönenberger, Determining symplectic fillings from planar open books. J. Sympl. Geom. **5**(1), 19–41 (2007)

Part III
Mathematical Biology

Understanding Locomotor Rhythm in the Lamprey Central Pattern Generator

Nicole Massarelli, Allan Yau, Kathleen Hoffman, Tim Kiemel and Eric Tytell

Abstract The lamprey central pattern generator (CPG) for locomotion consists of a collection of neurons in the spinal cord that is responsible for producing the rhythmic neural activity used for swimming. Mechanoreceptors in the margin of the spinal cord, called edge cells, detect the bending of the body and provide sensory feedback for the CPG. Thus, edge cells are essential for the CPG's ability to respond to perturbations. To investigate the CPG's response to perturbations during swimming, we compute entrainment ranges for stochastic bending signals where Gaussian band-limited white noise is added on top of a sinusoidal signal. Experimentally, the lamprey spinal cord was bent back-and-forth to entrain the CPG's rhythm, and then Gaussian band-limited white noise was added to the sensory stimulus. Correspondingly, we also developed mathematical models of the CPG circuit. Using the same stimuli in the models as was used in the experiment, we examine which properties of the CPG circuit are related to the observed experimental results.

Keywords Locomotion · Sensory feedback · Central pattern generator · Stochastic perturbation

---

N. Massarelli · K. Hoffman (✉)
Department of Mathematics and Statistics, University of Maryland Baltimore County, Baltimore, MD 21250, USA
e-mail: khoffman@umbc.edu

N. Massarelli
e-mail: massa2@umbc.edu

A. Yau
Tufts University, Medford, MA 02155, USA
e-mail: allan.yau@tufts.edu

T. Kiemel
Department of Kinesiology, University of Maryland, College Park, MD 20742, USA
e-mail: kiemel@umd.edu

E. Tytell
Department of Biology, Tufts University, Medford, MA 02155, USA
e-mail: eric.tytell@tufts.edu

G. Letzter et al. (eds.), *Advances in the Mathematical Sciences*, Association for Women in Mathematics Series 6, DOI 10.1007/978-3-319-34139-2_6

Mathematics Subject Classification 92B25 · 93E03 · 37N25

1 Introduction

Locomotion is a complex behavior that involves interaction between sensory inputs, control circuits in the central nervous system, and the musculoskeletal system. Most animal locomotion requires rhythmic oscillations of appendages, such as the side-to-side motion of a fish's tail. Therefore, much of the work on neural control of locomotion has focused on central pattern generator (CPG) circuits [10] that can produce the muscle activity pattern for these rhythmic behaviors. The defining characteristic of a CPG is that it can produce its rhythm when artificially isolated [19], despite lacking its normal sensory input and descending control. To date, CPGs have been found to be involved in locomotion in both vertebrates and invertebrates that swim [5, 9, 23, 32], walk [1, 15, 26], and fly [36]. An interesting property of the CPG is entrainment: in the presence of a rhythmic stimulus close to its intrinsic frequency, the CPG will match the stimulus frequency. This phenomenon can be measured experimentally in the lamprey [12, 22]. Tytell and Cohen measured entrainment ranges for an isolated lamprey spinal cord, when it was bent at various points along the spinal cord [29].

The locomotor CPG is part of a complex feedback loop. CPGs provide feed forward signals that activate muscles to bend the body. The body shape feeds back into the CPG *via* proprioceptors. We focus on a particular proprioceptor in the lamprey: the edge cell. Grillner discovered that edge cells, which are located on the margin of the spinal cord of lamprey, act as stretch receptors [12, 27]. Edge cells are a primary mechanism for sensory input into the lamprey CPG. After perturbations to the body, edge cells contribute to adjustments of the CPG signal. Edge cells are essential to the entrainment effect, and more generally, are keys in the CPG's ability to adapt its frequency to its environment.

For effective locomotion, animals must cope with many different types of variability, yet we know relatively little about how CPGs respond to perturbations. We study how the CPG responds to both deterministic and stochastic perturbations by measuring entrainment ranges experimentally and computationally. We have previously studied deterministic entrainment in response to sinusoidal bending, both experimentally [29] and computationally [20]. We now investigate how stochastic perturbations to sinusoidal bending affect the entrainment of the lamprey CPG. Due to the presence of noise, there will be variability in the experimental CPG signal, and in the computational CPG model as well. To determine entrainment in the stochastic case, we measure entrainment with circular vector strength of CPG bursts to quantify how closely the CPG frequency matches the forcing frequency. Using this measure, we compute stochastic entrainment ranges, for a range of bending frequencies, to compare with the deterministic ranges. From experimental bending and simulations, we see that the CPG is robust to noise and is able to entrain to the underlying sinusoidal bending frequency.

2 Experimental Results

Following the deterministic entrainment procedure [29], we perform similar experiments on the lamprey spinal cord to determine how perturbations affect the entrainment of the CPG. Between 20 and 50 segments of the spinal cord and notochord were dissected out of adult lampreys (*Ichthyomyzon unicuspis*) and placed in a bath with physiological saline containing between 0.5–2.0 mM glutamate, maintained at 8 degrees C. A key property of the lamprey CPG is that even excised from the body, when placed in a bath of excitatory amino-acid D-glutamate, the CPG produces a similar motor output produced as that by the intact animal during swimming [31]. Figure 1a illustrates the experimental setup of the lamprey CPG where the small triangles indicate where the spinal cord is pinned down. The rostral end corresponds to the left end of Fig. 1a and the caudal end corresponds to the right end. Note that the pins leave only a few segments at the caudal end of the spinal cord free to bend while the rostral most end is immobile. The data produced by the experiment is a recording of the voltage of the signal produced from one side of a single segment of the spinal cord. Motor output is recorded with suction electrodes placed on the ventral roots [11, 21]. This is referred to as "fictive" swimming [31]. The larger arrows in Fig. 1 represent the glass suction electrodes that record the electrical signals from the ventral roots where electrodes 1 and 2 are on the left side of the body and electrode 3 is on the right side. These signals correspond to electrical activity produced by the CPG that would then innervate muscles used for swimming. Figure 1b illustrates the

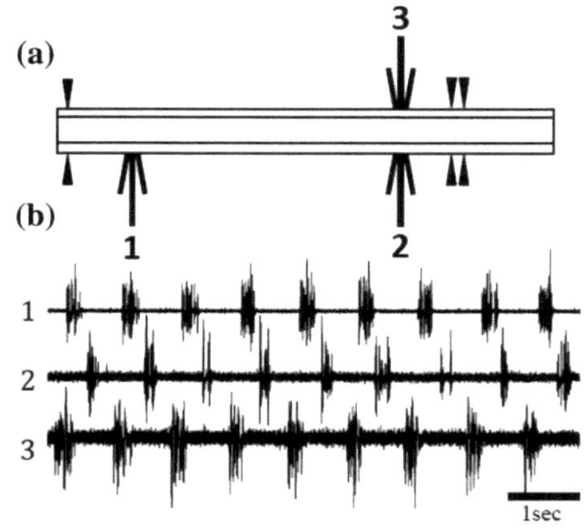

Fig. 1 Experimental recordings from ventral roots along the excised lamprey spinal cord in a bath of neurotransmitter. Figure **a** shows the experimental recording configuration. Glass suction electrodes are used to record from ventral roots along the notochord while bending the spinal cord back-and-forth. The *three large arrows* indicate three suction electrodes placed at different locations along the spinal cord. The *smaller triangles* indicate where the spinal cord was pinned down in the bath. Figure **b** shows a sample recording of the ventral root signals from segments along the spinal cord at positions indicated by the arrows in **a**, for a stationary spinal cord without bending

three signals recorded by the electrodes at various points along the spinal cord. This recording is the intrinsic CPG signal, that is, no bending stimulus was applied to elicit the CPG response. Note the recordings from electrodes 1 and 3 are bursting in anti-phase since they are taken directly opposite one another. Also there is a phase lag between recordings 1 and 2 because the electrodes are at different locations along the same side of the spinal cord. Periodically there are bursts in the data, which correspond to groups of action potentials. To analyze the data, we first determine when spikes and bursts of activity occur using the same methods used by Tytell and Cohen [29]. Any time the recording crosses a selected threshold, a spike time is recorded and burst times are calculated as the center of a cluster of spikes. Thus, we can see the intrinsic CPG frequency, f_0, from the burst frequency in the recordings.

To entrain the CPG to a different frequency f, a computer controlled motor is used to sinusoidally bend the spinal cord and notochord at a single point. An illustration of where the bending is applied to the spinal cord can be seen in Fig. 2g and an illustration of the bending signal is shown in Fig. 2a. To test the effects of perturbations, low-pass filtered Gaussian noise, with frequency components below 10 Hz, is added on to a steady motion that would normally entrain the CPG (see Fig. 2d). Comparing the bursting frequency during steady bending with the bursting frequency during noisy bending will allow us to determine how the CPG responds to perturbations. For the deterministic forcing, entrainment is determined by a straightforward comparison of the forcing frequency to the frequency of the CPG bursts. When the CPG is entrained, the bursts in each recording should occur with nearly the same frequency as the bending frequency since the CPG has altered its intrinsic frequency to match the bending stimulus. That is, the CPG now produces a signal at the bending stimulus frequency f instead of its natural frequency f_0. However, for the stochastic bending, the time between bursts will have more variability. To quantify the variability in burst times, we compute the circular vector strength of the burst phases. The phase of each burst is computed relative to the deterministic sinusoidal bending signal. We use the R statistic described in detail in Sect. 4 to quantitatively characterize the entrainment of the CPG.

Here we present the experimental results for one particular bending frequency to illustrate the effects of noisy bending on CPG entrainment. Figure 2 summarizes the comparison between the experimentally determined deterministic and stochastic entrainment for bending at 1.6 Hz. Figure 2a illustrates the sinusoidal bending along with the ventral root recordings, which show the CPG response to bending. A plot of burst frequency, or one over the time between bursts, is depicted in Fig. 2b where the bending frequency is 1.6 Hz, indicated with a solid black line. The three different color dots represent burst frequencies for the recordings at three different locations along the spinal cord, indicated by the arrows in Fig. 2g. The blue dots correspond to electrode 1, the red dots correspond to electrode 2, and yellow dots correspond to electrode 3. If the frequency of the bursts matches the frequency of the prescribed bending signal, then we conclude the CPG is entrained. We also plot, in Fig. 2c, the relative phases for each spike (phase of CPG spikes relative to the bending cycle) detected in the recording. For each recording, the spikes are centered around a constant phase in the bending cycle. This also indicates that the CPG is entrained

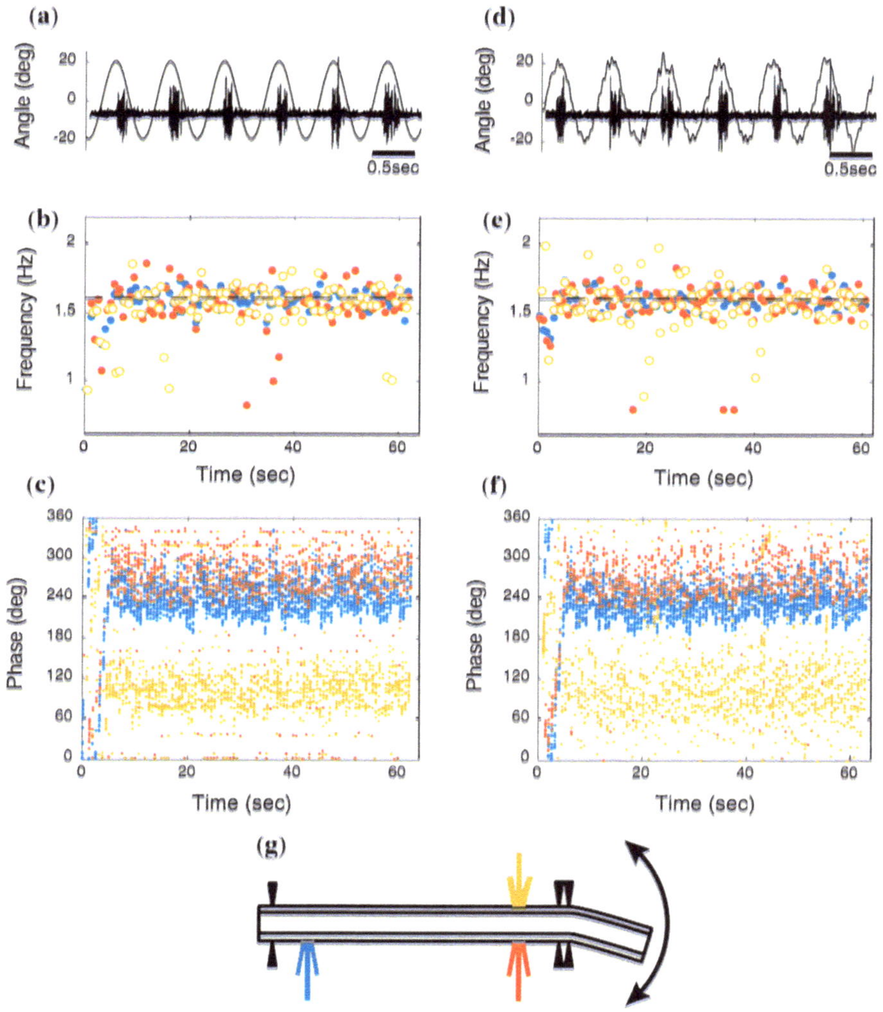

Fig. 2 Comparison of a deterministic sinusoidal forcing signal (**a-c**) and noisy sinusoidal forcing (**d-f**) applied to the last segment. *Blue dots* represent electrode 1, *red dots* represent electrode 2, and *yellow dots* represent electrode 3. Figures **a**, **d** show the relationship of the sinusoidal bending to a single ventral root recording. Figures **b**, **e** show that the oscillators have entrained to the 1.6 Hz forcing frequency. Figures **c**, **f** show the phase of the spikes relative to the forcing frequency. The noisy bending produces figures very similar to deterministic bending, indicating the lamprey CPG is robust to noise. Figure **g** shows the location of the recordings (*colored arrows*) and bending applied to the caudal (*right*) end of the spinal cord

to the forcing frequency $f = 1.6\,\text{Hz}$. When low-pass filtered Gaussian white noise is added to the bending signal, we still see bursts of activity in the CPG recording but there is more variability in the frequency. Again we compute the frequency and plot these for each cycle. Figure 2e shows the burst frequencies are more variable when noise is added to the bending signal. Despite the added noise, the burst frequencies are still clustered around $f = 1.6$ Hz. Figure 2f contains a plot of relative spike phases for the three electrode recordings. Note the spikes occur at roughly the same phase relative to the bending signal throughout the recording. This supports the conclusion that the CPG is entrained and bursts are occurring with the same frequency as the bending signal. These entrainment results hold for a range of bending frequencies and the exact range will be discussed in Sect. 4. Our analysis indicates that the entrainment of the lamprey CPG is robust to low-pass filtered Gaussian white noise as a perturbation.

3 Mathematical Modeling Results

Locomotor CPGs are commonly modeled as chains of coupled oscillators where each oscillator represents an anatomical segment of the spinal cord. These oscillators can then be modeled with various degrees of biological detail [3, 4, 6, 7, 17]. We simulate the experimental conditions described in Sect. 2 using the derived phase model in [20] to represent the lamprey CPG. Massarelli et al. [20] have studied deterministic entrainment ranges for the derived phase model. The derived phase model refers to the method of reducing a more detailed neural model, where individual neurons are modeled, to a phase model, where segments are modeled by a single variable which denotes its phase. The model represents each segment of the spinal cord with one variable corresponding to the phase of the ith oscillator, θ_i. We assume that each segment oscillates at an inherent frequency f_0. Each segment sends and receives input from the other oscillators through intersegmental coupling. This coupling is derived from a biologically detailed neural model, originally studied by Buchanan [2] and Williams [33], which describes connections between different cell types within a CPG segment.

We briefly review the derivation of the phase model from the neural model studied by Massarelli et al. [20]. This model was originally studied, without edge cell connections, by Buchanan [2] and Williams [33]. The cell classes and connections for one neural model oscillator are shown in Fig. 3. The neural model consisting of n oscillators, with forcing applied to the mth oscillator is represented by equations of the form

$$\dot{v}_{ij} = -G_R v_{ij} + G_T^j(1 - v_{ij}) + \sum_{k=1}^{n}\sum_{l=1}^{6} \alpha_{i-k}^{lj} G_0^{lj} h(v_{kl})(V_{\text{syn}}^l - v_{ij}) \tag{1}$$

$$+ \delta_{im}\alpha_f \sum_{s=1}^{2} G_f^{sj} h(v_{\text{ec}}^s(\theta_f))(V_{\text{syn,ec}}^{sj} - v_{ij}), \qquad \text{for } i = 1, \ldots, n;\ j = 1, \ldots, 6,$$

$$\dot{\theta}_f = \omega_f \tag{2}$$

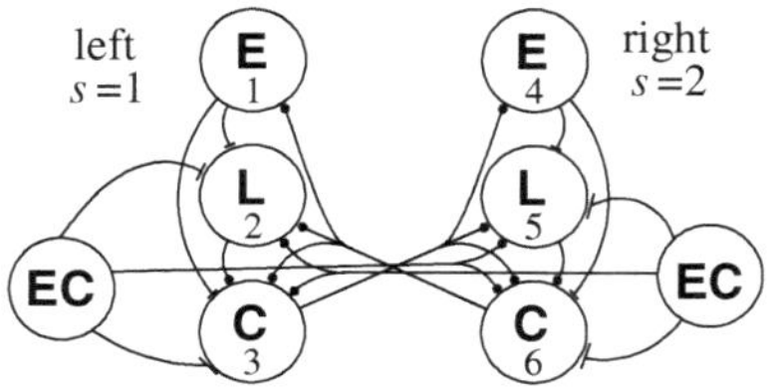

Fig. 3 Schematic of cell classes within one oscillator in the neural model described in [20]. The cell classes are excitatory interneurons (E), lateral inhibitory interneurons (L), crossed inhibitory interneurons (C), and edge cells (EC). Cells are labeled with their respective indices in the model and $s = 1$ or 2 denotes the left and right side of the oscillator respectively. Connections ending in bars are excitatory connections and connections ending with dots are inhibitory connections. Edge cells are only active in the segment where bending occurs

where

$$h(x) = \eta \log\left(1 + e^{\frac{x}{\eta}}\right)$$

is a smooth threshold function and

$$v_{\text{ec}}^{s}(\theta_f) = (-1)^s \sin(2\pi\theta_f)$$

defines the voltage of the edge cells on side s where θ_f is the phase of the forcer with frequency ω_f. The state variables v_{ij} denote the voltage of cell j in oscillator i. The first term, G_R denotes the resting conductance which drives the voltage to 0 while the second term, G_T, denotes the tonic excitatory conductance which drives the voltage towards 1. The double summation in (1) represents the connections between cells within oscillator i (intrasegmental connections) and also connections to the same cells in the other oscillators (intersegmental connections) where α_{i-k}^{lj} denotes the strength of the connection between cell j in oscillator i and cell l in oscillator k. Within the double summation, G_0^{lj} denotes the maximal synaptic conductance between cell types l and j. The function $h(x)$ is a smooth version of the threshold function used by Buchanan [2] and Williams [33]. As the parameter η goes to zero, (3) approaches the piecewise-linear version of h used in the original model. The parameter V_{syn}^{l} represents the reversal potential. The second single summation, multiplied by δ_{im}, represents the edge cell connections, which are only nonzero at the mth oscillator where forcing is applied. The strength of the edge cell connections is represented by α_f. Similarly to the other connections, G_{f}^{sj} denotes the synaptic conductance between the edge cells and cell j and $V_{\text{syn,ec}}$ denotes the reversal potential of the edge cells. The voltage of the edge cells, denoted by v_{ec}^{s} where $s = 1$ denotes the left side of the segment and $s = 2$ denotes the right side, is defined as a function of the forcing phase θ_f. We assume intersegmental coupling and forcing strength are small and thus (1) represents a chain of weakly coupled oscillators. The specific choice of coupling strengths α_{i-k} are defined later in (5) and (6). For a complete description of the neural model and parameters, along with the derivation of a phase model, please see [20].

Using the theory of weakly coupled oscillators [4, 24, 25], we approximate the neural connections with connections defined by phase differences, denoted in our model by H^{lj}. First, we numerically compute phase response curves (PRCs) for the neural model. PRCs measure the effect of a small perturbation, at a specified phase, on the phase of a single cell within the segment after several cycles. Thus, PRCs are a function of phase and illustrate the change in phase which would result if a perturbation occurred at that time. We then use PRCs and the standard techniques of phase reduction and averaging (see [13, 14, 18]) as applied to coupled oscillators by Varkonyi et al. [30]. The connection strength of an input from oscillator j to oscillator i is denoted by α_{i-j}. The following equations represent the derived phase oscillators comprising the CPG model:

$$\dot{\theta}_i = f_0 + \sum_{\substack{k=1 \\ k \neq i}}^{n} \sum_{j=1}^{6} \sum_{l=1}^{6} \alpha_{i-k}^{lj} H^{lj}(\theta_k - \theta_i) + \delta_{im} \left(\alpha_f \sum_{s=1}^{2} \sum_{j=1}^{6} H_f^{sj}(\theta_f - \theta_i) + \sigma\xi \right), \tag{3}$$

$$\dot{\theta}_f = f \tag{4}$$

for $i = 1, \ldots, n$, where forcing is applied to oscillator m with frequency f. The forcing phase, θ_f, is defined by (4). The triple summation represents the all-to-all coupling between segments within the chain. The double summation in (3) represents the contribution from the edge cells, where α_f is the strength of the connection. This term is multiplied by δ_{im}, the Kronecker delta function, which ensures the term is only nonzero when $i = m$. That is, we assume that the edge cells are only activated on the single segment where forcing is applied. Experimental evidence shows the existence of short and long connections among oscillators with decreasing connection strength for long connections [16], so we follow Varkonyi et al. [30] and choose intersegmental connection strengths which decay exponentially with length. Several experimental results suggest that there is asymmetry in coupling strengths, but the exact ratio is not known [16, 34, 35]. The exponential coupling strength is given by

$$\alpha_k = A_a e^{(-|k|/\lambda_a)} \tag{5}$$

for ascending connections of length $k = i - j$ and

$$\alpha_k = A_d e^{(-|k|/\lambda_d)} \tag{6}$$

for descending connections of length $k = i - j$ [30]. Here A_a, A_d, represent the amplitude of the coupling for ascending connections and descending connections. The scaling factors λ_a, λ_d, similarly control the relationship of the strength to the length of connections $|k|$. For these simulations, we use $A_a = 0.6$, $A_d = 0.04$, $\lambda_a = 0.75$, and $\lambda_d = 4$. For a complete description of the derived phase model (3) and its derivation see [20]. To this model, we add Gaussian band-limited white noise, ξ, to the forcing connection to represent the stochastic bending signal in the experiments. We vary the standard deviation of the added noise, denoted by σ, and compute entrainment ranges for various forcing frequencies f. Note $\sigma\xi$ is also multiplied by δ_{im} because we are only adding noise to the oscillator where edge cells are active.

To model the experimental results from Sect. 2, a chain of coupled phase oscillators was simulated under the same conditions as the biological experiments. Mathematically, the relative phase of the oscillators, ϕ_i, is computed by taking the difference $\theta_f - \theta_i$. With smooth sinusoidal forcing, constant relative phases indicate the CPG is entrained. That is, the oscillators in the chain all have the same frequency as the forcer θ_f. For sinusoidal forcing with low-pass filtered Gaussian white noise, represented in our model by the last term in (3), the relative phases will not be constant because of the noise added to the model. Figure 4 compares the simulations with deterministic and stochastic sinusoidal bending. Figure 4a illustrates the deterministic sinusoidal forcing signal. For a set range of frequencies, this signal will entrain the computational CPG for $\alpha_f = 3$. Figure 4b, c illustrate the entrainment of the CPG forced at oscillator 10. In Fig. 4c, the frequency is plotted for the entire length of the simulation for oscillators 9 and 10. Since the CPG is entrained, the frequency of both oscillators is the same as the forcing frequency (and is the same for all oscillators in the chain). When all the oscillators have the same frequency, the phases of the oscillators relative to the forcing signal will be constant, as seen in Fig. 4b. We can also force the chain of oscillators at different locations and we still see entrainment, as illustrated by Fig. 4d, where relative phases are plotted for forcing at the fifth oscillator.

We use the same types of plots for the stochastic sinusoidal forcing to determine the effects of perturbations on entrainment. To replicate the experimental results, we choose a noisy sinusoidal bending signal with same amount of noise relative to the amplitude of the underlying sinusoidal signal. That is, we choose the standard deviation of the noise relative to the amplitude of the sinusoid so that the ratio α_f/σ matches the experimental bending values. Figure 4e shows the noisy sinusoidal bending signal for our simulation, Fig. 4g plots the frequency of oscillators 9 and 10, and Fig. 4f shows the relative phase of oscillators 4, 8, 9 and 10 in the chain as a function of time. The computational CPG still appears entrained to the forcing frequency f despite the addition of noise. Figure 4g shows that the cycle frequency for oscillator 9 matches almost exactly with the forcing frequency while the frequency of oscillator 10 varies around f. Thus, the noisy forcing signal has a noticeable impact on the tenth oscillator where the forcing is applied, but not on the rest of the chain. Figure 4f shows the relative phases vary slightly around a mean phase consistent with the relative phase from the deterministic bending results, depicted in Fig. 4b. The relative phases also illustrate that the phase lag between segments is maintained throughout the simulation. More importantly, the simulations show that the noisy forcing on the tenth oscillator is reflected in the plot of its relative phase, but the noise drastically decreases in even the ninth oscillator. The relative phases of the remaining oscillators in the chain are also mostly unaffected. Thus, while noisy forcing creates more variability in the oscillator where bending is applied, the CPG signal from the remaining oscillators closely resembles the deterministic signal.

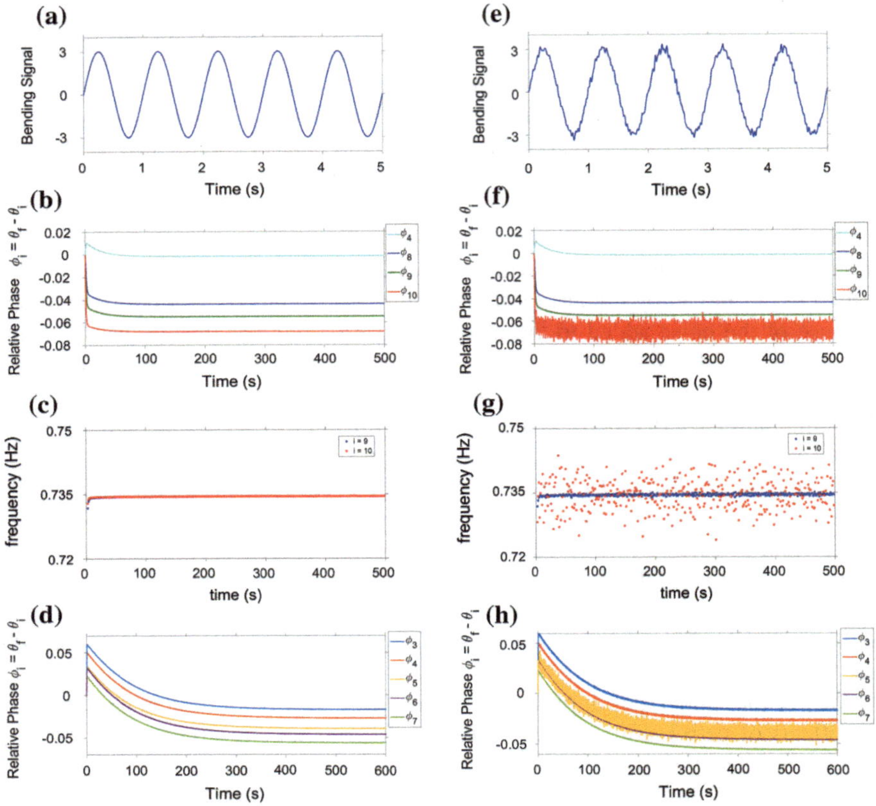

Fig. 4 Simulation of the derived phase model with sinusoidal (**a–d**) and noisy sinusoidal (**e–h**) forcing. Figure **a** shows the sinusoidal forcing signal applied to the tenth oscillator. Figure **c** plots the cycle frequency for oscillators 9 and 10 during the entire bending signal. Figure **b** shows the relative phases oscillators 4, 8, 9, and 10 with sinusoidal forcing applied to the tenth oscillator where $f - f_0 = 0.0005$. Figure **d** shows the relative phases oscillators 3 through 7 with sinusoidal forcing applied to the fifth oscillator where $f - f_0 = 0.0005$. Figure **e** illustrates the noisy sinusoidal forcing signal applied to the oscillator. Figure **g** plots the cycle frequency for oscillators 9 and 10 throughout the entire bending signal. Figure **f** shows the relative phases of oscillators 4, 8, 9, and 10 forced at $m = 10$ with $f - f_0 = 0.0005$, forcing strength $\alpha_f = 3$ and noise level $\sigma = 0.15$. Figure **h** shows the relative oscillators 3 through 7 forced at $m = 5$ with $f - f_0 = 0.0005$, forcing strength $\alpha_f = 3$ and noise level $\sigma = 0.15$

To further investigate how the noisy signal propagates along the chain, we simulate our model with noisy forcing applied to the middle of the chain at oscillator 5. Figure 4h shows the relative phases for oscillators 3 through 7 with the same noisy sinusoidal bending signal shown in Fig. 4e. We see the noise is still the largest where the forcing is applied at θ_5. Since forcing is applied to the middle of the chain, we can now see the difference in the oscillators above and below the forced oscillator. The relative phases of oscillators 3 and 4 are more noisy than the relative phases of oscillators 6 and 7. Our simulations, based on phase oscillators, indicate that the CPG sends information very close to the unperturbed signal when noisy sensory input is received. That is, the noise is mostly present in the oscillator being forced and the remaining oscillators still maintain the expected phase relationship. Thus, the CPG signal still produces a neural signal for steady swimming. Our results in this section for one forcing frequency $f = f_0 + 0.0005$ suggest that the CPG is strongly robust in the presence of noisy perturbations of sinusoidal bending. In the next section, we discuss how the model responds to different forcing frequencies and different levels of noise.

4 Entrainment Ranges

So far we have described experimental results for noisy sinusoidal bending in Sect. 2 and computational modeling of noisy sinusoidal bending in Sect. 3. These are two illustrative examples of the CPG remaining entrained to the underlying sinusoidal bending signal in the presence of noise. To summarize the effects of noise across a range of bending frequencies, we compute entrainment ranges for both the experimental bending data and the computational CPG model. For deterministic sinusoidal bending, entrainment corresponds to roughly constant relative phases. For the stochastically forced CPG, both the experimental and computational CPG signal will be more variable. Thus, to characterize entrainment for noisy sinusoidal bending, we calculate the circular mean and circular vector strength for the spike phase. Since our data is periodic and variable, we need to use circular statistics [8]. For spike phases, x_i, the circular mean is computed by the following

$$\bar{x} = \frac{1}{n}\sum_{i=1}^{n}\begin{pmatrix}\sin(2\pi x_i)\\ \cos(2\pi x_i)\end{pmatrix} \tag{7}$$

$$\bar{\theta} = \arctan(\bar{x}) \tag{8}$$

where $\bar{x}$ is the resultant vector and $\bar{\theta}$ denotes the average phase of the data. To compute circular variance we take

$$R = ||\bar{x}|| \tag{9}$$

where R denotes the circular vector strength. The circular mean, $\bar{\theta}$, represents the average phase of the CPG signal. For experimental data, $\bar{\theta}$ is the average phase of all the spikes in the CPG recording. These phases are plotted in Fig. 2c, f. For computational data, $\bar{\theta}_i$ is the average phase difference between θ_i and θ_f. To quantify the variability of these phases, we compare R where an R value closer to 1 means that there is a tight distribution of spike phases and that the spinal cord is entrained. A smaller value of R indicates there is more variability in the phases when the spikes

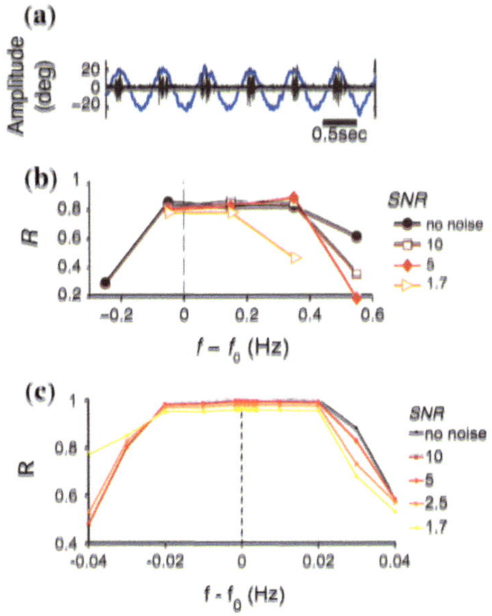

Fig. 5 Experimental and computational results indicate that the CPG is highly robust to noise. Figure **a** shows a sample recording showing entrainment of the CPG activity (*black*) to a noisy bending signal (*blue*) with SNR of 10. Figures **b** and **c** show the experimental (**b**) and computational (**C**) entrainment results for signals with a range of SNR values. R is the resultant vector length; $R > 0.8$ indicates entrainment

occur, and below some minimum value of R (approximately 0.8 in our results) the CPG is not entrained.

Figure 5 shows the R statistic plotted as a function of the entrainment range $f - f_0$ for different values of the signal-to-noise ratio (SNR). SNR is calculated by dividing the amplitude of the sinusoidal bending signal by the standard deviation of the Gaussian band-limited white noise which is added to the signal. Figure 5a plots a sample recording with noisy sinusoidal bending with SNR= 10. Figure 5b shows values of the R statistic as a function of forcing frequency obtained experimentally. Figure 5c shows the same plot using simulation of the derived phase model of the CPG with noisy sinusoidal forcing. In Fig. 5b, the R values are close to 1 for several SNRs when the forcing frequency is close to CPG's base frequency. However, as $|f - f_0|$ gets larger, the noise has more effect on the CPG signal and the R value decreases. For the computation model, the R-statistic is high for all levels of noise when $f - f_0$ lies in the deterministic entrainment range, but outside of this range the R statistic decreases. This is illustrated in Fig. 5c. Thus, when the computational CPG is entrained, it is highly robust to noise added to the forcing signal. Figure 5 illustrates that the lamprey CPG is highly resistant to noise, with no effect observed until the signal to noise ratio decreases to close to 1. The effects of noise are more subtle in the computational model in the sense that the noise is mostly seen in the oscillator where the noisy bending occurs, seen in Fig. 4f in oscillator 10. Also, the entrainment range for the derived phase model is much smaller (an order of magnitude less) than the experimental entrainment range. This is due to the assumptions of our model and our choice of coupling strength. In the derived phase model, entrainment ranges scale with the coupling strength. For example, if the intersegmental coupling was ten times stronger, the entrainment range would be ten times larger. Thus, the derived phase model still captures the important qualitative behavior of the experimental entrainment data.

5 Discussion

Vertebrate locomotion is a complex system involving the interaction of the locomotor CPG, muscle, body, and environment. The CPG also receives feedback from proprioceptors that sense how the body bends. We focus on the proprioceptive feedback from edge cells, mechanosensory cells on the margin of the lamprey spinal cord. We experimentally and computationally bend the spinal cord of the lamprey with a noisy sinusoidal signal and determine the range of forcing frequencies for which the CPG entrains to the forcer. To visualize the effect of the added noise on the entrainment of the CPG, we plotted burst frequencies and relative spike phases. Figures 2 and 4 compare the burst frequency and relative phases for experimental data and model simulations. From these comparisons, it appears the CPG is robust to noise during entrainment. In both the computational model and the experimental data, the burst frequency remains close to the forcing frequency f. To further quantify these results, we computed entrainment ranges as the signal-to-noise ratio increased. We define entrainment for noisy sinusoidal bending with the R statistic, where values above 0.8 indicate entrainment. The threshold 0.8 was chosen by comparing the experimental CPG recordings with the bending signal. When the forcing frequency is in the deterministic entrainment range, entrainment ranges for both experimental data and the derived phase model show that entrainment is mostly unaffected across a range of SNRs. The R statistic only begins to drop below 0.8 when the forcing frequency is much different than the base frequency (i.e., $|f - f_0|$ is large) and the SNR gets close to 1. Both our experimental data and our computational model indicate the CPG is highly robust to noisy sinusoidal bending in terms of maintaining entrainment.

As presented in this paper, we have experimental data for CPG output in response to bending and also simulated CPG output from a phase model. For this model, we chose a specific type of intersegmental coupling, namely nonuniform coupling asymmetry. This means that we choose ascending and descending coupling strengths so that for some connection lengths, ascending connections are stronger than descending ones, but for other lengths, ascending connections are weaker than descending. Previously, Massarelli et al. [20] computed deterministic entrainment ranges for both the neural model (1) and the derived phase model (3) and (4). The models best captured the qualitative properties of entrainment from experimental data when nonuniform coupling asymmetry was used [20]. This gave important insight into how intersegmental strengths vary along the spinal cord. The derived phase model, with nonuniform coupling asymmetry, also captures the stochastic bending results presented here. This further validates our model and supports our original claim that the lamprey CPG exhibits nonuniform coupling asymmetry for intersegmental connection strengths. One limitation of our model is that we assume edge cell inputs only affect a single segment. Anatomically, we know that edge cell axons may extend over multiple segments, with their axons going primarily rostrally (towards the head) [28]. Functionally, however, we do not know how strong these long connections are. Further work will be necessary to establish the effects of more distributed edge cell inputs.

Lamprey locomotion is part of a complex closed-loop system involving the CPG and sensory feedback from edge cells. To determine how these two components interact is challenging, because the CPG produces the signal to the muscles that bend the body, which in turn activates edge cells that feed back into the CPG. We know that during deterministic sinusoidal bending the edge cells provide sensory feedback which allows the CPG to alter its frequency to match the bending frequency f. To determine the effects of edge cell input on the CPG, we perturb the system and measure the resulting CPG signal. The differences between the deterministic and stochastic output help to illuminate how the edge cells affect the CPG signal and how entrainment is acheived. For forcing at the end of the chain, we saw our model closely resembled the experimental entrainment results (Fig. 5). In our derived phase model, we can easily vary where the forcing is applied in the chain. An interesting result from the noisy bending simulations arose from forcing the chain at the middle oscillator θ_5. We saw the noise from the forcing signal had a larger effect on oscillators above θ_5 and a smaller effect on the oscillators below. This result is especially interesting because we choose nonuniform coupling strengths, defined in (5) and (6), which means that short ascending connections are stronger than short descending connections but for longer connections the descending strengths are larger. Thus, the short connections may determine how much influence the sensory information from the edge cells has on the other oscillators in the chain. This result is supported by the relative phases from the experimental recordings plotted in Fig. 2c, f. Electrode 1 is farther away from the point of bending than electrodes 2 and 3 and the distribution of spikes in Fig. 2f for electrode 1 appear less variable. Note all of the recordings are above the point of bending so we cannot compare the effects of noisy bending on oscillators above and below the segment where forcing is applied. Thus, our model gives insight into how differences could arise based on differences intersegmental connection strengths. These modeling results are important because it is difficult to measure individual connection strengths experimentally. We also see that when noisy bending is applied to the 10th oscillator in the chain, the noise is greatly reduced in the rest of the oscillators. These results illustrate how noisy input into the CPG is filtered before it propagates to other segments in the spinal cord.

Our stochastic entrainment analysis characterizes some of the effects of noisy sinusoidal bending on the lamprey CPG. An alternative approach uses a harmonic transfer function (HTF), which fully characterizes the effects of small perturbations of a stable periodic system in the frequency domain. In our case, the periodic system corresponds to sinusoidal bending that entrains the CPG's rhythm and the perturbations are the noise added to bending. An HTF is an extension of the frequency response function (FRF), which fully characterizes small perturbations around a stable fixed point in the frequency domain. The FRF describes how sinusoidal input at any frequency f produces sinusoidal output at the same frequency, specifically it describes how gain (the ratio of output amplitude divided by input amplitude) and phase (the phase shift of the output relative to the input) vary across frequency. The idea behind the HTF is that for a periodic system, a perturbation at input frequency f produces outputs at multiple frequencies $f + kf_0$, where f_0 is the gait frequency and k is any integer. The HTF uses gain and phase to describe this input-output mapping for each k. We are currently applying this analysis to experimental data like that presented above.

Acknowledgments The authors wish to acknowledge that this work was partially funded by NSF Grant DBI-RCN 1062052, NSF Grant BCS-123011. This material is based upon work supported by, or in part by, the U.S. Army Research Laboratory and the U.S. Army Research Office under contract/grant number W911NF-14-1-0268.

References

1. A. Borgmann, S.L. Hooper, A. Büschges, Sensory feedback induced by front-leg stepping entrains the activity of central pattern generators in caudal segments of the stick insect walking system. J. Neurosci. **21**(9), 2972–2983 (2009)
2. J.T. Buchanan, Identification of interneurons with contralateral, caudal axons in the lamprey spinal cord: Synaptic interactions and morphology. J. Neurophys. **47**, 961–975 (1982)
3. A.H. Cohen, G.B. Ermentrout, T. Kiemel, N. Kopell, K.A. Sigvardt, T.L. Williams, Modeling of intersegmental coordination in the lamprey central pattern generator for locomotion. Trends Neurosci. **15**, 434–438 (1992)
4. A.H. Cohen, P.J. Holmes, R.H. Rand, The nature of the coupling between segmental oscillators of the lamprey spinal generator for locomotion: A mathematical model. J. Exp. Biol. **116**, 345–369 (1982)
5. A.H. Cohen, P. Wallén, The neuronal correlate of locomotion in fish. 'Fictive swimming' induced in an in vitro preparation of the lamprey. Exp. Brain Res. **41**, 11–18 (1980)
6. Ö. Ekeberg, A combined neuronal and mechanical model of fish swimming. Biol. Cybern. **69**, 363–374 (1993)
7. Ö. Ekeberg, S. Grillner, Simulations of neuromuscular control in lamprey swimming. Phil. Trans. Roy. Soc. Lond. B **354**(1385), 895–902 (1999)
8. N.I. Fisher. *Statistical Analysis of Circular Data*. (Cambridge University Press, cambridge, 1995)
9. S. Grillner, On the generation of locomotion in the spinal dogfish. Exp. Brain Res. **20**, 459–470 (1974)
10. S. Grillner, The motor infrastructure: from ion channels to neuronal networks. Nat. Rev. Neurosci. **4**, 573–586 (2003)
11. S. Grillner, A. McClellan, C. Perret, Entrainment of the spinal pattern generators for swimming by mechanosensitive elements in the lamprey spinal cord in vitro. Brain Res. **217**, 380–386 (1981)
12. S. Grillner, T. Williams, P.-Å. Lagerbäck, The edge cell, a possible intraspinal mechanoreceptor. Science **223**(4635), 500–503 (1984)
13. J. Guckenheimer, P. Holmes. *Nonlinear Oscillations, Dynamical Systems, and Bifurcations of Vector Fields*. (Springer, Berlin, 1990)
14. F.C. Hoppensteadt, E.M. Izhikevich, *Weakly Connected Neural Networks* (Springer, New York, 1997)
15. O. Kiehn, Locomotor circuits in the mammalian spinal cord. Ann. Rev. Neurosci. **29**(1), 279–306 (2006)

16. T. Kiemel, K.M. Gormley, L. Guan, T.L. Williams, A.H. Cohen, Estimating the strength and direction of functional coupling in the lamprey spinal cord. J. Comput. Neurosci. **15**, 233–245 (2003)
17. N. Kopell, G.B. Ermentrout, T.L. Williams, On chains of oscillators forced at one end. SIAM J. Appl. Math. **51**, 1397–1417 (1991)
18. Y. Kuramoto, *Chemical Oscillations, Waves, and Turbulence* (Springer, Berlin, 1984)
19. E. Marder, D. Bucher, Central pattern generators and the control of rhythmic movements. Curr. Biol. **11**(23), 986–996 (2001)
20. N. Massarelli, G. Clapp, K. Hoffman, T. Kiemel. Entrainment ranges for chains of forced neural and phase oscillators. J. Math. Neurosci. **6**(6), (2016). doi:10.1186/s13408-016-0038-9
21. A.D. McClellan, Brainstem command system for locomotion in the lamprey: localization of descending pathways in the spinal cord. Brain Res. **457**(2), 338–349 (1988)
22. A.D. McClellan, K. Sigvardt, Features of entrainment of spinal pattern generators for locomotor activity in the lamprey. J. Neurosci. **8**, 133–145 (1988)
23. D.L. McLean, M.E. Higashijima, J.R. Fetcho, A topographic map of recruitment in the spinal cord. Nature **446**(7131), 71–75 (2007)
24. J. Neu, Large populations of coupled chemical oscillators. SIAM J. Appl. Math **38**(2), 305–316 (1980)
25. J. Neu, The method of near-identity transformations and its applications. SIAM J. Appl. Math **38**(2), 189–208 (1980)
26. K.G. Pearson, S. Rossignol, Fictive motor patterns in chronic spinal cats. J. Neurophysiol. **66**(6), 1874–1887 (1991)
27. G. Viana di Prisco, P. Wallen, S. Grillner, Synaptic effects of intraspinal stretch-receptor neurons mediating movement-related feedback during locomotion. Brain Res. **530**, 161–166 (1990)
28. C.M. Rovainen, Synaptic interactions of identified nerve cells in the spinal cord of the sea lamprey. J. Comp. Neurol. **154**, 189–206 (1974)
29. E.D. Tytell, A.H. Cohen, Rostral versus caudal differences in mechanical entrainment of the lamprey central pattern generator for locomotion. J. Neurophys. **99**(5), 2408–2419 (2008)
30. P.L. Várkonyi, T. Kiemel, K.A. Hoffman, A.H. Cohen, P. Holmes, On the derivation and tuning of phase oscillator models for lamprey central pattern generators. J. Comp. Neurosci. **25**(2), 245–261 (2008)
31. P. Wallen, T. Williams, Fictive locomotion in the lamprey spinal cord in vitro compared with swimming in the intact and spinal animal. J. Physiol. **347**, 225–239 (1984)
32. J.C. Weeks, Neuronal basis of leech swimming: separation of swim initiation, pattern generation, and intersegmental coordination by selective lesions. J. Neurophysiol. 45(4), 698–723 (1981)
33. T.L. Williams, Phase coupling by synaptic spread in chains of coupled neuronal oscillators. Science **258**, 662–665 (1992)
34. T.L. Williams, K.A. Sigvardt, Intersegmental phase lags in the lamprey spinal cord: Experimental confirmation of the existence of a boundary region. J. Comput. Neurosci. **1**, 61–67 (1994)
35. T.L. Williams, K.A. Sigvardt, N. Kopell, G.B. Ermentrout, M.P. Remler, Forcing of coupled nonlinear oscillators: Studies of intersegmental coordination in the lamprey locomotor central pattern generator. J. Neurophys. **64**, 862–871 (1990)
36. D.M. Wilson, The central nervous control of flight in a locust. J. Exp. Biol. **38**(2), 471–490 (1961)

Applications of Knot Theory: Using Knot Theory to Unravel Biochemistry Mysteries

Candice Reneé Price

Abstract Although knots have been used since the dawn of humanity, the mathematical study of knots is only a little over 100 years old. Not only has knot theory grown theoretically, the fields of physics, chemistry, and molecular biology have provided many applications of mathematical knots. In this expository paper, we provide an overview of some connections between knot theory and DNA–protein interaction, outlining specifics of the biological mechanisms of DNA replication while providing an overview of related knot invariants. This work is based on an oral presentation given by the author at the Association of Women Mathematicians Research Symposium April 11, 2015.

Keywords Knots · Knot invariants · DNA · Topoisomerase · Recombinase

Mathematics Subject Classification 57M25 · 92B05

1 Knots and Links

A **knot** is defined as a closed, non-intersecting curve in $\mathbb{R}^3$. Formally it is a proper embedding of a circle in three dimensions (we call a mapping $f : X \to Y$ an embedding if the restriction mapping $f' : X \to f(X)$ is a homeomorphism) (Fig. 1).

Candice presented this work at the special session "Research from the Cutting EDGE." Candice was a mentor in the 2012 EDGE (Enhancing Diversity in Graduate Education) Program and received her doctorate in Mathematics shortly thereafter, also in 2012. The goal of EDGE is to strengthen the ability of women to successfully complete their graduate programs in the Mathematical Sciences. Please see the preface for more information about the EDGE Program and its founders.

C. Reneé Price (✉)
Department of Mathematics and Statistics, Sam Houston State University,
1806 Avenue J, Huntsville, TX 77340, USA
e-mail: candice.price@shsu.com; candice.price@shsu.edu

G. Letzter et al. (eds.), *Advances in the Mathematical Sciences*, Association for Women in Mathematics Series 6, DOI 10.1007/978-3-319-34139-2_7

Fig. 1 Examples of simple alternating knots

A **knot projection** is the two-dimensional image of the three-dimensional knot projected onto $\mathbb{R}^2$. At each double point in the projection (a crossing involving only two line segments), it is not clear which portion of the knot crosses over and which crosses under. In order to take care of this, gaps are left in the projection to indicate **overcrossing** and **undercrossings.** It is known that problematic intersections (see Fig. 2) can be avoided so that all intersections correspond to double points. A knot projection drawn with these criteria is called a **knot diagram**. It is always possible to find a plane so that the projection has a finite number of double points. An **alternating knot** is a knot which possesses a knot diagram in which crossings alternate between undercrossings and overcrossings.

An **n-component link** is an embedding of a disjoint union of n copies of S^1 into $\mathbb{R}^3$ (Fig. 3). We will take the view that a knot can be seen as a 1-component link. Thus, we will use the term *links* to refer to links of at least one component, while *knots* will refer only to links of one component.

A **link projection** is the two-dimensional image of the three-dimensional link projected onto a plane. Just as with knot diagrams, a **link diagram** is a link projection drawn with the same criteria.

Knots and links are studied through their diagrams. Links that have diagrams that can be drawn using a finite number of polygonal circuits (i.e., closed paths) in three-dimensional space are called **tame** (Fig. 4). All other links are known as **wild**. Most applications of knot theory concern only tame links, so we will only focus on this class of links.

We say two links, K_1 and K_2, are **equivalent** if there is an ambient isotopy between them. An **ambient isotopy** can be described as a continuous deformation from one

Fig. 2 Ambiguous and problematic intersections not allowed in knot diagrams

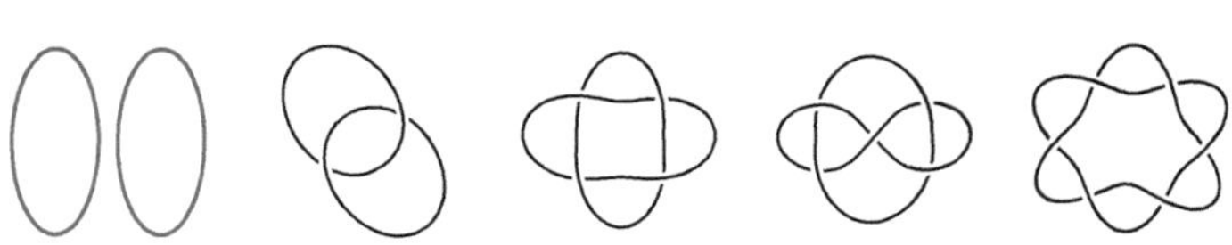

Fig. 3 Examples of simple 2 component links

Fig. 4 A polygonal projection and a smooth projection of the knot with three crossings has eight possible knot diagrams, two are shown here

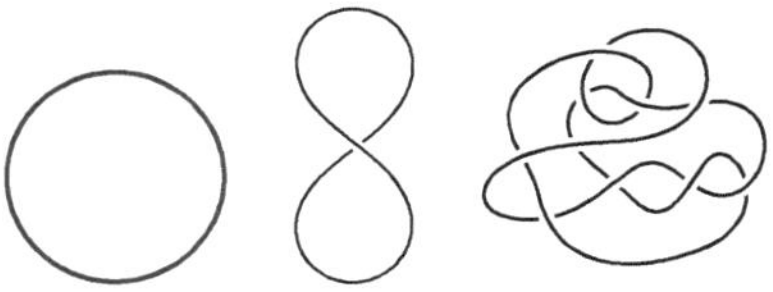

Fig. 5 Examples of equivalent trivial knots

link diagram (K_1) to the other (K_2) (Fig. 5). It allows us to stretch, bend and twist the link however we would like; we just cannot cut it. Mathematically, two links, K_1 and K_2, are **ambient isotopic** if there is an isotopy $h : \mathbb{R}^3 \times [0, 1] \to \mathbb{R}^3$ such that h_i is a homeomorphism such that $h(s, i) = h_i(s)$ where $h(K_1, 0) = h_0(K_1) = K_1$ and $h(K_1, 1) = h_1(K_1) = K_2$ [8]. If two knots are equivalent, we refer to these knots as knots of the same **knot type**, K, where K is the equivalence class under this equivalence relation.

In the 1930s, Kurt Reidemeister proved that an ambient isotopy may be described as a finite sequence of three moves, called **Reidemeister moves**, as discussed in Theorem 1.1.

Theorem 1.1 ([20]) *Two link diagrams L_1 and L_2 are equivalent if and only if they can be obtained from one another by a finite sequence of planar isotopies and the three moves: twist, poke, and slide (Figs. 6, 7, 8).*

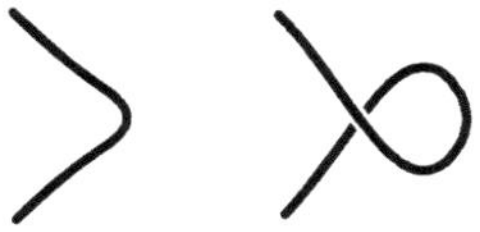

Fig. 6 R_1 move: "twist." This move allows us to put in or take out one crossing in a knot diagram

Fig. 7 R_2 move: "poke." This move allows us to either add or remove two crossings from a knot diagram

Fig. 8 R_3 move: "slide." This move allows us to slide one strand of the knot from one side of a crossing to the other side of the crossing

While Reidemeister has provided a tool for checking if two knots are the same knot type, performing these moves on a complex knot diagram can become very tedious. Thus, they are not a very efficient way of distinguishing knots. We now utilize certain topological properties of links to help discover equivalent knots and links.

1.1 Knot and Link Invariants

While Reidemeister moves are helpful to see if two links are equivalent, they are not as useful when showing that two links are not equivalent. Link invariants are utilized to show inequivalence between two link diagrams. A **link invariant** is a specific quality of a knot or link type that does not change its value under ambient isotopy. Thus, if two links are equivalent then their invariants are equal. Unfortunately, for the majority of invariants, the other direction is not usually true: equal invariant values for two link diagrams do not imply equivalent links.

One basic example of a link invariant is the minimum crossing number. The **minimum crossing number** is the number of crossings in the minimal diagram of the knot. That is, we minimize the number of crossings over all knot diagrams in that equivalence class or knot type (Fig. 9). A knot diagram with the minimal number of crossings is a **minimum regular diagram** of the knot.

Some invariants keep count of the number of topological changes made to a link diagram. For example, given a knot diagram, exchange locally overcrossings and undercrossings. This type of alteration may change the knot type. The **unknotting number** is the least number of crossing changes in a diagram of a knot needed to get to the trivial knot, minimized over all diagrams (Fig. 10).

Fig. 9 Examples of minimum regular diagrams of the first five knots

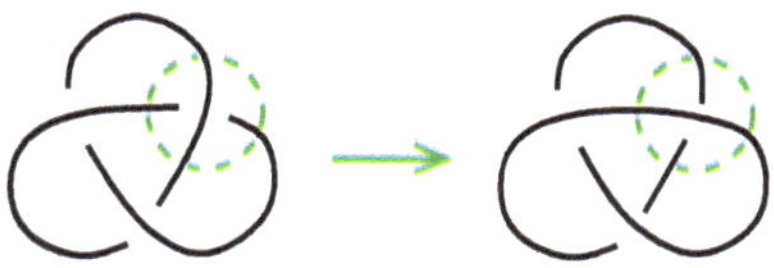

Fig. 10 Example of a topological change: crossing change. This example shows that the unknotting number of the knot 3_1 is 1

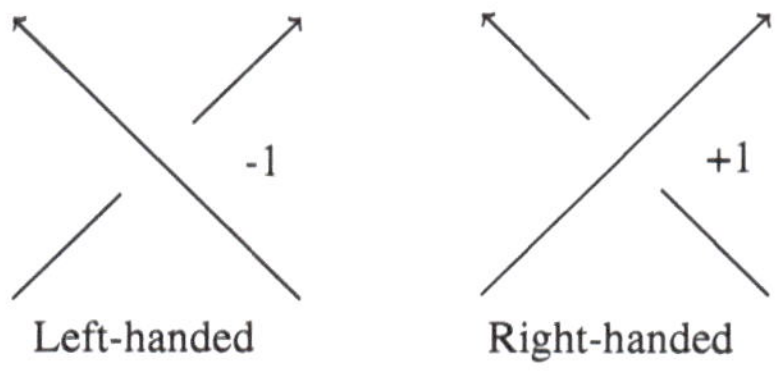

Fig. 11 Given an orientation, we can assign negative and positive crossings

The **linking number** is a link invariant for links of two or more components. It is calculated using the **crossing sign convention** (seen in Fig. 11) as follows take the sum of the crossing signs of each crossing between the different components of the link and dividing by two.

While the previous invariants give numerical quantities, other invariants can associate a polynomial to a knot type: the Alexander polynomial, the Jones polynomial, and the HOMFLY-PT polynomial [2, 12, 15]. Additionally, invariants such as Khovanov Homology and Knot Floer Homology associate a chain complex of abelian groups to a knot diagram [16, 19].

There are certain proteins, topoisomerase and recombinase, that can change the topology of DNA. **Topoisomerases** are proteins that can change the topological shape of DNA while keeping the genetic code unchanged. This action is essential to regulating supercoiling in DNA, unknotting and unlinking DNA, and preventing cell death [1, 4]. **Recombinases** are proteins that cut two segments of DNA and recombine them in some manner allowing for genetic diversity. These changes can inhibit or aid biological processes involving the structure of DNA including replication and transcription. The local actions of both proteins can be modeled using knot theory; therefore, applications of knot theory to problems involving these proteins have been extensively studied [6, 9, 11, 21–23].

2 Biology Background

A crucial advancement in molecular biology was made when the structure of DNA was determined by James Watson and Francis Crick in 1953 using results from Rosalind Franklin [17]. Its structure revealed how DNA can be replicated and provided clues about how a molecule of DNA might encode directions for

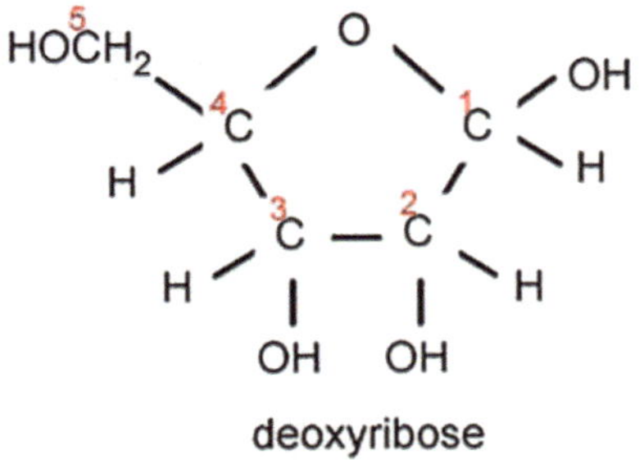

Fig. 12 Sugar ring made of five carbon atoms. Courtesy of [25]

producing proteins [1]. Nucleic acids consist of a chain of linked units called **nucleotides**. Each nucleotide contains a deoxyribose, a sugar ring made of five carbon atoms which are numbered as seen in Fig. 12. This sugar ring then forms bonds to a single phosphate group between the third and fifth carbon atoms of adjacent sugar rings (Fig. 13). The backbone of a DNA strand is made from alternating phosphate groups and sugar rings. The four bases found in DNA are Adenine (A), Thymine (T), Cytosine (C), and Guanine (G). The shapes and chemical structure of these bases allow hydrogen bonds to form efficiently between A and T and between G and C. These bonds, along with base stacking interactions, hold the DNA strands together [1]. Each base is attached to the first carbon atom in the sugar ring to complete the nucleotide (Fig. 13).

The bond between the sugar and the phosphate group give a direction to DNA strands. The asymmetric ends of the strands are called the $5'$ (**five prime**) and $3'$ (**three prime**) ends, with the $5'$ end having a phosphate group attached to the fifth carbon atom of the sugar ring and the $3'$ end with a terminal hydroxyl group attached to the third carbon atom of the sugar ring (Fig. 13). The direction of the DNA strands are read from $5'$ to $3'$. In a double helix the direction of one strand is opposite to the direction of the other strand: the strands are **antiparallel** [1].

Besides the standard linear form, a molecule of DNA can take the form of a ring known as **circular DNA**. One way to model circular DNA mathematically is as an annulus, **R**, an object that is topologically equivalent to $S^1 \times [-1, 1]$. The axis of **R** is $S^1 \times \{0\}$. With this view, we neglect the chemical properties of DNA and focus only on the topological structure. Due to this, we often model DNA as a rod, tube, or string in our schematics.

In this model we can choose an orientation for the axis of **R** and use the same orientation on $\partial\mathbf{R}$; thus, the axis and boundary curves of **R** have a parallel orientation. Note that this is a different convention than the biology/chemistry orientation. We use geometric invariants twist and writhe, denoted **Tw** and **Wr**, to describe the structure of the circular DNA molecule. **Writhe** can be determined by viewing the axis of **R** as a spatial curve and is measured as the average value of the sum of the positive and negative crossings of the axis of **R** with itself, averaged over all projections [18]. The sign convention for a crossing is given in (Fig. 11). **Twist** is defined as the amount that one of the boundary curves of **R** twists around the axis of **R** [4].

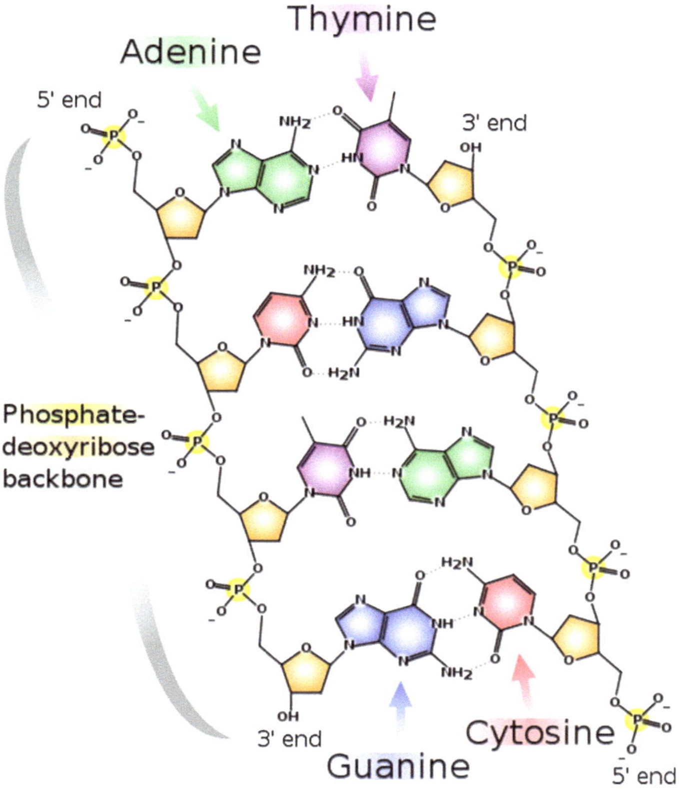

Fig. 13 Deoxyribonucleic acid. Using the direction convention given to DNA strands, we read this sequence as *ACTG*, or equivalently *CAGT*. Courtesy of [26]

One relationship between **Tw** and **Wr** is expressed in the following law:
Law 2.1 (Conservation Law [13])

$$\mathbf{Lk}(\mathbf{R}) = \mathbf{Tw}(\mathbf{R}) + \mathbf{Wr}(\mathbf{R})$$

where $\mathbf{Lk}(\mathbf{R})$ is the linking number of the oriented link formed by the two boundary curves of **R** with a parallel orientation.

We say that a DNA molecule is **supercoiled** when $\mathbf{Wr} \neq 0$. Native circular DNA appears negatively supercoiled under an electron microscope, i.e., $\mathbf{Wr} < 0$ (Fig. 14) [4].

Recall that the structure of DNA is a double-stranded helix, where the four bases are paired and stored in the center of this helix. While this structure provides stability for storing the genetic code, Watson and Crick noted that the two strands of DNA would need to be untwisted in order to access the information stored for replica-

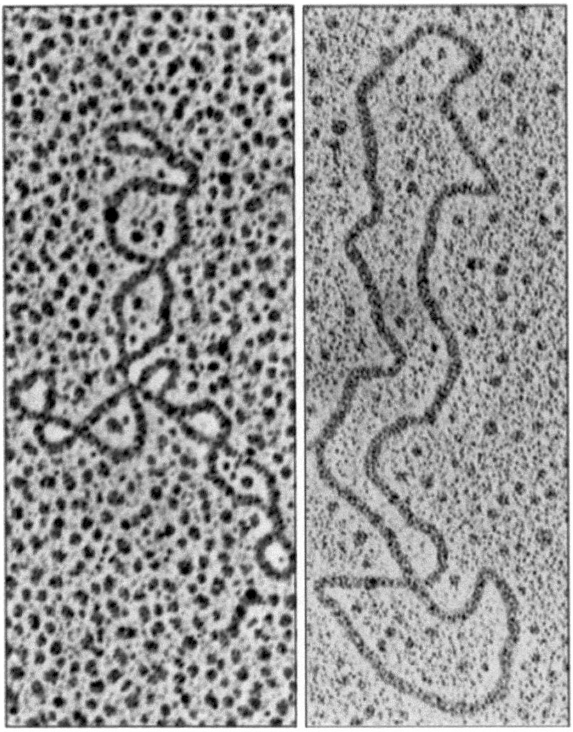

Fig. 14 Two examples of supercoiled DNA seen through an electron microscope. Reproduced with permission from [14]

tion [1]. They also foresaw that there should be some mechanism to overcome this problem.

2.1 Replication

There are three main topological forms that circular DNA can take: supercoiled, knotted, catenated (linked DNA molecules) or a combination of these. DNA is kept as compact as possible when in the nucleus, and these three states help or hinder this cause. However, when transcription or replication occurs, DNA must be accessible [24].

DNA replication is the process that starts with one DNA molecule and produces two identical copies of that molecule. During replication, the DNA molecule begins to unwind at a specific location and starts the synthesis of the new strands at this location, forming **replication forks** (Fig. 15, left). The DNA ahead of the replication fork becomes positively supercoiled, while DNA behind the replication fork becomes entangled, creating pre-catenanes, a state where the DNA molecules are beginning to form linked DNA molecules (Fig. 15, center). A topological problem occurs at the

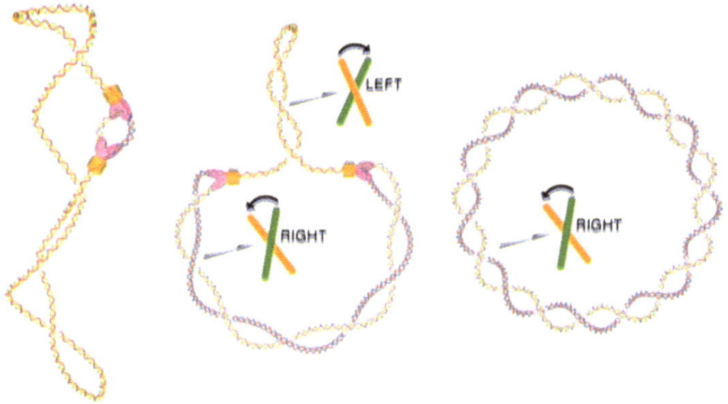

Fig. 15 Topological changes to DNA during replication of circular DNA. The process of replication begins with negatively supercoiled DNA. The replication forks are shown in *purple* and gold. Partially replicated DNA molecule: the replicated portions of the DNA are interwound with positive (*right-handed*) crossings, creating a pre-catenane, while the remaining unreplicated DNA is still negatively (*left-handed*) supercoiled. Completely replicated DNA shown as a DNA catenane with positive (*right-handed*) crossings. Used with permission from [27]

end of replication, when daughter chromosomes must be fully disentangled before the cell can split (Fig. 15, right) [24]. A protein, topoisomerase, plays an essential role in resolving this problem.

2.2 *Topoisomerase*

Topoisomerases are proteins that are involved in the packing of DNA in the nucleus and in the unknotting and unlinking of DNA links that can result from replication and other biological processes. These proteins bind to either single-stranded or double-stranded DNA and cut the phosphate backbone of the DNA. A **type I topoisomerase** cuts one strand of a DNA double helix allowing for the reduction or the introduction of stress (Fig. 16). Such stress is introduced or needed when the DNA strand is supercoiled or uncoiled during replication or transcription. **Type II topoisomerase** cuts both phosphate backbones of one DNA double helix, passes another DNA double helix through it, and then reseals the cut strands (Fig. 17). This action does not change the chemical composition and connectivity of DNA, but potentially changes its topology.

Thus the action of topoisomerase can be modeled as a crossing change, as illustrated in Fig. 11.

A biological issue that can arise from knotted DNA is during the replication of knotted DNA (Fig. 18). If the DNA molecule is knotted pre-replication, the resulting product after DNA replication is a DNA link (catenane). Thus, the two DNA molecules are not able to separate into two new cells, therefore causing problems in

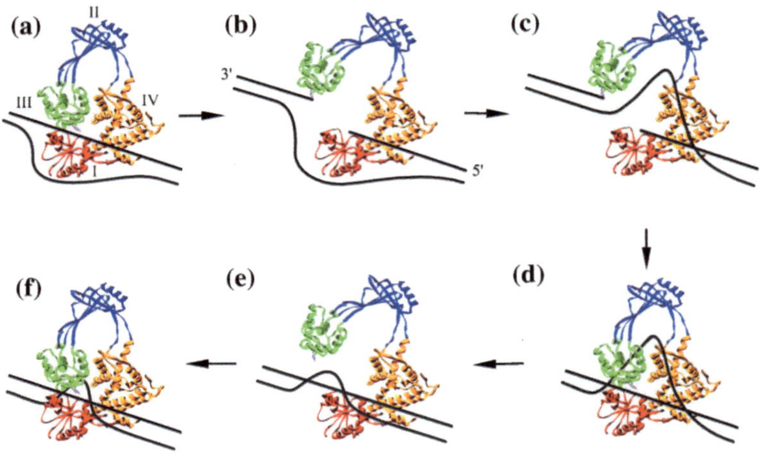

Fig. 16 Schematic of topoisomerase I action. Used with permission from [7]

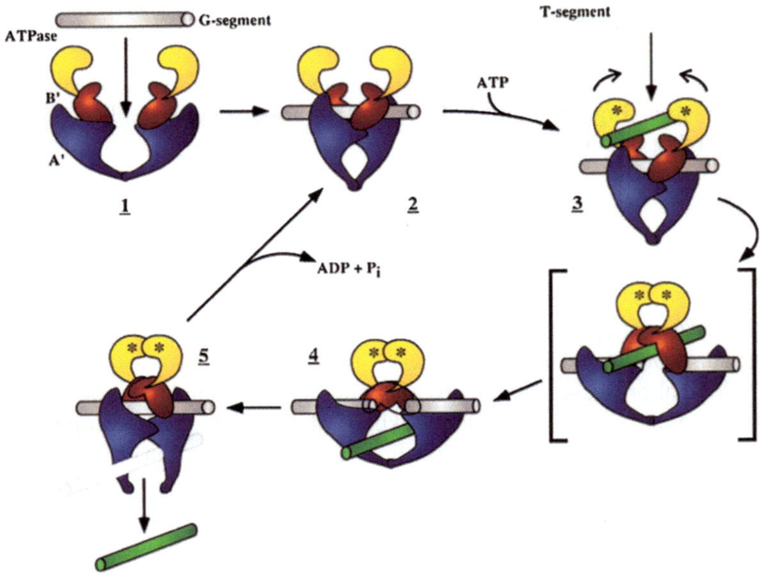

Fig. 17 Schematic of topoisomerase II action where the double-stranded DNA is modeled as a rod. Used with permission from [5]

the replication process and often leading to cell death [10]. These knots can be removed via the action of topoisomerase II. But how do the knots arise? Besides being a possible result of DNA packing [3], they can also be the product of recombinase action.

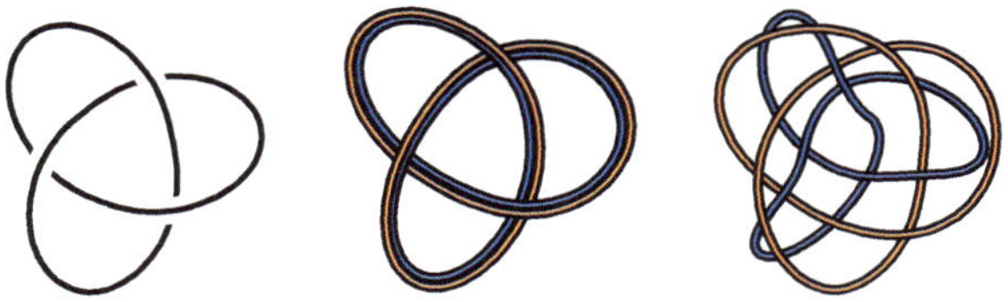

Fig. 18 Replication of knotted DNA: Assuming that the DNA strand is knotted (*left*), through replication that knot is doubled (*middle*). Notice the new topology of the DNA is a link which cannot be separated into two cells which is needed for replication (*right*)

2.3 Recombinase

Recombination is a process involving the genetic exchange of DNA where DNA sequences are rearranged by proteins known as **recombinases** [1]. **Site-specific recombination** is an operation on DNA molecules where recombination proteins, **site-specific recombinases**, recognize short specific DNA sequences on the recombining DNA molecules. First, two sequences from the same or different DNA molecule are drawn together. The recombinase then introduces a break near a specific site, known as a **recombination site**, on the double stranded DNA molecule. The protein then recombines the ends in some manner and seals the break (Fig. 19).

The DNA sequence of a recombination site can be used to give an orientation to this site. When two sites are oriented in the same direction, the sites are called **direct repeats** (Fig. 20). Recombinase action on direct repeats normally results in a change in the number of components (Fig. 21). If the two sites are oriented in opposite directions, the sites are called **inverted repeats** (Fig. 22). The action of

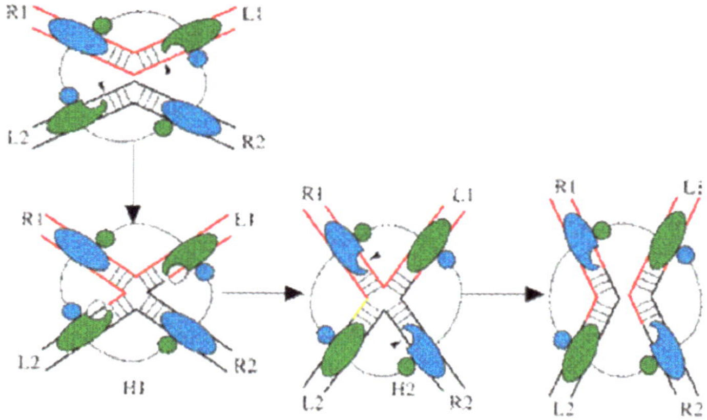

Fig. 19 An example of a site-specific recombinase mechanism where the protein breaks one strand of the double helix, recombines it and then does the same with the other strand

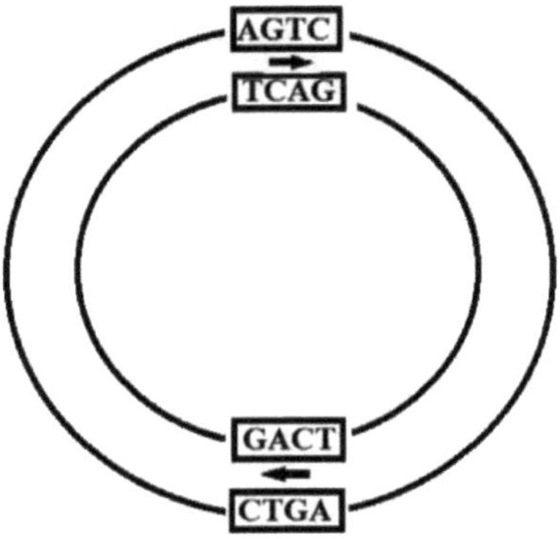

Fig. 20 Directed Repeats

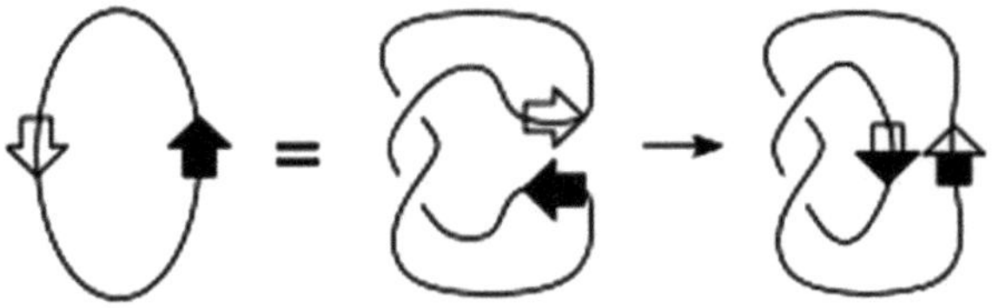

Fig. 21 Recombinase action on direct repeats

Fig. 22 Inverted Repeats

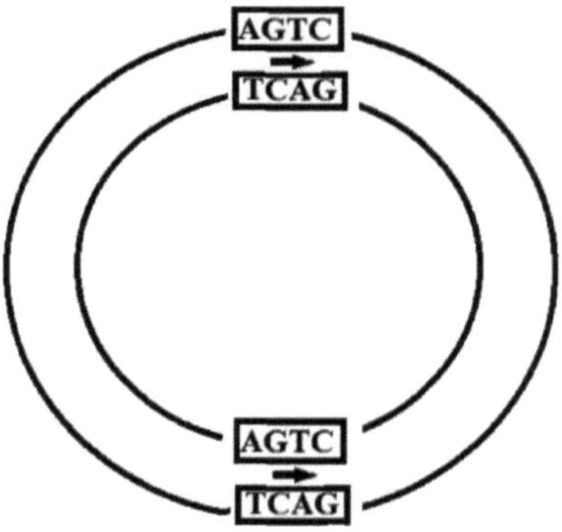

a recombinase on inverted repeats normally results in no change in the number of components (Fig. 23).

Thus DNA can be visualized as a complicated knot, due to recombinase action or packing, which must be unknotted by topoisomerase action in order for replication or transcription to occur. Hence, it is with no surprise that there lie connections between mathematical knot theory and biology. By thinking of DNA as a knot, we can use

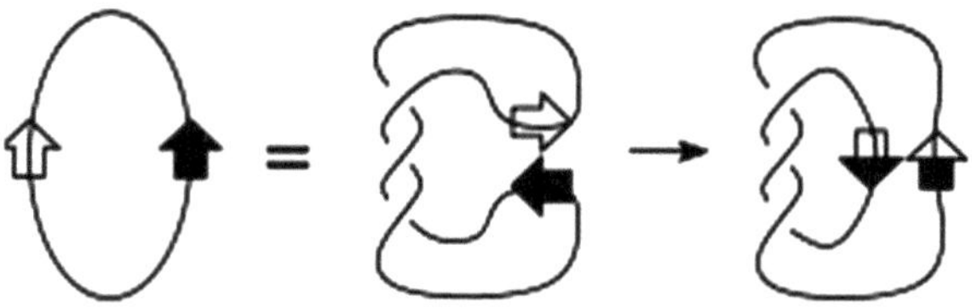

Fig. 23 Recombinase action on inverted repeats

knot theory to estimate the difficulty in unknotting the DNA. This can help estimate properties of the proteins involved in knotting and unknotting DNA and therefore unraveling many mysteries of biochemistry.

References

1. B. Alberts, D. Bray, K. Hopkins, A. Johnson, J. Lewis, M. Raff, K. Roberts, P. Walter, *Essential Cell Biology*, 2nd edn. (Garland Science/Taylor & Francis Group, Abingdon, 2003)
2. J.W. Alexander, Topological invariants of knots and links. Trans. Amer. Math. Soc. **30**(2), 275–306 (1928)
3. J. Arsuaga, M. Vazquez, P. McGuirk, S. Trigueros, D.W. Sumners, Joaquim Roca, DNA knots reveal a chiral organization of DNA in phage capsids. Proc. Natl. Acad. Sci. U. S. A **102**(26), 9165–9169 (2005)
4. A.D. Bates, A. Maxwell, *DNA Topology* (Oxford University Press, Oxford, 2005)
5. J.M. Berger, S.J. Gamblin, S.C. Harrison, J.C. Wang, Structure and mechanism of DNA topoisomerase ii. Nature (1996)
6. D. Buck, C.V. Marcotte, Tangle solutions for a family of DNA-rearranging proteins. Math. Proc. Cambridge Philos. Soc. **139**(1), 59–80 (2005)
7. J. Champoux, DNA topoisomerases: structure, function, and mechanism. Annu. Rev. Biochem. **70**, 369–413 (2001)
8. P.R. Cromwell, *Knots and Links* (Cambridge University Press, Cambridge, 2004)
9. I.K. Darcy, Biological distances on DNA knots and links: applications to XER recombination. J. Knot Theory Ramifications, 10(2):269–294 (2001). Knots in Hellas '98, Vol. 2 (Delphi)
10. R.W. Deibler, J.K. Mann, L.S. De Witt, L. Zechiedrich, Hin-mediated DNA knotting and recombining promote replicon dysfunction and mutation. BMC Mol. Biol. **8**(1), 44 (2007)
11. C. Ernst, D. Sumners, A calculus for rational tangles: applications to DNA recombination. Math. Proc. Camb. Phil. Soc **108**, 489–515 (1990)
12. P. Freyd, D. Yetter, J. Hoste, W.B.R. Lickorish, K. Millett, A. Ocneanu, A new polynomial invariant of knots and links. Bull. Amer. Math. Soc. (N.S.), 12(2):239?246 (1985)
13. F.B. Fuller, Decomposition of the linking number of a closed ribbon: a problem from molecular biology. Proc. Nat. Acad. Sci. U.S.A. **75**(8), 3557–3561 (1978)
14. J. Hardin, G.P. Bertoni, L.J. Kleinsmith, *Becker's World of the Cell*, 8th edn. (Benjamin Cummings, San Francisco, 2010)
15. V.F.R. Jones, A polynomial invariant for knots via von Neumann algebra. Bull. Amer. Math. Soc. (N.S.) **12**, 103–111 (1985)
16. M. Khovanov, A categorification of the Jones polynomial. Duke Math. J. **101**(3), 359–426 (2000)
17. A. Klug, Rosalind franklin and the discovery of the structure of DNA. Nature, 219:808–844

18. K. Murasugi, *Knot theory & its applications*. Modern Birkhäuser Classics. Birkhä user Boston Inc., Boston, MA, (2008). Translated from the 1993 Japanese original by Bohdan Kurpita, Reprint of the 1996 translation [MR1391727]
19. P. Ozsváth, Knot Floer Homology. Advanced Summer School in Knot Theory. International Center for Theoretical Physics (2009)
20. K. Reidemeister, *Knotentheorie* (Springer, Berlin, 1974). Reprint
21. D.W. Sumners, The role of knot theory in DNA research, in *Geometry and topology (Athens, Ga., 1985)*, vol. 105 of *Lecture Notes in Pure and Appl. Math.*, pp. 297–318. Dekker, New York (1987)
22. D.W. Sumners, Untangling DNA. Math. Intell. **12**(3), 71–80 (1990)
23. M. Vazquez, D.W. Sumners, Tangle analysis of Gin site-specific recombination. Math. Proc. Camb. Philos. Soc. **136**(3), 565–582 (2004)
24. J.C. Wang, *Untangling the Double Helix*, DNA Entanglement and the Action of the DNA Topoisomerases (Cold Spring Harbor Laboratory Press, Cold Spring Harbor, 2009)
25. Wikipedia. Deoxyribose— Wikipedia, the free encyclopedia, (2012). Accessed 20 May 2012
26. Wikipedia. DNA — Wikipedia, the free encyclopedia, (2012). Accessed 20-May-2012
27. G. Witz, A. Stasiak, DNA supercoiling and its role in DNA decatenation and unknotting. Nucl. Acids Res. **38**(7), 2119–2133 (2010)

Metapopulation and Non-proportional Vaccination Models Overview

Mayteé Cruz-Aponte

This article is dedicated to my beloved friend Dr. Lukasz Adam Koscielski (1985–2015) cheers to you for a brief but brilliantly well lived life. Kocham Cie i Tesknie za Toba bardzo.

Abstract Influenza viruses are a major cause of morbidity and mortality worldwide. The 2009 influenza pandemic not only brought to our attention the strengths and weaknesses of the public health system but also changed their priorities. Vaccination still is the most powerful allay for preventing or mitigating influenza outbreaks or other diseases. In this article, we summarize our findings that arose from two different published research articles (see Cruz-Aponte et al. BMC Infect. Dis. 11(1), 207, (2011), [22], Herrera-Valdez et al. Math. Biosci. Eng. (MBE) 8(1), 21–48, (2011), [36]) that were presented at the Association of Woman in Mathematics (**AWM**) 2015 symposium at the University of Maryland College Park. The first one is a metapopulation model we constructed using the data from México's 2009 epidemic patterns characterized by three peaks. These peak patterns were theoretically investigated via models that incorporate México's general trends of land transportation, public health measures, and the academic calendar trends of that year. After studying many mathematical models that incorporated vaccination into the modeling, we were not satisfied with the simplification approaches that usually took place. Vaccinating only the susceptible individuals or vaccinating a fraction of the population was not realistic when supplies or daily administration capacity was considered. Hence, in the second project we presented a SIR-like model that explicitly takes into account vaccine supply and the number of vaccines administered per day and places data-informed limits on these parameters. The model that we refer to as non-proportional vaccination model is a theoretical improvement that provides more accurate predictions of the mitigating effects of vaccination than the typical proportional model. For some parameter regimes, proportional and non-proportional models behave the

M. Cruz-Aponte (✉)
Department of Mathematics and Physics, University of Puerto Rico at Cayey,
205 Ave Antonio R Barcelo, Cayey 00736-9997, Puerto Rico
e-mail: maytee.cruz@upr.edu

G. Letzter et al. (eds.), *Advances in the Mathematical Sciences*, Association for Women in Mathematics Series 6, DOI 10.1007/978-3-319-34139-2_8

same, especially when the vaccination supplies were depleted but for others there were significant changes that predicted earlier or longer epidemics as we discuss further. Both of our models can be easily modified to be used by government and medical officials to create preparedness plans based on specific constraints.

Keywords Biomathematics · Dynamical systems and simulations

Mathematics Subject Classification 92B05 · 37M05

1 Introduction

It is well known that most epidemiological models are based on the compartmental epidemic models developed by Kermack and McKendrick [38, 39]. Kermack and McKendrick considered a fixed population with only three disjoint compartments representing epidemiological stages at time t of an individual: susceptible, infected, or recovered. Extensive work has been done by several authors for example, the work of Brauer [8–10] has focused on the study and analysis of compartmental models starting from the simple model and adjusting the compartments according to the behavior of a particular disease modeled. Epidemiologists now focus on complex models to adjust to real life situations, not only to generalize the spread of disease but also to include behavioral aspects of a disease and mechanisms that affect the time course of a given epidemic. For instance, the goal of our first project was to construct a metapopulation model to understand the patterns of México's 2009 epidemic using the country's terrestrial transportation trends, social distancing mechanism imposed by the government, and the impact of the academic calendar. Complex models are difficult to analyze mathematically and are computationally expensive, but they are necessary to study the possible scenarios that can lessen the catastrophic effects of a disease spread in the population. Many works have been published that include metapopulation models [4, 37, 40, 61]. Rvachev's (1985) work [61] was the first attempt to apply the method at a global scale modeling the spread of influenza from city to city around the world during the 1968–1969 A/H3N2 (Hong Kong Flu) pandemic. The work of Arino focused on metapopulation models as well [3–7, 40].

Part of this article presents a metapopulation model which started as a simple compartmental model and evolved into a complex model including transportation across the different states in México in the context of the 2009 A/H1N1 epidemic outbreak [36]. Our article considers terrestrial transportation only to construct the network connectivity of the 32 states of México. The models in this paper are largely based upon the work of Hyman and LaForce [37] who modeled the spread of influenza between cities of the United States. However, we modify their models for the constraints of México and the 2009 pandemic. The second part of the paper focuses on a theoretical improvement on vaccination approaches in mathematical models in general.

Vaccination is one of the most effective tools health professionals have to prevent or mitigate outbreaks, and even eradicate diseases from the population [70, 71]. Seasonal vaccines are developed based on predictions about which influenza strains are likely to be circulating [60, 70] and distributed prior to and during the influenza season. However, when the guess is slightly off and the vaccine strains and the circulating strains or novel mutations of distinct strains are circulating, epidemics can emerge unexpectedly. Vaccination programs are highly effective in decreasing hospitalizations and deaths in high risk groups like the elderly and small children [33, 50, 59]. Vaccination prior to the initial outbreak of an epidemic or potential pandemic is not possible most of the time [56, 62, 70]. However, seasonal diseases such as influenza pandemics are characterized by multiple outbreaks separated by months or years, which can allow time for strategy development by heath officials such as development and administration of vaccination to mitigate the burden and morbidity [26, 32, 43, 47, 72]. Instead of using the general oversimplification approaches of vaccination in the literature, our goal is to theoretically improve an SIR-type model to explicitly take into account vaccine supply and the amount of vaccines that can realistically be administered per day to provide more accurate predictions of the mitigating effects of vaccination.

In this overview, we present a modification of a vaccination model that is an improvement on the classical and typical models of vaccination and presents the marked difference between both models. We called this model the non-proportional vaccination model [22]. The importance on addressing this marked difference is that as we improved the precision of modeling approaches, public health officials can prepare better strategies to contain an epidemic specifically to lessen the burden in the already saturated health care system. Vaccine distribution worldwide or even within a country is not democratic, resources are limited and many countries already struggle with insufficient doses of vaccines, as well as limited medical supplies, facilities, and personnel [25, 46, 55]. Health care facilities and their medical staff are limited and can only offer a maximum number of vaccines per day that can be small relative to the population size depending on the country [1, 13, 15, 57, 58, 67]. Vaccination resources and distribution affect the size and dynamics of influenza outbreaks, hence understanding how this happens is crucial to outbreak preparedness [70]. Hence, better modeling approaches are needed to study and examine the effectiveness of vaccination in mitigating outbreaks and the goal of our model is to base educated assumptions about supply and administration of vaccines paying close attention to vaccine stockpiles, the population that receive the vaccines and the time frame of vaccine administration (Fig. 1). Previous studies modeling vaccination during a pandemic assume that vaccines are administered only to susceptible individuals which is not the case [11, 17, 18, 27, 28, 42, 52]. First, medical professionals are rarely able to determine an individual's epidemiological status (i.e., susceptible, infected, recovered) prior to vaccination. Second, laboratory testing of individuals seeking vaccination is not required [14]. Finally, individuals may not be aware of their own epidemiological status, either because they were asymptomatic or are unsure of the virus that afflicted them in the first place. Hence, our modeling approach takes into

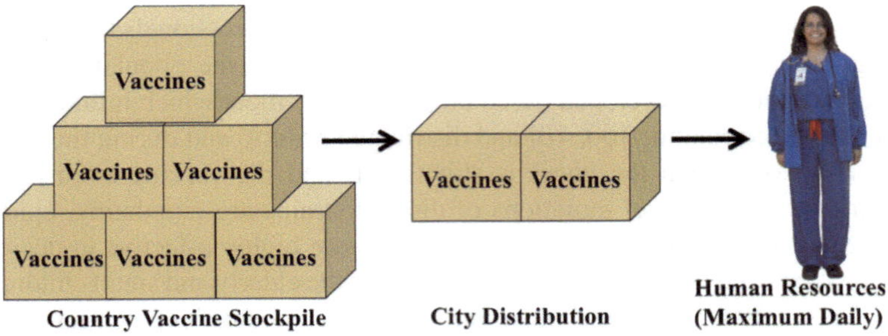

Fig. 1 Vaccination stockpile distribution scheme. Starting with a given stockpile per country that will be distributed per city and then distribute by medical personnel daily for a given period of time

consideration these factors and not only puts constraints on these parameters but also administers vaccine to any individual that is not visibly symptomatic or confirmed.

The rest of this article is organized as follows: In Sect. 2 we present the methodology for each model. In Sect. 2.1 we describe the Metapopulation model constructed to address the events that shaped the A-H1N1 2009 epidemic in México and in Sect. 2.2 the methodology of our non-proportional vaccination model is presented. The non-proportional vaccination model takes into consideration factors that affect vaccination campaigns such as starting day, duration, stockpile size, and daily administration capacity. In Sect. 3 simulations and results are presented separately (Sect. 3.1 for the Metapopulation model and Sect. 3.2 for the Non-proportional model). Section 4 of the article presents overall final remarks of both of our research articles.

The following published papers were summarized in this article:

- Herrera-Valdez, M.A., Cruz-Aponte, M. and Castillo-Chavez, C. Multiple outbreaks for the same pandemic: Local transportation and social distancing explain the different "waves" of A-H1N1pdm cases observed in México during 2009. (2011) Mathematical Biosciences and Engineering (MBE), 8(1):21–48 [36].
- Cruz-Aponte, M., McKiernan, E.C., Herrera-Valdez, M.A. Mitigating effects of vaccination on influenza outbreaks given constraints in stockpile size and daily administration capacity. (2011) BMC Infectious Diseases 11:1, 207 [22].

2 Methodology

In this article, in Sect. 2, we will briefly describe the methodology used in both projects presented. Section 2.1, presents the metapopulation model and describes the mechanisms imposed in the model to replicate the patterns of México's 2009 epidemic. We constructed a networking approach for the 32 states of México using the country's terrestrial transportation trends. We constructed a modulation function to control the probability of an effective infection by adding social distancing mechanisms imposed by the government and the school academic calendar that alter the spread of the disease. In Sect. 2.2 the non-proportional vaccination model will present the theoretical improvements we implemented. We take into account vaccine supplies and the amount of vaccines that can realistically be administered per day to provide more accurate predictions of the mitigating effects of vaccination. These variable factors can be changed according to the country's resources available to predict accurately the outcomes of any vaccine campaign that can potentially be put into place. For a more detailed methodology refer to [22, 36].

2.1 Metapopulation Model

The main focus of this model is to explore the role of social distancing, school closures, transportation patterns, and vaccination policies on the time course of México's epidemic. México is divided into 31 states and the Federal District (DF) that contains México City. For the purpose of the connectivity network and based on the terrestrial transportation patterns of the country, these 32 regions are regarded as nodes in a star-shaped, weighted graph with all nodes connected to DF, but not directly connected to each other. Regions are regarded as strongly and weakly connected to D.F. (México City) as illustrated in Fig. 2, based on transportation patterns (for more details see [36]). The infection rate is assumed to change as a function of social distancing measures, behavioral changes induced by media alert, and school closures as modulated in Fig. 4A. The rates of infection and recovery periods are assumed depending on the region [4] while the interactions between individuals within a given region are assumed to be the same across all states. Populations in the states that compound the influenza corridor are assumed to be more susceptible to the disease, they are also assumed to recover slower than the rest [41]. For the parameter values used in the simulations refer to Table 1 on Sect. 3 (Simulations and Results).

2.1.1 Mathematical Model

The population of each region is divided into distinct epidemiological classes per city k: S_k, E_k, C_k, U_k, R_k, and V_k representing susceptible, incubating, infected and confirmed, infected but not confirmed, recovered, and vaccinated, respectively. The infection rate for city k, β_k, represents the mean infection probability per contact $\lambda_k(t)$ defined as

$$\lambda_k(t) = g(t)\frac{\beta_k}{N_k}\left(C_k + U_k + \mu_E E_k\right) \tag{1}$$

where the parameter μ_E takes values between 0 and 1 modeling a decrease in the infectivity of individuals who are within the incubation period. The contact rate is modulated by a function $g(t)$ to capture social distancing and school closures at specific dates during the year. The modulation of the infection rate was defined using a combination of sigmoid functions as illustrated in Fig. 4A. The system of coupled

Fig. 2 The Mexican states that contributed with more than half of the total cases during the initial spread of A/H1N1 up to June 4, 2009 are shown in *dark gray*. The remaining states (*light gray*) were the main contributors to secondary outbreaks later in the year. The *red dots* mark states in the influenza corridor [2]

ordinary differential equations is defined as follows

$$\dot{S}_k = (Q_k - \lambda_k)\, S_k + \sum_{i \neq k} Q_{i1} S_i - \varepsilon(t) - \nu_{S_k} \tag{2}$$

$$\dot{E}_k = (Q_k - \alpha)\, E_k + \sum_{i \neq k} Q_{i1} E_i + \lambda_k S_k + \varepsilon(t) - \nu_{E_k} \tag{3}$$

$$\dot{C}_k = (Q_k - \sigma_k - \delta_C)\, C_k + \sum_{i \neq k} Q_{i1} C_i + \alpha p E_k \tag{4}$$

$$\dot{U}_k = (Q_k - \sigma_k - \delta_U)\, U_k + \sum_{i \neq k} Q_{i1} U_i + \alpha(1-p) E_k - \nu_{U_k} \tag{5}$$

$$\dot{R}_k = Q_k R_k + \sum_{i \neq k} Q_{i1} R_i + \sigma_k C_k + \sigma_k U_k - \nu_{R_k} \tag{6}$$

$$\dot{V}_k = Q_k V_k + \sum_{i \neq k} Q_{i1} V_i + \nu_{S_k} + \nu_{I_k} + \nu_{U_k} + \nu_{R_k}, \tag{7}$$

$$\dot{w}_k = -\left(\nu_{S_k} + \nu_{E_k} + \nu_{U_k} + \nu_{R_k}\right) \tag{8}$$

$$\dot{D}_k = \delta_C C_k + \delta_U U_k \tag{9}$$

with variables w and D representing, respectively, the available vaccine stockpile and disease-induced deaths. All population numbers are in thousands of individuals; the time is in days with t_0 equal to January 1, 2009. The term $Q_k = Q_{k0} - \sum\{Q_{0i} : i = 1, \ldots, M, i \neq k\}$ denotes the proportion of people traveling from region k to DF minus the proportion of people returning to region k.

2.2 Non-proportional Vaccination Model

Understanding how vaccination resources can affect the size and dynamics of influenza outbreaks is crucial to outbreak preparedness constructing a model that computationally addresses all the real time components that affect a vaccination campaign to develop better strategies and test if they are capable to contain an epidemic. Classical modeling that examines the effectiveness of vaccination in mitigating a disease are based on unrealistic assumptions on the supply and administration capabilities of vaccines. Most of the time it is assumed that the vaccine stockpiles are large relative to the size of the population or the financial capabilities of a particular country [31, 45, 52, 73]. The majority of models in the literature assume that vaccines are administered only to susceptible individuals [11, 17, 18, 27, 28, 42, 52].

2.2.1 Mathematical Model

The non-proportional vaccination model distributes vaccines to the populations S, I_U, and R_C, susceptible, infected unconfirmed and recovered individuals respectively. Individuals from I_U and R_C are those who become infected, but seek vaccination either because they are asymptomatic or they are unaware of the specific viral strain of their previous illness. Since the length of time simulated is no longer than a year, it is assumed that recovered populations and vaccinated individuals have total protection against the virus that does not wane in time. Vaccines that go to individuals in populations I_U and R_C are considered wasted, since immunity was already acquired. Other compartments of the model are V_U to keep track of vaccinated individuals from I_U that will develop symptoms since they already acquired the virus and the vaccine will not protect them against it and will follow its course at the same rate, and V_{SC} to track those vaccinated from populations that are susceptible S and recovered R_C that will not develop symptoms over time. The mean infection probability per contact is given by

$$\lambda(S, I, t) = b\frac{S}{N}\left[I_C + \alpha(I_U + V_U)\right], \tag{10}$$

The system of ordinary differential equations is defined as follows:

$$\dot{S} = -\lambda(S, I, t) - v_S(t) \tag{11}$$

$$\dot{I}_C = p\lambda(S, I, t) - (c + \delta)I_C \tag{12}$$

$$\dot{I}_U = (1-p)\lambda(S,I,t) - (c+\delta)I_U - \nu_U(t) \tag{13}$$
$$\dot{V}_U = \nu_U(t) - (c+\delta)V_U \tag{14}$$
$$\dot{R_C} = cI_C - \nu_R(t) \tag{15}$$
$$\dot{R_U} = c(I_U + V_U) \tag{16}$$
$$\dot{V}_{SC} = \nu_S(t) + \nu_R(t) \tag{17}$$
$$\dot{D} = \delta(I_C + I_U + V_U) \tag{18}$$

where p is the probability of being infected and confirmed, c is the rate of recovery of infected individuals (1/recovery time), δ is the infection-related death rate, b represents the mean probability of infection per contact.

3 Results and Simulations

In this section, we present a summary of our main results and simulations of each research work starting with the metapopulation modeling approach. Using the metapopulation model with all the mechanisms imposed to mimic the patterns of México's 2009 epidemic, we worked on several simulations to explain how the mechanism took place and the significance of it. For instance, the shape of the epidemic is characterized by three "waves", the first two waves were interrupted by social distancing policies, the closing of schools in the summer, and altered by delays in transportation. Our imposed modeling approach did confirm that the structure of the network that mimicked the transportation system contributed significantly to the generation of the three outbreaks; but, more than that, our approach showed that social distancing and school closures have a delaying effect in the total epidemic.

In the non-proportional vaccination model we presented simulations to compare classical vaccination models and our non-proportional model that is a theoretical improvement. We took into account vaccine supplies and the amount of vaccines that can realistically be administered per day to provide more accurate predictions of the mitigating effects of vaccination. All these variable factors can be changed according to the country's resources available to predict accurately the outcomes of any vaccine campaign that can potentially be put into place. Looking closely at the simulations, there are key regimes where the two models differ. Depending on the case longer epidemics were predicted or distinct final size of epidemics were seen. This was not the case in all regimes specifically when the vaccines were depleted in both models before the peak of the epidemic both models behave the same (see Fig. 7).

Table 1 Parameters for the A-H1N1 2009 pandemic metapopulation SEUCR model

Parameter	Value	Description	Reference
α^{-1}	2 days	Incubation period	[49]
σ_k^{-1}	7 days	Recovery period for State k	[20, 53]
μ_I	0.5	Reduction factor for infectivity during incubation	Estimated, [52]
β_k	0.95	Mean infection probability per contact for State k	Estimated, [53]
δ_x	10^{-6}	Influenza-induced death rate for $x = \{C, U\}$	Estimated, [53]
p	[0.1, 0.3]	Probability of confirmed case	Estimated
$\hat{\nu}$	$[1, 60] \times 10^3$/day	Maximum vaccines per day	Estimated, [24, 44]
F	$[500, 1000] \times 10^3$/day	Thousands of people traveling to/from D.F. per day	[51]

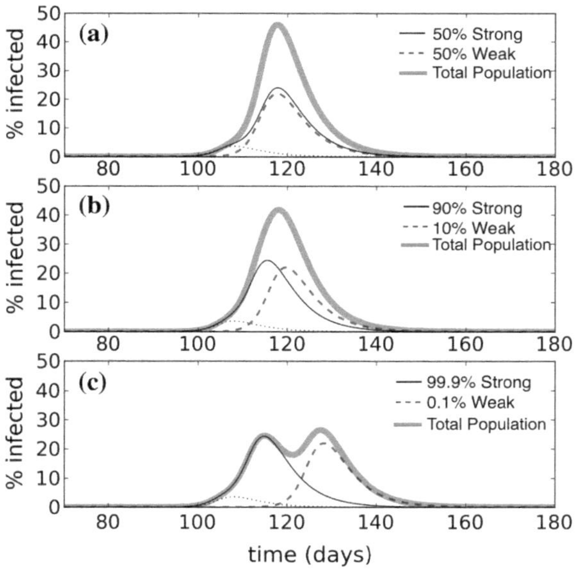

Fig. 3 Influence of transportation on the time course of the epidemic outbreak. The *solid* and *dashed curves* are, respectively, the total of infectious people in strongly and weakly connected populations. The *dotted line* is the epidemic curve in the originating state, Veracruz. The *thick gray curve* represent the total population of infected individuals in all cities. Graph A shows simulations in which strongly and weakly connected populations contribute, nearly the same, to the traffic through México City. Graphs B and C show cases in which the contribution of strongly connected populations is large relative to the weakly connected contribution

3.1 Metapopulation Model

There are several factors that are responsible for the shape of an epidemic such as social distancing or school breaks (summer or winter) among others. We focus specifically on the shape of the 2009 AH1N1 pandemic that occurred in México and is the center of attention of our model. First, in the case of the terrestrial transportation of individuals among states it is a delay mechanism but it does not drive the generation of multiple outbreaks. The spread of A-H1N1 was not uniform across México, not all states suffered the burden simultaneously as the data showed later on. To test if the network connectivity and transport was the driving force of these multiple outbreaks we tested different scenarios in our simulations.

In Fig. 3 it is evident that transportation alone does not contribute to the generation of multiple waves. Figure 3 is the case in which all Mexican states contribute to the flow into and out of DF nearly proportionally to their population size ($q = 0.5$) which represent that 50 % of the population that comes from the strongly connected states and the other 50 % from the weakly connected states. The small delay between the strongly and weakly connected is mainly due to the small difference between the contributions of strongly and weakly connected states and from the slight changes from infection rates and recovery periods (assumed for the different populations). If the traffic weight q for the Mexican states in the strongly connected subset is increased, the delays between the peaks become larger and what it seams to be a two-wave pattern is evident when unrealistically it is assumed that 999 out of every 1000 individuals traveling through México City come from the strongly connected states as seen in Fig. 3C. In general, the delay is an increasing function of q. However, for the delay in the weakly connected regions to be similar to the delay of the secondary "wave" observed in the epidemic, the contribution of the weakly connected states to the total traffic has to be negligible. Therefore, transportation may contribute to the creation of a delay between peaks of confirmed cases for different states, but the connectivity between DF and the 31 Mexican states alone does not drive multiple waves of the same outbreak.

In Fig. 4 the infectivity rate is modulated by the introduction of a function $g(t)$ depending on time and events that affected the contact among individuals. These events were two main ones: social distancing government imposed measures and school closures. Some of these events are: the first peak dropped is reached after the Mexican government imposed social distancing measures and school closures. The second peak dropped marks the end of the school year. To test the possibility that social distancing and school closures had an impact on the epidemic, changes in contact are simulated using sigmoid functions that are time-dependent modulation of the infection rates in each state, λ_k, $k = 0, \ldots, 31$. The combined functions decrease and increase the rate of infection λ_k, at specific points in time (see Fig. 4A). In Fig. 4 the decrease and increase of the infection rates representing the sudden changes in transmission that may occur during closure and reopening of schools and social distancing measures suggest that the implementation of measures that decrease the contact rates in combination with the school calendar have a significant mitigating

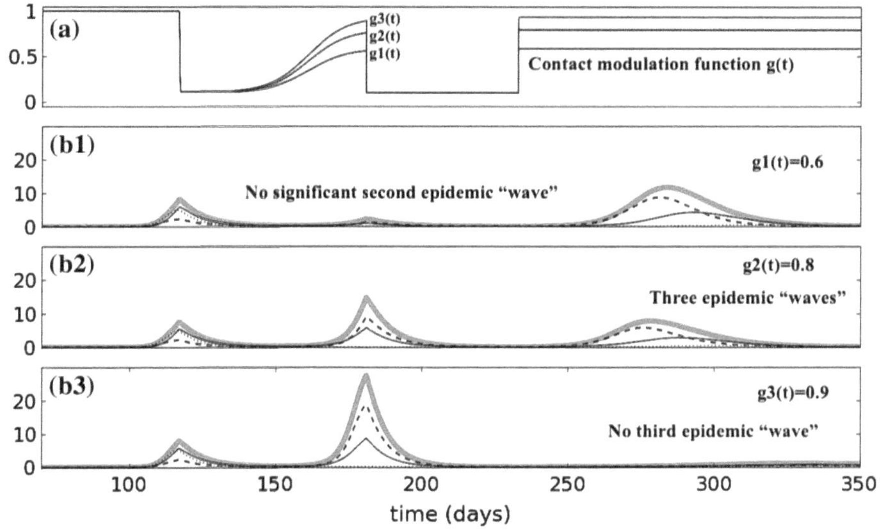

Fig. 4 Social distancing and school closures can create multiple outbreaks. Panel A shows different modulations of the infection rate after behavioral changes occurred. Panels B1, B2 and B3 show different time courses of the epidemic depending on the modulation of the contact rate. Panels B1, B2, and B3 correspond respectively to the functions $g1(t)$, $g2(t)$, and $g3(t)$. The *thick curves* are the sum of all infected individuals. The strongly and weakly connected populations are shown in *solid* and *dashed black lines*, respectively. The epidemic starts on day 78 in the states of Veracruz and Oaxaca

effect on the spread of the influenza in 2009. The patterns shown by the strongly and weakly connected states in the different scenarios shown in Fig. 4 reflect aspects of the epidemic that are driven by our function $g(t)$ that can suppress future outbreaks if the infectivity rate is not recovered 100 %. We assume by means of individuals being careful and taking measures to protect themselves such as using mask or distancing themselves if they are contagious. For instance, if the infection rate recovers to a small proportion of what it was at first (Fig. 4A), function $g1(t) = 0.6$, infections would occur at a very low rate after the intervention. In this case, the model predicts that there would be two large "waves" during the year with one outbreak of small amplitude during the summer (Fig. 4B1).

In conclusion, transport and the partition of the population into weakly and strongly connected states induces a delay in the dynamics. The weakly connected states have an epidemic outbreak that occurs after the strongly connected states have had an epidemic. The modulation of the infection rate by social distancing, school closures, and the academic calendar is enough to explain the emergence of the multiple waves of infection. Early arrival of the vaccine will have had a significant impact on the time course of the epidemic if they were available: at the beginning of the summer, or prior to the start of the new school semester. As vaccines become available later into the start of the epidemic there are more wasted vaccines. The intervention at the beginning of the April outbreak did mitigate the spread of the disease, but as a consequence generate two more waves and hence determined the shape of the data collected.

3.2 Non-proportional Vaccination Model

The non-proportional model is constructed in a way that allows the simulation of specific limits on the total number of vaccines available, the number that can be administered per day to a single population, the relative supply to different epidemiological classes, and the effects of the timing and duration of vaccination campaigns. Wherever possible, the values of these parameters were based on real and simulated data. In Table 2 all the parameters used in simulations are presented.

To compare the non-proportional model and the classical proportional models in the literature Fig. 5 shows the effects of vaccination in the proportional and non-proportional models for different campaign starts and a fix population size of 10^8. Vaccination occurs at a rate of 1 % of the eligible population per day for the pro-

Table 2 Parameters used in simulations for the non-proportional vaccination model

Parameter	Range	Description	Source
p	0.2 or 0.65	Probability of being confirmed	[12, 29, 48]
α	0.5	Relative infectiousness of unconfirmed class	[12, 21, 30]
t_a	20, 50, or 80 days	Start of vaccination campaign	Set to occur 10, 40, or 70 days after t_0
t_b	Variable	End of vaccination campaign	Depends on campaign start and duration
t_d	Variable	Depletion of vaccine stockpile	Depends on stockpile size set $\bar{v}$
t_0	Day 10	Starting point of the epidemic	Arbitrary
b	0.476 or 0.346	Mean probability of infection per contact	Adjusted as function of p so R_0=2.0[a]
c	1/7 (1/days)	Rate of recovery	[12, 20, 34]
N	10^8	Total population size	e.g. México [64, 69]
$\bar{v}$	30×10^6	Vaccine stockpile size	Based on 30 % coverage; [35, 46, 54]
$\bar{v}_D$	10^5–10^7	Maximum number of vaccines per day	Based on clinic data [1, 13, 15, 57, 58, 67]
k	0.001–0.1	Proportion of eligible vaccinated per day	[17, 23, 27, 43]
δ	10^{-6}(1/days)	Infection-related death rate	Based on U.S. viral surveillance data [63]

[a]See seasonal/pandemic R_0 values [16, 66]

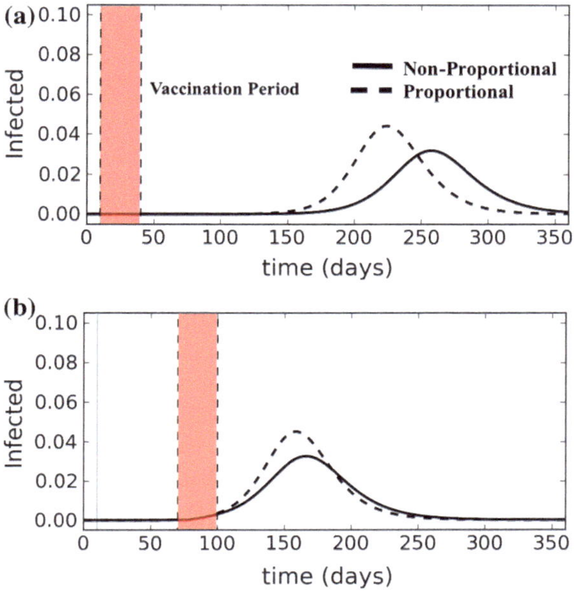

Fig. 5 Effects of vaccination in the proportional and non-proportional models for different campaign starts. The graphs show the proportion of infected people as a function of time. The population size is 10^8. The epidemic starts on day $t_0 = 10$ (*solid vertical gray line*). The vaccination campaign (*shaded pink region*) begins on day 10 A, or 70 B, and lasts 28 days. Vaccination occurs at a rate of 1 % of the eligible population per day (proportional; $k = 0.01$), or at a maximum of 10^6 vaccines per day (non-proportional, $\bar{v}_D = 10^6$)

portional model that is comparable to a maximum of 10^6 vaccines per day on our non-proportional model approach. The epidemic starts on day 10 and the vaccination campaign lasts approximately a month. Figure 5A shows a vaccination campaign that starts simultaneously with the epidemic; this can be the case of a seasonal influenza where there were not enough vaccines produced in advance. In Fig. 5B, the vaccination campaign (shaded pink region) starts 70 days before the start of the epidemic; this can be the case of an epidemic of a novel strain where the vaccines were produced quickly. It can be observed that in both scenarios the proportional model predicts epidemics that occur days prior to the epidemics predicted with our model. Though not very noticeable, these epidemics (the proportional ones) are more morbid but are shorter in terms of duration. This is more noticeable in Fig. 7 where we compare these patterns more closely.

Similarly in Fig. 6 there is a comparison of the proportional and non-proportional models when the stockpile is depleted. The initial population size is 10^8 people with a 40 day vaccination campaign that guarantees that the vaccine stockpile is depleted. Vaccination occurs at a rate of 1 % of the eligible population per day (proportional; $k = 0.01$), or at a maximum of 10^6 vaccines per day (non-proportional, $\bar{v}_D = 10^6$). The vaccination campaign is initiated on day 20 for Fig. 5A, or 80 for Fig. 5B. In this case, both models behavior (proportional and non-proportional) are relatively the same. Hence, there is a regime where both models are comparable depending on the rate the vaccination is administered, the campaign duration and if the stockpile is depleted.

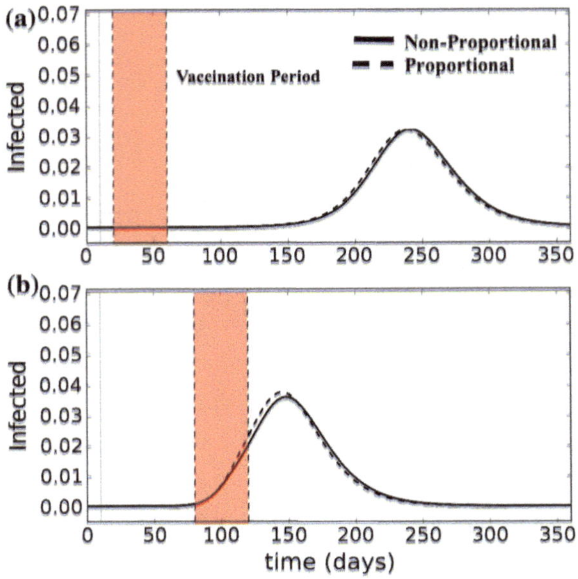

Fig. 6 Comparison of the proportional and non-proportional models when the stockpile is depleted. The initial population size is 10^8 people. The vaccination campaign is initiated on day 20 (A), or 80 (B), and lasts 40 days such that all the vaccines are used in both models. Start (t_a)and stop (t_b) times of the campaign are indicated by *dashed vertical lines*. Vaccination occurs at a rate of 1 % of the eligible population per day (proportional; $k = 0.01$), or at a maximum of 10^6 vaccines per day (non-proportional, $\bar{v}_D = 10^6$)

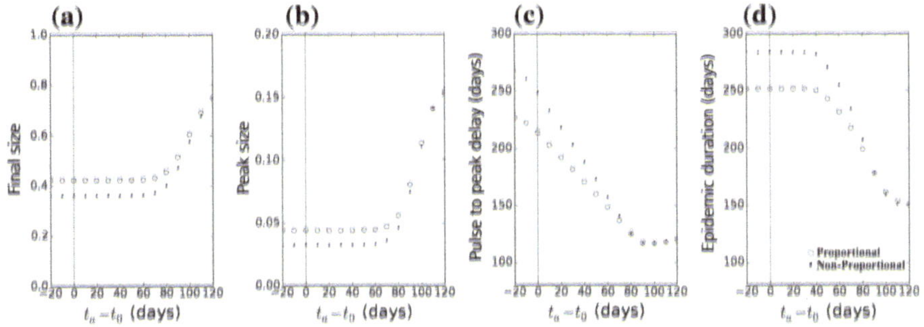

Fig. 7 Effects of vaccination in the two models for different administration rates and campaign durations. Epidemic measures are shown for proportional (*open circles*) and non-proportional (*filled dots*) models. Final size A, peak size B, peak time C, and epidemic duration D are plotted as a function of the difference between the vaccination start time (t_a) and the onset of the initial outbreak (t_0; *solid gray line*). The vaccination campaign durations and daily administration rates are as follows: 28 day campaign with k = 0.01 or $\bar{v}_D = 10^6$

Although the models behave similarly when examined under a conservative vaccination regime, when a moderate regime is put in place; with the population size to 10^8, vaccination rate of 1 % for the proportional model and 10^6 vaccines per day for the non-proportional model; important differences between the models on all four measures (Fig. 7A–D) are revealed. For early vaccination starts, final and peak sizes are smaller, while peak times and epidemic durations are larger in the non-proportional than the proportional model. As discussed previously, these differences result from the higher level of vaccine coverage achieved in the non-proportional relative to the proportional model. When vaccination starts later, due to the increasing number of vaccinated individuals from populations I_U and R_C, the difference between the models decreases until the models converge on most measures. Interestingly, with respect to epidemic duration, the two models not only converge, but reverse their respective relationship: epidemic durations are slightly smaller in the non-proportional model for very late vaccination start times (Fig. 7D).

In terms of limited number of vaccines administered per day: The key observation prompting the development of the model presented here was that most existing models of vaccination distribute vaccines based on a proportion of the eligible population. Vaccination clinics operate with a finite number of medical professionals for a finite number of hours, however, it is clear that distribution happens in practice based on the number of vaccines that can be administered per day. Pandemic preparedness plans devised by county health departments often calculate the necessary length of vaccination campaigns using a formula based on daily administration capacity. Therefore, we model vaccination by placing a limit on the number of daily vaccines (non-proportional model).

In terms of total vaccines administered: We predicted, based on the solutions of the equations representing the proportional and non-proportional models, that the different decays in the vaccinable population would lead to distinct epidemic dynamics. The non-proportional model always administers a larger total number of vaccines, which results in smaller and later, but sometimes longer, epidemics than in the proportional model. If the vaccination continues until the stockpile is depleted, the same total number of vaccines are administered in each model, and the epidemics produced are very similar in time course and severity.

In terms of the difference in epidemic duration: One of the largest differences between the two models when different total numbers of vaccines are administered is the epidemic duration. There is an increased coverage of the population in the non-proportional model, which allows the epidemic to develop more slowly, but can also cause it to last longer than predicted by the proportional model.

4 Final Remarks

In the metapopulation deterministic model, this work supports the view that the three epidemic "waves" are actually the same epidemic wave that has been interrupted by different factors such as social distancing, the academic calendar, and connectivity of cities by transportation dynamics. These results support the fact that massive governmental intervention measures did mitigate the spread of diseases but in reality they are costly economically. However, if they can be applied long enough for an intervention such as a vaccination campaign to be ready or medical supplies and treatment to be available it can lessen the public health burden. In the case of the 2009 AH1N1 influenza pandemic specifically in México: the first two waves were interrupted by social distancing policies, the closing of schools in the summer, and altered by delays in transportation effectiveness. To summarize, our modeling approach confirmed that México's transportation structure and the movement of individuals in this network contributed significantly to the generation of the three outbreaks; but more importantly social distancing and school closures have a delaying effect in the spread of the epidemic. It is important to point out that the third one was significantly larger since no significant interventions took place. When we look closely at an epidemic outbreak and see it as a network, an unprotected population is likely to suffer from further epidemic outbreaks that can be more morbid in comparison to the initial outbreak wave if no further mitigation strategies are put in place. Governments can use a strategy based on this knowledge about the possible delays induced in an epidemic outbreak to initially mitigate the spread of influenza or any other disease while resources become available, but an alarming alternative is that if the resources are not available when the full outbreak occurs, the consequences can be significant or even catastrophic depending on the severity of the disease in question. The A-H1N1 virus that caused the 2009 pandemic was mild in terms of mortality and by the time the third wave was happening, the authorities were not worried about containing the burden since people themselves were taking precautionary measures such as wearing masks and using hand sanitizers. But this might not happen in future pandemics as different influenza viruses emerge by mutations and recombinations like the 2009 pandemic which was caused by a novel form of the virus having portions of avian, porcine, and human influenza viruses [19]. There is a need for more resources to increase the capacity of mass production of vaccines and treatment in preparation for a possibly more severe influenza epidemic in the near future.

In the case of the non-proportional vaccination modeling approach, based on the numerical solutions of the equations representing the proportional and non-proportional vaccination models, different decays in the vaccinable population represent distinct epidemic dynamics depending on distinct vaccination scenarios in which the campaign duration and daily administration limit are imposed. Marked differences between the two models are evident in the epidemic duration when different amounts of vaccines are administered. This arises from the increased immunity of the population in the non-proportional model, which allows the epidemic to progress slowly causing it to last days or even weeks longer than predicted by the traditional proportional model. Models that can give more accurate predictions about the length of epidemics will allow health care professionals and medical facilities to prepare accordingly. The importance of this non-proportional model is that it constitutes a theoretical improvement over existing models, since it includes accurate data of available resources that can be used as parameter choices, vaccination of multiple epidemiological classes, a reasonable vaccine stockpile, limits on the number of vaccines administered per day, and ways to estimate wasted resources that can be adapted to any particular scenario. In particular, the non-proportional method of vaccine administration implemented in our model provides accurate predictions of the mitigating effects of vaccination. Public Health officials can use our non-proportional

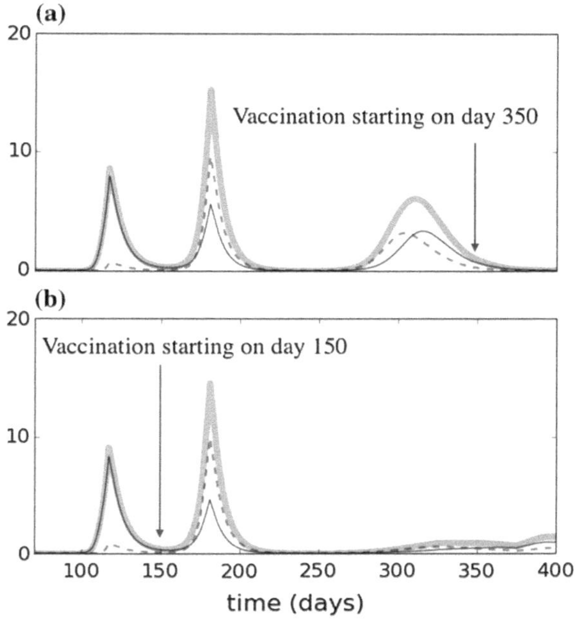

Fig. 8 Time course of the epidemic with vaccination before the second wave (**a**) and during the third wave as in the case of México during the third wave (**b**) of the 2009 pandemic. Vaccination of a maximum of 100,000 individuals per day from a stockpile of 30 million. The vertical axes represents the percentage of the population. **a** Starting vaccination times at day 350, corresponding to December 16, and **b** at day 150, corresponding to May 30 start of the summer. These simulations were performed assuming that a single outbreak starting on the first week of April around day 100 occurred

model as a tool to create preparedness plans for specific communities based on their available resources.

While the A-H1N1 pandemic was in full effect and the WHO had a phase 5 alert it was evident that the potential supply of vaccines was not going to be sufficient for the world population and at it bests will be more than 900 million [24]; that is, perhaps enough to cover 10–15% of the current world population [68]. The first 650 thousand vaccines from an estimated 30 million vaccines, arrived in México on November 23, 2009 [44]. However, by the beginning of January 2010, the Secretariat of Health in México had approximately 13 million vaccines available, of which only 1.5 million had been administered to the general population [65]. Looking closely at our metapopulation model of México, we can see in Fig. 8b that if vaccines where available before the second outbreak was full blown, the morbidity of the epidemic would have been controlled. In México vaccines were put in effect near the end of November around day 350 (Fig. 8a) when the third and final outbreak was already ending meaning that the majority of the vaccines used were really wasted.

Hence, while this work might answer some important questions for metapopulation models and realistic measures of mitigation effects on vaccination campaigns, there are still many questions regarding how to establish a regime of parameter values, initial conditions and different factors such as population size and number of trials that need to be explored further.

Acknowledgments I would like to thank the organizers of the Mathematical Biology session at the AWM Research Symposium for inviting me to give a talk: Dr. Erika T. Camacho and Dr. Talitha

Washington. Also, I would like to thank my collaborators in the work presented at the symposium and summarize in this article Dr. Marco A. Herrera-Valdez, Dr. Erin C. McKiernan and Dr. Carlos Castillo Chavez for their support and their valuable discussions and feedback. Last but not least, thanks to my Figure 1 model register nurse Iris Aldecoa from Scottsdale Healthcare.

References

1. K. Aaby, R.L. Abbey, J.W. Herrmann, M. Treadwell, C.S. Jordan, K. Wood, Embracing computer modeling to address pandemic influenza in the 21st century. J. Public Health Manag. Pract. **12**(4), 365–372 (2006)
2. R. Acuña-Soto, Death records from historical archives: a valuable source of epidemiological information, *Mathematical and Statistical Estimation Approaches in Epidemiology* (Springer, Dordrecht, 2009), pp. 189–194
3. J. Arino, Diseases in Metapopulations, vol. 11. Series in Contemporary Applied Mathematics (2009), pp. 65–123. (Also CDM Preprint Series report 2008-04)
4. J. Arino, P. van den Driessche, A multi-city epidemic model. Math. Popul. Stud. **10**(3), 175–193 (2003)
5. J. Arino, P. van den Driessche, The basic reproduction number in a multi-city compartmental epidemic model. Lect. Notes Control Inf. Sci. **294**, 135–142 (2003)
6. J. Arino, J.R. Davis, D. Hartley, R. Jordan, J.M. Miller, P. van Den Driessche, A multi-species epidemic model with spatial dynamics. Math. Med. Biol. **22**(2), 129 (2005)
7. J. Arino, F. Brauer, P. Van den Driessche, J. Watmough, J. Wu, A model for influenza with vaccination and antiviral treatment. J. Theor. Biol. **253**(1), 118–130 (2008)
8. F. Brauer, Compartmental models in epidemiology. Math. Epidemiol. **1945**, 19–79 (2008)
9. F. Brauer, Mathematical epidemiology is not an oxymoron. BMC Public Health **9**, 1–11 (2009)
10. F. Brauer, C. Castillo-Chavez, *Mathematical Models in Population Biology and Epidemiology* (Springer, New York, 2001)
11. F. Carrat, J. Luong, H. Lao, A.V. Sallé, C. Lajaunie, H. Wackernagel, A 'small-world-like' model for comparing interventions aimed at preventing and controlling influenza pandemics. BMC Med. **4**, 26 (2006)
12. F. Carrat, E. Vergu, N.M. Ferguson, M. Lemaitre, S. Cauchemez, S. Leach, A.J. Valleron, Time lines of infection and disease in human influenza: a review of volunteer challenge studies. Am. J. Epidemiol. **167**(7), 775–785 (2008)
13. Centers for Disease Control and Prevention. Large-scale vaccination clinic output and staffing estimates: an example (2009)
14. Centers for Disease Control and Prevention. Influenza symptoms and laboratory diagnostic procedures. Accessed March 2011
15. B.H. Cho, K.A. Hicks, A.A. Honeycutt, N. Hupert, O. Khavjou, M. Messonnier, M.L. Washington, A tool for the economic analysis of mass prophylaxis operations with an application to H1N1 influenza vaccination clinics. J. Public Health Manag. Pract. **17**(1), E22–E28 (2011)
16. G. Chowell, M.A. Miller, C. Viboud, Seasonal influenza in the United States, France, and Australia: transmission and prospects for control. Epidemiol. Infect. **136**(06), 852–864 (2008)
17. G. Chowell, C. Viboud, X. Wang, S.M. Bertozzi, M.A. Miller, Adaptive vaccination strategies to mitigate pandemic influenza: Mexico as a case study. PLoS One **4**(12), e8164 (2009)
18. M.L. Ciofi degli Atti, S. Merler, C. Rizzo, M. Ajelli, M. Massari, P. Manfredi, C. Furlanello, G. Scalia Tomba, M. Iannelli, N. Ahmed, Mitigation measures for pandemic influenza in Italy: an individual based model considering different scenarios. PLoS One **3**, e1790 (2008)
19. B.J. Coburn, B.G. Wagner, S. Blower, Modeling influenza epidemics and pandemics: insights into the future of swine flu (H1N1). BMC Med. **7**, 30 (2009)
20. R.B. Couch, J.A. Kasel, Immunity to influenza in man. Annu. Rev. Microbiol. **37**(1), 529–549 (1983)

21. R.B. Couch, R.G. Douglas Jr., D.S. Fedson, J.A. Kasel, Correlated studies of a recombinant influenza-virus vaccine. III. Protection against experimental influenza in man. J. Infect. Dis. **124**(5), 473–480 (1971)
22. M. Cruz-Aponte, E. McKiernan, M.A. Herrera-Valdez, Mitigating effects of vaccination on influenza outbreaks given constraints in stockpile size and daily administration capacity. BMC Infect. Dis. **11**(1), 207 (2011)
23. J.M. Epstein, D.M. Goedecke, F. Yu, R.J. Morris, D.K. Wagener, G.V. Bobashev, Controlling pandemic flu: the value of international air travel restrictions. PLoS One **2**(5), 401 (2007)
24. M. Falco, Cdc: production of H1N1 flu lagging (2009). http://nats.sct.gob.mx/nats/sys/index.jsp?i=3
25. D.S. Fedson, Pandemic influenza and the global vaccine supply. Clin. Infect. Dis. **36**(12), 1552–1561 (2003)
26. D.S. Fedson, Preparing for pandemic vaccination: an international policy agenda for vaccine development. J. Public Health Policy **26**(1), 4–29 (2005)
27. N.M. Ferguson, D.A.T. Cummings, C. Fraser, J.C. Cajka, P.C. Cooley, D.S. Burke, Strategies for mitigating an influenza pandemic. Nature **442**(7101), 448–452 (2006)
28. A. Flahault, E. Vergu, L. Coudeville, R.F. Grais, Strategies for containing a global influenza pandemic. Vaccine **24**(44–46), 6751–6755 (2006)
29. A. Flahault, X. de Lamballerie, T. Hanslik, N. Salez, Symptomatic infections less frequent with H1N1pdm than with seasonal strains. PLoS Curr. **1**, RRN1140 (2009)
30. H.M. Foy, M.K. Cooney, I.D. Allan, J.K. Albrecht, Influenza B in households: virus shedding without symptoms or antibody response. Am. J. Epidemiol. **126**(3), 506–515 (1987)
31. T.C. Germann, K. Kadau, I.M. Longini Jr., C.A. Macken, Mitigation strategies for pandemic influenza in the United States. PNAS **103**(15), 5935–5940 (2006)
32. W.P. Glezen, Herd protection against influenza. J. Clin. Virol. **37**(4), 237–243 (2006)
33. P.A. Gross, A.W. Hermogenes, H.S. Sacks, J. Lau, R.A. Levandowski, The efficacy of influenza vaccine in elderly persons. Ann. Intern. Med. **123**(7), 518–527 (1995)
34. F.G. Hayden, R. Fritz, M.C. Lobo, W. Alvord, W. Strober, S.E. Straus, Local and systemic cytokine responses during experimental human influenza A virus infection. Relation to symptom formation and host defense. J. Clin. Investig. **101**(3), 643–649 (1998)
35. Health Industry Distributors Association. 2008–2009 influenza vaccine production and distribution (2009)
36. M.A. Herrera-Valdez, M. Cruz-Aponte, C. Castillo-Chavez, Multiple outbreaks for the same pandemic: local transportation and social distancing explain the different "waves" of A-H1N1pdm cases observed in México during 2009. Math. Biosci. Eng. (MBE) **8**(1), 21–48 (2011)
37. J.M. Hyman, T. Laforce, Modeling the spread of influenza among cities, *Biomathematical Modeling Applications for Homeland Security* (Society for Industrial and Applied Mathematics, Philadephia, 2003), pp. 215–240
38. W.O. Kermack, A.G. McKendrick, Contributions to the mathematical theory of epidemics. Proc. R. Soc. Lond. **115**, 700–721 (1927)
39. W.O. Kermack, A.G. McKendrick, Contributions to the mathematical theory of epidemics III. Further studies of the problem of endemicity. Bull. Math. Biol. **53**(1), 89–118 (1991)
40. K. Khan, J. Arino, W. Hu, P. Raposo, J. Sears, F. Calderon, C. Heidebrecht, M. Macdonald, J. Liauw, A. Chan et al., Spread of a novel influenza A (H1N1) virus via global airline transportation. N. Engl. J. Med. **361**(2), 212 (2009)
41. C.D. Kozul, K.H. Ely, R.I. Enelow, J.W. Hamilton, Low-dose arsenic compromises the immune response to influenza a infection in vivo. Environ. Health Perspect. PubMed **117**(9), 1441–1447 (2009)
42. B.Y. Lee, S.T. Brown, P. Cooley, J.J. Grefenstette, R.K. Zimmerman, S.M. Zimmer, M.A. Potter, R. Rosenfeld, W.D. Wheaton, A.E. Wiringa et al., Vaccination deep into a pandemic wave: potential mechanisms for a "third wave" and the impact of vaccination. Am. J. Prev. Med. **39**(5), e21–e29 (2010)

43. V.J. Lee, G.G. Fernandez, M.I. Chen, D. Lye, Y.S. Leo, Influenza and the pandemic threat. Singap. Med. J. **47**(6), 463–470 (2006)
44. F. Libenson, Llegaron al Edomex 66 mil vacunas contra AH1N1 (2009). http://elinformantemexico.com/index.php/noticias/llegaron-al-edomex-66-mil-vacunas-contra-ah1n1-franklin-libenson-violante.html
45. I.M. Longini Jr., A. Nizam, S. Xu, K. Ungchusak, W. Hanshaoworakul, D.A.T. Cummings, M.E. Halloran, Containing pandemic influenza at the source. Science **309**(5737), 1083–1087 (2005)
46. Macroepidemiology of Influenza Vaccination Study Group. The Macro-epidemiology of influenza vaccination in 56 countries, 1997–2003. Vaccine 23(44), 5133–5143 (2005)
47. M.A. Miller, C. Viboud, M. Balinska, L. Simonsen, The signature features of influenza pandemics-implications for policy. N. Engl. J. Med. **360**(25), 2595 (2009)
48. A.S. Monto, J.S. Koopman, I.M. Longini Jr., Tecumseh study of illness. XIII. Influenza infection and disease, 1976–1981. Am. J. Epidemiol. **121**(6), 811–822 (1985)
49. M.R. Moser, T.R. Bender, H.S. Margolis, G.R. Noble, A.P. Kendal, D.G. Ritter, An outbreak of influenza aboard a commercial airliner. Am. J. Epidemiol. **110**(1), 1 (1979)
50. K.L. Nichol, Efficacy and effectiveness of influenza vaccination. Vaccine **26**, D17–D22 (2008)
51. North American transportation statistics database (2014). http://nats.sct.gob.mx/nats/sys/index.jsp?i=3
52. M. Nuño, G. Chowell, A.B. Gumel, Assessing the role of basic control measures, antivirals and vaccine in curtailing pandemic influenza: scenarios for the US, UK and the Netherlands. J. R. Soc. Interface **4**(14), 505 (2007)
53. M. Nuño, G. Chowell, X. Wang, C. Castillo-Chavez, On the role of cross-immunity and vaccines on the survival of less fit flu-strains. Theor. Popul. Biol. **71**(1), 20–29 (2007)
54. Oliver Wyman Group and Program for Appropriate Technology in Health. Influenza vaccine strategies for broad global access, key findings and project methodology (2007)
55. H. Oshitani, T. Kamigaki, A. Suzuki, Major issues and challenges of influenza pandemic preparedness in developing countries. Emerg. Infect. Dis. **14**(6), 875–880 (2008)
56. M.T. Osterholm, Preparing for the next pandemic. N. Engl. J. Med. **352**(18), 1839–1842 (2005)
57. Peterborough County-City Health Unit. Pandemic Influenza Plan, Annex A: Mass Vaccination Plan (2010)
58. F.B. Phillips, J.P. Williamson, Local health department applies incident management system for successful mass influenza clinics. J. Public Health Manag. Pract. **11**(4), 269 (2005)
59. J. Ramet, C. Weil-Olivier, W. Sedlak, Influenza vaccination: the pediatric perspective. Vaccine **25**(5), 780–787 (2007)
60. C.A. Russell, T.C. Jones, I.G. Barr, N.J. Cox, R.J. Garten, V. Gregory, I.D. Gust, A.W. Hampson, A.J. Hay, A.C. Hurt et al., Influenza vaccine strain selection and recent studies on the global migration of seasonal influenza viruses. Vaccine **26**, D31–D34 (2008)
61. L.A. Rvachev, I.M. Longini Jr., A mathematical model for the global spread of influenza. Math. Biosci. **75**(1), 3–22 (1985)
62. K. Stohr, M. Esveld, Will vaccines be available for the next influenza pandemic? Science **306**(5705), 2195–2196 (2004)
63. W.W. Thompson, D.K. Shay, E. Weintraub, L. Brammer, N. Cox, L.J. Anderson, K. Fukuda, Mortality associated with influenza and respiratory syncytial virus in the United States. JAMA: J. Am. Med. Assoc. **289**(2), 179 (2003)
64. U.S. Census Bureau. International data base (IDB). Accessed March 2011
65. B. Valadez, Aplicadas solo 10% de las dosis contra el AH1N1 SSA (2009). http://www.milenio.com/node/368812
66. C. Viboud, T. Tam, D. Fleming, A. Handel, M.A. Miller, L. Simonsen, Transmissibility and mortality impact of epidemic and pandemic influenza, with emphasis on the unusually deadly 1951 epidemic. Vaccine **24**(44–46), 6701–6707 (2006)
67. M. Washington, Evaluating the capability and cost of a mass influenza and pneumococcal vaccination clinic via computer simulation. Med. Decis. Mak. **29**(4), 414–423 (2009)

68. Wikipedia. World population 1800–2100 (2010). http://en.wikipedia.org/wiki/file:world-population-1800-2100.png
69. Wikipedia. List of countries by population. http://en.wikipedia.org/wiki/list_of_countries_by_population. Accessed 7 Mar 2011
70. World Health Organization. WHO guidelines on the use of vaccines and antivirals during influenza pandemics (2004). www.who.int/entity/csr/resources/publications/influenza/11_29_01_a.pdf
71. World Health Organization. Strengthening pandemic influenza preparedness and response (2005). www.who.int/csr/disease/influenza/a58_13-en.pdf
72. World Health Organization. Pandemic influenza preparedness and response (2009). http://whqlibdoc.who.int/publications/2009/9789241547680_eng.pdf
73. Y. Yang, J.D. Sugimoto, M.E. Halloran, N.E. Basta, D.L. Chao, L. Matrajt, G. Potter, E. Kenah, I.M. Longini Jr., The transmissibility and control of pandemic influenza A (H1N1) virus. Science **326**(5953), 729 (2009)

Controlling a Cockroach Infestation

Hannah Albert, Amy Buchmann, Laurel Ohm, Ami Radunskaya and Ellen Swanson

Abstract The cockroach is one of the world's most prolific and resilient pests, with over 3,500 species worldwide. It is important to understand the growth and adaptive mechanisms of cockroach colonies in order to safely control these populations. We present a continuous time, age-structured population model of the Blattella germanica cockroach that includes the application of pesticides and the development of resistant subpopulations. The resulting system of differential equations is then used to optimize treatment strategies using analytical and heuristic optimization techniques. While the model shows that the roach-free equilibrium is always unstable, the strategic application of pesticides can keep populations low, even when a drug-resistant subpopulation develops.

Amy presented this work at the special session "Research from the Cutting EDGE." Amy is the 60th member of the EDGE (Enhancing Diversity in Graduate Education) Program to receive her doctorate in Mathematics. The goal of EDGE is to strengthen the ability of women to successfully complete their graduate programs in the Mathematical Sciences. Please see the preface for more information about the EDGE Program and its founders.
Ami co-organized the special session "Research from the Cutting EDGE." Ami is also a co-director for the EDGE Program, together with Ulrica Wilson from Morehouse College.

H. Albert
Department of Applied Mathematics, Illinois Institute of Technology, Chicago, IL 60616, USA
e-mail: halbert@hawk.iit.edu

A. Buchmann (✉)
Department of Mathematics and Center for Computational Science, Tulane University, New Orleans, LA 70118, USA
e-mail: abuchman@tulane.edu

L. Ohm
School of Mathematics, University of Minnesota, Minneapolis, MN 55455, USA
e-mail: ohmxx039@umn.edu

A. Radunskaya
Department of Mathematics, Pomona College, Claremont, CA 91711, USA
e-mail: aradunskaya@pomona.edu

E. Swanson
Department of Mathematics, Centre College, Danville, KY 40422, USA
e-mail: ellen.swanson@centre.edu

G. Letzter et al. (eds.), *Advances in the Mathematical Sciences*, Association for Women in Mathematics Series 6, DOI 10.1007/978-3-319-34139-2_9

Keywords Mathematical model · Blattella germanica · Optimal pesticide application

Mathematics Subject Classification 92

1 Introduction

Despite its diminutive size (1.3–1.6 cm long) [1], *Blattella germanica* is one of the most common and resilient pests worldwide, costing over a billion dollars in pest removal services in the US each year [1]. The nocturnal scavenger is frequently found in human establishments where food sources are readily available, especially restaurants, food processing centers, hotels, individual homes, and apartment complexes. Their flat bodies afford them easy access to human dwellings through cracks in doors and walls, and allow them to hide easily in narrow crevices during daylight hours.

In addition to ruining foods by leaving foul-smelling secretions wherever they walk, the German cockroach can also serve as a pathogenic vector [5, 22], and its skin can produce potent allergens [4, 16]. Pesticides represent one of the quickest and most effective ways of controlling cockroach outbreaks, as even the cleanest house can still harbor the little brown pests. However, rampant pesticide use can also negatively affect human health, especially at high levels of toxicity or if accidentally ingested [13]. Thus the use of pesticides should balance the certainty of cockroach elimination with the level of toxicity introduced into the surrounding environment. Ideally, the minimum amount of pesticide required to control the cockroach population will be applied [13].

This paper builds on several previous models. A system of difference equations is used in [9] to describe the effect of pesticides on roaches at various stages in their development, resulting in a discrete time, age-structured model of the roach populations. In their model, Larter et al. divide the roach population into twenty stages, or "stadia." Discrete time and continuous time models are compared by Wu et al. in [21]. The focus of this paper is to consider the competition between two cockroach species, *Blattella germanica* and *B. bisignata*, as well as the effect of circadian cycles on the population dynamics. The authors of [21] find no significant differences between the results of the discrete and continuous models. Our goal is to determine optimal pesticide treatment strategies and, therefore, we choose to use the simplest models that will give us useful information. Our approach is to use a continuous time model, with a minimal number of age groups. The model described in this paper represents the population of *Blattella germanica* subdivided into roaches that are resistant and non-resistant to the pesticide. We consider two mechanisms of pesticide resistance acquisition: developmental—acquired over the course of the cockroach's lifetime via repeated nonlethal levels of exposure to a particular pesticide, and genetic—resistant adults likely pass down the cellular ability to metabolize certain pesticides to their offspring [10, 11, 19, 20]. A system of ordinary differential equations models the roach population and is presented in Sect. 2. The goal is to determine a treatment that requires the smallest amount of pesticide necessary to eliminate the roach population. The ability to achieve and maintain this roach-free state is explored in Sect. 3.1 through a stability analysis. The optimal pesticide treatment determined by a genetic algorithm is presented in Sect. 3.2. The treatment schedule varies the amount of pesticide applied and frequency of pesticide application.

2 The Mathematical Model

In order to analyze the effects of pesticide application on *Blattella germanica* proliferation, we separate the cockroach population into four categories—eggs, adult roaches, pesticide-resistant eggs, and pesticide-resistant adult roaches. These categories allow us to account for differences in mobility between eggs and adults, and differences in pesticide toxicity between resistant and non-resistant populations. According to this structure, the egg stage includes only the eggs, while the adult stage includes both juveniles, or nymphs, and mature individuals. Most pesticides in common use are applied to surfaces and transmitted to the insects as they walk on the surfaces. Both nymphs and adults are mobile, and can be poisoned by walking through the areas where pesticide has been applied. However, the eggs remain unharmed in the female cockroach's *ootheca*—a hard, protective casing enclosing the eggs [13]. Hence, pesticides applied to surfaces will affect these two groups differently. The roach population groups in the model do not differentiate between juveniles and adults, or between males and females. The death rate of the adult population is taken to be an average of these subgroups (see Sect. 2.4 for details).

Resistance acquisition arises via two possible mechanisms: as an evolutionary byproduct, or as a continuous process requiring exposure over time. Hence, as a simplification, we divide the population into resistant and susceptible subpopulations, with corresponding low and high kill rates associated with pesticide application. We assume that the acquisition of resistance depends on both the length of exposure and pesticide dosage, and thus can be expressed as a function of the population size and the concentration of pesticides in a given area. The model accounts for delayed resistance acquisition following continuous exposure by allowing non-resistant adult roaches to move to the resistant population following prolonged, low-dosage exposure to pesticides.

In addition to considering the time evolution of the four subpopulations, we consider the amount of pesticide present at any given time as a fifth dynamic state variable in the model. We consider various pesticide application scenarios, with differing pesticide concentration per area per application, and differing time periods between applications. We do a stability analysis and use a genetic algorithm to determine the most effective combination of pesticide strength and application scheduling that results in lowest accumulated toxicity while still eliminating the cockroach population in the long run.

2.1 The Basic Model

The populations of susceptible eggs and adults at time t are represented by $E(t)$ and $A(t)$, respectively. After the application of pesticide, non-resistant adults can join the resistant sub-population, represented by $A_R(t)$. The model assumes that the classification of eggs (resistant or not) depends on the classification of the female mate. For example, resistant female roaches produce only resistant eggs, represented in the model by $E_R(t)$. The pesticide present at time t is represented by $P(t)$. The model equations are

$$\frac{dE}{dt} = -gE + b_A A \tag{1a}$$

$$\frac{dA}{dt} = \left(1 - \frac{A + A_R}{k}\right) gE - \left(1 + \frac{A + A_R}{k}\right) d_A A - f_A(P)A - f_Z(P)A \tag{1b}$$

$$\frac{dE_R}{dt} = -gE_R + b_R A_R \tag{1c}$$

$$\frac{dA_R}{dt} = \left(1 - \frac{A + A_R}{k}\right) gE_R - \left(1 + \frac{A + A_R}{k}\right) d_A A_R - f_R(P)A_R + f_Z(P)A \tag{1d}$$

$$\frac{dP}{dt} = T(t) - d_P P \tag{1e}$$

Each population evolves according to a growth term and a death term. The growth terms are linear functions of the populations, with rates derived from the literature (see Table 2). The birth rates for the non-resistant and resistant populations are given by b_A and b_R, respectively. Eggs become adults at rates that depend on both the length of the egg stage, as well as the survival rate of the eggs: these rates are combined into the parameter g. The growth rate of the new adult population decreases as the total population size approaches the carrying capacity k. Similarly, the death rate for the adult population, denoted by d_A, is assumed to increase as the adult population approaches the carrying capacity k due to increased competition for resources.

The amount of pesticide present at a given time t is governed by the $\frac{dP}{dt}$ equation, where $T(t)$ represents the rate at which pesticide is applied at time t, and d_P is the decay constant for the pesticide. The decay rate is calculated as $d_P = \frac{1}{\tau}\ln(2)$, where τ is the pesticide half-life (see Table 1). The effect of pesticide on the resistant and non-resistant adults is represented by $f_R(P)$ and $f_A(P)$, respectively. The function $f_Z(P)$ describes the rate at which susceptible insects develop resistance to the pesticide.

2.2 The Effect of Pesticides

We account for the effects of pesticide application on the non-resistant and resistant adult cockroach populations by including nonlinear functions of the amount of pesticide present, $f_A(P)$ and $f_R(P)$, respectively. These kill rates allow for the evolutionary acquisition of immunity. Presumably, more resistant cockroaches survive each pesticide application, resulting in a larger ratio of resistant to non-resistant cockroaches in the mating pool and increasing the chance that the next batch of eggs contains genetically resistant roaches. We assume that these functions are increasing, saturating functions of the amount of pesticide, with maximum kill rate c_A and c_R for the non-resistant and resistant populations, respectively (see Fig. 1). In the absence of any calibration data, both of these maximum kill rates are set to 1 in the current implementation. The dose-dependent pesticide kill rates are given by the functions

$$f_A(P) = \frac{c_A P}{s_A \cdot k + P} \quad , \qquad f_R(P) = \frac{c_R P}{s_R \cdot k + P} \tag{2}$$

where P is in micrograms, and c_A and c_R are in units of day^{-1}. The 50 % lethal dose of pesticide per individual, or LD_{50}, is the amount of pesticide per cockroach that will result in a 50 % mortality among non-resistant (s_A) and resistant roaches (s_R).

Various studies [10, 11, 19, 20] detail the possibility of susceptible insects acquiring immunity via upregulation of transcription genes for metabolic detoxification. As the amount of pesticide increases beyond a certain threshold, the ability of the individual to acquire the necessary mutations that result in resistance decreases due to the toxicity. This could be described by a function of the form

$$f_Z(P) = \alpha P e^{-\beta P} \tag{3}$$

which is shown in Fig. 1. The parameter α controls the maximum rate of resistance acquisition, and the inverse of parameter β describes the pesticide dose resulting in the maximum resistance.

2.3 Pesticide Parameters

The pesticide parameters are listed in Table 1 and described below. For toxins such as cypermethrin—a typical active ingredient in household pesticides such as RAID—the 50 % lethal dose for susceptible adult cockroaches is $s_A = 0.049\,\mu g$ per cockroach [14, 17]. Resistant cockroaches, on the other hand, experience 50 % mortality at a dose of $s_R = 0.24\,\mu g$ per cockroach [14, 17]. As an ad hoc value, we take the LD_{10} value for cypermethrin—the toxin dosage resulting in 10 % insect mortality—to be the dosage at which the surviving roaches experience the highest rate of acquired metabolic resistance. Since the LD_{10} is given as the dose per roach, we assume that maximum resistance acquisition occurs as the cockroach population approaches carrying capacity k, 10,000 roaches [8]. Thus $\beta = 1/[0.01\,\mu g$ per cockroach $\times 10{,}000$ roaches$] = 0.01\,\mu g^{-1}$ [18]. For α, which is proportional to the maximum rate of resistance acquisition per μg of pesticide applied, we use a 20-fold increase in resistance over a 4-day exposure period to calculate the maximum rate. We account for the dosage dependence by dividing the fivefold increase per day by

Table 1 Pesticide parameters

Name	Description	Value	Units	Source
s_A	LD_{50}: non-resistant adults	0.049	µg per roach	[14, 17]
s_R	LD_{50}: resistant adults	0.24	µg per roach	[14, 17]
α	$\propto$ maximum rate of resistance gain	$0.05 * \beta * e$	Per $(\mu g)^2$ Per day	[8, 18]
β	Inverse of max resistance dose	0.01	Per µg	[8, 20]
d_P	Pesticide decay rate	$\frac{\ln(2)}{5}$	Per day	[2, 7]

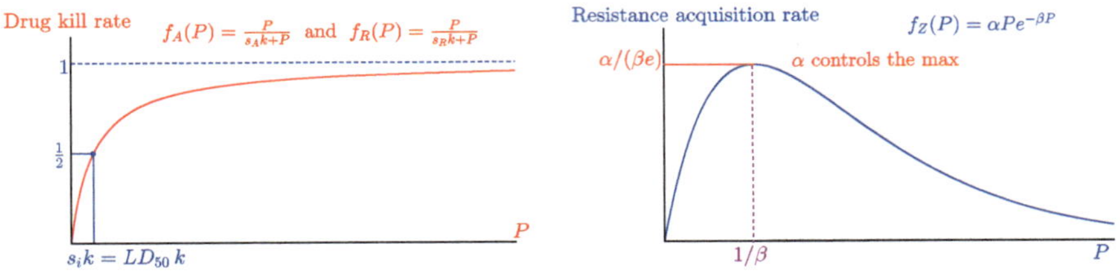

Fig. 1 The effect of pesticide on cockroach survival ($f_A(P)$ and $f_R(P)$ *Left Panel*) and the development of resistance ($f_Z(P)$ *Right Panel*). Units are per individual

the LD_{10} dose multiplied by the carrying capacity for the cockroaches: [maximum rate of acquisition]= $\frac{20}{4}$ days^{-1}/[0.01 µg per cockroach $\times$10,000 roaches] = 0.05 [µg $\cdot$ days]$^{-1}$ [20]. Taking α to be $0.05 * \beta * e$, gives a maximum resistance development rate of 0.05 when the dosage of pesticide applied is 0.01 µg per cockroach, or the LD_{10}.

2.4 Population Growth Parameters

Table 2 gives the population growth parameter values used in the model and the derivation of these values is described below. The calculation of the per capita birth rate assumes that 50 % of the adult cockroaches are female and adult *Blattella germanica* females are capable of reproduction for a three-month period [12] during a total lifespan of 160 days [6]. A typical female produces one ootheca per month [3] containing, on average, 35 viable eggs, for a total of three reproductive events and 105 hatching eggs per 160-day lifetime of a female. The average birth rate for the total adult population is $0.5\frac{105}{160} = 0.33$ eggs per adult cockroach per day, assuming that the birth rate of resistant and non-resistant eggs are identical. For the sake of simplicity, the model considers only the eggs that eventually hatch, so that the natural death rate of unhatched eggs is incorporated into the parameter, g. Hence, the 35 eggs per reproductive cycle represent the average number of viable larvae rather than actual eggs per ootheca. The base life span of the roaches is the average

Table 2 Parameter values used in the model

Name	Description	Value	Units	Source
b_A	Birth rate: non-resistant eggs	0.33	Per individual per day	[3, 6, 12]
b_R	Birth rate: resistant eggs	0.33	Per individual per day	[3, 6, 12]
g	Maturation rate of eggs	$\frac{1}{21}$	Per individual per day	[15]
d_A	Base death rate of adults	$\frac{1}{140}$	Per individual per day	[6]
k	Carrying capacity	10,000	Individuals	[8]
c_A, c_R	Maximum kill rate of non-resistant and resistant populations due to pesticide	1	Per day	ad hoc value

of the life spans reported in [6]. The carrying capacity is based on an apartment complex with 10 apartments, using the peak density observed by Koehler et al. [8] of approximately 1,000 *Blattella germanica* per apartment.

3 Results

The goal of this project is to determine a treatment plan that simultaneously minimizes the amount of pesticide and the roach population. The ideal case would result in no roaches. The stability of the system (1) at the roach-free equilibrium is analyzed in Sect. 3.1. A genetic algorithm, described in Sect. 3.2, determines the best treatment plan to minimize both the roaches and the pesticide.

3.1 Stability Analysis

We note that, in the absence of pesticides, when the entire population is susceptible, the model has a non-zero steady state at

$$E^* = \frac{b_A}{g} A^* \quad A^* = k \frac{b_A - d_A}{b_A + d_A}. \tag{4}$$

For our choice of parameter values (see Table 2), $A^* = k\frac{137}{143}$, so the value of k is very close to this steady state population, justifying the term "carrying capacity". This scenario is depicted in Fig. 2. It is reasonable to assume that, in the presence of

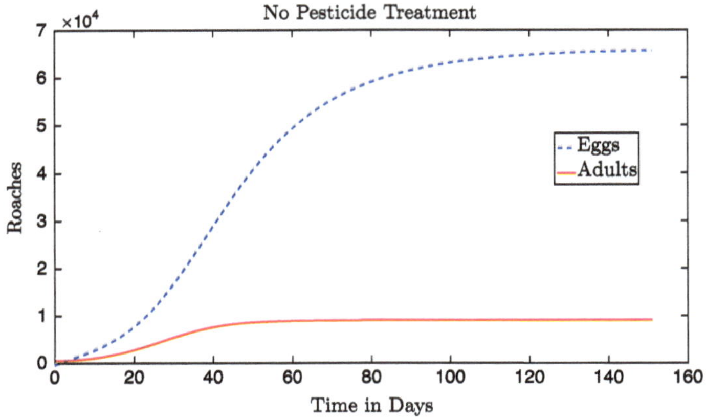

Fig. 2 Without treatment, the susceptible adult population reaches its maximum value, as long as the initial value is strictly positive. The zero equilibrium is an unstable steady state without treatment as long as $b_A > d_A$

a roach infestation and before treatment is applied, the non-resistant population has reached this steady state, and the resistant population is zero.

We are interested in driving the roach populations to zero through the application of pesticide, so we analyze the stability of the zero-roach equilibrium. Since the differential equation for the pesticide decouples from the other four equations, we consider the equations for the adult and egg populations for a given value of P. A linear stability analysis of the four-dimensional system describing the evolution of the susceptible and resistant cockroach populations around the roach-free equilibrium: $(0, 0, 0, 0)$ gives the characteristic polynomial

$$p(\lambda) = [(g + \lambda)(d + f_R(P) + \lambda) - bg][(g + \lambda)(d + f_A(P) + f_Z(P) + \lambda) - bg],$$

where, to simplify the notation, we write $b = b_R = b_A$ and $d = d_A$. The roots of this polynomial can be expressed as:

$$\lambda = \frac{-(g + \hat{d}) \pm \sqrt{(g + \hat{d})^2 - 4g(\hat{d} - b)}}{2}$$

where $\hat{d} = d + f_R(P)$ for the first two roots, and $\hat{d} = d + f_A(P) + f_Z(P)$ for the third and fourth roots. In the absence of pesticide, when $f_R(P) = f_A(P) = f_Z(P) = 0$, the roots of the characteristic polynomial will all be negative as long as $d > b$, i.e., as long as the death rate of the cockroaches is greater than the birth rate. This result is intuitively clear—and the condition will never be satisfied if a roach infestation is occurring. The expression also tells us that the roach-free equilibrium will be stable in the presence of pesticide as long as

$$d + f_R(P) > b \quad \text{and} \quad d + f_A(P) + f_Z(P) > b.$$

Since $f_R(P) < f_A(P) + f_Z(P)$ for all values of P, we know that the zero equilibrium will be *stable*, and the roach population will die out, as long as $f_R(P) > b - d$. This condition translates to:

$$P > \frac{(b-d)s_R k}{1-b+d}. \tag{5}$$

The differential equation for P can be solved explicitly if we assume that a constant rate of pesticide is given for one hour in any given day. Thus, $T(t)$ is non-zero only from $t = 0$ to $t = 1/24 \mod 1$, and during these intervals it is a constant multiple of a fixed increment, l. These assumptions make sense, since pesticide is typically applied consistently in discrete increments over a relatively short time period. Therefore, we assume that a pesticide application over an interval of N days can be described by a vector of length N: $T(t) = (T_1, T_2, T_3, \ldots, T_N)$, where $T_i \in \{0, 1, 2, \ldots\}$ is the *intensity* of the treatment on a given day i, and the *amount* given on that day is lT_i for a fixed increment, l (see Fig. 3).

We can analyze the long term behavior of the roach population if we assume a regular treatment schedule: one treatment every n days, where the amount of pesticide applied in each treatment is a constant, lT. We can solve the differential equation for P with initial condition $P(0) = 0$. We assume that the first treatment is given on day 1, the next on day $n+1$, the next on day $2n+1$, and so on. Each treatment lasts one hour ($\frac{1}{24}$th of a day), and has treatment intensity, T. For this analysis, we rescale the pesticide amounts so that $l = 1$, i.e., the intensity, T, is equal to the amount of pesticide given per unit time.

$$P(t) = \begin{cases} \frac{T}{d_P}\left(1 - e^{-d_P t}\right) & 0 \le t < 1/24 \\ P(1/24)e^{-d_P(t-1/24)} & 1/24 \le t < n \\ \frac{T}{d_P}\left(1 - e^{-d_P(t-n)}\right) + P(n)e^{-d_P(t-n)} & n \le t < n + 1/24 \\ P(n+1/24)e^{-d_P(t-(n+1/24))} & n + 1/24 \le t < 2n \\ \vdots & \end{cases}$$

As $t \to \infty$, this solution approaches a stable periodic solution which varies from P_{low} to P_{high} (see Fig. 4). The period of this stable periodic solution is n days. We can find the minimum value of this periodic solution by equating the expressions for P_{low} at the beginning and end of the cycle (set $t = 0$ at the beginning, or low point, of the cycle)

$$P_{high} = \frac{T}{d_P} - \frac{T}{d_P}e^{-d_P/24} + P_{low}e^{-d_P/24}, \quad P_{low} = P_{high}e^{-d_P(n-1/24)}$$

$$\Rightarrow P_{low} = \frac{T}{d_P}\frac{e^{d_P/24} - 1}{e^{d_P n} - 1}.$$

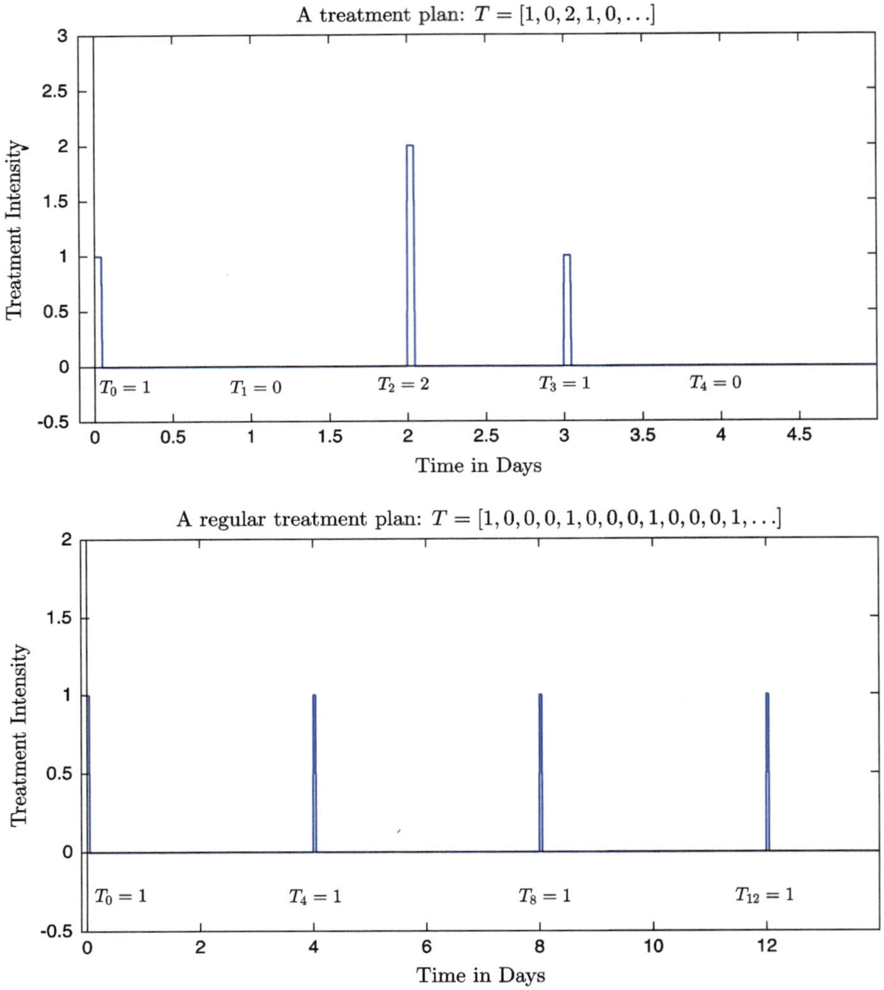

Fig. 3 Two treatment plans and their descriptions as vectors: a treatment plan where the intervals between treatments and the amount of treatment varies (*top graph*), and a treatment where both the amount and the number of days between treatments is constant (*lower graph*)

From Eq. 5 the roach population will die out if $P_{low} > \frac{(b-d)s_R k}{1-b+d}$. This gives a bound on n, the number of days in the cycle (as a function of a fixed pesticide amount, T)

$$n < \frac{1}{d_P} \ln \left[\frac{T}{d_P} \left(e^{d_P/24} - 1 \right) \frac{1-b+d}{(b-d)sk} + 1 \right]. \tag{6}$$

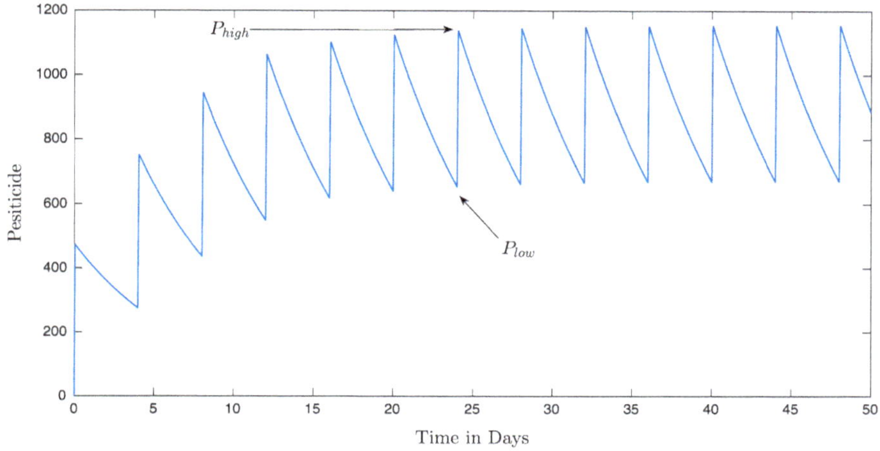

Fig. 4 Pesticide applied regularly every n days results in pesticide levels that converge to a stable periodic cycle. The maximum (P_{high}) and minimum (P_{low}) of this cycle can be determined analytically

For the parameter values listed in Tables 1 and 2, with $IT = \dfrac{500}{24}$ μg/day, this critical value is $n_{crit} \approx 2.6$. This analysis shows that the roach population can be eradicated with a one-hour treatment every 2 days. In fact, $n < n_{crit}$ is a sufficient but not

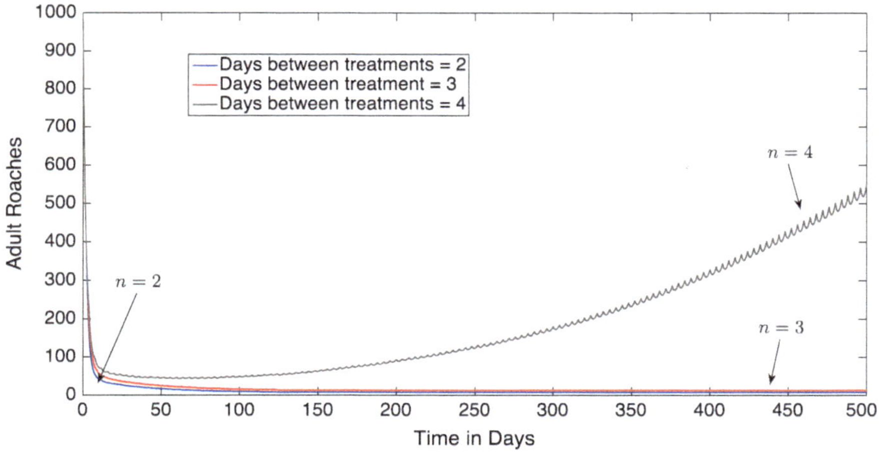

Fig. 5 For one-hour treatments of a fixed amount given every n days, there is a critical value of n above which the roach population will escape control. For our parameter set, treatment every 2.6 days or more often is sufficient to keep the roach population down according to Eq. (6). In fact, this theoretical lower bound is too strict: treatment every 3 days drives the roach population to zero eventually, while treatment every 4 days does not control the population

necessary condition for the eradication of the roaches, since it was calculated using a "worse case scenario," where all roaches are resistant. Simulations show that, in fact, a treatment every 3 days drives the roach population to zero, while a treatment every 4 days is not sufficient to control the population (see Fig. 5).

3.2 Genetic Algorithm

The stability analysis in Sect. 3.1 gives the optimal treatment plan assuming pesticide is applied in one-hour treatments given every n days. There are many treatment plans that are not regular. In order to determine a treatment that does not require a pesticide application at regular intervals, we run simulations in MATLAB and use a genetic algorithm to find an optimal treatment plan.

Treatment intensities $T(t) = (T_1, T_2, T_3, \ldots, T_N)$ are randomly selected where $T_i \in \{0, 1, 2\}$ and the constant multiplier is $l = \frac{500}{24}\,\mu\text{g/day}$. Treatment plans are generated for a 120-day period, and simulations using these treatment plans are run to determine their effects on the roach population. A genetic algorithm is implemented with the aim of finding a treatment plan that minimizes the objective function

$$Round(A(N) + A_R(N)) + \frac{1}{P_{max}} \int_0^N T(t)dt \tag{7}$$

where $Round(A(N) + A_R(N))$ is the total number of adult roaches at the end of the simulation, P_{max} is the total amount of pesticide applied in the treatment plan given by $T(t) = (2, \ldots, 2)$, and $\int T(t)dt$ is the total amount of pesticide applied over N days. In the implementation discussed here, $N = 120$. Notice that the first term in the objective function is a positive integer, and the second term is a real number between 0 and 1. Therefore, the primary concern is finding plans that eradicate all roaches, and the secondary concern is minimizing the amount of pesticide used.

In each generation of the genetic algorithm, 10 simulations are run, and the objective function is computed for each treatment. The two best (minimizing) treatment plans are selected for the subsequent generation. In addition, two mutations of these plans are also created by reassigning each T_i with 20 % probability. The remaining six treatment plans that compromise the next generation are randomly generated.

The treatment plan shown in Fig. 6 (top left) was obtained from 400 generations of the genetic algorithm. High doses are applied initially, followed by multiple days without treatment and days with a low dose before another high dose is applied. Treatments are occasionally applied over consecutive days and there are also consecutive days with no treatment applied. Figure 7 (left panel) shows the resulting adult and egg population and a comparison of the adult populations resulting from the regular treatment ($n = 3$) and the treatment determined by the genetic algorithm

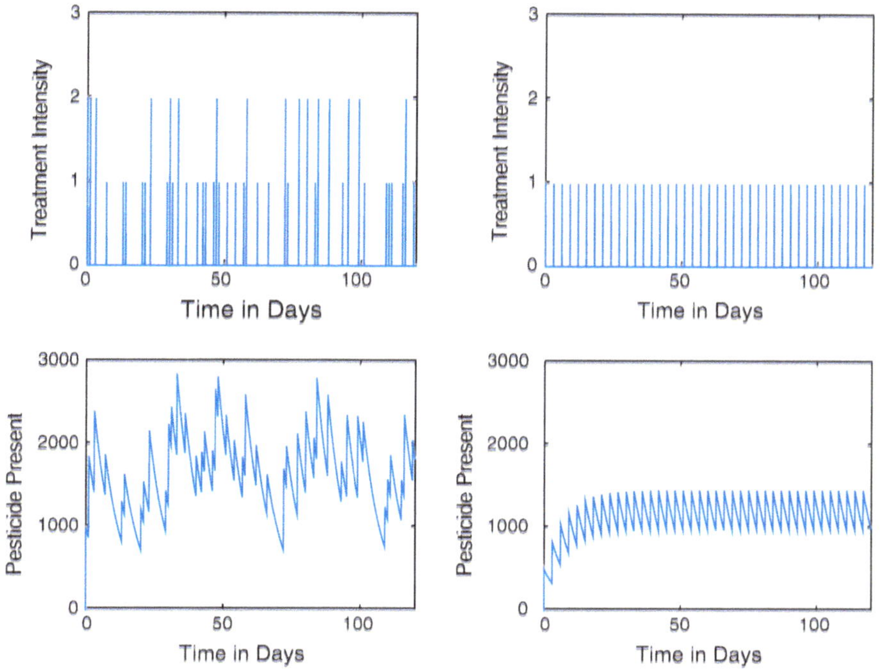

Fig. 6 A treatment plan that emerges from 400 generations of the genetic algorithm (*top left*) and the amount of pesticide present $P(t)$ resulting from that treatment plan (*bottom left*), and the regular treatment plan with $n = 3$ (*top right*) and the resulting pesticide present $P(t)$ (*bottom right*)

(right panel). The adult population of roaches is eradicated very quickly. In fact, the treatment plan generated from the genetic algorithm kills off the adult population quicker than the regular treatment plan (Fig. 7 right panel). While applying pesticide every 3 days will eradicate the roaches eventually, the adult roach population is still above zero after 120 days of treatment.

Though the optimal treatment plan is able to kill off roaches faster than the regular treatment plan, it also uses more pesticides (Fig. 6). The pesticide levels applied in the regular treatment plan converge to a stable periodic cycle that stays below 1200 μg. The pesticide levels in the optimal treatment plan spike up near 3000 μg multiple times throughout the 120 day period. The objective function used in the genetic algorithm prioritized the complete eradication of roaches over limiting pesticide levels. In the future, different choices for the objective function could be used to find effective treatment plans.

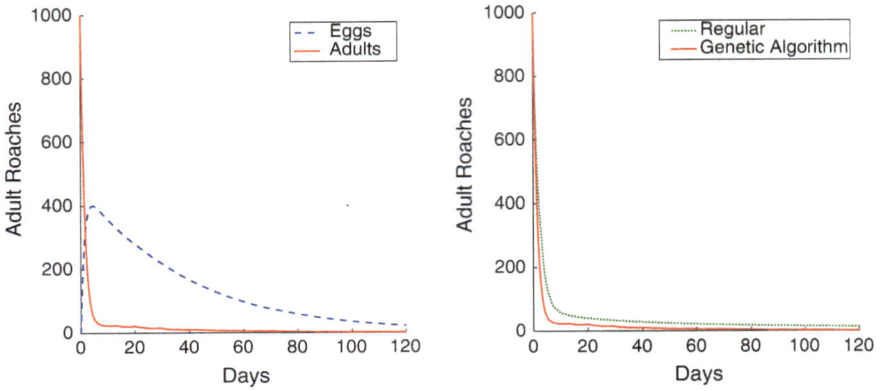

Fig. 7 The adult roach and egg populations (*left panel*) resulting from the treatment plan determined by the genetic algorithm, and a comparison of the adult populations resulting from both the regular treatment plan with $n = 3$ and the genetic algorithm (*right panel*). The treatment plan obtained from the genetic algorithm is very effective and kills off 95 % of the adult population within the first 10 days

4 Discussion

In this study, we explored an age-structured model of a cockroach population that includes the application of one pesticide, and the development of a resistant population. We give a formula for calculating the periodic treatment regimen that keeps the population near zero. Using a heuristic optimization technique, we suggest other nonperiodic treatments that are more effective in reducing the cockroach population while minimizing the toxic effect of pesticides.

The model and analysis presented here is intended as a preliminary description of a possible approach to the design of pesticide treatment regimes. We see this model as the first step in a more comprehensive study that would take into account practical constraints due to the location of the infestation, the toxicities of the pesticide, options for multiple treatments, and environmental fluctuations. In particular, we see several directions for the future development of this modeling approach.

The model should be validated using experimental data. Fluctuations in parameter values such as the birth and death rates due to seasonal changes or geographical location should be noted. It is possible that model refinements, such as the distinction between the nymph and the adult stages, need to be included in order to accurately model the time evolution of the total population.

Pesticides vary in efficacy, toxicity, and the promotion of resistance. The application of multiple pesticides should be explored in order to minimize the growth of resistant populations and to mitigate toxicity to humans and other species. In particular, the effect of nontoxic strategies such as the installation of sound-emitting devices and simple eradication by thorough cleaning, should be included in the model as a first-line strategy to lower the initial roach population.

Cockroaches are ubiquitous pests whose presence in residences and food preparation areas can have serious consequences. The heuristic optimization technique described in this paper can be easily adapted to include a variety of treatments and other objective functions. The use of a validated mathematical model to run in silico experiments could result in suggested treatment protocols that could increase overall health by minimizing the roach population while keeping toxic pesticides at minimal levels.

Acknowledgments This paper grew out of a workshop in differential equations that was part of the EDGE 2013 summer program. We would like to thank the other members of the workshop: Jessica Poole, Kara Keller, Yeng Xiong, Karamatou Yacoubou Djima and Professor Eirini Poimenidou for their work in the early stages of the project. We also thank Professor Elzie McCord for his invaluable information on the biology of cockroaches, and the effects of pesticides. The workshop was supported by a grant from the NSF, DMS 1136857. We would also like to thank the Association for Women in Mathematics for organizing the 2015 AWM Symposium and the EDGE Foundation, with support from the NSF, for continuing support of this research project.

References

1. X. Bonnefoy, H. Kampen, K. Sweeney, *Public Health Significance of Urban Pests* (World Health Organization, Copenhagen, 2008)
2. N.P.I. Center, Pesticide fact sheet: cypermethrin, *Oregon State University and United States Environmental Protection Agency, Environmental and Molecular Toxicology* (1998)
3. P. Cornwell, The Cockroach: A laboratory insect and an industrial pest (book) Publisher Hutchinson Year 1968 Author Cornwell, P.B. Volume I Date-Added 2014-08-21 22:11:08 +0000 Date-Modified 2014-08-21 22:14:48 +0000 Local Files Remote URLs The Cockroach: A laboratory insect and an industrial pest, vol. I. Hutchinson (1968)
4. F. de Blay, J. Sanchez, G. Hedelin, A. Perez-Infante, A. Vérot, M. Chapman, G. Pauli, Dust and airborne exposure to allergens derived from cockroach (*blattella germanica*) in low-cost public housing in strasbourg (france). J. Allergy Clin. Immunol. Elsevier **99**(1), 107–112 (1997)
5. R. Fotedar, U.B. Shriniwas, A. Verma, Cockroaches (*blattella germanica*) as carriers of microorganisms of medical importance in hospitals. Epidemiol. Infect. Camb. Univ. Press **107**(01), 181–187 (1991)
6. C. Gemeno, G.M. Williams, C. Schal, Effect of shelter on reproduction, growth and longevity of the german cockroach, blattella germanica (dictyoptera: Blattellidae). Eur. J. Entomol. **108**, 205–210 (2011)
7. D. Jones, Environmental fate of cypermethrin. Environmental Monitoring and Pest Management, Department of Pesticide Regulation, Sacramento, CA 95814, (1998)
8. P.G. Koehler, R.S. Patterson, R.J. Brenner, German cockroach (*orthoptera: blattellidae*) infestations in low-income apartments. J. Econ. Entomol. **80**(2), 446–450 (1987)
9. R. Larter, P. Chadwick, Use of a general model to examine control procedures for a cockroach population. Res. Popul. Ecol. **25**, 238–248 (1983)
10. P. Mamidala, S.C. Jones, O. Mittapalli, Metabolic resistance in bed bugs. Insects **2**(1), 36–48 (2011)
11. J.R. Misra, M.A. Horner, G. Lam, C.S. Thummel, Transcriptional regulation of xenobiotic detoxification in Drosophila. Genes Dev. **25**(17), 1796–1806 (2011)
12. C.D.M. Müller-Graf, E. Jobet, A. Cloarec, C. Rivault, M. van Baalen, S. Morand, Population dynamics of host-parasite interactions in a cockroach-oxyuroid system. OIKOS **95**, 431–440 (2001)

13. M.K. Rust, J.M. Owens, D.A. Reierson, *Understanding and Controlling the German Cockroach* (Oxford University Press, Oxford, 1995)
14. C. Schal, Sulfluramid resistance and vapor toxicity in field-collected German cockroaches (dictyoptera: Blattellidae). J. Med. Entomol. **29**(2), 207–215 (1992)
15. C. Schal, G.L. Holbrook, J.A. Bachmann, V.L. Sevala, Reproductive biology of the german cockroach, blattella germanica: juvenile hormone as a pleiotropic master regulator. Arch. Insect Biochem. Physiol. **35**(4), 405–426 (1997)
16. C. Schou, P. Lind, E. Fernandez-Caldas, R.F. Lockey, H. Løwenstein, Identification and purification of an important cross-reactive allergen from american (periplaneta americana) and german (blattella germanica) cockroach. J. Allergy Clin. Immunol. Elsevier **86**(6), 935–946 (1990)
17. J.G. Scott, D.G. Cochran, B.D. Sigfried, Insecticide toxicity, synergism, and resistance in the german cockroach (*dictyoptera: blattellidae*). J. Econ. Entomol. **83**(5), 1698–1703 (1990)
18. S. Toft, A.P. Jensen, No negative sublethal effects of two insecticides on prey capture and development of a spider. Pestic. Sci. **52**(3), 223–228 (1998)
19. S.M. Valles, Toxicological and biochemical studies with field populations of the german cockroach, blattella germanica. Pestic. Biochem. Physiol. Elsevier **62**(3), 190–200 (1998)
20. L. Willoughby, H. Chung, C. Lumb, C. Robin, P. Batterham, P.J. Daborn, A comparison of drosophila melanogaster detoxification gene induction responses for six insecticides, caffeine and phenobarbital. Insect Biochem. Mol. Biol. **36**(12), 934–942 (2006)
21. H.H. Wu, H.J. Lee, S.B. Horng, L. Berec, Modeling population dynamics of two cockroach species: effects of the circadian clock, interspecific competition and pest control. J. Theor. Biol. **249**, 473–486 (2007)
22. L. Zurek, C. Schal, Evaluation of the german cockroach (*blattella germanica*) as a vector for verotoxigenic *escherichia coli* f18 in confined swine production. Vet. Microbiol. Elsevier **101**(4), 263–267 (2004)

The Impact of Violence Interruption on the Diffusion of Violence: A Mathematical Modeling Approach

Shari A. Wiley, Michael Z. Levy and Charles C. Branas

Abstract Public health approaches to interrupting infectious disease transmission have yet to be informed by traditional deterministic models of contagion. We investigate this gap in current violence prevention research by introducing a Susceptible–Transmitter–Victim Epidemic model, based on the classic Susceptible–Infectious–Recovered differential equation model, to explore the impact of violence interruption on the diffusion of violence. Uncertainty and sensitivity analysis are done using Latin hypercube sampling. Based on sensitivity analysis results, model predictions appear to be overestimating annual gun assault cases, where the mean estimate of the gun assault rate at equilibrium is double the average gun assault rate over the past decade. Several key parameters are identified as significant to gun assault predictions and may account for model imprecision. Scenario analysis is also done to determine the effectiveness of violence interruption programs. Results suggest that targeting all potential violence transmitters can reduce gun violence three times more than an intervention that only targets gun-owning individuals, indicating the importance of taking a holistic approach to violence interruption and prevention. Our results also suggest that having individuals in the population transmitting violence, whether or not they are participating in gun violence, is sufficient to sustain a gun violence epidemic.

Keywords Contagious violence · Infectious disease model · Violence interruption · Gun violence

Mathematics Subject Classification 92B05

S.A. Wiley (✉) · M.Z. Levy · C.C. Branas
Department of Biostatistics and Epidemiology, University of Pennsylvania, Philadelphia, PA 19104, USA
e-mail: shari.wiley@hamptonu.edu

M.Z. Levy
e-mail: mzlevy@mail.med.upenn.edu

C.C. Branas
e-mail: cbranas@mail.med.upenn.edu

S.A. Wiley
Department of Mathematics, Hampton University, Hampton, VA 23668, USA

G. Letzter et al. (eds.), *Advances in the Mathematical Sciences*, Association for Women in Mathematics Series 6, DOI 10.1007/978-3-319-34139-2_10

1 Introduction

Numerous violence prevention programs are moving toward a broader public health contagion paradigm to understand and interrupt community violence. The novelty of these paradigms is their use of infectious disease prevention concepts to interrupt and prevent community violence. An example is the Cure Violence program, which has outlined approaches analogous to preventing or containing epidemic disease outbreaks for intervening on and preventing violence through the use of community-based "interrupters", or people who are inserted in potentially violent situations in an effort to break the contagion effect or escalation of violence between individuals [1]. This model also has a prevention component that includes creating employment and education opportunities for high-risk individuals, and using outreach to address social and group norms regarding attitudes toward violence [1, 2].

Trauma-informed, hospital-based interventions have also been effective in violence interruption and prevention. The hospital-based model identifies repeat violent crime trauma victims admitted to hospitals and provides victims with an outlet for aggression through family or group therapy as well as providing substance abuse treatment, follow-up psychosocial care and trauma recovery assistance [2–4].

However, unlike these above-mentioned approaches that interrupt infectious disease diffusion, these violence prevention paradigms have yet to be informed by traditional deterministic mathematical models of contagion. We attempt to investigate this connection by formulating a mathematical model of contagion and applying it to the spread and interruption of gun violence in Philadelphia.

Adaptations of mathematical, infectious disease models to study social contagion are common. Most relevant is a study by Patten and Arboleda-Flórez [5], where they used an SIR epidemic model to study crowd violence. Results from their study indicated that the length of time "infectious" individuals remained in the crowd impacted the probability of violence transmission. They also discussed the role that alcohol consumption may play on increasing the probability of violence transmission.

Additional examples of studies that used mathematical models to describe social contagion include the diffusion of alcohol consumption among peers [6–8], rumors [9, 10], bulimia in college females [11], and posttraumatic stress in children [12]. Studies using spatial techniques to study contagious violence have also been conducted [13, 14], as well as the use of a game theory modeling design to determine the effectiveness of gun control on reducing violence [15]. We expand on these previous quantitative approaches to develop an epidemic model of contagious violence inclusive of gun ownership subpopulations and violence interruption and prevention paradigms.

Literature on the contagious nature of violence has identified a multitude of ways that violence and aggression can be transmitted between individuals [1, 13, 16–20]. In a study on the effects of adolescent exposure to community violence, Kelly [17] found that some adolescents feel social pressure to participate in violence when violence is prevalent in their surrounding community. In addition, violence victimization among adolescents can also lead to the subsequent transmission of violence, where the victims seek out the security that being in a gang provides [17]. Crooks et al. [18] found that, particularly in adolescent males, exposure to child abuse can result in a child perpetrating community violence in the future. Other studies reveal that observing violence through television and video game outlets can also lead to participation in subsequent violence [1, 19, 20].

We capture this complexity of violence transmission by adopting the classic susceptible–infectious–recovered (SIR) modeling framework to describe the impact of interrupting community violence transmission on the incidence of gun assaults. There is evidence in the literature that links gun possession to increased risk of gun violence victimization [21–26]. In order to incorporate this into our model, we include subpopulations that consist of non-gun owners, as well as legal and illegal gun owners to explore legal and illegal gun ownership as factors related to gun violence transmission. We then use the gun assault victim population to estimate changes in the occurrence of violence over time.

2 Methods

2.1 STV Model: Susceptible–Transmitter–Victimized Epidemic Model for Violence Transmission

We introduce a Susceptible–Transmitter–Victimized (STV) epidemic model to explore the impact of violence interruption on contagious violence. We define our susceptible class, S, as individuals that are both vulnerable to adopting a culture of violence and vulnerable to violence victimization. The transmitter population, T, are individuals that are currently engaged in a culture of violence. Violence transmitters can infect a susceptible in one of two ways: (1) they can influence others to adopt a violent lifestyle with a transmission efficiency of β, in which case the susceptible moves into the transmitter class; or (2) violence transmitters can assault a susceptible with a firearm at a transmission rate of α, after which the susceptible moves to the victimized class, V. We assume the transmission rate of violence, β, is a function of social and environmental factors such as peer pressure, exposure to violence at home or in ones neighborhood, and socioeconomical factors.

Violence transmitters can recover to the susceptible class as a result of violence interruption efforts at a rate of σ. We assume there is no immunity from violence transmission or violence victimization. Thus, individuals are either susceptible, violence transmitters, or victims of gun violence. We assume that violence transmitters are victimized by gun violence and move to the victimized class at a rate of Γ. After a gun assault victim recovers, they either return to the susceptible class at rate $q\gamma$ or become transmitters at rate $(1-q)\gamma$, where γ is the recovery rate and q the proportion of victims that become susceptible after recovery. We assume that victims can move to the transmitter class after recovery as a result of their exposure to violence

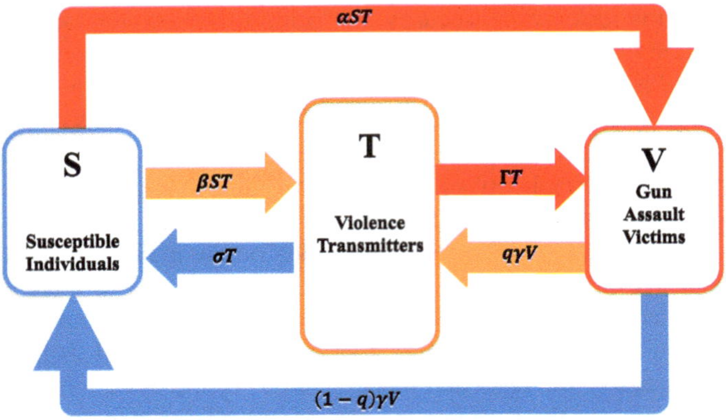

Fig. 1 Flow diagram for STV model

or prior attitudes toward violence. We give a conceptual framework of our model in Fig. 1.

We make model assumptions consistent with the classic SIR compartmental model and variations [27–30]. First, we assume homogeneous mixing, which implies that each susceptible has an equal probability of an infectious contact with any violence transmitter. Thus, the transmission efficiency of violent behavior and gun violence is assumed to be frequency-dependent and is represented by a mass action term [28–30]. We also assume that the overall population, $S + T + V = K$, remains constant throughout the duration of the epidemic, where K is the total population.

The rate that individuals move in an out of compartments correspond to exponentially distributed waiting times [29, 30]. Thus, $\frac{1}{\sigma}$ is the average time that it takes an individual to transfer from the transmitter population to the susceptible population as a result of a violence interruption campaign, $\frac{1}{\Gamma}$ is the average time it takes to be victimized by gun assault as a result of the lifestyle associated with being a violence transmitter, and $\frac{1}{\gamma}$ is the average time it takes to recover from gun assault victimization and transfer back to either the susceptible or transmitter population.

2.2 *STV Model with Gun Ownership Subpopulations*

To better model gun violence, we add subpopulations based on gun ownership to our STV model. We consider legal gun owners (LGO) as individuals that legally own and use a firearm, illegal gun owners (IGO) are individuals that illegally own or use a firearm, and non-gun owners (NGO) are individuals that do not own a firearm. Since we are including a legal gun ownership population, our total population consists of individuals that are 21 years old or older. With these additional subpopulations, our three compartment model becomes a nine-compartment model.

Our three susceptible populations are given by:

$$\frac{dS_N}{dt} = -\beta S_N(T_N + T_L + T_I) - S_N(\alpha_L T_L + \alpha_I T_I) + (1 - q_N)\gamma V_N + \sigma_N T_N \quad (1)$$
$$\frac{dS_L}{dt} = -\beta S_L(T_N + T_L + T_I) - S_L(\alpha_L T_L + \alpha_I T_I) + (1 - q_L)\gamma V_L + \sigma_L T_L \quad (2)$$
$$\frac{dS_I}{dt} = -\beta S_I(T_N + T_L + T_I) - S_I(\alpha_L T_L + \alpha_I T_I) + (1 - q_I)\gamma V_I + \sigma_I T_I. \quad (3)$$

In Eqs. 1–3, the first term is the per capita rate at which a susceptible becomes infected with violence. Similarly, the second term is the per capita rate at which a susceptible becomes victimized by gun violence. The last term is the fraction of recovered gun assault victims that return to a respective susceptible population. We assume that the gun assault transmission rate is different for legal and illegal gun owners. Therefore, α_L is the gun assault transmission rate for LGO and α_I is the gun assault transmission rate for illegal gun owners.

Similarly, we assume the violence interruption rate may be different for each subpopulation, allowing us to account for violence intervention programs that target transmitters that are likely to be gun owners. Thus, σ_N, σ_L and σ_I are the violence interruption rates for the NGO, LGO and IGO populations, respectively.

Since we assume that our total population remains constant, S_N, S_I, and S_L represent proportions of the total population, K. If X_N, X_L and X_I represent the population size of the susceptible non-gun owner, legal gun owners and illegal gun owners respectively, then

$$S_N = \frac{X_N}{K}, \quad S_L = \frac{X_L}{K} \text{ and } S_I = \frac{X_I}{K}.$$

When the population is free of contagious violence, that is, when $T_N = T_L = T_I = 0$, the sum of the susceptible populations is the total population, $S_N + S_L + S_I = K$.

The equations for the transmitter populations are given by:

$$\frac{dT_N}{dt} = \beta S_N(T_N + T_L + T_I) + q_N\gamma V_N - \alpha_L T_N T_L - \alpha_I T_N T_I - \sigma_N T_N \quad (4)$$
$$\frac{dT_L}{dt} = \beta S_L(T_N + T_L + T_I) + q_L\gamma V_L - \Gamma T_L - \sigma_L T_L \quad (5)$$
$$\frac{dT_I}{dt} = \beta S_I(T_N + T_L + T_I) + q_I\gamma V_I - \Gamma T_I - \sigma_I T_I. \quad (6)$$

Similar to the equations for the susceptible population, T_N, T_L and T_I represent proportions of the total population, K. Letting Y_N, Y_L and Y_I be the subpopulation sizes of the transmitting non-gun owners, legal gun owners and illegal gun owners, respectively, then

$$T_N = \frac{Y_N}{K}, \quad T_L = \frac{Y_L}{K} \text{ and } T_I = \frac{Y_I}{K}.$$

The first terms of Eqs. 4–6 represent newly infected individuals coming from their respective susceptible population. The second terms represent the fraction of recovered gun assault victims that become transmitters after recovery. The last terms of the rate of change equations for the transmitter populations represent violence transmitters that recover and move to their respective susceptible population.

Lastly, the victim classes are given by:

$$\frac{dV_N}{dt} = S_N(\alpha_L T_L + \alpha_I T_I) + T_N(\alpha_L T_L + \alpha_I T_I) - \gamma V_N. \tag{7}$$

$$\frac{dV_L}{dt} = S_L(\alpha_L T_L + \alpha_I T_I) + \Gamma T_L - \gamma V_L. \tag{8}$$

$$\frac{dV_I}{dt} = S_I(\alpha_L T_L + \alpha_I T_I) + \Gamma T_I - \gamma V_N. \tag{9}$$

We define Z_N, Z_L and Z_I as the population sizes of the victimized non-gun owners, victimized legal gun owners, and the victimized illegal gun owners. Then,

$$V_N = \frac{Z_N}{K}, \ V_L = \frac{Z_L}{K} \text{ and } V_I = \frac{Z_I}{K}.$$

The first terms of the rate of change equations for the victim populations are incidence of gun assaults per unit time of the susceptible non-gun owners, legal gun owners, and illegal gun owners, respectively. Similarly, the second term in Eq. 7 is the incidence of gun assaults per unit time of the violence transmitting non-gun owner population. The second terms in Eqs. 8 and 9 are the fraction of violence transmitting legal and illegal gun owners that are eventually victimized by gun violence as a result of their lifestyle. The last term in each of Eqs. 7–9 are recovered gun assault victims that recover either to their respective susceptible population or to their respective violence transmitter population. We give a full description of model parameters in Table 1.

2.3 *Uncertainty and Sensitivity Analysis via the Latin Hypercube Sampling Method*

We use the Latin Hypercube Sampling (LHS) method to conduct uncertainty and sensitivity analysis. The LHS method is a stratified sampling method that samples with memory. This sampling method is a highly efficient as it reduces the number of simulations needed to sufficiently conduct uncertainty and sensitivity analysis (compared to a Monte Carlo sampling method) [31, 32]. This approach was first introduced for infectious disease models by Blower and Dowlatabadi [31].

We sample 10 of our 12 model parameters from uniform distributions and the remaining two parameters, α_L and α_I from a triangle distribution. For parameters where data were available, maximum and minimum values were selected based on

Table 1 Model parameter description

Parameter	Explanation	Parameter	Explanation
S_N	Proportion of susceptible NGO	β	Transmission efficiency of transmitters
S_L	Proportion of susceptible LGO	α_L	Transmission efficiency of violence by LGO
S_I	Proportion of susceptible IGO	α_I	Transmission efficiency of violence by IGO
T_N	Proportion of transmitter NGO	σ_N	Per capita rate of violence intervention for NGO transmitters
T_L	Proportion of transmitter LGO	σ_L	Per capita rate of violence intervention for LGO transmitters
T_I	Proportion of transmitter IGO	σ_I	Per capita rate of violence intervention for IGO transmitters
V_N	Proportion of NGO gun assault victims	γ	Recovery rate of gun assault victims
V_L	Proportion of LGO gun assault victims	Γ_L	Per capita rate that LGO transmitters become gun assault victims
V_I	Proportion of IGO gun assault victims	Γ_I	Per capita rate that IGO transmitters become gun assault victims
K	Total population $(S + T + V)$	q_L	Proportion recovered LGO gun assault victims that recover to transmitter class
q_N	Proportion recovered NGO gun assault victims that recover to transmitter class	q_I	Proportion recovered IGO gun assault victims that recover to transmitter class

NGO non-gun owners, *LGO* legal gun owners, *IGO* illegal gun owners, *GAV* gun assault victims the published data. However, published data on many of our parameters were not available in which case, model assumptions and initial conditions were used to try best estimate these parameter values. A discussion on how we estimated each parameter is given in the following section, and a summary of parameter values is given in Table 2.

We used the LHS function in the pse package included in the R programming software to generate our sampling space. We initially generated 200 samples for each of the 12 parameters. We discarded sample combinations that did not satisfy our model constraint (discussed below), which left us with 102 samples per parameter to use in our final analysis.

2.3.1 Initial Conditions for Model Simulation

We estimated the proportion of susceptible legal gun owners in Philadelphia through using data from a Public Health Management Corporation (PHMC) Southeastern Pennsylvania Household Health survey. As part of the survey, residents were asked if they owned a gun. The overall percentage of yes responses between 2000 and 2006 was 10 % and we use this number to represent the proportion of susceptible legal gun owners in Philadelphia.

To estimate the proportion of individuals that illegally own or use firearms, we focus on individuals that cannot legally purchase guns. Using a 2011 labor market study [33], we estimate the fraction of the population that are ex-felons to be 5 %. According to an NIJ study [34], on average, 37 % of arrestees have illegally possessed a firearm at some point. Combining these percentages, we estimate that 2 % of the population are illegal gun owners.

Without knowing the number of violence transmitters in the population, we assume that the initial transmitter proportion is 1 %, where $T_N = .003$, $T_L = .003$ and $T_I = .004$, and assume the initial victim populations are 0. Since all the proportions must sum to one, the initial population density of susceptible non-gun owners is $S_N(0) = 1 - S_L(0) - S_I(0) - T_N(0) - T_L(0) - T_I(0) = 87\,\%$. We summarize our initial condition values in Table 2.

2.3.2 Parameter Estimations for Model Simulation

For the gun assault recovery time, γ, we considered at several studies on hospital gunshot wound admissions [21, 40–42]. Kellermann et al. studied firearm injuries in three major US cities and reported a median recovery time for gunshot wounds of 3 days. Median recovery time of 6 days was reported in [40], where they studied firearm injuries in Sweden over an eight-year period. Feliciano et al. [41] studied 300 abdominal gunshot wounds in Houston, TX and reported a mean recovery time 7 days. Cowey et al. [42] reported a mean recovery times of 9 days for 187 patients in the UK. Taking these studies into account, for analysis we explore recovery times ranging from 3 days to 9 days.

Table 2 Parameter intervals, distribution function, and initial conditions for uncertainty analysis

Parm	Min	Max	Distribution	Var	I.C
q_N [35–37]	0.15	0.5	Uniform	S_N	0.87
q_L [35–37]	0.15	0.5	Uniform	S_L (PHMC)	0.1
q_I [35–37]	0.15	0.5	Uniform	S_I [33, 34]	0.02
β	β_0	2.4	Uniform	T_N	0.003
α_L	0	0.01	Triangular (peak = 0.003)	T_L	0.003
α_I	0	0.01	Triangular (peak = 0.008)	T_I	0.004
σ_N	0.5	2	Uniform	V_N	0
σ_L	0.5	2	Uniform	V_L	0
σ_I	0.5	2	Uniform	V_I	0
Γ_L [21–24]	0.04	0.05	Uniform		
Γ_I [38, 39]	0.1	0.14	Uniform		
γ [21, 40–42]	26	52	Uniform		

To estimate Γ_L, we use previous studies to estimate age range when legal gun owners are victimized [21–24]. From, these studies we found that the average age of gun violence victimization was between 40 and 44. Since the legal age to purchase a firearm is 21, this makes the average time as a LGO transmitter before victimization is 19–23 years.

We used Philadelphia Police Department homicide data from [38, 39] to estimate the rate at which IGO transmitters are victimized by gun violence. The average age of homicide victims in Philadelphia is approximately 31 years, which reduces to a mean wait time of 10 years. We also explored mean wait times as low as 7 years (28 years at time of shooting) to explore additional risk associated with individuals that illegally possess firearms.

To explore the effects of violence interruption on violence transmission, we explore successful violence interruption efforts that result in the recovery of a transmitter occurring within 6 months to two years of intervention. We determine the proportion of victims that recover to the transmitter class as result of their exposure to violence based on several studies on retaliatory violence. From [35], Kubrin and Weitzer reported that 17 % of gun-related homicides that occurred in St. Louis from 1985 to 1995 were a result of retaliation. In studies on adolescent attitudes toward violence retaliation after being assaulted, Wiebe et al. [36] and Copeland et al. [37] found that 47 and 16 %, respectively, of adolescent victims wished to retaliate. Based on these studies, we explore values between 15 and 50 %.

We do not have data for transmission efficiency of violence. However, to determine a lower bound for β, we choose a threshold value that will ensure that violence remains prevalent in the population, versus a violence-free population. To do this, we assume that β is greater than the sum of rates moving out of each of the violence transmitter populations. This ensures that the transmission efficiency of violence is sufficient to maintain the spread of violence irrespective of individuals leaving a transmitter class as a result of gun assault victimization or successful intervention. This gives us the model constraint,

$$\beta > \max\{\alpha_L + \alpha_I + \sigma_N, \Gamma + \sigma_L, \Gamma + \sigma_I\} = \beta_0. \tag{10}$$

For analysis, we sample β values subject to the constraint given in Eq. 10. Based on preliminary analysis, we chose an upper bound of 2.4 for β, to ensure the range of parameter values being sampled for our model produced realistic predictions of violence.

In the absence of data for the transmission rates of legal and illegal gun violence, we sample values between 0 and 0.01 for α_L and α_I using the triangle distribution. We consider transmission rates for illegal gun owners that are greater than the transmission rate for legal gun owners. Thus, we choose the peak values as 0.003 and 0.008 for α_L and α_I, respectively. During preliminary analysis, we also considered the reverse scenario, however, model predictions of the incidence of gun assaults were similar. We give a summary of the distribution functions and sampling intervals for model parameters in Table 2.

Based on the sensitivity analysis, we did a second LHS run sampling only key parameters, where non-key parameters were fixed at their means from the initial LHS. We used the least squares approach to determine the parameter values that most accurately describe the incidence of gun violence. We defined new gun assault cases at year k for NGOs by

$$s_N(k)(\alpha_L t_L(k) + \alpha_I t_I(k)) + t_N(k)(\alpha_L t_L(k) + \alpha_I t_I(k)), \tag{11}$$

new gun assault cases at year k for LGOs by

$$s_L(k)(\alpha_L t_L(k) + \alpha_I t_I(k)) + \Gamma_L t_L(k), \tag{12}$$

and new gun assault cases at year k for IGOs by

$$s_I(\alpha_L t_L(k) + \alpha_I t_I(k)) + \Gamma_I t_I(k), \tag{13}$$

where year $k \in \{1, 14\}$ corresponds with observed gun assault data from 2001 to 2013 inclusive. We defined total new gun assaults (TNGA) as the sum of Eqs. 11–13. Thus, for our model simulations, we selected the parameters that minimized

$$\sum_{k=1}^{14} (\mathit{predicted}\ \mathit{TNGA}(k) - \mathit{observed}\ \mathit{TNGA}(k))^2. \tag{14}$$

3 Results

3.1 Uncertainty Analysis Results from 102 Model Simulations

Using parameter values produced from the initial LHS (see Table 6 in Appendix), we ran 102 model simulations. We used time steps in years, and for each model run we predicted equilibrium prevalences for the overall transmitter population and gun assault incidences per year at equilibrium. Histograms for each outcome are given in Fig. 2.

The STV model converged to a positive equilibrium population for each of the 102 model simulations. Model predictions for gun assault cases per year at equilibrium ranged from 49 per capita (based on a population of 100,000) to 678 per capita, with mean 430 per capita (see Table 3). This prediction for gun assault incidences is double the average per capita gun assault rate for Philadelphia from 1998 to 2012, which was approximately 200 gun assaults per year.

Table 3 reflects results from 102 model simulations for the per capita gun assault rate per year at equilibrium, the equilibrium prevalence for the total susceptible population, and the equilibrium prevalence for the total violence transmitter.

The model also predicts that a large percentage of the population will eventually become violence transmitters, where the average prediction for equilibrium prevalence was 42 % ranging from 5 to 73 % (see Table 3). These large variations in model predictions is most likely a result of uncertainty, or incorrect estimation of many of the model parameters.

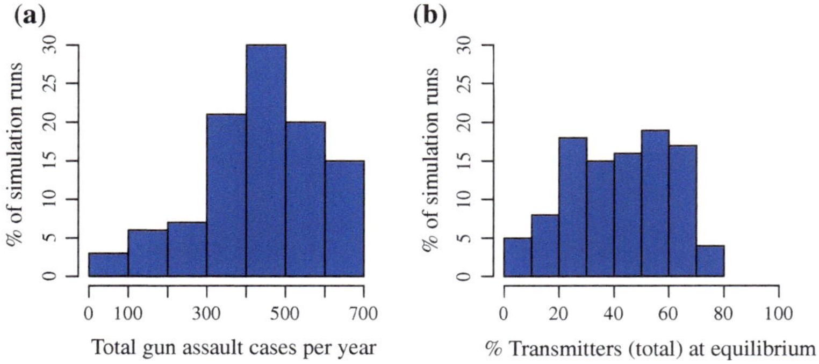

Fig. 2 Uncertainty analysis results for 102 model simulations. In (**a**), equilibrium per capita gun assault rates were calculated for each subpopulation then summed to determine overall gun assault rate for each of the 102 simulations. In (**b**), equilibrium prevalences were calculated for each of the violence transmitter subpopulations then summed to determine overall prevalence of violence transmitters for each of the 102 simulations

Table 3 Summary statistics from sensitivity analysis

	Mean	SD	25 %	50 %	75 %	Min	Max	95 % C.I.
Gun assault cases	430.38	144.29	354.57	434.39	531.57	48.57	677.82	(402.03, 458.72)
Susceptible (%)	58.76	17.97	44.01	58.33	73.64	27.58	96.22	(55.23, 62.29)
Transmitters (%)	42.23	17.97	27.35	42.66	56.98	4.78	73.41	(38.7, 45.76)

3.2 Model Sensitivity to Estimated Parameters

Partial rank correlation coefficients (PRCC) were calculated for each parameter sampled in the uncertainty analysis. Two-sided tests were used to determine the significance of PRCC values. We report three levels of significance $***P < 0.001$, $**P < 0.01$ and $*P < 0.05$. The PRCC values for significant model parameters with respect to the total per capita gun assault rate at equilibrium are given in Table 4.

The most significant parameter for predicting the per capita gun assault rate was the violence interruption rate for the NGO transmitter population with a PRCC of −0.94. Model predictions were also highly sensitive to the transmission efficiency of violence, which had a PRCC of −0.89. Targeted violence interruption in the LGO and IGO transmitter population were also strongly correlated to prediction imprecision of incidences of gun assaults, with PRCC values of −0.78 and −0.73, respectively.

Table 4 PRCC values from sensitivity analysis for the per capita gun assault rate at equilibrium

All GAV		LGO GAV		IGO GAV	
Parameter	PRCC	Parameter	PRCC	Parameter	PRCC
σ_N	−0.94***	σ_N	−0.93***	σ_N	−0.89***
β	0.89***	σ_L	−0.92***	σ_I	−0.85***
σ_L	−0.78***	β	0.88***	β	0.81***
σ_I	−0.73***	Γ_L	0.58***	Γ_I	0.51***
Γ_L	0.48***	q_L	0.2*		
Γ_I	0.35***				
α_L	0.25*				

***P<0.001
**P<0.01
*P<0.05
PRCC values reported for the overall per capita gun assault rate, per capita gun assault rate for legal gun owners, and per capita gun assault rate for illegal gun owners at equilibrium. Per capital gun assault rates are based on the estimated number of yearly gun assault victims at equilibrium

Several additional parameters were significantly correlated to predictions of the per capita gun assault rate. The rate at which IGO transmitters are victimized, Γ_I, had a PRCC value of 0.38, and the LGO transmission rate of gun violence, α_L, had a PRCC value of 0.34. The proportion of IGO victims that recover to the transmitter population (q_I), the rate at which LGO transmitters are victimized (Γ_L) and the proportion of NGO that recover to the NGO transmitter population (q_N) had PRCC values of 0.28, 0.23 and 0.18, respectively.

For our second LHS run, we fixed parameter values at their mean (see Table 6 in the appendix) from the first LHS and only varied key parameters. We considered key parameters as parameters with PRCC greater than 0.7 in absolute value, which included β, σ_N, σ_L, and σ_I. We used the results from our second LHS run of key parameters to minimize Eq. 14. The values for the key parameters that minimized Eq. 14 were $\beta = 2.1$, $\sigma_N = 1.92$, $\sigma_L = 1.62$ and $\sigma_I = 0.87$. For these values, the STV model predicted a equilibrium gun assault rate of 204 and mean gun assault rate of 154 over 14 years compared to the observed mean gun assault rate of 201.

The results from the second LHS run are listed in Table 7 in the Appendix.

3.3 STV Model Simulations with Violence Interruption

To discuss the impact of violence interruption strategies on gun violence outcomes, we estimate per capita gun assault rates over a 30-year period under different scenarios of violence interruption interventions. As an initial starting point we use key parameter values that minimize Eq. 14. Based on the sensitivity analysis, β, σ_N, σ_L, and σ_I were identified as key parameters. We use mean values from the initial LHS run for all other parameters for our simulations. We estimate per capita gun assault rates under different scenarios of violence intervention efforts that target LGO transmitters only, IGO transmitters only, both LGO and IGO transmitters and all violence transmitters.

For violence interruption efforts focused on LGO transmitters only, a 10 % increase in violence interruption and intervention efforts result in a 12 % decline in incidences of gun assaults after 10 years, and a 20 % increase in the rate of violence intervention results in a 22 % decline in total incidences of gun assaults after 10 years. Similarly, when intervention efforts target IGO transmitters, a 10 % increase in intervention efforts result in a 9 % decline in total gun assaults over 10 years and an 17 % decline for a 20 % increase in violence intervention efforts.

When both gun-owning populations are targeted, a 10 % increase in violence intervention efforts result in a 21 % decline in gun assaults, and a 20 % increase in violence intervention efforts result in 37 % decline in gun assaults. However, when all transmitter populations are targeted, the decline in gun assault cases per year is more than tripled. A 10 % increase in violence intervention rates result in a 41 % reduction in gun assaults over 10 years, and a 20 % increase in violence intervention rates result in a 91 % decline in gun assaults over 10 years. We illustrate the impact

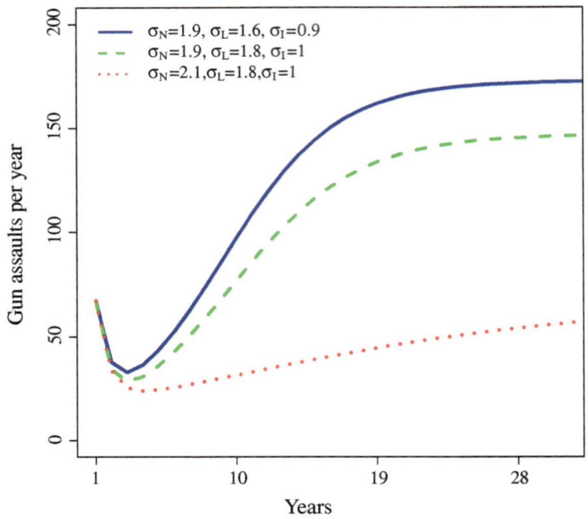

Fig. 3 Impacts of violence interruption on the incidence of gun assaults. This figure depicts effects of violence intervention efforts on total incidences of gun assaults (per 100,000 people) per year over 30 years. The *solid, blue line* represents baseline values, $\gamma = 44.35$, $q_N = .5$ $q_L = .51$, $q_I = .49$, $\Gamma_L = .05$, $\Gamma_I = .13, \beta = 2.1$, $\alpha_L = 0.004, \alpha_I = .006$, $\sigma_N = 1.92$, $\sigma_L = 1.62$, $\sigma_I = 0.87$, where the per capita gun assault rate is 172.17 at 30 years. The *dashed, green line* depicts a 10 % increase in violence intervention efforts in the LGO and IGO transmitter populations only ($\sigma_N = \sigma_L = \sigma_I = 1.9$ and all other parameters left at their baseline values), resulting in a decrease in the per capita gun assault rate to 76.9 at 30 years. The *dotted, red line* represents a 10 % increase from baseline values in violence intervention efforts ($\sigma_N = 2.11$, $\sigma_L = 1.78$, $\sigma_I = 0.96$, and all other parameters remained fixed at their baseline value) across all transmitter populations. The resulting per capita gun assault rate at 30 years is 55.58

of targeted intervention efforts in gun-owning versus non-gun-owning populations in Fig. 3 and Table 5.

In Fig. 4, we illustrate equilibrium prevalences for the total susceptible and transmitter populations at different rates of violence interruption. At the baseline values (parameter values that minimized Eq. 14), the transmitters account for 13 % of the population at equilibrium resulting in a per capita gun assault of 173. For violence intervention rates 10 % above baseline, the proportion of transmitters in the population at equilibrium is reduced by more than 50–4.6 % corresponding to a per capita gun assault rate of 64. For a 20 % increase in violence intervention efforts, the STV system converges to the violence free state, where the susceptible and transmitter equilibrium populations 100 and 0 %, respectively.

Table 5 Percent changes in violence interruption rates versus percent change in gun assault rates

% Change in violence intervention rate	% Change in total gun assault victims			
	LGO	IGO	GO	All T
5	−7	−5	−11	−41
10	−12	−9	−21	−67
15	−18	−13	−30	−83
20	−22	−17	−37	−91

In this table, we show changes in overall gun assault rates for different levels of increased violence intervention efforts. The LGO, IGO, GO, and All T columns represents percent change in overall gun assault rate when only LGO transmitters, IGO transmitters, both legal and illegal gun owner transmitters or all transmitters are targeted, respectively. Reference rates for the percent change are the baseline values for violence intervention efforts, $\sigma_N = 1.92$, $\sigma_L = 1.62$, and $\sigma_I = 0.87$. All other parameter values are same as in Fig. 3

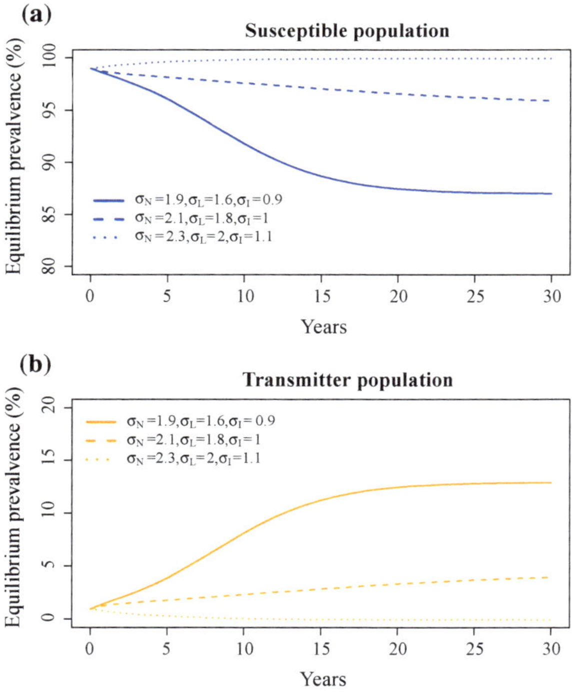

Fig. 4 **The impact of violence interruption on violence transmitter prevalence. a** and **b** illustrate the equilibrium prevalence for the susceptible and transmitter populations, respectively, under different scenarios of violence intervention efforts. In both figures, the *solid line* corresponds to violence intervention efforts at their baseline values ($\sigma_N = 1.9$, $\sigma_L = 1.6$ and $\sigma_I = 0.9$). The *dashed line* corresponds with violence intervention efforts 10 % above their baseline values and that *dotted line* corresponds to violence intervention efforts 20 % above baseline values. All other parameters remained fixed at their baseline values (see Fig. 3)

4 Discussion and Conclusions

Our model builds upon public health contagion approaches in providing a general modeling framework for the diffusion of violence that is adaptable to multiple urban communities. Consistent with violence prevention programs, results support the idea that interrupting violence transmission and reducing the number of transmitting individuals can have a significant impact on reducing violent events like gun assaults. However, large variations in model predictions of future gun assault cases indicate that further model development is needed before results can be used to inform violence prevention and interruption programs.

Several factors may be contributing to unreliable model predictions of yearly gun assault cases at equilibrium. First, 4 of the 12 model parameters have significant PRCC magnitudes greater than 0.5, viz., the rates of violence intervention for non-gun owners, legal gun owners, and illegal gun owners, and the transmission efficiency of violence. Additionally, the rate of intervention for non-gun owners and the transmission efficiency of violence both have PRCC magnitudes greater than 0.9 (0.95 and 0.93 respectively), indicating the STV model is extremely sensitive to these parameter estimations. Moreover, the estimates for the 4 key parameters mentioned above were not informed by empirical data as it was unavailable. Thus, parameter estimations are unreliable and potentially leading to overestimations in per capita gun assault rates. In order to make more accurate predictions on the occurrence of gun violence, further empirical data could be generated for these key parameters.

Second, one of our primary model assumptions is perfect mixing (homogeneity), where all susceptible individuals in the study region have an equal probability of becoming a transmitter or a gun assault victim. However, this may not be a suitable assumption given that susceptibility to becoming a violence transmitter is dependent on an individual's exposure to violence [1, 16–20]. This assumption may account for the large discrepancy with respect to targeted violence intervention rates, and the models sensitivity to the NGO transmitter violence interruption rate, where disparity in the proportions of NGO compared to LGO and IGO may provide imprecise results on the effectiveness of targeted campaigns within the gun-owning populations. Incorporating individual and spatial heterogeneities into the model, where we could consider high-risk individuals and high-risk neighborhoods, could result in more precise predictions on violence diffusion and provide more insight on the impacts of violence interruption and prevention programs. Although our model results do indicate targeting only illegal gun owners would not have a overwhelming impact on the incidence of gun assaults, more empirical data is needed to make further conclusions. For instance, more observation based estimations of gun violence transmission rates for legal and illegal gun owners could lead to more reliable predictions on the impact legal versus illegal gun ownership.

Lastly, our model does not include age structure. Age structure is typically included in infectious disease models that describe childhood diseases such as measles, rubella and chickenpox, as these diseases affect children at a much higher rate than adults [28, 30]. Community violence is similar in that manner as it affects adolescents and young adults at a much higher rate than the adult population. Individuals between the ages of 12 and 24 accounted for 45 % of all violent, nonfatal firearm injuries in 2013 (rate of 53 per 100,000), compared to 35 % in the 25–40 age group (rate of 39 per 100,000) [43]. Including additional compartments to capture this disproportion in gun violence among adolescents and young adults is necessary to develop a more accurate model of contagious violence.

This study introduces a general modeling framework that can be used to elucidate impacts of violence interruption and prevention programs on the occurrence of contagious community gun violence. Many violence prevention programs, including more recent programs that have adopted a violence interruption paradigm [1–4], consist of components that intervene on potential violence transmitters (e.g., individuals victimized by violence) in order to stop the transmission of violence. Our model results support this approach by indicating that the most effective method for reducing gun violence are interventions that target all individuals at risk of becoming violence transmitters irrespective of whether or not individuals have immediate access to a firearm.

Acknowledgments The authors would like to thank the reviewers for their very thorough comments that were used to improve the manuscript.

This work was supported by the National Institute of Alcohol Abuse and Alcoholism, http://www.niaaa.nih.gov/, grant number R01AA020331-01A1 and the Center for Disease Control and Prevention, http://www.cdc.gov/, Center grant R49CE002474.

The authors declare that no competing financial interests exist.

Appendix: Additional Tables

See Tables 6 and 7.

Table 6 LHS parameter results

Run	q_N	q_L	q_I	β	α_L	α_I	σ_N	σ_L	σ_I	γ	Γ_I	Γ_L
1	0.133	0.032	0.653	2.061	0.005	0.006	1.171	1.306	1.261	41.266	0.132	0.040
2	0.117	0.352	0.398	2.327	0.006	0.004	1.629	1.134	1.681	48.241	0.126	0.042
3	0.783	0.297	0.768	2.103	0.003	0.007	0.579	1.201	0.639	49.791	0.123	0.043
4	0.487	0.562	0.543	2.096	0.008	0.005	0.684	0.856	1.479	44.056	0.127	0.044
5	0.497	0.458	0.853	1.494	0.005	0.007	1.404	1.366	0.894	44.444	0.121	0.047
6	0.237	0.923	0.588	1.571	0.003	0.008	0.676	1.224	1.426	44.521	0.134	0.045
7	0.312	0.102	0.993	2.333	0.008	0.008	0.669	0.834	1.644	45.839	0.113	0.045
8	0.018	0.142	0.758	2.298	0.003	0.002	0.661	1.261	0.579	44.986	0.136	0.040
9	0.658	0.438	0.182	2.389	0.001	0.002	0.849	0.706	0.871	43.746	0.133	0.045

(continued)

Table 6 (continued)

Run	q_N	q_L	q_I	β	α_L	α_I	σ_N	σ_L	σ_I	γ	Γ_I	Γ_L
10	0.873	0.057	0.683	2.375	0.008	0.005	1.336	0.661	1.156	39.484	0.119	0.049
11	0.908	0.158	0.297	1.612	0.005	0.004	0.504	1.351	0.759	49.481	0.134	0.043
12	0.898	0.357	0.318	1.522	0.003	0.009	0.571	0.609	1.321	45.761	0.128	0.046
13	0.638	0.558	0.163	1.983	0.005	0.003	1.899	0.924	1.771	48.086	0.131	0.043
14	0.057	0.068	0.613	1.962	0.007	0.006	1.666	1.719	1.494	37.159	0.113	0.044
15	0.878	0.302	0.663	2.144	0.005	0.006	0.706	0.511	1.591	38.476	0.114	0.049
16	0.302	0.863	0.723	2.138	0.006	0.007	0.714	1.899	1.419	48.861	0.127	0.048
17	0.448	0.938	0.062	1.767	0.005	0.005	0.796	0.579	1.284	40.104	0.137	0.045
18	0.868	0.618	0.342	2.264	0.004	0.004	1.831	0.699	1.584	41.421	0.135	0.041
19	0.773	0.628	0.658	1.304	0.002	0.009	1.239	1.044	1.081	43.281	0.123	0.048
20	0.388	0.182	0.338	2.011	0.003	0.007	1.486	1.486	1.389	40.181	0.117	0.048
21	0.102	0.818	0.497	1.928	0.001	0.009	0.969	0.631	0.669	51.496	0.123	0.046
22	0.843	0.648	0.788	1.675	0.007	0.008	1.359	0.916	0.594	44.289	0.132	0.044
23	0.367	0.077	0.247	2.208	0.004	0.009	0.879	0.826	0.924	47.931	0.128	0.043
24	0.052	0.988	0.438	2.236	0.005	0.005	1.096	0.894	1.674	42.041	0.112	0.045
25	0.933	0.072	0.037	1.913	0.005	0.006	1.261	0.684	0.939	41.111	0.130	0.049
26	0.618	0.177	0.603	1.823	0.003	0.008	0.691	0.639	1.186	50.489	0.135	0.049
27	0.427	0.728	0.923	1.837	0.004	0.003	0.954	1.471	1.134	42.584	0.139	0.044
28	0.333	0.237	0.743	1.829	0.000	0.002	1.179	1.081	1.501	41.964	0.110	0.040
29	0.593	0.748	0.042	2.151	0.007	0.006	1.674	1.336	0.976	51.806	0.111	0.040
30	0.453	0.753	0.583	1.619	0.009	0.008	0.609	0.991	1.336	37.469	0.113	0.047
31	0.988	0.543	0.783	1.640	0.007	0.001	0.789	0.901	0.706	51.574	0.112	0.040
32	0.518	0.873	0.983	2.026	0.006	0.005	0.736	1.696	0.549	46.149	0.139	0.046
33	0.923	0.692	0.068	2.005	0.004	0.005	1.951	1.539	1.381	46.769	0.112	0.049
34	0.122	0.933	0.112	1.556	0.001	0.001	0.901	1.186	0.729	44.134	0.135	0.044
35	0.062	0.273	0.328	1.893	0.003	0.004	1.591	0.714	1.089	43.049	0.137	0.047
36	0.588	0.778	0.477	2.228	0.008	0.006	1.029	1.179	0.789	45.916	0.121	0.049
37	0.978	0.403	0.968	1.844	0.002	0.008	1.771	0.571	0.834	45.064	0.122	0.044
38	0.273	0.888	0.333	2.067	0.003	0.005	0.999	1.591	1.036	40.724	0.117	0.048
39	0.342	0.492	0.573	2.032	0.003	0.006	0.744	0.849	1.374	42.661	0.111	0.046
40	0.748	0.603	0.393	1.900	0.004	0.007	1.254	1.006	0.796	47.621	0.114	0.042
41	0.848	0.323	0.808	2.354	0.005	0.005	1.006	1.014	1.726	46.536	0.114	0.041
42	0.318	0.548	0.618	2.341	0.005	0.007	1.314	0.819	1.231	49.404	0.114	0.048
43	0.698	0.833	0.158	2.110	0.003	0.007	1.944	1.629	1.921	51.186	0.127	0.049
44	0.928	0.758	0.633	2.397	0.002	0.001	1.419	1.479	0.969	46.071	0.121	0.046
45	0.022	0.432	0.492	1.949	0.004	0.003	0.526	1.614	1.096	40.336	0.112	0.048
46	0.958	0.048	0.558	2.221	0.002	0.008	1.854	1.299	1.554	36.771	0.128	0.047
47	0.247	0.138	0.858	2.131	0.001	0.006	0.729	1.419	1.411	40.646	0.137	0.041
48	0.177	0.217	0.748	1.290	0.005	0.007	0.909	0.519	0.946	38.709	0.126	0.046
49	0.098	0.928	0.953	2.159	0.002	0.003	1.201	1.951	1.366	49.714	0.133	0.049
50	0.222	0.328	0.833	2.082	0.004	0.007	1.441	1.854	1.779	47.001	0.118	0.043
51	0.953	0.282	0.678	2.046	0.004	0.007	0.931	0.804	0.661	51.651	0.121	0.046

(continued)

Table 6 (continued)

Run	q_N	q_L	q_I	β	α_L	α_I	σ_N	σ_L	σ_I	γ	Γ_I	Γ_L
52	0.798	0.768	0.798	1.956	0.003	0.005	1.501	1.141	1.464	49.094	0.119	0.043
53	0.713	0.188	0.422	2.186	0.006	0.002	1.516	1.284	1.456	40.879	0.113	0.042
54	0.738	0.338	0.142	1.801	0.003	0.005	1.704	0.729	0.751	37.546	0.112	0.049
55	0.573	0.513	0.052	2.292	0.007	0.004	0.924	1.051	1.111	38.321	0.134	0.048
56	0.407	0.948	0.192	1.851	0.007	0.002	1.569	0.999	1.689	43.901	0.131	0.046
57	0.042	0.883	0.242	2.200	0.003	0.005	1.224	0.984	0.819	47.156	0.112	0.047
58	0.758	0.518	0.357	1.402	0.009	0.006	0.939	0.969	0.526	44.831	0.137	0.046
59	0.558	0.553	0.698	2.383	0.007	0.008	0.519	1.944	1.876	46.459	0.116	0.045
60	0.888	0.523	0.643	1.907	0.006	0.007	0.834	1.846	0.556	38.399	0.139	0.042
61	0.853	0.417	0.432	2.256	0.006	0.003	1.044	0.864	1.734	43.436	0.138	0.050
62	0.158	0.713	0.453	1.935	0.008	0.010	1.734	0.939	1.006	44.366	0.120	0.047
63	0.292	0.843	0.098	2.242	0.004	0.007	0.991	1.441	1.809	50.256	0.123	0.042
64	0.077	0.793	0.812	2.179	0.006	0.008	1.546	0.909	1.636	42.429	0.138	0.043
65	0.458	0.333	0.863	1.970	0.002	0.009	1.074	0.691	0.901	45.529	0.129	0.049
66	0.858	0.128	0.087	1.990	0.004	0.005	0.759	1.681	1.801	45.606	0.136	0.048
67	0.833	0.958	0.933	2.284	0.005	0.007	1.014	1.981	1.756	47.079	0.133	0.046
68	0.403	0.168	0.568	1.486	0.005	0.005	1.381	1.194	1.141	43.204	0.133	0.047
69	0.778	0.052	0.217	2.312	0.003	0.005	1.599	0.796	0.984	37.081	0.120	0.046
70	0.242	0.743	0.177	2.123	0.005	0.006	1.366	1.959	0.631	46.691	0.135	0.047
71	0.007	0.268	0.467	2.319	0.003	0.009	1.659	1.171	0.841	41.731	0.132	0.043
72	0.432	0.028	0.623	1.409	0.006	0.007	0.826	0.789	0.931	50.721	0.116	0.043
73	0.983	0.858	0.638	1.514	0.002	0.007	0.871	0.751	0.879	48.939	0.123	0.042
74	0.643	0.668	0.092	1.998	0.004	0.006	0.886	1.809	1.846	43.514	0.124	0.042
75	0.347	0.763	0.928	2.348	0.005	0.008	1.351	0.736	1.651	46.381	0.124	0.047
76	0.733	0.773	0.442	1.472	0.003	0.006	1.021	0.759	1.276	38.554	0.131	0.041
77	0.513	0.232	0.002	1.228	0.002	0.008	1.104	0.504	0.654	42.894	0.125	0.041
78	0.753	0.708	0.893	1.942	0.005	0.007	1.711	0.669	1.306	40.491	0.138	0.041
79	0.072	0.808	0.232	1.668	0.006	0.005	1.456	0.526	1.066	50.179	0.140	0.047
80	0.202	0.913	0.873	2.172	0.006	0.006	1.824	0.841	1.209	38.786	0.118	0.041
81	0.112	0.528	0.077	1.599	0.004	0.004	0.556	0.871	1.104	46.226	0.117	0.044
82	0.147	0.738	0.012	1.683	0.007	0.008	0.856	0.744	0.541	48.319	0.124	0.045
83	0.653	0.398	0.222	1.872	0.004	0.004	0.564	1.554	1.021	48.784	0.124	0.049
84	0.142	0.472	0.133	1.976	0.004	0.007	0.511	0.654	1.696	41.189	0.130	0.041
85	0.553	0.503	0.407	1.921	0.004	0.009	1.764	0.886	0.804	51.419	0.131	0.043
86	0.603	0.678	0.282	2.075	0.006	0.007	1.479	1.921	0.849	42.196	0.130	0.050
87	0.598	0.878	0.713	1.444	0.001	0.003	1.396	1.359	1.299	37.004	0.114	0.047
88	0.048	0.278	0.312	2.088	0.005	0.007	0.751	1.494	1.246	37.391	0.128	0.043
89	0.092	0.422	0.688	2.277	0.004	0.004	1.681	1.576	1.164	48.396	0.130	0.042
90	0.463	0.812	0.472	2.250	0.003	0.004	1.741	1.066	1.059	44.211	0.120	0.046
91	0.232	0.007	0.207	2.369	0.004	0.005	1.299	1.929	1.351	51.729	0.129	0.047
92	0.543	0.247	0.608	2.215	0.004	0.006	1.869	1.644	1.471	49.326	0.134	0.043
93	0.012	0.318	0.032	2.117	0.004	0.008	1.321	1.329	1.291	42.506	0.115	0.049
94	0.743	0.633	0.778	1.395	0.004	0.006	0.541	1.089	0.961	48.706	0.119	0.042
95	0.913	0.658	0.347	2.194	0.008	0.009	0.699	1.816	0.571	39.096	0.122	0.043

(continued)

Table 6 (continued)

Run	q_N	q_L	q_I	β	α_L	α_I	σ_N	σ_L	σ_I	γ	Γ_I	Γ_L
96	0.628	0.383	0.482	2.271	0.007	0.005	1.156	0.586	1.989	36.616	0.113	0.047
97	0.683	0.683	0.362	2.306	0.001	0.006	1.846	1.344	0.676	41.654	0.136	0.045
98	0.673	0.263	0.703	2.361	0.006	0.008	1.141	1.734	1.951	49.869	0.125	0.042
99	0.668	0.198	0.578	2.040	0.004	0.008	1.216	1.824	1.606	39.174	0.135	0.044
100	0.993	0.312	0.708	2.165	0.002	0.006	1.974	1.779	1.014	44.599	0.135	0.046
101	0.307	0.673	0.302	1.857	0.006	0.009	1.111	1.599	0.519	38.631	0.131	0.044
102	0.217	0.538	0.628	1.780	0.005	0.003	1.194	1.674	0.564	39.794	0.126	0.041
Means												
	0.501	0.506	0.493	1.984	0.004	0.006	1.171	1.166	1.179	44.346	0.125	0.045

Table 7 LHS results for key parameters

Run	β	σ_N	σ_L	σ_I
1	1.837	1.573	0.608	1.468
2	1.692	0.537	0.657	1.117
3	2.347	1.488	1.887	1.073
4	1.799	1.042	0.627	0.632
5	2.315	1.232	1.988	1.163
6	2.397	0.512	1.248	0.573
7	2.177	1.623	1.823	1.952
8	2.082	1.452	1.373	0.657
9	1.547	1.518	0.917	1.183
10	2.095	1.567	1.782	0.527
11	2.133	1.607	1.042	1.808
12	2.139	1.163	0.557	0.532
13	1.793	0.777	0.667	1.032
14	1.535	1.173	1.367	0.863
15	2.057	1.812	0.682	1.577
16	2.089	0.892	1.863	0.583
17	1.724	1.018	1.538	0.752
18	1.862	0.807	1.512	1.258
19	1.812	0.932	1.218	0.838
20	2.221	1.508	1.518	1.978
21	2.309	1.667	0.988	0.777
22	2.353	1.448	0.873	1.897
23	2.359	0.532	1.893	1.933
24	2.114	0.547	1.298	1.617
25	1.661	1.147	0.507	1.387
26	1.944	1.782	0.583	0.958
27	1.327	1.242	1.097	0.642

(continued)

Table 7 (continued)

Run	β	σ_N	σ_L	σ_I
28	2.045	1.698	0.902	1.002
29	2.284	0.613	0.568	1.653
30	1.453	1.272	1.387	0.613
31	1.680	0.757	0.892	0.588
32	2.296	0.843	1.413	1.377
33	1.698	0.578	1.278	0.978
34	2.170	0.598	1.393	1.962
35	2.183	1.583	1.948	0.792
36	2.334	0.978	1.647	1.157
37	1.900	0.823	1.397	0.853
38	1.831	1.218	1.377	0.733
39	1.371	1.252	0.647	0.807
40	2.196	1.323	0.733	1.248
41	2.365	1.712	1.593	1.552
42	1.881	0.667	0.613	1.373
43	2.107	1.778	1.353	0.892
44	1.994	1.347	0.752	1.508
45	2.227	1.542	1.952	0.618
46	1.629	1.127	0.618	1.302
47	2.189	1.363	1.228	1.012
48	1.673	0.858	0.838	0.728
49	1.610	1.478	0.718	0.603
50	1.755	1.208	1.587	0.917
51	2.063	1.393	1.718	0.547
52	2.013	1.302	0.728	1.583
53	2.340	0.652	1.002	0.698
54	1.963	1.107	0.603	1.677
55	2.164	1.613	0.897	1.613
56	1.938	1.143	1.448	1.008
57	2.372	1.433	1.962	1.423
58	1.705	0.743	1.038	1.252
59	2.302	0.688	0.692	1.028
60	1.529	1.093	0.573	1.042
61	1.604	1.157	1.157	0.677
62	1.384	0.938	0.828	0.973
63	1.560	1.373	1.252	1.407
64	2.158	1.262	1.143	1.788
65	2.258	1.133	0.632	0.797
66	2.391	1.498	1.843	1.153

(continued)

Table 7 (continued)

Run	β	σ_N	σ_L	σ_I
67	2.032	1.708	1.657	1.367
68	1.573	0.568	1.032	1.383
69	2.252	1.087	0.998	1.883
70	2.202	0.863	0.848	1.742
71	2.240	1.292	1.857	1.442
72	2.145	0.907	1.147	0.828
73	2.070	1.387	0.598	0.743
74	2.384	0.588	1.667	1.317
75	2.026	1.903	1.427	1.867
76	1.787	0.988	1.343	1.353
77	1.761	1.238	0.978	0.718
78	1.950	1.353	1.903	1.177
79	2.271	0.752	0.703	0.932
80	1.875	0.968	1.778	0.738
81	2.126	0.748	1.613	0.983
82	2.051	1.603	1.603	1.823
83	1.931	1.577	1.327	0.708
84	2.007	1.317	0.907	1.192
85	1.415	0.998	0.593	1.058
86	1.434	0.562	0.522	0.902
87	1.925	1.048	1.528	0.787
88	1.711	0.958	0.738	1.222
89	1.472	1.008	0.578	1.282
90	2.019	1.188	0.502	1.688
91	2.321	0.502	0.677	1.873
92	2.290	1.688	0.863	1.698
93	2.246	1.643	1.083	0.593
94	1.743	1.413	1.183	0.667
95	1.956	0.917	1.583	1.752
96	2.076	1.873	1.938	1.562
97	2.265	1.403	1.772	1.123
98	2.151	1.077	1.462	1.812
99	1.686	1.587	1.107	1.228
100	2.214	1.512	0.958	1.433
101	2.233	1.377	0.777	0.692
102	2.101	1.917	1.623	0.873
103	2.120	1.722	0.767	0.963
104	1.648	1.333	0.887	0.713

(continued)

Table 7 (continued)

Run	β	σ_N	σ_L	σ_I
105	2.277	1.192	1.758	0.897
106	1.906	0.637	1.417	1.427
107	2.208	1.768	0.797	1.308
108	1.642	1.482	1.407	1.393
109	1.868	1.853	0.792	1.492
110	2.038	1.312	0.948	0.993
111	2.328	1.113	1.383	1.542
112	1.749	0.902	1.103	0.578
113	2.378	0.603	1.323	1.212
114	1.982	1.212	1.827	1.597
115	1.510	1.492	1.222	1.103
116	1.912	1.442	1.232	0.647
117	1.818	1.653	1.438	0.703

References

1. G. Slutkin, Violence is a contagious disease, in *Contagion of Violence: Workshop Summary* (2012)
2. D.M. Patel, M.A. Simon, R.M. Taylor et al., Contagion of Violence: Workshop Summary (National Academies Press, Washington, 2013)
3. C. Cooper, D.M. Eslinger, P.D. Stolley, Hospital-based violence intervention programs work. J. Trauma-Inj. Infect. Crit. Care **61**, 534–540 (2006)
4. N. Karraker et al., Violence is preventable: a best practices guide for launching and sustaining a hospital-based program to break the cycle of violence. Office of Victims of Crime, Office of Justice Programs, US Department of Justice, Washington, DC (2011)
5. S. Patten, J. Arboleda-Florez, Epidemic theory and group violence. Soc. Psychiatry Psychiatr. Epidemiol. **39**, 853–856 (2004)
6. F. Sánchez, X. Wang, C. Castillo-Chávez, D.M. Gorman, P.J. Gruenewald, Drinking as an epidemica simple mathematical model with recovery and relapse, *Therapists Guide to Evidence-Based Relapse Prevention* (Academic Press, New York, 2007), p. 353
7. A. Cintrón-Arias, F. Sánchez, X. Wang, C. Castillo-Chavez, D.M. Gorman, and P.J. Gruenewald. The role of nonlinear relapse on contagion amongst drinking communities. In *Mathematical and statistical estimation approaches in epidemiology*, pages 343–360. Springer, 2009
8. L. Almada, E. Camacho, R. Rodriguez, M. Thompson, L. Voss, Deterministic and small-world network models of college drinking patterns. Technical report, AMSSI Technical Report (2006)
9. L. Zhao, H. Cui, X. Qiu, X. Wang, J. Wang, SIR rumor spreading model in the new media age. Phys. A: Stat. Mech. Appl. **392**, 995–1003 (2013)
10. L.M. Bettencourt, A. Cintrón-Arias, D.I. Kaiser, C. Castillo-Chávez, The power of a good idea: quantitative modeling of the spread of ideas from epidemiological models. Phys. A: Stat. Mech. Appl. **364**, 513–536 (2006)
11. B. González, E. Huerta-Sánchez, A. Ortiz-Nieves, T. Vázquez-Alvarez, C. Kribs-Zaleta, Am i too fat? Bulimia as an epidemic. J. Math. Psychol. **47**, 515–526 (2003)
12. B. Pfefferbaum, Posttraumatic stress disorder in children: a review of the past 10 years. J. Am. Acad. Child Adolesc. Psychiatry **36**, 1503–1511 (1997)

13. A.M. Zeoli, J.M. Pizarro, S.C. Grady, C. Melde, Homicide as infectious disease: using public health methods to investigate the diffusion of homicide. Justice Q. **31**, 609–632 (2014)
14. C. Loftin, Assaultive violence as a contagious social process. Bull. N. Y. Acad. Med. **62**, 550 (1986)
15. R. Taylor, A game theoretic model of gun control. Int. Rev. Law Econ. **15**, 269–288 (1995)
16. L.R. Huesmann, The contagion of violence: the extent, the processes, and the outcomes, in *Social and Economic Costs of Violence: Workshop Summary* (2012), pp. 63–69
17. S. Kelly, The psychological consequences to adolescents of exposure to gang violence in the community: an integrated review of the literature. J. Child Adolesc. Psychiatr. Nurs. **23**, 61–73 (2010)
18. C.V. Crooks, K.L. Scott, D.A. Wolfe, D. Chiodo, S. Killip, Understanding the link between childhood maltreatment and violent delinquency: what do schools have to add? Child Maltreatment **12**, 269–280 (2007)
19. L.R. Huesmann, J. Moise-Titus, C.-L. Podolski, L.D. Eron, Longitudinal relations between children's exposure to TV violence and their aggressive and violent behavior in young adulthood: 1977–1992. Dev. Psychol. **39**, 201 (2003)
20. L.R. Huesmann, Nailing the coffin shut on doubts that violent video games stimulate aggression: comment on Anderson et al. Psychol. Bull.**136**(2), 179–181 (2010)
21. A.L. Kellermann et al., Injuries due to firearms in three cities. N. Engl. J. Med. **335**, 1438–1444 (1996)
22. P. Cummings, T.D. Koepsell, D.C. Grossman, J. Savarino, R.S. Thompson, The association between the purchase of a handgun and homicide or suicide. Am. J. Public Health **87**, 974–978 (1997)
23. A.L. Kellermann et al., Gun ownership as a risk factor for homicide in the home. N. Engl. J. Med. **329**, 1084–1091 (1993)
24. K.M. Grassel, G.J. Wintemute, M.A. Wright, M. Romero, Association between handgun purchase and mortality from firearm injury. Inj. Prev. **9**, 48–52 (2003)
25. C.C. Branas, T.S. Richmond, D.P. Culhane, T.R. Ten Have, D.J. Wiebe, Investigating the link between gun possession and gun assault. Am. J. Public Health **99**, 2034 (2009)
26. C.D. Phillips, O. Nwaiwu, D.K. McMaughan Moudouni, R. Edwards, S.-H. Lin, When concealed handgun licensees break bad: criminal convictions of concealed handgun licensees in texas, 2001–2009. Am. J. Public Health **103**, 86–91 (2013)
27. W. Kermack, A. McKendrick, Contributions to the mathematical theory of epidemicsi. Bull. Math. Biol. **53**, 33–55 (1991)
28. M.J. Keeling, P. Rohani, *Modeling Infectious Diseases in Humans and Animals* (Princeton University Press, Princeton, 2008)
29. H.W. Hethcote, The mathematics of infectious diseases. SIAM Rev. **42**, 599–653 (2000)
30. F. Brauer, C. Castillo-Chavez, *Mathematical Models in Population Biology and Epidemiology* (Springer, New York, 2011)
31. S.M. Blower, H. Dowlatabadi, Sensitivity and uncertainty analysis of complex models of disease transmission: an HIV model, as an example. Int. Stat. Rev./Revue Internationale de Statistique **62**, 229–243 (1994)
32. S. Blower, D. Hartel, H. Dowlatabadi, R. Anderson, R. May, Drugs, sex and HIV: a mathematical model for New York city. Philos. Trans. R. Soc. Lond. Ser. B: Biol. Sci. **331**, 171–187 (1991)
33. J. Schmitt, K. Warner, Ex-offenders and the labor market. Work. USA **14**, 87–109 (2011)
34. S.H. Decker, S. Pennell, A. Caldwell, Illegal firearms: access and use by arrestees (US Department of Justice, Office of Justice Programs, National Institute of Justice, Washington, DC, 1997)
35. C.E. Kubrin, R. Weitzer, Retaliatory homicide: concentrated disadvantage and neighborhood culture. Soc. Probl. **50**, 157–180 (2003)
36. D.J. Wiebe, M.M. Blackstone, C.J. Mollen, A.J. Culyba, J.A. Fein, Self-reported violence-related outcomes for adolescents within eight weeks of emergency department treatment for assault injury. J. Adolesc. Health **49**, 440–442 (2011)

37. N. Copeland-Linder, S.B. Johnson, D.L. Haynie, S.-E. Chung, T.L. Cheng, Retaliatory attitudes and violent behaviors among assault-injured youth. J. Adolesc. Health **50**, 215–220 (2012)
38. Research, P. P. D. and Section, P. U. S. Murder analysis Philadelphia police department 2007–2010 (2011). http://www.phillypolice.com/assets/crime-maps-stats/PPD.Homicide.Analysis.2007-2010.pdf. Accessed 3 Sept 2014
39. Research, P. P. D. and Section, P. U. S. Murder/shooting analysis 2012 (2011). http://www.phillypolice.com/assets/crime-maps-stats/PPD-Homicide-Analysis-2011-vs-2012.pdf. Accessed 3 Sept 2014
40. L. Boström, B. Nilsson, A review of serious injury and death from gunshot wounds in Sweden: 1987–1994. Eur. J. Surg. **165**, 930–936 (1999)
41. D.V. Feliciano, J. Burch, V. Spjut-Patrinely, K.L. Mattox, G.L. Jordan Jr., Abdominal gunshot wounds. An urban trauma center's experience with 300 consecutive patients. Ann. Surg. **208**, 362 (1988)
42. A. Cowey, P. Mitchell, J. Gregory, I. Maclennan, R. Pearson, A review of 187 gunshot wound admissions to a teaching hospital over a 54-month period: training and service implications. Ann. R. Coll. Surg. Engl. **86**, 104 (2004)
43. For Disease Control, C. and (CDC), P. Web-based injury statistics query and reporting system (WISQARS) (2003). http://www.cdc.gov/ncipc/wisqars. Accessed 3 Mar 2015

Part IV
Probability and Stochastic Processes

Cramér's Theorem is Atypical

Nina Gantert, Steven Soojin Kim and Kavita Ramanan

Abstract The empirical mean of n independent and identically distributed (i.i.d.) random variables $(X_1, \ldots, X_n)$ can be viewed as a suitably normalized scalar projection of the n-dimensional random vector $X^{(n)} \doteq (X_1, \ldots, X_n)$ in the direction of the unit vector $n^{-1/2}(1, 1, \ldots, 1) \in \mathbb{S}^{n-1}$. The large deviation principle (LDP) for such projections as $n \to \infty$ is given by the classical Cramér's theorem. We prove an LDP for the sequence of normalized scalar projections of $X^{(n)}$ in the direction of a generic unit vector $\theta^{(n)} \in \mathbb{S}^{n-1}$, as $n \to \infty$. This LDP holds under fairly general conditions on the distribution of X_1, and for "almost every" sequence of directions $(\theta^{(n)})_{n\in\mathbb{N}}$. The associated rate function is "universal" in the sense that it does not depend on the particular sequence of directions. Moreover, under mild additional conditions on the law of X_1, we show that the universal rate function differs from the Cramér rate function, thus showing that the sequence of directions $n^{-1/2}(1, 1, \ldots, 1) \in \mathbb{S}^{n-1}$, $n \in \mathbb{N}$, corresponding to Cramér's theorem is atypical.

Keywords Large deviations · Projections · High-dimensional product measures · Cramér's theorem · Rate function

Mathematics Subject Classification 60F10 (primary) · 60D05 (secondary)

N. Gantert
Faculty for Mathematics, Technical University of Munich,
Boltzmannstr. 3, 85748 Garching, Germany
e-mail: gantert@ma.tum.de

S.S. Kim
Division of Applied Mathematics, Brown University,
182 George Street, Providence, RI 02192, USA
e-mail: steven_kim@brown.edu

K. Ramanan (✉)
Division of Applied Mathematics, Brown University,
182 George Street, Providence, RI 02912, USA
e-mail: kavita_ramanan@brown.edu

G. Letzter et al. (eds.), *Advances in the Mathematical Sciences*, Association for Women in Mathematics Series 6, DOI 10.1007/978-3-319-34139-2_11

1 Introduction

Let $X^{(n)} = (X_1, \ldots, X_n)$ be a sequence of n independent and identically distributed (i.i.d.) $\mathbb{R}$-valued random variables with common distribution $\gamma \in \mathcal{P}(\mathbb{R})$, the space of probability measures on $\mathbb{R}$. A fundamental probabilistic question is how the empirical mean of $X^{(n)}$ behaves as the length of the sequence n increases. From a geometric perspective, the empirical mean is a suitably normalized version of (the scalar component of) the projection of the n-dimensional vector $X^{(n)}$ in the direction of the unit vector $\iota^{(n)}$, defined by

$$\iota^{(n)} \doteq \frac{1}{\sqrt{n}}(\overbrace{1, 1, \ldots, 1}^{n \text{ times}}) \in \mathbb{S}^{n-1}. \tag{1}$$

In other words, we can write

$$W_\iota^{(n)} \doteq \frac{1}{\sqrt{n}} \langle X^{(n)}, \iota^{(n)} \rangle_n = \frac{1}{n} \sum_{i=1}^{n} X_i , \tag{2}$$

where $\langle \cdot, \cdot \rangle_n$ denotes the Euclidean inner product. With some abuse of terminology, for $x \in \mathbb{R}^n$ and $v \in \mathbb{S}^{n-1}$, we hereby write the "*projection* of x in the direction v" to refer to the scalar component $\langle x, v \rangle_n \in \mathbb{R}$ (rather than the vector $\langle x, v \rangle_n v \in \mathbb{R}^n$). Then, the expression (2) indicates that questions on the empirical mean for large n can be rephrased in a geometric language as questions on suitably normalized projections of high-dimensional random vectors.

The classical Cramér's theorem characterizes the large deviations behavior of (2), the empirical mean of i.i.d. random variables, as $n \to \infty$. In particular, if $X_1 \sim \gamma$ has some finite exponential moments, in the sense that

$$\exists\, t_0 > 0 \text{ s.t. } \forall\, |t| < t_0, \quad \Lambda(t) \doteq \log \mathbb{E}[e^{tX_1}] < \infty, \tag{3}$$

then we have the limit

$$\lim_{n \to \infty} \frac{1}{n} \log \mathbb{P}(W_\iota^{(n)} \geq x) = -\Lambda^*(x),$$

where * denotes the *Legendre transform*,

$$\Lambda^*(x) \doteq \sup_{t \in \mathbb{R}} \{tx - \Lambda(t)\}. \tag{4}$$

We refer to [17, Sect. 12] for a review of the Legendre transform (also known as the *convex conjugate*).

Given the geometric view of empirical means given by (2), it is natural to investigate analogs of Cramér's theorem for normalized projections in directions $\theta^{(n)} \in \mathbb{S}^{n-1}$ other than $\iota^{(n)}$. Such projections correspond to *weighted* means,

$$W_\theta^{(n)} \doteq \frac{1}{\sqrt{n}} \langle X^{(n)}, \theta^{(n)} \rangle_n = \frac{1}{n} \sum_{i=1}^{n} X_i \sqrt{n} \theta_i^{(n)}. \tag{5}$$

Our main result is an LDP for $(W_\theta^{(n)})_{n\in\mathbb{N}}$ for almost every (in a sense that is specified below) sequence of directions $\theta = (\theta^{(1)}, \theta^{(2)}, \dots)$. In particular, we show that the associated rate function does not depend on θ, and that it differs from the Cramér rate function Λ^*. That is, the sequence of directions $(\iota^{(n)})_{n\in\mathbb{N}}$ corresponding to Cramér's theorem is "atypical"!

Remark 1 While the LDP for (5) is novel, the corresponding law of large numbers (LLN) and central limit theorem (CLT) for weighted sums are well known. For example, a weak LLN follows from Chebyshev's inequality, and a CLT follows from the Lindeberg condition (see, e.g., [11, Sect. VIII.4, Theorem 3]).

The outline of this note is as follows. In Sect. 2, we state our main results and discuss their relation to prior work. In Sect. 3, we prove the claimed LDP. In Sect. 4, we establish that Cramér's theorem is atypical, and also comment on a generalization that is considered in [13].

2 Main Results

We first set some notation. Suppose the random variables $X_1, X_2, \dots$ are all defined on a common probability space $(\Omega, \mathcal{F}, \mathbb{P})$. Let $\|\cdot\|_n$ denote the Euclidean norm on $\mathbb{R}^n$. Write σ_{n-1} for the unique rotation invariant probability measure on $\mathbb{S}^{n-1}$, the unit sphere in $\mathbb{R}^n$. Let $\mathbb{S} \doteq \prod_{n\in\mathbb{N}} \mathbb{S}^{n-1}$, and let $\pi_n : \mathbb{S} \to \mathbb{S}^{n-1}$ be the coordinate map such that for $\theta = (\theta^{(1)}, \theta^{(2)}, \dots) \in \mathbb{S}$, we have $\pi_n(\theta) = \theta^{(n)}$. Let σ be a probability measure on (the Borel sets of) $\mathbb{S}$ such that

$$\sigma \circ \pi_n^{-1} = \sigma_{n-1}, \quad n \in \mathbb{N}. \tag{H1}$$

The generic example to keep in mind that satisfies (H1) is the product measure $\sigma = \bigotimes_{n\in\mathbb{N}} \sigma_{n-1}$, in which case the projection directions $\theta^{(n)}$, $n \in \mathbb{N}$, are independent under σ. However, our results allow for more general dependencies; for more discussion on σ and the condition (H1), see Remark 3.

For σ-a.e. $\theta \in \mathbb{S}^{n-1}$, we prove a large deviation principle for the sequence $(W_\theta^{(n)})_{n\in\mathbb{N}}$ with a rate function that does not depend on θ. We refer to [9] for general background on large deviations. In particular, recall the following definition:

Definition 1 The sequence of probability measures $(\mu_n)_{n\in\mathbb{N}} \subset \mathcal{P}(\mathbb{R})$ is said to satisfy a *large deviation principle (LDP)* with a *rate function* $\mathbb{I} : \mathbb{R} \to [0, \infty]$ if $\mathbb{I}$ is lower semicontinuous, and for all Borel measurable sets $\Gamma \subset \mathbb{R}$,

$$-\inf_{x\in\Gamma^\circ} \mathbb{I}(x) \le \liminf_{n\to\infty} \tfrac{1}{n} \log \mu_n(\Gamma^\circ) \le \limsup_{n\to\infty} \tfrac{1}{n} \log \mu_n(\bar{\Gamma}) \le -\inf_{x\in\bar{\Gamma}} \mathbb{I}(x),$$

where Γ° and $\bar{\Gamma}$ denote the interior and closure of Γ, respectively. Furthermore, $\mathbb{I}$ is said to be a *good rate function* if it has compact level sets.

We say the sequence of $\mathbb{R}$-valued random variables $(\xi_n)_{n\in\mathbb{N}}$ satisfies an LDP if the sequence of laws $(\mu_n)_{n\in\mathbb{N}}$ given by $\mu_n = \mathbb{P}\circ\xi_n^{-1}$ satisfies an LDP.

In particular, for empirical means of i.i.d. random variables, we recall the following classical result, due to [5, 7].

Theorem 1 **(Cramér)** *Let* $(X_n)_{n\in\mathbb{N}}$ *be an i.i.d. sequence such that* (3) *holds, and let* $\iota = (\iota^{(1)}, \iota^{(2)}, \dots)$ *be defined as in* (1). *Then the sequence* $(W_\iota^{(n)})_{n\in\mathbb{N}}$ *of* (2) *satisfies an LDP with the good rate function* $\mathbb{I}_\iota$*, given by*

$$\mathbb{I}_\iota(w) \doteq \Lambda^*(w) = \sup_{t\in\mathbb{R}}\{tw - \Lambda(t)\}. \tag{6}$$

Let $\nu \in \mathcal{P}(\mathbb{R})$ denote the standard one-dimensional Gaussian measure. In the sequel, we assume the following condition on Λ, the logarithmic moment generating function (log mgf) of $X_1 \sim \gamma$:

$$\forall t \in \mathbb{R}, \quad \int_{\mathbb{R}} |\Lambda(tu)|^4 \nu(du) < \infty. \tag{H2}$$

Note that (H2) is stronger than even requiring the exponential moment condition in (3) to hold with $t_0 = \infty$. For an absolutely continuous γ with density, a sufficient condition for (H2) is that the decay of the tail of the density is strictly faster than exponential, in the following sense:

Lemma 1 *Suppose that* γ *has density* f*, and that there exist* $p \in (1,\infty)$ *and constants* $0 < C_1, C_2, C_3 < \infty$ *such that for* $|x| > C_1$*, we have*

$$f(x) \le C_2 e^{-C_3|x|^p}.$$

Then there exists some constant $C < \infty$ *such that* Λ*, the log mgf of* γ*, satisfies the following upper bound for all* $t \in \mathbb{R}$*:*

$$\Lambda(t) \le C|t|^{p/(p-1)} + C.$$

Moreover, this implies that Λ *satisfies the condition* (H2).

Proof By Young's inequality applied to the conjugate exponents p and $\frac{p}{p-1}$, for $\varepsilon > 0$ and $t, y \in \mathbb{R}$,

$$ty \le \left(\varepsilon^{-1/p}|t|\right)\left(\varepsilon^{1/p}|y|\right) \le \tfrac{p-1}{p}\varepsilon^{-1/(p-1)}|t|^{p/(p-1)} + \frac{\varepsilon|y|^p}{p}.$$

In the following, let C absorb all constants, and note that for $0 < \varepsilon < C_3 p$ and $t \in \mathbb{R}$,

$$\begin{aligned}\Lambda(t) &= \log \int_{|y|\le C_1} e^{ty} f(y)dy + \log \int_{|y|>C_1} e^{ty} f(y)dy \\ &\le C_1|t| + \log C_2 + \log \int_{\mathbb{R}} e^{ty} e^{-C_3|y|^p} dy \\ &\le C_1|t|^{p/(p-1)} + C_1 + \tfrac{p-1}{p}\varepsilon^{-1/(p-1)}|t|^{p/(p-1)} - \tfrac{1}{p}\log(C_3 p - \varepsilon) + 1 \\ &= C|t|^{p/(p-1)} + C.\end{aligned}$$

From the preceding inequalities, since the Gaussian measure ν has finite moments of every order, it is clear that Λ satisfies the integrability condition (H2). □

We define the following analog of the log mgf in the case of weighted sums,

$$\Psi(t) \doteq \int_{\mathbb{R}} \Lambda(tu)\frac{1}{\sqrt{2\pi}}e^{-u^2/2}du, \quad t \in \mathbb{R}. \tag{7}$$

Our first main result is the following.

Theorem 2 (Weighted LDP) *Assume* (H1) *and* (H2). *Then, for* σ*-a.e.* $\theta \in \mathbb{S}$, *the sequence* $(W_\theta^{(n)})_{n\in\mathbb{N}}$ *of* (5) *satisfies an LDP with the convex good rate function* $\mathbb{I}_\sigma$, *given by*

$$\mathbb{I}_\sigma(w) \doteq \Psi^*(w) = \sup_{t\in\mathbb{R}}\{tw - \Psi(t)\}. \tag{8}$$

The proof of Theorem 2 is given in Sect. 3, with intermediate steps established in Sects. 3.1 and 3.2, and the proof completed in Sect. 3.3.

In principle, the rate function $\mathbb{I}_\sigma$ of Theorem 2 could depend on the particular choice of θ, but our result shows that the rate function is the same for σ-a.e. θ. In the case where σ is the product measure $\sigma = \bigotimes_{n\in\mathbb{N}} \sigma_{n-1}$, this follows immediately from the Kolmogorov zero-one law. That is, let $\mathcal{T}_n$ be the sigma-algebra generated by $(\theta^{(k)})_{k\ge n}$, and let

$$\mathcal{T} \doteq \bigcap_{n=1}^{\infty} \mathcal{T}_n \tag{9}$$

denote the *tail sigma-algebra* induced by $(\theta^{(1)}, \theta^{(2)}, \dots)$. The rate function $\mathbb{I}_\sigma$ is measurable with respect to $\mathcal{T}$, and the Kolmogorov zero-one law states that $\mathcal{T}$ is trivial under the product measure. Hence, $\mathbb{I}_\sigma$ coincides for σ-a.e. $\theta \in \mathbb{S}$. However, our claim holds for general σ satisfying (H1). In particular, Example 2(b) in Sect. 3.1 gives an example of σ such that $\theta^{(1)}, \theta^{(2)}, \dots$ are highly dependent, $\mathcal{T}$ is not trivial, and hence, the lack of dependence of the rate function $\mathbb{I}_\sigma$ on θ is not a priori obvious.

Given the σ-a.e. statement of Theorem 2, it is natural to ask what happens on the set of measure zero in $\mathbb{S}$ where the stated LDP does not hold. In particular, our second main result Theorem 3 shows that under certain additional conditions on Λ, the sequence of directions ι associated with Cramér's theorem is exceptional, in the sense that Cramér's rate function $\mathbb{I}_\iota$ differs from the universal rate function $\mathbb{I}_\sigma$. For the following theorem, we assume γ is symmetric, or specifically:

$$\forall t \in \mathbb{R}, \quad \Lambda(t) = \Lambda(-t). \tag{H3}$$

Theorem 3 (Atypicality) *Assume Λ satisfies* (H3)*, and let $\mathbb{I}_\iota$ and $\mathbb{I}_\sigma$ be given by* (6) *and* (8)*, respectively.*

(a) *If $\Lambda \circ \sqrt{\cdot}$ is concave on $\mathbb{R}_+$, then $\mathbb{I}_\sigma(w) \geq \mathbb{I}_\iota(w)$ for all $w \in \mathbb{R}$.*
(b) *If $\Lambda \circ \sqrt{\cdot}$ is convex on $\mathbb{R}_+$, then $\mathbb{I}_\sigma(w) \leq \mathbb{I}_\iota(w)$ for all $w \in \mathbb{R}$.*
(c) *If $\Lambda \circ \sqrt{\cdot}$ is concave or convex, but not linear, on $\mathbb{R}_+$, then $\mathbb{I}_\sigma(w) = \mathbb{I}_\iota(w) < \infty$ if and only if $w = 0$.*

The proof of Theorem 3 is given in Sect. 4.

We now provide some sufficient conditions (established in [1]) for the convexity or concavity conditions of Theorem 3 to hold.

Proposition 1 *Assume the exponential moment condition* (3) *and the symmetry condition* (H3).

(i) *Suppose $\gamma \neq \delta_0$, the Dirac mass at 0. Define $\varphi : \mathbb{N} \to \mathbb{R}$ by*

$$\varphi(k) \doteq (2k+1)\frac{\mathbb{E}[|X_1|^{2k}]}{\mathbb{E}[|X_1|^{2k+2}]}, \quad k \in \mathbb{N}.$$

If φ is non-decreasing (resp., non-increasing), then $\Lambda \circ \sqrt{\cdot}$ is concave (resp., convex) on $\mathbb{R}_+$.

(ii) *Suppose γ has density f such that* $\log f \circ \sqrt{\cdot}$ *is concave (resp., convex) on $\mathbb{R}_+$. Then $\Lambda \circ \sqrt{\cdot}$ is concave (resp., convex) on $\mathbb{R}_+$.*

Proof Part (i) is established in Theorem 7 of [1]. Part (ii) follows from applying Theorem 12 of [1] with their f replaced by our $f \circ \sqrt{\cdot}$, and noticing that the integrability of $f \circ \sqrt{\cdot}$ follows from the fact that f has finite first moment, due to the exponential moment condition of (3). □

Example 1 Suppose γ is the *generalized normal distribution* with location 0, scale $\alpha > 0$, and shape $\beta > 1$; that is, $\gamma = \mu_{\alpha,\beta}$, where

$$\mu_{\alpha,\beta}(dx) \doteq \frac{1}{2\alpha\Gamma(1+\frac{1}{\beta})} e^{-(|x|/\alpha)^\beta} dx$$

It follows from Lemma 1 that $\mu_{\alpha,\beta}$ satisfies (H2), which implies (3). It is also easy to see that $\mu_{\alpha,\beta}$ satisfies (H3). Thus, the conditions of Proposition 1 are satisfied. It follows immediately from Proposition 1(ii) that for $\beta \geq 2$ (resp., for $\beta \leq 2$), $\Lambda \circ \sqrt{\cdot}$ is concave (resp., convex). In fact, for $\beta \neq 2$, the concavity (resp., convexity) is strict.

The preceding example suggests the particular role of the Gaussian, which corresponds to $\beta = 2$. In particular, $\gamma = \mu_{\alpha,2}$ for some $\alpha > 0$ if and only if $\Lambda \circ \sqrt{\cdot}$ is linear. Thus, we could interpret the conditions of Theorem 3 as evaluating whether our distribution of interest is "more" or "less" log-concave than the Gaussian. We also have the following result in the Gaussian case (i.e., when $\gamma = \mu_{\alpha,2}$), which holds for *all* θ as opposed to just for σ-a.e. θ.

Proposition 2 *Suppose $\gamma = \mu_{\alpha,2}$ for some $\alpha > 0$. Then, for all $\theta \in \mathbb{S}$, the sequence $(W_\theta^{(n)})_{n\in\mathbb{N}}$ satisfies an LDP with the good rate function $\Psi^*(w) = \Lambda^*(w) = (w/\alpha)^2$, where Λ^* is defined in* (4) *with Λ the log mgf of the Gaussian with mean 0 and variance $\alpha^2/2$.*

Proof This follows from the fact that for all $n \in \mathbb{N}$, the Gaussian measure on $\mathbb{R}^n$ is spherically symmetric, and hence, for any $\theta^{(n)} \in \mathbb{S}^{n-1}$, the law of $\langle X^{(n)}, \theta^{(n)}\rangle_n$ is the same as the law of $\langle X^{(n)}, \iota^{(n)}\rangle_n$. Thus, the LDP for $(W_\theta^{(n)})_{n\in\mathbb{N}}$ follows from the classical Cramér's theorem for empirical means of i.i.d. Gaussians, for which the rate function can be easily computed to be $\Lambda^*(w) = (w/\alpha)^2$. □

Remark 2 It is not clear whether a converse of Proposition 2 holds. That is, whether $\mathbb{I}_\sigma \equiv \mathbb{I}_\iota$ if and only if γ is Gaussian. As one possible approach in this direction, it would be sufficient to show that for any measure γ satisfying both (H2) and (H3) (and possibly some additional natural conditions), the function $\Lambda \circ \sqrt{\cdot}$ must be either concave or convex.

Aside from the sequence of Cramér directions $\iota \in \mathbb{S}$, another natural sequence of directions to consider is the sequence of canonical basis vectors, $e_1 = (e_1^{(1)}, e_1^{(2)}, \dots) \in \mathbb{S}$, where

$$e_1^{(n)} \doteq (1, \overbrace{0, \dots, 0}^{n-1 \text{ times}}) \in \mathbb{S}^{n-1}.$$

Then $W_{e_1}^{(n)} = X_1/\sqrt{n}$ for all n. The following result states that under certain tail conditions, such normalized projections yield a trivial LDP, again with a rate function different from $\mathbb{I}_\sigma$.

Proposition 3 *Assume the following condition (which is stronger than* (H2)*):*

$$\exists\, C < \infty,\ r \in [0, 2) \text{ such that } \forall\, t \in \mathbb{R},\quad \Lambda(t) \le C(1 + |t|^r). \tag{H2$'$}$$

Then the sequence $(W_{e_1}^{(n)})_{n\in\mathbb{N}}$ satisfies an LDP with the trivial good rate function χ_0 given by

$$\chi_0(x) \doteq \begin{cases} 0, & x = 0; \\ \infty, & x \neq 0. \end{cases}$$

Proof Consider the limit log mgf associated with the Gärtner–Ellis theorem (recalled for convenience later in Theorem 4). For all $t \in \mathbb{R}$,

$$\Lambda_n(t) \doteq \frac{1}{n} \log \mathbb{E}[\exp(tnW_{e_1}^{(n)})] = \frac{1}{n} \log \mathbb{E}[\exp(t\sqrt{n}X_1)] = \frac{1}{n}\Lambda(t\sqrt{n}) \le \frac{1}{n}(C|t|^r n^{r/2} + C).$$

Since the exponent r of (H2′) satisfies $r < 2$ by assumption, we have $\lim_{n\to\infty} \Lambda_n(t) = 0$ for all $t \in \mathbb{R}$. Thus, by the Gärtner–Ellis theorem, the sequence $(W_{e_1}^{(n)})_{n\in\mathbb{N}}$ satisfies an LDP with good rate function $0^* = \chi_0$. □

2.1 Relation to Prior Work

There is a wealth of literature on large deviations for weighted sums, but our work seems to be the first to emphasize the unique position of Cramér's theorem in the geometric setting. Moreover, it appears that none of the existing literature is readily adaptable to our particular problem. We offer a partial (but inevitably, incomplete) survey of existing results.

In the somewhat classical works of Book [2, 3], we can find asymptotics bounds for quantities of the form

$$\mathbb{P}\left(\frac{\sum_{k=1}^n a_{nk} X_k}{\sum_{k=1}^n a_{nk}} > c\right),$$

where $(a_{nk})_{k\le n, n\in\mathbb{N}}$ is a triangular array of weights such that $\sum_{k=1}^n a_{nk}^2 = 1$ for all n. However, this does not address our setting because if we let $a_{nk} = \theta_k^{(n)}$, we have $\sum_{k=1}^n a_{nk}^2 = 1$, but this only yields tail bounds of the form $\mathbb{P}(W_\theta^{(n)} > cn^{-1/2}\sum_{k=1}^n \theta_k^{(n)})$, as opposed to the desired asymptotics for $\mathbb{P}(W_\theta^{(n)} > c)$. Furthermore, Book does not establish an LDP or identify a rate function.

In a more recent line of work, consider [14], where on their p. 932, their λ and ν correspond to our n and k, respectively. For $Z \sim N(0,1)$, we have the following correspondence:

$$\begin{aligned} a_j(n) &= \mathbf{1}_{\{j\le n-1\}} \frac{1}{n}\sqrt{n}\theta_j^{(n)}, \quad j,n\in\mathbb{N}; \\ \sum_{j=0}^{\infty} a_j(n)^k &\approx \sum_{j=0}^{n-1} \frac{1}{n^k} z_j^k \overset{(\varsigma\text{-a.e.})}{\approx} \frac{\mathbb{E}[|Z|^k]}{n^{k-1}} \doteq \frac{a_k}{n^{k-1}}, \quad k,n\in\mathbb{N}; \\ \phi(n) &= n, \quad n\in\mathbb{N}. \end{aligned}$$

Suppose that the sequence $(a_k)_{k\in\mathbb{N}}$ (which depends on the particular choice of weights $a_j(n)$, $j,n\in\mathbb{N}$) satisfies the following condition (from p. 932 of [14]):

$$\lim_{k\to\infty} |a_k|^{1/k} < \infty. \tag{10}$$

The main result of [14] is that for a sequence of i.i.d. random variables $(X_k)_{k\in\mathbb{N}}$ with cumulants $(c_k)_{k\in\mathbb{N}}$, if condition (10) holds, then the sequence of weighted means $\frac{1}{n}\sum_{j=1}^n a_j(n)X_j$, $n\in\mathbb{N}$, satisfies an LDP with rate function χ^*, the Legendre transform of $\chi(t) \doteq \sum_{k=2}^{\infty} \frac{a_k c_k}{k!} t^k$. However, the finiteness condition (10) does not hold in our setting of $a_k = \mathbb{E}[|Z|^k]$, since the following limit is infinite:

$$\lim_{k\to\infty} \mathbb{E}[|Z|^k]^{1/k} = \sqrt{2} \lim_{k\to\infty} \Gamma(\tfrac{k+1}{2})^{1/k} = \infty.$$

Therefore, the weighted mean LDP of [14] does not apply in our setting.

Yet more recently, [16] proves an LDP for weighted empirical means similar to (5), except with weights that are uniformly bounded (in n). Our results correspond to *unbounded* weights $\sqrt{n}\theta_i^{(n)}$ which are not covered by their results. Similarly, [6] proves an LDP for empirical means of certain bounded functionals, which again fails to apply to our unbounded weights.

In the context of information theory, [8] states an LDP for sums of the form $\frac{1}{n}\sum_{i=1}^n \rho(x_i, Y_i)$, where $(x_i)_{i\in\mathbb{N}}$ are "weights," $(Y_i)_{i\in\mathbb{N}}$ is a sequence of random variables satisfying certain mixing properties, and $\rho : \mathcal{X} \times \mathcal{Y} \to \mathbb{R}_+$ for Polish spaces $\mathcal{X}$ and $\mathcal{Y}$. The LDP is stated in the form of a *generalized asymptotic equipartition property* for "distortion measures." However, note that ρ is assumed to be nonnegative, so a function like $\rho(x, y) = xy$ (corresponding to projections) does not fit within the setting of [8]. Moreover, their weights $(x_i)_{i\in\mathbb{N}}$ are assumed to be a realization of a stationary ergodic process, which is not the case for our weights $\sqrt{n}\theta^{(n)}$ that are drawn from the scaled sphere $\sqrt{n}\mathbb{S}^{n-1}$. This lends our work a geometric rather than information-theoretic interpretation.

The paper [12], co-authored by the first and third authors of this work, also analyzes weighted sums of i.i.d. random variables, but there the emphasis is on sums of subexponential random variables, rather than the weights themselves.

The most closely related work to our own is the recent work of [4], which gives strong large deviations (i.e., refined asymptotics) for weighted sums of i.i.d. random variables and i.i.d. weights, conditioned on the weights. Our weights $\sqrt{n}\theta_i^{(n)}$ are not i.i.d., but in Sects. 3.1 and 3.2, we prove that Theorem 2 can be reduced to an LDP for the sequence $(\widehat{W}_z^{(n)})_{n\in\mathbb{N}}$ defined in (12), which is an i.i.d. weighted sum, conditional on given weights. With some additional calculations from this point, the rate function $\mathbb{I}_\sigma$ of Theorem 2 could then be deduced from the conditional LDP of [4], stated in their Theorem 1.6 with rate function defined in their Eq. (1.13). Note that condition (iii) of their Theorem 1.6 has two parts, but our integrability condition (H2) corresponds only to their first part; in fact, it follows from Lemma 4 that their second part follows from our condition (15), which is weaker than (H2), and thus, need not be assumed separately. Moreover, our research (completed independently) differs due to our emphasis on a geometric point of view; as a consequence, we can explicitly identify a rate function $\mathbb{I}_\sigma$ and highlight the atypical position occupied by Cramér's theorem.

Lastly, the method we use is a simplification of those developed in a companion paper [13], where we consider normalized projections of certain *non*-product measures, as well as projections in random directions.

3 The σ-Almost Everywhere LDP

3.1 The Surface Measure on $\mathbb{S}^{n-1}$

In this section, we recall a convenient representation for a random vector distributed according to the surface measure on $\mathbb{S}^{n-1}$, in order to obtain (13), which reduces σ-a.e. statements into more tractable statements about Gaussian random variables. Let $\mathbb{A} \doteq \prod_{n\in\mathbb{N}} \mathbb{R}^n$ denote the space of infinite triangular arrays. That is, $z \in \mathbb{A}$ is of the form $z = (z^{(1)}, z^{(2)}, \dots)$ where $z^{(n)} \in \mathbb{R}^n$ for all $n \in \mathbb{N}$. Let $\mathcal{R} : \mathbb{A} \to \mathbb{A}$ be the map such that for $z \in \mathbb{A}$, the nth row of $\mathcal{R}(z)$ is

$$[\mathcal{R}(z)]^{(n)} \doteq \frac{z^{(n)}}{\|z^{(n)}\|_n}.$$

Let $\bar{\pi}_n : \mathbb{A} \to \mathbb{R}^n$ denote the nth row map such that $\bar{\pi}_n(z) = z^{(n)}$. Let ν denote the Gaussian measure on $\mathbb{R}$, and let $\nu^{\otimes n}$ denote the standard Gaussian measure on $\mathbb{R}^n$.

Lemma 2 *If $\zeta \in \mathcal{P}(\mathbb{A})$ is such that*

$$\zeta \circ \bar{\pi}_n^{-1} = \nu^{\otimes n}, \quad n \in \mathbb{N}, \tag{11}$$

then $\sigma \doteq \zeta \circ \mathcal{R}^{-1}$ satisfies (H1). *Conversely, if $\sigma \in \mathcal{P}(\mathbb{S})$ satisfies* (H1), *then there exists some $\zeta \in \mathcal{P}(\mathbb{A})$ satisfying* (11) *such that $\sigma = \zeta \circ \mathcal{R}^{-1}$.*

Proof Both results are merely a restatement of the well known fact that if $Z^{(n)}$ has the n-dimensional standard Gaussian distribution, then $Z^{(n)}/\|Z^{(n)}\|_n$ is uniformly distributed on the unit sphere $\mathbb{S}^{n-1}$, and independent of $\|Z^{(n)}\|_n$. □

Note that Lemma 2 states that for any given $\sigma \in \mathcal{P}(\mathbb{S})$, we can find a corresponding $\zeta \in \mathcal{P}(\mathbb{A})$. Fix such a pair (σ, ζ). Now, for $z \in \mathbb{A}$, define

$$\widehat{W}_z^{(n)} \doteq \frac{1}{n} \sum_{i=1}^{n} X_i \, z_i^{(n)}. \tag{12}$$

Then, given $W_\theta^{(n)}$ as defined in (5), and any good rate function $\mathbb{I} : \mathbb{R} \to [0, \infty]$, Lemma 2 implies that

$$\begin{aligned} &\sigma\left(\theta \in \mathbb{S} : (W_\theta^{(n)})_{n\in\mathbb{N}} \text{ satisfies an LDP with good rate function } \mathbb{I}\right) \\ &\quad = \zeta\left(z \in \mathbb{A} : (\tfrac{\sqrt{n}}{\|z^{(n)}\|_n} \widehat{W}_z^{(n)})_{n\in\mathbb{N}} \text{ satisfies an LDP with good rate function } \mathbb{I}\right). \end{aligned} \tag{13}$$

In addition, Lemma 2 yields a large class of examples of σ satisfying (H1), constructed via ζ satisfying (11). We specify two such examples below.

Example 2 (a) Consider the completely independent case, where the elements $Z_i^{(n)}$, $i = 1, \ldots, n$, $n \in \mathbb{N}$, are all independent; then the law of $\mathcal{R}(Z)$ is the product measure $\sigma = \bigotimes_{n \in \mathbb{N}} \sigma_{n-1}$, where each row $\theta^{(n)}$ of θ is independent under σ. As previously noted, the tail sigma-algebra $\mathcal{T}$ induced by the rows (defined in (9)), is trivial in this case due to the Kolmogorov zero-one law.

(b) Alternatively, consider the following highly dependent case: let $\zeta \in \mathcal{P}(\mathbb{A})$ satisfy (11) such that for ζ-a.e. $z \in \mathbb{A}$, we have $z_i^{(n)} = z_i^{(m)}$ for all $i \in \mathbb{N}$ and $m, n \geq i$ (i.e., constant within columns). Then, let $\sigma = \zeta \circ \mathcal{R}^{-1}$, so that σ satisfies (H1) by Lemma 2. In this case, there is strong dependence across rows which precludes a claim regarding triviality of the tail sigma-algebra $\mathcal{T}$ induced by the rows. In fact, consider the event

$$A \doteq \left\{\theta \in \mathbb{S} : \lim_{n \to \infty} \sqrt{n}\theta_1^{(n)} > 0\right\}$$

Note that A is measurable with respect to $\mathcal{T}$. However, we also have due to the strong law of large numbers (ζ-a.e., as stated precisely in (14)),

$$\sigma(A) = \zeta\left(z \in \mathbb{A} : \lim_{n \to \infty} \sqrt{n}\, z_1^{(n)}/\|z^{(n)}\|_{n,2} > 0\right) = \zeta\left(z \in \mathbb{A} : z_1^{(1)} > 0\right) = \frac{1}{2}.$$

That is, $\mathcal{T}$ is non-trivial, and so $\mathbb{I}_\sigma$ cannot *a priori* be declared as σ-a.e. constant through a simple analysis of the tail sigma-algebra.

Remark 3 We assume the condition (H1) not in an attempt to be as general as possible, but rather to point out that the universality of the rate function is a genuinely interesting phenomenon. Specifically, if we only consider the independent case of Example 2(a), then the fact that $\mathbb{I}_\sigma$ is "universal" (in that it does not depend on θ) is a consequence of the fact that the tail sigma-algebra $\mathcal{T}$ is trivial. However, Example 2(b) shows that universality of the rate function is a more general phenomenon that holds even when $\mathcal{T}$ is non-trivial. The condition (H1) only imposes constraints on the "marginal" distribution of the nth row of the array θ, and imposes no restrictions on the dependence across different rows $\theta^{(n)}$, $n \in \mathbb{N}$. In fact, for $Z \sim \zeta$ satisfying (11), the elements of Z need not even be *jointly* Gaussian in order for the law of $\mathcal{R}(Z)$ to satisfy (H1).

3.2 Exponential Equivalence

As a consequence of Lemma 2 and the equality in (13), we can replace σ-a.e. statements about $W_\theta^{(n)}$, $n \in \mathbb{N}$, with ζ-a.e. statements about $(\sqrt{n}/\|z^{(n)}\|_n)\widehat{W}_z^{(n)}$, $n \in \mathbb{N}$. In this section, we go further and explain why in the large deviations setting, we can ignore the contribution of the multiplicative factor $\sqrt{n}/\|z^{(n)}\|_n$. That is, we show that such a factor yields an exponentially equivalent sequence, defined as follows.

Definition 2 Let $(\xi_n)_{n\in\mathbb{N}}$ and $(\tilde{\xi}_n)_{n\in\mathbb{N}}$ be two sequences of $\mathbb{R}$-valued random variables such that for all $\delta > 0$,

$$\limsup_{n\to\infty} \frac{1}{n} \log \mathbb{P}(|\xi_n - \tilde{\xi}_n| > \delta) = -\infty;$$

then $(\xi_n)_{n\in\mathbb{N}}$ and $(\tilde{\xi}_n)_{n\in\mathbb{N}}$ are said to be *exponentially equivalent*.

Proposition 4 ([9]) *If $(\xi_n)_{n\in\mathbb{N}}$ is a sequence of random variables that satisfies an LDP with good rate function $\mathbb{I}$, and $(\tilde{\xi}_n)_{n\in\mathbb{N}}$ is another sequence that is exponentially equivalent to $(\xi_n)_{n\in\mathbb{N}}$, then $(\tilde{\xi}_n)_{n\in\mathbb{N}}$ satisfies an LDP with good rate function $\mathbb{I}$.*

Lemma 3 *Let $(\xi_n)_{n\in\mathbb{N}}$ be a sequence of random variables that satisfies an LDP with a good rate function $\mathbb{I}$. Let $(a_n)_{n\in\mathbb{N}}$ be a deterministic sequence such that $a_n \to 1$ as $n \to \infty$, and let $(\tilde{\xi}_n)_{n\in\mathbb{N}}$ be another sequence defined by:*

$$\tilde{\xi}_n = a_n \xi_n, \quad n \in \mathbb{N}.$$

If $\mathbb{I}$ is quasiconvex—that is, if the set $\{x \in \mathbb{R} : \mathbb{I}(x) \in (-\infty, c)\}$ is convex for all $c \in \mathbb{R}$—then $(\xi_n)_{n\in\mathbb{N}}$ and $(\tilde{\xi}_n)_{n\in\mathbb{N}}$ are exponentially equivalent.

Proof For $\varepsilon > 0$, let $N_\varepsilon < \infty$ be such that for all $n \geq N_\varepsilon$, we have $|1 - a_n| < \varepsilon$. For $n \geq N_\varepsilon$ and any $\delta > 0$,

$$|\tilde{\xi}_n - \xi_n| \geq \delta \quad \Leftrightarrow \quad |\xi_n| \cdot |1 - a_n| \geq \delta \quad \Rightarrow \quad |\xi_n| \geq \tfrac{\delta}{\varepsilon}.$$

Because $\mathbb{I}$ is lower semicontinuous and has compact level sets, it achieves its global minimum at some (not necessarily unique) $\bar{x} \in \mathbb{R}$. Fix $\delta > 0$ and let $\varepsilon > 0$ be small enough such that $|\bar{x}| < \frac{\delta}{\varepsilon}$. Then,

$$\begin{aligned}
\limsup_{n\to\infty} \frac{1}{n} \log \mathbb{P}(|\tilde{\xi}_n - \xi_n| > \delta) &\leq \limsup_{n\to\infty} \frac{1}{n} \log \mathbb{P}(|\xi_n| \geq \tfrac{\delta}{\varepsilon}) \\
&\leq - \inf_{|x|\geq\delta/\varepsilon} \mathbb{I}(x) \\
&= - \min\left[\mathbb{I}(\tfrac{\delta}{\varepsilon}),\ \mathbb{I}(-\tfrac{\delta}{\varepsilon})\right].
\end{aligned}$$

The second inequality follows from the LDP for $(\xi_n)_{n\in\mathbb{N}}$. The last equality follows from the fact that if a quasiconvex function has a global minimizer $\bar{x}$, then it is non-increasing for $x < \bar{x}$, and non-decreasing for $x > \bar{x}$ [15, Lemma 1]. Hence, since the rate function $\mathbb{I}$ is quasiconvex and has a global minimizer $\bar{x}$ which satisfies $|\bar{x}| < \delta/\varepsilon$, it follows that if $x \geq \delta/\varepsilon$ (resp., $x \leq -\delta/\varepsilon$), then we have $\mathbb{I}(x) \geq \mathbb{I}(\delta/\varepsilon)$ (resp., $\mathbb{I}(x) \geq \mathbb{I}(-\delta/\varepsilon)$). Lastly, take the limit as $\varepsilon \to 0$, and use the compactness of the level sets of $\mathbb{I}$ to conclude that $\mathbb{I}(\frac{\delta}{\varepsilon}) \to +\infty$ and $\mathbb{I}(-\frac{\delta}{\varepsilon}) \to +\infty$. This proves the required exponential equivalence. □

Fix ζ satisfying (11). Due to the strong law of large numbers, we have that for ζ-a.e. $z \in \mathbb{A}$,

$$\frac{\sqrt{n}}{\|z^{(n)}\|_n} = \left(\frac{1}{n}\sum_{i=1}^{n}(z_i^{(n)})^2\right)^{-1/2} \xrightarrow{n\to\infty} 1. \tag{14}$$

Thus, we are in a prime position to apply Lemma 3, which motivates the analysis of an LDP for $(\widehat{W}_z^{(n)})_{n\in\mathbb{N}}$.

3.3 Proof of the LDP for $(W_\theta^{(n)})_{n\in\mathbb{N}}$

We aim to prove an LDP for the sequence $(\widehat{W}_z^{(n)})_{n\in\mathbb{N}}$; that is, an LDP for sums of independent but not identically distributed random variables (where the lack of identical distribution comes from the inhomogeneous weights $z_i^{(n)}$ within the sum). The Gärtner–Ellis theorem (recalled below) is well suited for such an LDP.

Theorem 4 (Gärtner-Ellis) *Let $(\xi_n)_{n\in\mathbb{N}}$ be a sequence of $\mathbb{R}$-valued random variables. Suppose that the* limit log mgf $\bar{\Lambda} : \mathbb{R} \to [0,\infty)$ *defined by*

$$\bar{\Lambda}(t) \doteq \lim_{n\to\infty} \frac{1}{n} \log \mathbb{E}[e^{tn\xi_n}]$$

is finite and differentiable at all $t \in \mathbb{R}$. Then $(\xi_n)_{n\in\mathbb{N}}$ satisfies an LDP with the convex good rate function $\bar{\Lambda}^*$, the Legendre transform of $\bar{\Lambda}$.

For a proof of Theorem 4, we refer to [10, Theorem V.6], which also includes a more general version of the Gärtner–Ellis theorem that applies even if $\bar{\Lambda}$ is finite for only *some* $t \in \mathbb{R}$ (under mild additional conditions).

The following lemma establishes a property of Ψ that will be used in the application of the Gärtner–Ellis theorem.

Lemma 4 *Suppose that*

$$\forall t \in \mathbb{R}, \quad \int_{\mathbb{R}} |\Lambda(tu)|\, \nu(du) < \infty. \tag{15}$$

Then, the function Ψ of (7) is differentiable on $\mathbb{R}$.

Proof For each $t \in \mathbb{R}$, differentiability of Ψ at t follows from the differentiability of $t \mapsto \Lambda(tu)$ for all $u \in \mathbb{R}$, and an application of the dominated convergence theorem with the dominating function

$$g_t(u) \doteq |\Lambda'((t-1)u)u| + |\Lambda'((t+1)u)u|, \quad u \in \mathbb{R}.$$

Indeed, fix $t \in \mathbb{R}$ and for each $\delta \in (-1, 1)$ and $u \in \mathbb{R}$, define the difference quotient $R_{t,\delta}(u) \doteq [\Lambda((t+\delta)u) - \Lambda(tu)]/\delta$. Then,

$$|R_{t,\delta}(u)| \leq \sup\left\{|\Lambda'((t+\alpha)u)u| : \alpha \in [-1, 1]\right\} \leq g(u),$$

where the last inequality uses the fact that $t \mapsto u\,\Lambda'(tu)$ is monotone. To show that g_t is integrable, first note that the convexity of Λ implies that for $u, s \in \mathbb{R}$,

$$\Lambda(su) - \Lambda(0) \leq \Lambda'(su)\,su \leq \Lambda(2su) - \Lambda(su),$$

and hence,

$$|\Lambda'(su)su| \leq |\Lambda(0)| + |\Lambda(su)| + |\Lambda(2su)|.$$

Since, by the assumption (15), for every $s \in \mathbb{R}$, the right-hand side is an integrable function of u, it follows that g_t is also integrable for every $t \in \mathbb{R}$. □

Proof of Theorem 2 Due to Lemma 2 (in particular, its consequence, (13)), it suffices to prove a ζ-a.e. LDP for the sequence $((\sqrt{n}/\|z\|_n)\widehat{W}_z^{(n)})_{n\in\mathbb{N}}$, where $\widehat{W}_z^{(n)}$ is defined as in (12). Due to Lemma 3 and the limit (14), it suffices to prove a ζ-a.e. LDP for $(\widehat{W}_z^{(n)})_{n\in\mathbb{N}}$. To this end, we consider the Gärtner–Ellis limit log mgf for the sequence $(\widehat{W}_z^{(n)})_{n\in\mathbb{N}}$. For every $n \in \mathbb{N}$ and $t \in \mathbb{R}$, we have due to the independence of X_i, $i = 1, \ldots, n$,

$$\Lambda_{n,z}(t) \doteq \frac{1}{n}\log \mathbb{E}\left[\exp\left(tn\widehat{W}_z^{(n)}\right)\right] = \frac{1}{n}\log\prod_{i=1}^{n}\mathbb{E}\left[\exp\left(tX_i z_i^{(n)}\right)\right] = \frac{1}{n}\sum_{i=1}^{n}\Lambda(tz_i^{(n)}). \tag{16}$$

We first claim that for ζ-a.e. $z \in \mathbb{A}$, the Gärtner–Ellis limit log mgf, the limit of (16), satisfies, for each $t \in \mathbb{R}$,

$$\lim_{n\to\infty}\Lambda_{n,z}(t) = \int_{\mathbb{R}}\Lambda(tu)\nu(du) = \Psi(t), \tag{17}$$

with Ψ as defined in (7).

We proceed by proving the following modified claim (obtained by interchanging the quantifiers in our original claim): for each $t \in \mathbb{R}$, for ζ-a.e. $z \in \mathbb{A}$, the expression (17) holds. Note that if z were an i.i.d. sequence instead of a triangular array, our modified claim would follow from the usual strong law of large numbers. However, the strong LLN does not necessarily extend to empirical means of rows of i.i.d. random variables in a triangular array (see, e.g., [18, Example 5.41]). On the other hand, if the common distribution of the i.i.d. elements (in our case, each of the random variables $\Lambda(tz_i^{(n)})$, $i = 1, \ldots, n$, $n \in \mathbb{N}$) has finite fourth moment, then the strong LLN follows from a standard weak LLN and Borel–Cantelli argument [18, p. 113, (i)]. Due to our assumption (H2), it follows that for all $t \in \mathbb{R}$, for ζ-a.e. $z \in \mathbb{A}$, the limit (17) holds.

Next, we aim to interchange the quantifiers to establish the original claim. Note that for each $n \in \mathbb{N}$, $\Lambda_{n,z}$ of (16) is a convex function (since it is the sum of convex functions). Now, let $T \subset \mathbb{R}$ be countable and dense. Then, it follows from countable additivity that for ζ-a.e. $z \in \mathbb{A}$, the convex functions $\Lambda_{n,z}(t)$ converge pointwise as $n \to \infty$ to $\Psi(t)$, for all t in the dense subset $T \subset \mathbb{R}$. Hence, the convex analytic considerations of [17, Theorem 10.8] imply that the pointwise convergence of $\Lambda_{n,z}(t)$ to $\Psi(t)$ holds for all $t \in \mathbb{R}$. That is, for ζ-a.e. $z \in \mathbb{A}$, for all $t \in \mathbb{R}$, the limit (17) holds, proving our original claim.

Since (H2) holds, $\Psi(t) < \infty$ for all $t \in \mathbb{R}$ and, because (15) follows trivially from (H2), Lemma 4 implies that Ψ is differentiable on $\mathbb{R}$. Therefore, by the Gärtner–Ellis Theorem (Theorem 4), for ζ-a.e. $z \in \mathbb{A}$, the sequence $(\widehat{W}_z^{(n)})_{n\in\mathbb{N}}$ satisfies an LDP with good rate function Ψ^*. □

4 Atypicality

In this section, we compare the rate function $\mathbb{I}_\sigma$ with the Cramér rate function $\mathbb{I}_\iota$. We first use Jensen's inequality to compare the associated log mgfs Ψ and Λ.

Lemma 5 *Assume* (H3), *and let* Λ *and* Ψ *be defined as in* (3) *and* (7), *respectively.*

(a) *If* $\Lambda \circ \sqrt{\cdot}$ *is concave on* $\mathbb{R}_+$, *then* $\Psi(t) \leq \Lambda(t)$ *for all* $t \in \mathbb{R}$.
(b) *If* $\Lambda \circ \sqrt{\cdot}$ *is convex on* $\mathbb{R}_+$, *then* $\Psi(t) \geq \Lambda(t)$ *for all* $t \in \mathbb{R}$.
(c) *If* $\Lambda \circ \sqrt{\cdot}$ *is concave or convex, but not linear, on* $\mathbb{R}_+$, *then* $\Lambda(t) = \Psi(t)$ *if and only if* $t = 0$.

Proof We begin with part (a). Let ν be the standard Gaussian distribution, and let $Z \sim \nu$ be a standard Gaussian random variable. Then, for all $t \in \mathbb{R}$, we have

$$\begin{aligned} \Psi(t) &= \mathbb{E}[\Lambda(tZ)] \\ \text{(symmetry)} \quad &= \mathbb{E}\left[\Lambda\left((t^2Z^2)^{1/2}\right)\right] \\ \text{(Jensen)} \quad &\leq \Lambda\left(\mathbb{E}[t^2Z^2]^{1/2}\right) \\ &= \Lambda(t). \end{aligned}$$

Similar calculations can be used to establish part (b). As for part (c), recall that in Jensen's inequality, equality holds if and only if either: (i) $\Lambda \circ \sqrt{\cdot}$ is linear; or (ii) the underlying random variable is almost surely constant. Note that (i) is not the case by assumption. As for (ii), this holds if and only if t^2Z^2 is almost surely constant, which is the case if and only if $t = 0$. □

Before we prove the theorem, we recall some basic facts about the log mgf of $X_1 \sim \gamma$. Let the *domain* of a function $f : \mathbb{R} \to \mathbb{R}$ be the set $D_f \doteq \{x \in \mathbb{R} : f(x) < \infty\}$. For a set $D \subset \mathbb{R}$, let D° denote the interior of D.

Lemma 6 *Let $\Lambda(t) = \log \mathbb{E}[e^{tX_1}]$ be the log mgf of some random variable X_1. Then,*

1. *Λ is lower semicontinuous;*
2. *Λ is smooth in D°_Λ;*
3. *Λ is convex.*

Furthermore, if X_1 is non-degenerate (i.e., not a.s. constant), then

4. *Λ is strictly convex in D°_Λ;*
5. *Λ^* is differentiable in $D^\circ_{\Lambda^*}$;*
6. *for $x \in D^\circ_{\Lambda^*}$, the maximum in the definition of the Legendre transform is uniquely attained—that is, the following quantity is well defined:*

$$t_x \doteq \arg\max\{tx - \Lambda(t)\}. \tag{18}$$

Proof These are mostly standard, but we provide sketches of the proofs. For 1., lower semicontinuity follows from Fatou's lemma. For 2., smoothness follows from interchanging differentiation and expectation. Convexity in 3. and strict convexity in 4. follow from Hölder's inequality. As for 5., it is classical that if a function is lower semicontinuous and strictly convex in the interior of its domain, then its Legendre transform is differentiable in the interior of its domain (see [17, Theorem 26.3]). Lastly, for 6., it is also classical that for $x \in D^\circ_{\Lambda^*}$, we have $t_x = (\Lambda^*)'(x)$ (see [17, Theorem 26.5]). □

Proof of Theorem 3 Assume without loss of generality that X_1 is non-degenerate. If it were degenerate, then due to the symmetry condition (H3), the law of X_1 must be that $\gamma = \delta_0$, in which case $\Lambda = \Psi = 0$. Therefore, $\mathbb{I}_\sigma$ and $\mathbb{I}_\iota$ are both equal to the characteristic function at 0 (which is equal to 0 at $w = 0$ and $+\infty$ for all other w), and the result is trivial.

Suppose $\Lambda \circ \sqrt{\cdot}$ is concave (the convex case is similar, but with inequalities reversed). Due to Lemma 5, we have $\Psi(t) \le \Lambda(t)$ for all $t \in \mathbb{R}$, which due to the definition of the Legendre transform implies that $\mathbb{I}_\sigma(w) = \Psi^*(w) \ge \Lambda^*(w) = \mathbb{I}_\iota(w)$ for all $w \in \mathbb{R}$, thus proving (a) (and (b) for the convex case).

Further assume the stronger condition of (c), that $\Lambda \circ \sqrt{\cdot}$ is concave but not linear. Then, for $w \in \mathbb{R}$ such that $\Lambda^*(w) < \infty$, let t_w be as in (18), which is well defined due to the non-degeneracy condition of Lemma 6. Then,

$$\begin{aligned} \mathbb{I}_\sigma(w) = \Psi^*(w) &\ge t_w w - \Psi(t_w) \\ &\ge t_w w - \Lambda(t_w) \\ &= \Lambda^*(w) = \mathbb{I}_\iota(w). \end{aligned}$$

Due to Lemma 5, the second inequality above is an equality if and only if $t_w = 0$, which occurs if and only if $(\Lambda^*)'(w) = 0$. Note that Λ is symmetric, so Λ^* is also symmetric (by definition of the Legendre transform). Moreover, the smoothness of Λ (see Lemma 6), implies the strict convexity of Λ^* within its domain (see [17, Theorem 26.3]). Thus, $(\Lambda^*)'(w) = 0$ if and only if $w = 0$. This yields the claim of part (c). □

Remark 4 In this paper, we address the "atypical" nature of the directions $\iota^{(n)} = (1, 1, \ldots, 1)$ associated with Cramér's theorem for large deviations of product measures. But in fact, the notions of atypicality and universal rate function extend beyond the product case. In particular, the companion paper [13] establishes LDPs for random projections of random vectors distributed according to the uniform measure on ℓ^p balls, again with a rate function that coincides for σ-a.e. sequence of directions, and the sequence of directions $\iota^{(n)} = (1, 1, \ldots, 1), n \in \mathbb{N}$, can be shown to be atypical in that setting as well.

Acknowledgments NG and KR would like to thank ICERM, Providence, for an invitation to the program "Computational Challenges in Probability," where some of this work was initiated. SSK and KR would also like to thank Microsoft Research New England for their hospitality during the Fall of 2014, when some of this work was completed. SSK was partially supported by a Department of Defense NDSEG fellowship. KR was partially supported by ARO grant W911NF-12-1-0222 and NSF grant DMS 1407504. The authors would like to thank an anonymous referee for helpful feedback on the exposition.

References

1. F. Barthe, A. Koldobsky, Extremal slabs in the cube and the Laplace transform. Adv. Math. **174**(1), 89–114 (2003)
2. S.A. Book, Large deviation probabilities for weighted sums. Ann. Math. Stat. **43**(4), 1221–1234 (1972)
3. S.A. Book, A large deviation theorem for weighted sums. Zeitschrift für Wahrscheinlichkeitstheorie und Verwandte Gebiete **26**(1), 43–49 (1973)
4. A. Bovier, H. Mayer, A conditional strong large deviation result and a functional central limit theorem for the rate function. ALEA Lat. Am. J. Probab. Math. Stat. **12**, 533–550 (2015)
5. H. Chernoff, A measure of asymptotic efficiency for tests of a hypothesis based on the sum of observations. Ann. Math. Stat. **27**, 1–22 (1956)
6. Z. Chi, Stochastic sub-additivity approach to the conditional large deviation principle. Ann. Probab. **29**(3), 1303–1328 (2001)
7. H. Cramér, Sur un nouveau théoréme-limite de la théorie des probabilités. Actualités Scientifiques et Industrielles **736**, 5–23 (1938)
8. A. Dembo, I. Kontoyiannis, Source coding, large deviations, and approximate pattern matching. IEEE Trans. Inf. Theory **48**(6), 1590–1615 (2002)
9. A. Dembo, O. Zeitouni, *Large Deviations Techniques and Applications*, 2 edn. (Springer, Berlin, 1998)
10. F. Den Hollander. *Large Deviations, Fields Institute Monographs*, vol. 14 (American Mathematical Society, Providence, 2008)
11. W. Feller, *An Introduction to Probability Theory and Its Applications*, vol. II (Wiley, New York, 1970)
12. N. Gantert, K. Ramanan, F. Rembart, Large deviations for weighted sums of stretched exponential random variables. Electron. Commun. Probab. **19**, 1–14 (2014)
13. N. Gantert, S.S. Kim, K. Ramanan, Large deviations for random projections of ℓ^p balls. Preprint (2015) arXiv:1512.04988
14. R. Kiesel, U. Stadtmüller, A large deviation principle for weighted sums of independent identically distributed random variables. J. Math. Anal. Appl. **251**(2), 929–939 (2000)
15. D.G. Luenberger, Quasi-convex programming. SIAM J. Appl. Math. **16**(5), 1090–1095 (1968)

16. J. Najim. A Cramér type theorem for weighted random variables. Electron. J. Probab. **7**(4), 1–32 (2002)
17. R.T. Rockafellar, *Convex Analysis* vol. 28 (Princeton University Press, Princeton, 1970)
18. J.P. Romano, A.F. Siegel, *Counterexamples in Probability and Statistics* (CRC Press, Boca Raton, 1986)

Counting and Partition Function Asymptotics for Subordinate Killed Brownian Motion

Sarah Bryant

Abstract We consider the subordinate killed Brownian motion process generated by first killing Brownian motion at some boundary point on a smooth bounded domain then subordinating by a Lévy time-clock. For classes of subordinators satisfying some growth requirements, we establish asymptotic growth for the eigenvalues associated to these processes. Using an abelian argument we are then able to prove first-term asymptotics for the trace of the heat semigroup, or partition function. For $\alpha/2$-stable subordinators we prove second-order term asymptotics of the partition function with constants dependent on volume and surface area of the boundary.

Keywords Subordinate killed Brownian motion · Counting function · Partition function · Asymptotics

Mathematics Subject Classification 60G51

1 Introduction and Preliminaries

In his celebrated 1965 paper "Can One Hear the Shape of a Drum?" Marc Kac [7] discussed a beautifully phrased problem with deep mathematical implications. His paper outlined connections between the eigenvalues for the Dirichlet Laplacian $-\Delta|_D$ on a domain and the geometry of the underlying domain. By Dirichlet Laplacian we mean the Laplacian with zero boundary condition. Two domains are said to be isospectral if they have the same eigenvalues, and isometric if they are congruent in

Sarah presented this work at the special session "Research from the Cutting EDGE." Sarah is the 28th member of the EDGE (Enhancing Diversity in Graduate Education) Program to receive her doctorate in Mathematics. The goal of EDGE is to strengthen the ability of women to successfully complete their graduate programs in the Mathematical Sciences. Please see the preface for more information about the EDGE Program and its founders.

S. Bryant (✉)
Department of Mathematics, Shippensburg University, 1871 Old Main Drive, Shippensburg, PA 17257, USA
e-mail: sarahbryant456@gmail.com

G. Letzter et al. (eds.), *Advances in the Mathematical Sciences*, Association for Women in Mathematics Series 6, DOI 10.1007/978-3-319-34139-2_12

the sense of Euclidean geometry. The title is then interpreted as follows: if one has perfect pitch and can hear the fundamental tones (eigenvalues) of a drum (domain), then can one hear the shape (geometry) of that drum? Or, succinctly, are there two non isometric domains that are isospectral? In the following years, for many domains the answer has been answered in the affirmative (we direct the interested reader [6, 10], among others).

In studying this problem, it is common to consider the information gained by the asymptotics of the counting function rather than the individual eigenvalues themselves. Our paper contributes to the existing literature by leveraging the relationship between the spectrum of killed Brownian motion and subordinate killed Brownian motion to prove asymptotic behavior of the counting function and a related probabilistic quantity, the partition function, for a large class of subordinate killed processes on bounded domains. The results prove that for the subordinate killed Brownian motion and its associated operator we "hear" the volume and surface area of the domain the same as in the Brownian motion case. Our result for second-term asymptotics is proven only for the $\alpha/2$-stable subordinators and certain smooth domains, and so is limited in its generality, but does prove constants dependent on the domain in terms of multiples of those in the corresponding result in the Brownian motion case. Future work will hopefully generalize in two directions, for larger classes of subordinators and for more general domains.

We will now introduce some definitions and enough background potential theory to present the main results.

Definition 1 (Subordinator) An increasing, one-dimensional Lévy process $T = (T_t : t \geq 0)$ taking values in $[0, \infty)$ with $T_0 = 0$ is called a *subordinator*. The Laplace transform of the law of T is given by $\mathbb{E}[e^{-\lambda T_t}] = e^{-t\varphi(\lambda)}$ for $\lambda > 0$. The function φ is called the *Laplace exponent* of T.

The Laplace exponent $\varphi : (0, \infty) \to \mathbb{R}$ can be written in the form

$$\varphi(\lambda) = a + b\lambda + \int_0^\infty (1 - e^{-\lambda t})\mu(dt). \tag{1}$$

Laplace exponents φ are known to be Bernstein functions satisfying $\varphi \in C^\infty(0, \infty)$ with $\varphi \geq 0$ and $(-1)^n D^n(\varphi) \leq 0$, for all $n \in \mathbb{N}$ (see [1], p. 53).

Let D be a bounded domain in $\mathbb{R}^d$. Given a Lévy process X in D we write $X^D = (X_t^D : t \geq 0)$ for the process killed upon leaving D and $X_\varphi = (X_{T_t} : t \geq 0)$ for the process subordinated by an independent subordinator T with Laplace exponent φ, as given above. If we first kill X in D and then subordinate, the resulting process is called a subordinate killed process. When T has Laplace exponent φ we denote this process by $(X^D)_\varphi = ((X^D)_{T_t} : t \geq 0)$. We will be considering this process in the case that X is Brownian motion, D is a bounded domain in $\mathbb{R}^d$ and φ satisfies some growth requirements.

For the sake of completion, we note here that the killed subordinate processes (subordinate first, then kill) have a much more complicated spectral theory than the subordinate killed processes. These processes are closely related to the theory of nonlocal PDEs with Dirichlet condition. When the underlying process is Brownian motion and T is an $\alpha/2$-stable subordinator the killed subordinate process is the rotationally invariant α-stable process in D. In [12], the authors include a nice discussion of the differences between subordinate killed and killed subordinate processes with regard to their trajectories and potential theory. We note that the first-term asymptotics of the partition function for α-stable processes on bounded domains was proven in [4]. The second-term asymptotics for the α-stable processes were proven for domains with certain smoothness conditions in $\mathbb{R}^d$ by Bañuelos and Kulczycki in [2]. The results in those cases rely on different approaches than those presented in this note, due to the more complicated nature of their spectral properties.

In contrast, description of the spectrum for the subordinate killed processes is quite straightforward. When D is a bounded domain $\mathbb{R}^d$ the spectrum for the generator of killed Brownian motion, that is, the operator $-\frac{1}{2}\Delta|_D$, is known to be discrete and may be represented by $\{\lambda_j\}_{j\geq 1}$. It is known that $\{\varphi(\lambda_j)\}_{j\geq 1}$ are then the eigenvalues for the subordinated killed process (see, for example, [11]) and the eigenfunctions of the two processes are identical. In other words, the subordinate killed process is generated by the operator $-\varphi(-\frac{1}{2}\Delta|_D)$.

Definition 2 **(Counting Function)** For a killed Lévy process X on a bounded domain $D \subset \mathbb{R}^d$ with associated spectrum of the infinitesimal operator given by $\{\lambda_j\}_{j\geq 1}$, we may define the counting function $N_D(\lambda) = \#\{j : \lambda_j \leq \lambda\}$.

The famous Weyl's theorem regarding asymptotics of the counting function (in the d-dimensional Brownian motion case) is given by

$$\lim_{\lambda\to\infty} \lambda^{-d/2} N_D(\lambda) = c_d|D|, \tag{2}$$

where $c_d = \frac{\omega_d}{(4\pi)^d}$ is a constant dependent only on dimension and $|D|$ represents the d-dimensional Lebesgue measure (volume). This is sometimes written as

$$N_D(\lambda) \sim c_d|D|\lambda^{d/2}, \qquad \lambda \to \infty. \tag{3}$$

A probabilistic connection to the counting function is found in the related partition function. We formulate the definition in terms of the subordinate killed Brownian motion.

Definition 3 **(Partition Function)** Let $p_D^{\varphi}(t, x, y)$ represent the transition density for the subordinate killed Brownian motion in D. This density may be written

$$p_D^{\varphi}(t, x, y) = \int_0^{\infty} p_D(t, x, y)\mu_t(ds),$$

where p_D is the transition density for Brownian motion in D and μ_t is the density of the subordinator. The partition function, also known as the trace of the heat semigroup, is defined as $Z_D^{\varphi}(t) = \int_D p_D^{\varphi}(t, x, x)dx$.

The remainder of the paper follows the following outline: in Sect. 2 we prove first-order counting function asymptotics for the subordinate killed Brownian motion in a domain $D \subset \mathbb{R}^d$ with finite volume. In Sect. 3 we prove first-order asymptotics for $Z_D^{\varphi}(t)$ under some assumptions on D and φ. In Sect. 4 we prove second-order asymptotics for $Z_D^{\varphi}(t)$ for the $\alpha/2$-stable subordinator.

2 First- and Second-Order Asymptotics of the Counting Function

We will first state a few results regarding the behavior of the Laplace exponent of the subordinator. We assume throughout that with D of finite volume, the subordinate killed process has discrete associated spectrum $\{\mu_j\}_{j\geq 1} = \{\varphi(\lambda_j)\}_{j\geq 1}$, where $\{\lambda_j\}_{j\geq 1}$ is the spectrum associated with the killed Brownian motion in bounded domain D. Observe that since φ is non-decreasing, when φ has a well-defined inverse $\psi : (0, \infty) \to \mathbb{R}$ the inverse ψ is also non-decreasing. Hence we may relate N_D and N_D^{φ} in the following manner.

$$\begin{aligned} N_D^{\varphi}(\lambda) &= \#\{j : \varphi(\lambda_j) \leq \lambda\} \\ &= \#\{j : \lambda_j \leq \psi(\lambda)\} \\ &= N_D(\psi(\lambda)). \end{aligned} \tag{4}$$

We claim that the above equalities hold for a large class of Laplace exponents, namely those which are not bounded. Notice in the case that $\lim\limits_{\lambda\to\infty} \varphi(\lambda) = C < \infty$, then there exists $\lambda_0 > 0$ such that $N_D^{\varphi}(\lambda) = \infty$ for all $\lambda > \lambda_0$. Hence consideration of growth behavior of $N_D^{\varphi}(\lambda)$ as $\lambda \to \infty$ is invalid.

It follows by the mean value theorem applied to φ' that if $\varphi(a) = \varphi(b)$ for $a < b$ then $\varphi(x) = \varphi(a)$ for all $x \geq a$. In other words, if the function is ever constant, it remains so. Thus if φ does not have a horizontal asymptote, then it is invertible. Moreover, in this case, the inverse ψ has domain $(0, \infty)$.

By the previous, and since $N_D^{\varphi} = N_D \circ \varphi^{-1}$, we have the following results on first- and second-order asymptotics of the subordinate killed Brownian motion.

Theorem 1 *For a bounded domain $D \subset \mathbb{R}^d$ and subordinate killed Brownian motion with unbounded Laplace exponent φ we have*

$$N_D^{\varphi}(\lambda) \sim c_d\,|D|\,[\psi(\lambda)]^{d/2}, \qquad \lambda \to \infty, \tag{5}$$

where $|D|$ is the volume of D and c_d is the same constant, dependent only on dimension, from (2).

Proof Follows from (2) and (4). □

The following second-order asymptotics for λ large hold for many smooth domains. See [5] for more information and references.

$$N_D(\lambda) = c_d|D|\lambda^{d/2} - \hat{c}_d|\partial D|\lambda^{(d-1)/2} + o(\lambda^{(d-1)/2}), \tag{6}$$

where $|D|$ is the volume of D and $|\partial D|$ its surface area (or $(d-1)$-dimensional Lebesgue measure of its boundary). It follows by (2) and (4) that for these domains we have the corresponding asymptotics for the counting function for the subordinate killed process.

Theorem 2 *On domains where (6) holds and Laplace exponent φ is unbounded, we have*

$$N_D^{\varphi}(\lambda) = c_d|D|[\psi(\lambda)]^{d/2} - \hat{c}_d|\partial D|[\psi(\lambda)]^{(d-1)/2} + o([\psi(\lambda)]^{(d-1)/2}). \tag{7}$$

Notice constants in (7) remain the same as in (6).

3 First-Order Asymptotics of the Partition Function

In this section we use results from Sect. 2, along with the Karamata Tauberian theorem, to prove the asymptotics for the partition function.

Definition 4 (Regularly Varying) A function f is regularly varying at infinity with index $\rho \geq 0$ if $\frac{f(\lambda x)}{f(x)} \to \lambda^{\rho}$ as $x \to \infty$ for all $\lambda > 0$.

When $\rho = 0$ in the above definition, we say f is slowly varying at infinity. Thus a function f is regularly varying at ∞ of index $\rho \geq 0$ if $f(x) = x^{\rho}\ell(x)$, for some function ℓ slowly varying at ∞.

Theorem 3 (Karamata Tauberian Theorem, see [3]**)**
Let $\Gamma(t)$ be the Gamma function, also known as the generalized factorial. Let U be a non-decreasing, right-continuous function on $\mathbb{R}$ with $U(x) = 0$ for all $x < 0$. Define its Laplace–Stieltjes transform

$$\widehat{U}(s) = \int_0^{\infty} e^{-sx} dU(x).$$

If ℓ varies slowly at ∞, $c \geq 0$, and $\rho \geq 0$, then the following are equivalent:

$$U(x) \sim \frac{c}{\Gamma(1+\rho)} x^{\rho}\,\ell(x), \qquad x \to \infty,$$
$$\widehat{U}(s) \sim cs^{-\rho}\,\ell(1/s), \qquad s \to 0+\,.$$

It is well-known that $\widehat{N_D^{\varphi}}(\lambda) = Z_D^{\varphi}(t)$. By applying the abelian part of the Karamata–Tauberian theorem we may now state asymptotics for $Z_D^{\varphi}(t)$.

By Theorem 1, we may apply Theorem 3 to find the asymptotics of $Z_D^{\varphi}(t)$ whenever $(\psi)^{d/2}$ is regularly varying at ∞. The following proposition allows us to write the main theorem with our hypothesis in terms of the Laplace exponent of the subordinator.

Proposition 1 *Given a function φ and its function inverse ψ, the following are equivalent.*

1. *$\varphi(\lambda)$ is regularly varying at ∞ with index $\rho > 0$.*
2. *$\psi(\lambda)$ is regularly varying at ∞ with index $\frac{1}{\rho} > 0$.*
3. *$(\psi(\lambda))^{d/2}$ is regularly varying at ∞ with index $\frac{1}{\rho} \cdot \frac{d}{2} > 0$.*

Proof See, for example, Proposition 1.3.6 in [3]. □

Theorem 4 *For a bounded domain $D \subset \mathbb{R}^d$ and subordinate killed Brownian motion with Laplace exponent φ regularly varying at ∞ with index $\rho \in (0, 1]$, then*

$$Z_D^{\varphi}(t) \sim c_d|D|\,\Gamma\left(1 + \frac{d}{2\rho}\right)\left[\psi\left(\frac{1}{t}\right)\right]^{d/2}, \qquad t \to 0^+.$$

Proof Follows from Theorems 1, 3, and Proposition 1. □

We remark now that the main result of this section allows us to write explicit formulas for partition function asymptotics for a wide class of subordinators, including the following

Example 1 For $\alpha/2$-stable subordinators with $0 < \alpha < 2$, we have $\varphi(\lambda) = \lambda^{\alpha/2}$ unbounded and inverse $\psi(\lambda) = \lambda^{2/\alpha}$. By Theorem 1,

$$N_D^{\varphi}(\lambda) \sim c_d|D|\lambda^{d/\alpha}, \qquad \lambda \to \infty.$$

In this case $\varphi(\lambda)$ is regularly varying at ∞ of index $\alpha/2$, hence by Theorem 4

$$Z_D^{\varphi}(t) \sim c_d|D|\Gamma\left(1 + \frac{d}{\alpha}\right)t^{\frac{-d}{\alpha}}, \qquad t \to 0.$$

Example 2 Similarly, the results apply to relativistic α-stable subordinators, i.e., those characterized by Laplace exponent $\varphi(\lambda) = (\lambda + \gamma)^{\alpha/2} - \gamma^{\alpha/2}$ $0 < \alpha < 2$, for some $\gamma > 0$. The associated inverse is $\psi(\lambda) = (\lambda + \gamma^{\alpha/2})^{2/\alpha} - \gamma$. Since φ is unbounded it follows

$$N_D(\varphi)(\lambda) \sim c_d|D|((\lambda + \gamma^{\alpha/2})^{2/\alpha} - \gamma)^{d/2}, \qquad \lambda \to \infty.$$

It is easily verified that $\varphi(\lambda)$ is regularly varying at ∞ if index $\alpha/2$ and by Theorem 4

$$Z_D^{\varphi}(t) \sim c_d|D|\Gamma(1+\frac{d}{\alpha})\left[\left(\frac{1}{t}+\gamma^{\alpha/2}\right)^{2/\alpha}-\gamma\right]^{d/2}, \qquad t \to 0.$$

4 Second-Order Asymptotics of the Partition Function for $\alpha/2$-Stable Subordinator

The Weyl–Berry conjecture on the second term in the asymptotics of $N_D(\lambda)$ involves the measure of the boundary. The interpretation is that for certain domains we "hear" the perimeter. On many smooth domains D in $\mathbb{R}^d$ (see [8, 9]), the second-order expansions for large λ have been shown to be of the form

$$N_D(\lambda) = c_d|D|\lambda^{d/2} - \hat{c}_d|\partial D|\lambda^{(d-1)/2} + o(\lambda^{(d-1)/2}), \tag{8}$$

where the constants depend only on dimension. On domains where (8) holds, we use the relationship between N_D and Z_D to prove the second-term expansion of the subordinate killed Brownian motion where the subordinator is a stable process.

Theorem 5 *Let T_t be an $\alpha/2$-stable subordinator, independent of d-dimensional Brownian motion B. For the subordinate killed process $X_t = (B^D)_{T_t}$, we have the following second-order expansion of the partition function for small $t > 0$ on the bounded domains D with $|\partial D| < \infty$ where the Weyl–Berry conjecture (6) holds*

$$Z_D^{\varphi}(t) = c_d\Gamma(d/\alpha+1)|D|t^{-d/\alpha} - \hat{c}_d\Gamma((d-1)/\alpha+1)|\partial D|t^{-(d-1)/\alpha} + o(t^{-(d-1)/\alpha}). \tag{9}$$

Proof This result does not follow from Theorem 3, but requires a different approach. We will proceed with an abelian proof, so named because we rely on the Laplace–Stieltjes integral form for Z_D^{φ}. We begin by performing integration by parts. Notice that the boundary term is 0. We then use the expansion (7) for $N_D^X(\lambda)$ with $f(\lambda) = o(\lambda^{(d-1)/\alpha})$. We repeatedly use the formula $\int_0^\infty s^r e^{-s} ds = \Gamma(r+1)$.

$$\begin{aligned} Z_D^{\varphi}(t) &= \int_0^\infty e^{-\lambda t} dN_D^X(\lambda) \\ &= t\int_0^\infty e^{-\lambda t} N_D^X(\lambda) d\lambda \\ &= t\int_0^\infty e^{-\lambda t} c_d|D|\lambda^{d/\alpha} d\lambda - t\int_0^\infty c_d'|\partial D|\lambda^{(d-1)/\alpha} e^{-\lambda t} d\lambda \\ &\qquad + t\int_0^\infty f(\lambda) e^{-\lambda t} d\lambda \\ &= I - II + III. \end{aligned} \tag{10}$$

$$\begin{aligned} I &= t\int_0^\infty e^{-\lambda t} c_d|D|\lambda^{d/\alpha} d\lambda \\ &= t\int_0^\infty e^{-s} c_d|D|(s/t)^{d/\alpha} ds/t \\ &= c_d|D|\Gamma(d/\alpha+1)t^{-d/\alpha}. \end{aligned} \tag{11}$$

A similar argument holds for II.

$$\begin{aligned} II &= t\int_0^\infty e^{-\lambda t} c_d'|\partial D|\lambda^{d-1/\alpha} d\lambda \\ &= c_d'\Gamma(\frac{d-1}{\alpha}+1)|\partial D|t^{-(d-1)/\alpha}. \end{aligned} \tag{12}$$

Let III $= g(t)$. Thus,

$$g(t) = t\int_0^\infty f(\lambda)e^{-\lambda t} d\lambda.$$

It remains to be shown that $g(t) = o(t^{-(d-1)/\alpha})$. That is, given $\varepsilon > 0$ there exists $t_0 = t_0(\varepsilon)$ such that $t^{(d-1)/\alpha} g(t) \le \varepsilon$ for all $t \le t_0$.

Since $f(\lambda) = o(\lambda^{(d-1)/\alpha})$, given $\varepsilon > 0$ we may choose $\lambda_0 = \lambda_0(\varepsilon)$ sufficiently large so that such that $\frac{f(\lambda)}{\lambda^{(d-1)/\alpha}} \le \frac{\varepsilon}{2\Gamma(\frac{d-1}{\alpha}+1)}$ for all $\lambda \ge \lambda_0$.

$$\begin{aligned} t^{\frac{d-1}{\alpha}} g(t) &= t^{\frac{d-1+\alpha}{\alpha}} \int_0^\infty f(\lambda)e^{-\lambda t} d\lambda \\ &= t^{\frac{d-1+\alpha}{\alpha}} \left(\int_0^{\lambda_0} f(\lambda)e^{-\lambda t} d\lambda + \int_{\lambda_0}^\infty f(\lambda)e^{-\lambda t} d\lambda\right) \\ &\le t^{\frac{d-1+\alpha}{\alpha}} \left(\sup_{[0,\lambda_0]} f(\lambda)(1-e^{-\lambda_0 t}) + \frac{\varepsilon}{2\Gamma(\frac{d-1}{\alpha}+1)} \int_{\lambda_0}^\infty \lambda^{\frac{d-1}{\alpha}} e^{-\lambda t} d\lambda\right) \\ &\le t^{\frac{d-1+\alpha}{\alpha}} \left(\sup_{[0,\lambda_0]} f(\lambda) + \frac{\varepsilon}{2\Gamma(\frac{d-1}{\alpha}+1)} t^{-\frac{d-1+\alpha}{\alpha}} \Gamma(\frac{d-1}{\alpha}+1)\right) \\ &= t^{\frac{d-1+\alpha}{\alpha}} \sup_{[0,\lambda_0]} f(\lambda) + \frac{\varepsilon}{2}. \end{aligned}$$

We now may choose $t_0(\varepsilon)$ sufficiently small so that for all $t \le t_o$ it follows $t^{\frac{d-1+\alpha}{\alpha}} \sup_{[0,\lambda_0]} f(\lambda) < \frac{\varepsilon}{2}$ and thus $t^{(d-1)/\alpha} g(t) \le \varepsilon$ for all $t \le t_0$, as desired.

□

References

1. D. Applebaum, *Lévy Processes and Stochastic Calculus*, vol. 93, Cambridge Studies in Advanced Mathematics (Cambridge University Press, Cambridge, 2004)
2. R. Bañuelos, T. Kulczycki, Trace estimates for stable processes. Probab. Theory Relat. Fields **142**, 313–338 (2008)
3. N. Bingham, C. Goldie, J. Teugels, *Regular Variation*, Encyclopedia of Mathematics and its Applications (Cambridge University Press, Cambridge, 1987)
4. R.M. Blumenthal, R.K. Getoor, On the distribution of first hits for the symmetric stable process. Trans. Am. Math. Soc. **99**, 540–554 (1961)
5. C. Brossard, Can one hear the dimension of a fractal. Commun. Math. Phys. **104**, 103–122 (1986)
6. C. Gordon, D. Webb, S. Wolpert, Isospectral plane domains and surfaces via Riemannian orbifolds. Invent. Math. **110**, 1–22 (1992)
7. M. Kac, Can one hear the shape of a drum? Am. Math. Mon. **73**(4), 1–23 (1966) (Part 2: Papers in Analysis)
8. N.V. Kuznecov, Asympototic distribution of eigenfrequencies of a plane membrane in the case of separable variables (Russian) Differencial'nye Uravnija **2** (1966) 1385–1402
9. R.B. Melrose, Weyl's conjecture for manifolds with concave boundary, Geometry of the Laplace operator (Proc. Sympos. Pure Math., Univ. Hawaii, Honolulu, Hawaii, 1979) Proc. Sympos. Pure Math., XXXVI, Am. Math. Soc., Providence, R.I. (1980) 257–274
10. J. Milnor, Eigenvalues of the Laplace operator on certain manifolds. Proc. Natl. Acad. U.S.A. **51** (1964) 542ff
11. R. Song, Z.-Q. Chen, Two-sided Eigenvalue estimates for subordinate processes in domains. J. Funct. Anal. **226**, 90–113 (2005)
12. R. Song, Z. Vondraček, Potential theory of subordinate killed Brownian motion in a domain. Probab. Theory Relat. Fields **125**, 578–592 (2003)

Part V
Statistics

A Statistical Change-Point Analysis Approach for Modeling the Ratio of Next Generation Sequencing Reads

Jie Chen and Hua Li

Abstract One of the key features of statistical change-point analysis is to estimate the unknown change-point locations for various statistical models imposed on the sample data. This analysis can be done through a hypothesis testing process, a model selection perspective, or a Bayesian approach, among other methods. Change-point analysis has a wide range of applications in research fields such as statistical quality control, finance and economics, climate study, medicine, genetics, etc. In this paper, a change-point analysis motivated by the modeling of genomic data will be provided. The high throughput next generation sequencing (NGS) technology is now frequently used in profiling tumor and control samples for the study of DNA copy number variants (CNVs). In particular, the ratio of the read count of the tumor sample to that of the control sample is popularly used for identifying CNV regions. To identify CNV regions is equivalent to finding change-points that potentially exist in the NGS reads ratio data. We present a change-point model and a Bayesian solution for the estimation of the change-point locations in NGS reads ratio data. Simulation studies of the proposed method indicate the effectiveness of the proposed method in identifying change-point locations. Applications of the proposed change point model for identifying boundaries of DNA copy number variation (CNV) regions using the next generation sequencing data of breast cancer/tumor cell lines and lung cancer cell line will be presented.

Keywords Change point analysis · DNA copy numbers · Next generation sequencing data

Mathematics Subject Classification 62F03 · 62F10 · 62F15 · 92D20

J. Chen (✉)
Department of Biostatistics and Epidemiology, Augusta University,
1120 15th Street, Augusta, GA 30912, USA
e-mail: jiechen@augusta.edu

H. Li
Stowers Institute for Medical Research, 1000 E 50th Street, Kansas City,
MO 64110, USA
e-mail: HUL@stowers.org

G. Letzter et al. (eds.), *Advances in the Mathematical Sciences*, Association for Women in Mathematics Series 6, DOI 10.1007/978-3-319-34139-2_13

1 Introduction

The innovative next generation sequencing (NGS) technology has made it possible to study much more biological research problems at the genomic level. Using the NGS technology, the entire targeted genome is broken into small pieces (called reads), and these reads are then ligated to adapters for massive parallel sequencing. Reads are then mapped and assembled back to the genome. The resulting outputs, the NGS data (consisting of mapped reads), are massive and requires special modeling approaches in order to retrieve the most scientifically interpretable results.

DNA copy number variations (CNVs) refer to genomic regions where DNA copy numbers are different from the normal copy number. For instance, the normal copy number for human being is 2 and one could see copy numbers of 1, 3, or 4, etc. in a person's different genomic regions, which indicates CNVs in those regions. Genetic variation exists in all human beings and takes various forms, thus giving the diversity of human beings, and CNV sizes can range from kilobases (kb) to megabases (Mb) [1]. Studies have shown that CNVs comprise about twenty percent of variation in gene expression [2]. Many studies [3–7] have concluded that variations in DNA copy numbers are common in cancer and other diseases. Therefore, locating correct CNV regions will help improve the development of medical diagnostic tools and treatment regimes for various diseases. Specifically, the ratio of NGS read count of the tumor sample to that of the control sample is popularly used for identifying CNV regions.

There are two approaches to applying NGS technology to the study of CNVs [8]. One approach is to use the NGS read counts in a tumor sample to study the CNVs without sequencing a control sample genome [4, 9, 10]. The read counts approach is based on the assumption [11] that the sequencing process is uniform in getting the reads from the target genome, and then the number of reads mapped to a region is expected to be proportional to the number of times the region appears in the DNA sample, and hence the copy number of any genomic region can be estimated by counting the number of reads aligned to that particular region. A recent work of detecting CNV regions using NGS read counts is given in [12]. The other approach is to apply NGS to both the test sample and control sample genomes and obtain sequencing reads (usually of 36 ĩ00 base pair (bp) long depending on which sequencing platform is used) for both the test sample and the control sample genomes. Then the ratio of the read counts can be used for the analysis of CNVs.

As observed by many authors, there are several steps involved in the analysis of either the read count data or the ratio of reads data for CNV detection, and these steps include raw reads processing, alignment of the reads for both the test sample and control sample genomes, the normalization of the data and GC content correction, CNV boundaries detection, and copy number estimation for each region between any two breakpoints. We concentrate on CNV boundary detection (or breakpoint detection) and copy number estimation in this work by using the NGS reads ratio data.

A few recent studies addressed the problem of identifying CNVs using ratio of reads data. A method called CNV-seq was developed to detect CNVs in reads ratio data using a sliding window approach by assuming that the reads ratio follows a Gaussian distribution [13]. An approach in which the CBS method (developed in [14] for array Comparative Genomics Hybridization (aCGH) data) was used to identify the boundaries of CNV regions in the ratio of reads was presented in [15] and a sliding window approach in detecting genomic regions that are hypothesized to contain CNVs was implemented. A method, called rSW-seq, for detecting CNVs in tumor versus control sample genome reads ratios using ad hoc thresholds was used in [16]. Xi et al. [17] used the Bayesian information criterion to detect CNVs using the reads ratio data. Undoubtedly, these methods have provided good strategies in identifying CNVs using the massive NGS reads ratio data. There are, however, still many unsolved problems and challenges in modeling the NGS data that are needed to be solved to advance our understanding of the complex biological systems.

The detection of the boundaries of CNV regions can be viewed as a statistical change point detection problem [18]. A statistical change point model based method was proposed to help detect CNV regions using the ratio of NGS reads in our earlier work [19]; moreover, simulation studies and more applications are added in this paper as an expansion of our earlier work [19]. The rest of the paper is organized as follows. Details of the proposed CNV detection method are given in Sect. 2. Simulation studies are present in Sect. 3 and applications of the methods to NGS reads ration data are provided in Sect. 4. Conclusions are given in Sect. 5.

2 The Approach

Let X_i be the ratio of the number of aligned reads of the test sample genome to that of the control sample genome at the ith bin along the chromosome under study, $i = 1, \ldots, n$, and n is the total number of bins. As many authors noted, the number of aligned reads in either the test sample or the control sample follows Poisson distributions, so the ratio X_i is indeed the ratio of a Poisson random variable to that of a truncated Poisson (no zeros) random variable. The exact probability distribution of this ratio is to-date unknown. However, according to the work of [15], when the local average read is greater than 80 in a given window length, the log base 2 of the reads ratio X_i is well approximated by a normal distribution. This assumption can be easily satisfied when the binning is applied to the reads. Now let Y_i represent the log base 2 of X_i, and we assume that Y_i follows approximately a normal distribution. Then the estimation of the boundaries of a CNV region becomes the problem of detecting breakpoints in the sequence Y_i.

2.1 The Method

We propose to use the mean and variance change model (MVCM) [20, 21] to model the log reads ratio Y_i for breakpoint detection. For simplicity, we first focus on how to detect one breakpoint k in the reads ratios of a certain window size M, modeled by MVCM. The estimation of a breakpoint can be stated as searching for an unknown bin index k such that the subsequence $Y_i, i = 1, \ldots, k$ can be viewed as a random sample taken from $N(\mu_1, \sigma_1^2)$ and the subsequence $Y_i, i = k+1, \ldots, M$ can be viewed as from $N(\mu_2, \sigma_2^2)$. As one of the major goals in change-point analysis is to estimate the true location of the change-point (breakpoint), the Bayesian approach has some advantage in its unique perspective in change point analysis. Being provided the posterior probability of any location as a possible change-point, researchers can make informed conclusions and interpretation of that location, instead of going through a hypothesis testing process. In this sense, knowing the posterior probability of a bin containing a possible CNV, the researchers will be able to further investigate that possible CNV with biological justification.

As all the reads of the test sample and control sample genomes are binned with the same width, it is straightforward to start with the assumption that the location of the breakpoint k is uniformly distributed along the chromosome, or the prior distribution, $p_0(k)$, of k can be taken to be

$$p_0(k) = \begin{cases} 1/(M-3), \textit{ for } k = 2, \ldots, M-2 \\ 0, \qquad\qquad \textit{otherwise}. \end{cases} \tag{1}$$

Note that the first and the last observations are assumed not to be breakpoint loci because when we start to search multiple breakpoints in the sequence, we will be using a sliding window to capture at most one breakpoint at a time and let the adjacent windows overlap one observation to take the edge effect out of the algorithm; we will add a zero before the first observation of the sequence so that the first observation of the sequence is essentially included in the search for breakpoints (see The Algorithm subsection). Within any segment, before or after the breakpoint, the binned reads ratios are assumed to be a sample from the same normal distribution, hence, it is natural to assign the prior distribution of the means μ_1 and μ_2, $p_0(\mu_1, \mu_2|k)$, to be proportional to some constant, or simply write

$$p_0(\mu_1, \mu_2|k) \propto \textit{constant}. \tag{2}$$

Now given the two means, μ_1 and μ_2, for the segments before and after the breakpoint k, we assume that the prior distribution of the variances σ_1^2 and σ_2^2 to be

$$p_0(\sigma_1^2, \sigma_2^2|\mu_1, \mu_2, k) \propto 1/\sigma_1^2\sigma_2^2. \tag{3}$$

With the priors given in (1)–(3), we can obtain the likelihood function given the sequence of reads ratio Y_i, i.e., $L(\mu_1, \mu_2, \sigma_1^2, \sigma_2^2, k)$, and hence obtain the joint posterior distribution of all the parameters, or $p_1(\mu_1, \mu_2, \sigma_1^2, \sigma_2^2, k)$, as

$$\begin{aligned} p_1(\mu_1, \mu_2, \sigma_1^2, \sigma_2^2, k) &\propto L(\mu_1, \mu_2, \sigma_1^2, \sigma_2^2, k) \\ &\quad \cdot p_0(\sigma_1^2, \sigma_2^2|\mu_1, \mu_2, k) \cdot p_0(\mu_1, \mu_2|k) p_0(k), \end{aligned} \tag{4}$$

where

$$
\begin{aligned}
L(\mu_1, \mu_2, \sigma^2, \sigma_1^2, k) &\triangleq L_1(\mu_1, \mu_2, \sigma_1^2, \sigma_2^2, k \mid X_i, i = 1, \ldots, M) \\
&\propto \left(\frac{1}{\sigma_1^2}\right)^{\frac{k}{2}} \exp\left\{-\frac{1}{2\sigma_1^2}\sum_{i=1}^{k}(X_i - \mu_1)^2\right\} \\
&\cdot \left(\frac{1}{\sigma_2^2}\right)^{\frac{n-k}{2}} \exp\left\{-\frac{1}{2\sigma_2^2}\sum_{i=k+1}^{M}(X_i - \mu_2)^2\right\}.
\end{aligned}
$$

From (4), the posterior probability, $p_1(k)$, of any location k being a possible breakpoint is obtained as [21]

$$
p_1(k) = p_1^*(k)/\sum_{j=2}^{M-2} p_1^*(j), \tag{5}
$$

for $k = 2, \ldots, M - 2$, where

$$
\begin{aligned}
p_1^*(j) = &\Gamma((j-1)/2)\Gamma((M-j-1)/2)j^{j/2-1} \\
&(M-j)^{(M-j)/2-1}\{j\sum_{i=1}^{j} Y_i^2 - (\sum_{i=1}^{j} Y_i)^2\}^{-(j-1)/2} \\
&\{(M-j)\sum_{i=j+1}^{M} Y_i^2 - (\sum_{i=j+1}^{M} Y_i)^2\}^{-(M-j-1)/2},
\end{aligned} \tag{6}
$$

and $\Gamma(a)$, $a > 0$, is the Gamma function, which is well defined in this case as k takes values from 2 to $M - 2$. Finally, in a window of size M, a location $\widehat{k}$ is an estimated location of the true breakpoint if the posterior probability (5) attains its maximum at $\widehat{k}$. Since the reads ratio data can be quite noisy, we suggest using a threshold value, p_t, starting from 0.55, to select k such that

$$
\widehat{k} = arg\{k : max(p_1(k)) \geq p_t\}. \tag{7}
$$

Furthermore, an approximate frequentist $(1 - \alpha)100\,\%$ confidence interval of the original reads ratio in each segment is obtained as

$$
(2^{\bar{Y} - z_{\frac{\alpha}{2}} S/\sqrt{M} + S^2/2}, 2^{\bar{Y} + z_{\frac{\alpha}{2}} S/M + S^2/2}), \tag{8}
$$

where $\bar{Y}$ and S^2 are, respectively, the sample mean and variance of the log reads ratio Y_i within the segment, M is the length of that segment, and $z_{\frac{\alpha}{2}}$ is the upper $(\alpha/2)$th percentile of the standard normal distribution.

2.2 The Implementation Algorithm

When searching for breakpoints in a long sequence, segmentation algorithms such as the binary segmentation [22], the circular binary segmentation [14], and dynamic programming [23], etc., have an obvious disadvantage of very low computational speed. For the NGS data, the sliding window approach is very appealing as a breakpoint is a local property in the neighborhood of the potential breakpoint, and hence a window of a much smaller size than the original length of the sequence is able to capture a breakpoint among the neighboring points. Our goal is to provide a user-friendly algorithm that can provide all potential breakpoints in a very minimal computing time. For the NGS data, our algorithm that combines a sliding window strategy with the above Bayesian posterior probability (5) is summarized in the following steps:

- *Step 1*: After preprocessing of the NGS data, get the reads ratio data and convert them to Y_i's. Select a window size M between 12 and 30 (it can go beyond 30 for a very long sequence). The window size will divide Y_i's of length n into $w = [n/M] + 1$ windows. For the jth window, denote the window size as W_j, $j = 1, \ldots, w$. We initialize the first window with size $W_1 = 0$, let $W_j = M$, for $j = 2, \ldots, w - 1$, and note that the last window is of size $W_w n - (w - 1) * M$.
- *Step 2*: The user will select a threshold value, p_t, from 0.55 to 0.99, depending on the research goal of getting more or less candidate breakpoints selected. When the threshold is increasing, the adjacent segments are merged together with less breakpoints and vice versa.
- *Step 3*: After steps (1) and (2), for the jth window with window size W_j, the algorithm automatically computes the posterior probability of each location being a potential breakpoint using (5) and (6), finds the maximum posterior probability within this window, and if it is greater than the threshold value p_t, identifies that location as the breakpoint location according to (7) for that window. This process will be done for all windows by the algorithm.
- *Step 4*: One can repeat steps (1)–(3) for different window sizes until the max posterior probability stabilizes.
- *Step 5*: The algorithm will finally return the number of breakpoints, their locations, etc., give the mean reference line of each segment between two breakpoints, provide the posterior probabilities within each window for all windows, and output a frequetist $(1 - \alpha)100\,\%$ confidence interval, according to (8) of the reads ratio for each segment.

3 Simulation Studies

To perceive how adaptive our Bayesian algorithm is towards the departure from normality, we carried out a series of simulation studies for the cases when the reads ratios are too noisy to follow normal distributions. As the sample mean reads ratio typically range from .35 to 3 or more with a very small variance (ranging from .009 to .5814) in most of the chromosomes in the three sequencing data sets (see Sect. 4 for details), we first simulated 1000 cases of normal observations using mean and variance parameters in the above mentioned ranges, the results are nearly as perfect as expected, i.e., the true changes are identified by the algorithm with very small error (in terms of the estimated mean square error (MSE). Partial results are given in Table 1). We also simulated 1000 cases where the reads ratios are distributed with skewness as the exponential, gamma, and Weibull distributions, respectively, and we restricted the location parameter for these distributions to stay in the range of .35–3. The results of the change loci identified by our algorithm with these three non-normal distributions are given in Tables 2, 3 and 4, respectively. In each table, $\bar{\hat{\tau}}$ stands for the average change of the estimated breakpoint out of the 1000 simulations and

Table 1 Simulation results for normal case

Before	After	Loci	Window size M					
$\mu_1(\sigma_1^2)$	$\mu_2(\sigma_2^2)$		12		20		32	
			$\bar{\hat{\tau}}$	MSE	$\bar{\hat{\tau}}$	MSE	$\bar{\hat{\tau}}$	MSE
.35(.003)	1(.01)	M/4	3.00	0.000	5.00	0.000	8.00	0.000
		M/2	6.00	0.000	10.0	0.000	15.0	0.000
		3M/4	9.00	0.000	15.0	0.000	24.0	0.000
.35(.003)	1(.01)	M/4	3.00	0.000	5.00	0.000	8.00	0.000
		M/2	6.10	0.589	10.0	0.166	15.2	0.236
		3M/4	9.00	0.000	15.0	0.000	24.0	0.000

Table 2 Simulation results for exponential case

Before	After	Loci	Window size M					
λ_1	λ_2		12		20		32	
			$\bar{\hat{\tau}}$	MSE	$\bar{\hat{\tau}}$	MSE	$\bar{\hat{\tau}}$	MSE
.35	1	M/4	5.48	16.29	8.47	51.32	13.4	136.6
		M/2	6.00	8.341	10.1	30.86	15.8	85.08
		3M/4	6.83	14.56	11.7	45.67	19.3	120.7
.35	1.5	M/4	4.81	12.15	7.63	36.27	10.9	73.49
		M/2	6.00	6.787	9.89	23.39	16.3	51.09
		3M/4	7.25	11.71	12.8	35.35	21.2	78.30
.35	2	M/4	4.49	9.881	6.98	27.49	10.3	56.61
		M/2	6.06	5.889	10.1	15.86	16.2	29.12
		3M/4	7.44	10.47	13.3	25.88	22.2	51.30
.35	3	M/4	4.07	6.959	6.16	16.50	8.95	25.35
		M/2	5.97	3.790	10.3	8.834	16.4	15.03
		3M/4	8.05	6.550	13.9	17.23	23.3	25.46

Table 3 Simulation results for gamma case

Before	After	Loci	Window size M					
α_1, β_1	α_2, β_2		12		20		32	
			$\bar{\hat{\tau}}$	MSE	$\bar{\hat{\tau}}$	MSE	$\bar{\hat{\tau}}$	MSE
1, .5	1.5, 1	M/4	4.53	7.966	6.53	25.54	9.44	46.13
		M/2	5.81	6.921	9.71	21.19	15.4	45.08
		3M/4	6.88	13.15	12.1	39.16	20.6	81.71
1, .5	4, .7	M/4	3.08	1.198	4.87	1.202	7.95	1.391
		M/2	5.94	9.663	9.72	36.94	15.1	109.8
		3M/4	8.44	3.325	14.5	5.745	23.7	4.328
1, .8	3, 1.2	M/4	4.51	11.48	6.93	37.59	10.2	95.15
		M/2	5.84	2.789	9.71	4.497	15.7	3.298
		3M/4	7.91	6.455	14.0	11.97	23.3	13.59
1.2, .9	3.1, 1.5	M/4	3.40	3.185	5.12	5.091	7.98	4.563
		M/2	5.83	2.780	9.74	4.396	15.9	3.028
		3M/4	7.98	6.189	13.9	12.69	23.3	14.10

Table 4 Simulation results for Weibull case

Before	After	Loci	Window size M					
k_1, λ_1	k_2, λ_2		12		20		32	
			$\bar{\hat{\tau}}$	MSE	$\bar{\hat{\tau}}$	MSE	$\bar{\hat{\tau}}$	MSE
1.5, 3.1	2.1, 3.5	M/4	3.42	2.967	5.27	4.817	8.06	4.015
		M/2	5.99	1.439	9.89	1.904	15.9	1.080
		3M/4	8.70	1.916	14.8	2.736	23.8	2.550
2, 3.3	2.6, 4	M/4	3.09	0.942	5.03	1.093	7.92	0.634
		M/2	5.93	0.563	9.88	0.443	15.8	0.701
		3M/4	8.80	0.876	14.8	0.800	23.9	0.525
2.1, 2.8	2.7, 3.5	M/4	3.17	1.592	5.02	2.161	7.93	1.802
		M/2	5.90	0.776	9.83	0.940	15.8	0.701
		3M/4	8.65	1.793	14.8	1.221	23.7	1.527
2.2, 3.4	1.8, 2.3	M/4	3.22	1.642	5.07	1.851	7.94	0.472
		M/2	5.93	0.476	9.93	0.478	15.9	0.455
		3M/4	8.87	0.532	14.9	0.587	23.9	0.595

MSE stands for the estimated mean square error of the estimated change location or $\widehat{MSE}(\hat{\tau})$. From Tables 2, 3 and 4, we see that the algorithm can perform reasonably well even when there are changes in the parameters of an underlying distribution which is not normal. Among the three non-normal cases, the algorithm performed the worst for the exponential case (most skewed), and performed the best for the Weibull distribution. If the magnitudes of the parameters are not large enough, the algorithm may not pick those exact change loci based on the observations. This is expected for any change-point search algorithm that is based on a parametric distribution assumption. The algorithm performs reasonably well in terms of the average of the estimated location being close to the true change location and the estimated mean square error being small. The simulation also indicates that for the dense data like the reads ratios, the small window sizes (12–20) are better to capture the local changes.

4 Applications to Tumor and Cancer Cell Lines

We applied our method to the sequencing data of three cell lines [15], namely, the breast tumor cell line HCC1954, the breast carcinoma cell line HCC1143, and the lung cancer cell line NCI-H2347. The data are downloaded from the National Center for Biotechnology Information, or NCBI, website. The data processing descriptions are given in [19].

We first applied our algorithm to the reads ratio data of the tumor cell line HCL1954 to the matched normal cell line BL1954 for the binned data with bin size of100K bp for clear illustration purpose. We summarize the result for chromosome 8 in Table 5. Then we provide Fig. 1 to showcase the output of the result on Chromosome 8 for threshold 0.70 and window size of 20. The breakpoints identified by this algorithm well matched the discoveries on this cell line in the literature (see Discussion).

After the identification of each segment with differential reads ratios, the log reads ratios within each segment are supposed to follow a Gaussian distribution if the data only contains normal random noise. To check if the segmented data actually can be viewed from a Gaussian distribution, the normal Q–Q plots are used. Specifically, Fig. 2 shows the normal Q–Q plots after 22 breakpoints are identified using the threshold of .70 and a window size of 20 for chromosome 8 binned at 100 Kbp. The 6th, 13th 15th, 17th, and 20th segments are quite deviate from normality, the 5th, 7th, 10th, 12th, 16th, 19th segments somewhat indicate normality for the middle 50 % of the data but have outliers and tail issues, while the rest 11 segments indicate normality. The departure from normality in some segments is due to the very noisy sequencing results observed during the data processing stage.

The algorithm also provides the location of every breakpoint, the mean of the reads ratios X_i's of each segment, and a confidence interval for the reads ratio of each segment based on X_i's with the user's choice of the confident level for each threshold value and each window size. For example, we provide the details for chromosome 8 in Table 5.

The availability of the 95 % individual confidence intervals in Table 5 on the mean ratios for all segments provides a good statistical interpretation of the estimated copy number ratio for all segments. As the upper bounds of the 95 % CIs for segments 1–4 and 23, are all less than 1, we are 95 % confident to state that the copy numbers in the sample is less than the copy numbers of the control sample, i.e., there is a deletion

Table 5 Detailed results for HCC-BL1954 Chromosome 8 at threshold value of $p_0 = 0.70$

Segment	Posterior position p_1		Segment region (bp)	Mean of ratio	CI of CN ratio
1			1–200000	0.4554	(0.4509, 0.4599)
2	0.8788	2	200000–25800000	0.4381	(0.4323, 0.4441)
3	0.9497	258	25800001–35800000	0.4383	(0.4312, 0.4454)
4	0.7654	358	35800001–37100000	0.4429	(0.4259, 0.4606)
5	0.9738	371	37100001–39300000	1.8286	(1.6610, 2.0131)
6	0.8297	393	39300001–53300000	1.0281	(0.9717, 1.0877)
7	0.8571	504	53300001–60500000	1.0092	(0.9947, 1.0240)
8	0.8022	576	60500001–69800000	1.0449	(1.0292, 1.0609)
9	0.9936	669	69800001–76500000	1.6818	(1.6426, 1.7218)
10	0.8045	736	76500001–77600000	2.0855	(1.9805, 2.1959)
11	0.8509	747	77600001–80100000	1.7146	(1.6640, 1.7667)
12	0.7387	772	80100001–86200000	1.2601	(1.1951, 1.3286)
13	0.8403	833	86200001–88000001	2.0332	(1.9260, 2.1463)
14	0.7809	850	88000001–93600000	2.4991	(2.3892, 2.6141)
15	0.7722	906	93600001–104200000	2.5510	(2.4862, 2.6175)
16	0.7603	1012	104200001–110700000	4.0092	(3.4669, 4.6362)
17	0.8148	1077	110700001–115000000	2.6971	(2.3733, 3.0651)
18	0.8672	1120	115000001–117400000	5.5511	(5.2919, 5.8229)
19	0.9921	1144	117400001–125400000	2.1128	(2.0179, 2.2122)
20	0.7321	1224	125400001–127900000	2.2792	(1.7966, 2.8914)
21	0.8916	1249	127900001–130200000	2.1260	(1.6837, 2.6846)
22	0.7842	1272	130200001–131500000	1.4422	(1.3934, 1.4928)
23	0.9968	1285	131500001–146364022	0.7757	(0.7641, 0.7875)

of copy number in each of those segments. As the 95 % CIs of segments 6 and 7 contains 1, it indicates that there are essentially no copy number changes in these two segments. Similarly, we conclude that the copy numbers in the sample is more than that of the control sample in the rest of the segments. After we look for more details, we conclude that for segments 8–9, 11–12, and 22, the copy number in the sample is less than 2 times of that in the control sample for each of these segments. Similarly, we state that the copy numbers in the sample is more than 2 times of that in the control sample for segments 17 and 1, and the copy number of segments 10, 13, 20, is about 2 times that of the control sample. Segments 14–15 and 21 have more than 2.5 times of the copy numbers in the test sample in comparison with the control sample. Furthermore, the copy number of segment 16 in the test sample is more than 4 times of that of the control sample. Finally, segment 18 of the test sample has more than 5 times of the DNA copy numbers than that of the control sample.

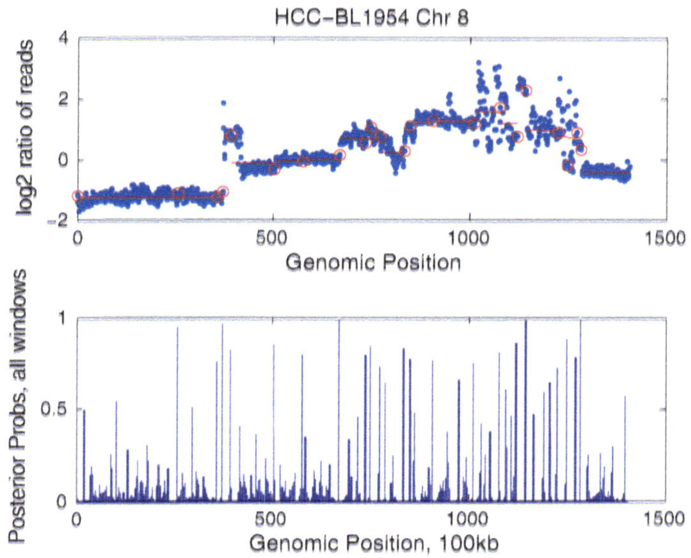

Fig. 1 *Upper panel* a scatter plot of the log base 2 reads ratios with breakpoints identified as *red circles* and *red horizontal line* as the mean of each segment between two identified breakpoints for the threshold of .70 and window size of 20. *Lower panel* posterior probabilities for each position within each window

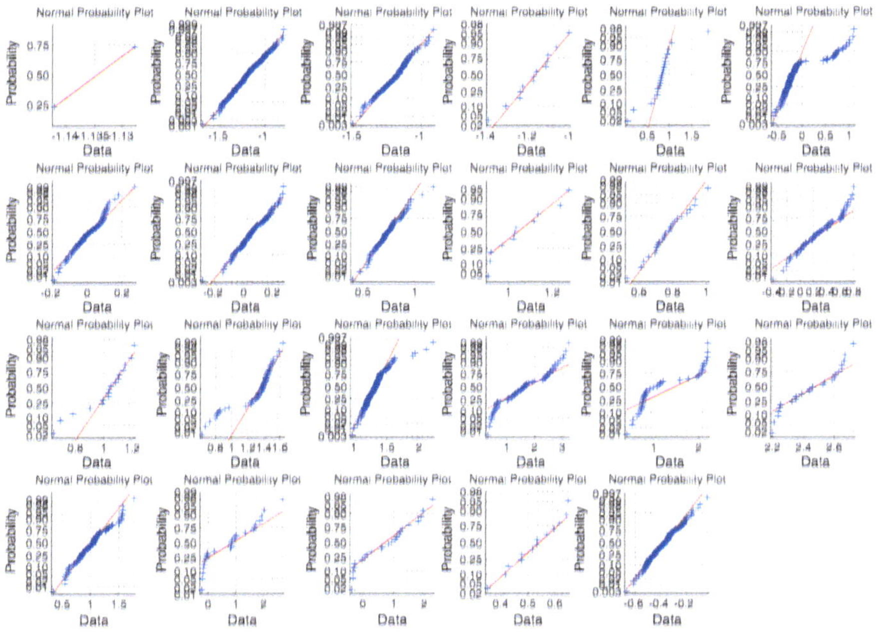

Fig. 2 Normal Q–Q plots for all the 23 segments after 22 breakpoints are identified

Moreover, this algorithm is sensitive and fast enough to search breakpoints for any bin sized reads ratio data. Figure 3 shows how the algorithm can capture the pattern of the reads ratio for different bin sized data using chromosome 11 of HCC-BL1954 as an example.

The overall pattern of the reads ratio changes is clearly shown in the two plots and our algorithm can identify the breakpoints in each case with meaningful results. The application of our algorithm to the other two datasets, HCC-BL1143 and NCI-H2347/2347BL, gives us the similar results and we provide two sets of figures for the CNV identification illustration and normal Q–Q plots for chromosome X of HCC-BL1143 (Fig. 4) and for chromosome 5 of NCI-H2347/2347BL (Fig. 5). For these two chromosomes, the breakpoints identified are in line with Chiang et al. 2009 and the normality checking shows reasonable results with minor exceptions in one or two segments.

We have used the commercially available biological database, BioBase [24], to closely look into biologically interpretable CNVs identified using our algorithm. The 23 segments of chromosome 8 from HCC-BL1954 listed in Table 2 contain many verified CNVs in the literature. We entail some of them here due to space limit.

In the deletion region ranging from 25,800 to 35,800 Kbp (segment 3), there are several well documented genes, for example, gene GTF2E2 localizes to a region of chromosome 8 associated with numerous cancers and Werner syndrome, making it a candidate gene for involvement in such disorders; the downregulation of mRNA associated with gene PPP2CB correlates with prostate cancer; and decreased expression of gene UNC5D in left ventricle is associated with end-stage dilated cardiomyopathy. In the region (with 2 times amplification in copy number) ranging from 73,600 to 74,700 Kbp (segment 10), gene TERF1 expression is upregulated in adrenal cortical cancers. In the region (1.5 times amplification in copy number) ranging from 77,200 to 83,300 Kbp (segment 11), gene TCEB1 with increased mRNA expression correlates with prostate cancer. In the region (2.5 times amplification in copy number) ranging from 85,000 to 90,600 Kbp (segment 14), gene E2F5 amplification and gene overexpression correlates with breast tumor; gene CA1 may promote tumor cell motility and contribute to tumor growth and metastasis, gene WWP1 is upregulated in breast cancer; etc. In the region (3 times amplification in copy number) ranging from 107,700 to 112,000 Kbp (segment 17), gene ABRA has several roles including that its associated mRNA expression is upregulated in type 2 diabetes and that its expression is increased in gastric cancer. In this same segment, gene EBAG9 is upregulated in breast, pancreatic and various other neoplasms, etc. In the most amplified region (5.5 times) ranging from 112,000 to 114,400 Kbp (segment 18), gene CSMD3 maps to a breakpoint region associated with autistic disorder and its mutation is associated with familial colorectal cancer. In the region (2 times amplification in copy number) ranging from 122,400 to 124,900 Kbp (segment 20), gene FAM83A related protein expression is upregulated in breast carcinoma and gene ATAD2 copy number variation correlate with ovarian cancer; gene amplification and overexpression correlates with aggressive cancers; increased expression correlates with poor survival in breast cancer; and increased expression act as a predictor marker in breast and lung cancers. In the region (2 times amplification) ranging from 124,900 to 127,200 Kbp

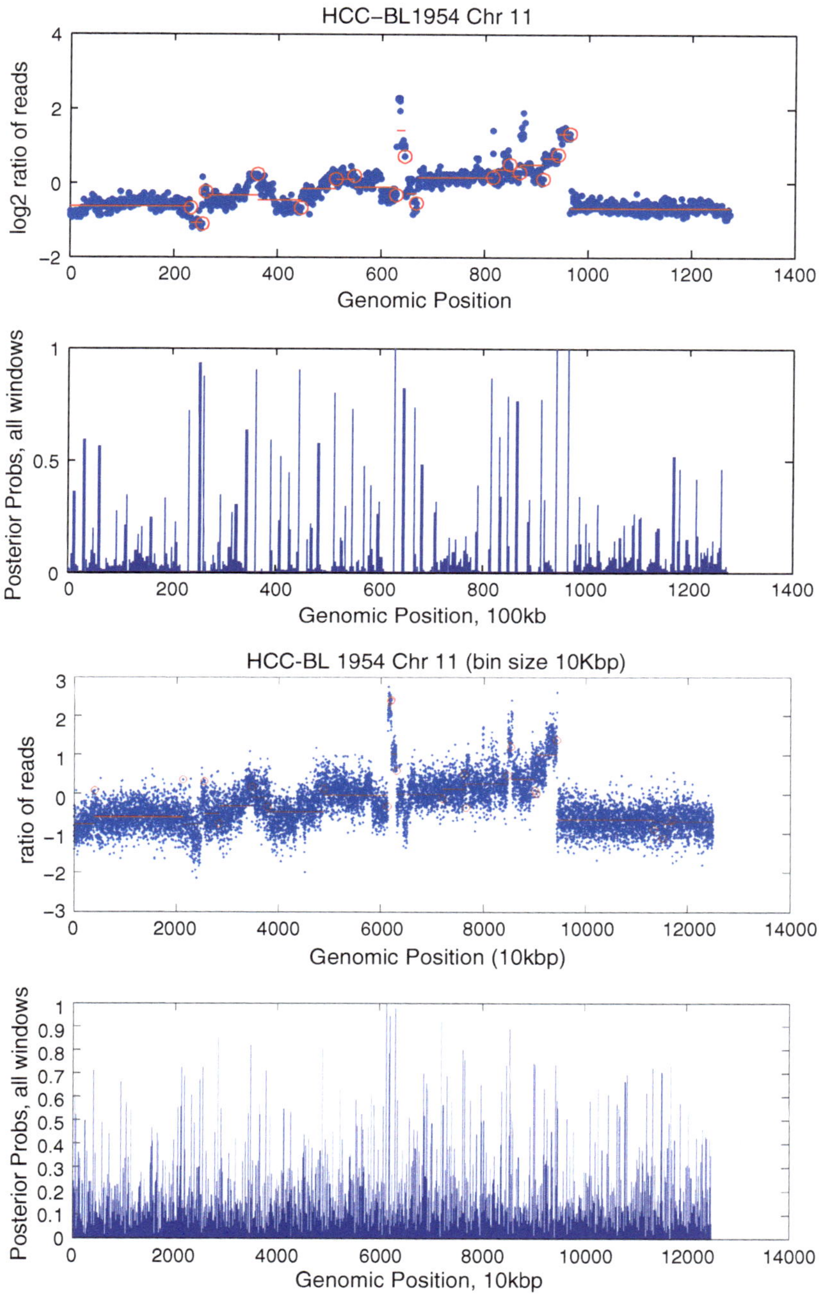

Fig. 3 *Upper panel* results for the log base 2 reads ratios data binned with size 100 K for chromosome 11 of HCC1954 with window size of 20 and threshold of .75. *Lower panel* results for the log base 2 reads ratios data binned with size 10 K for chromosome 11 of HCC1954 with window size of 20 and threshold of .75

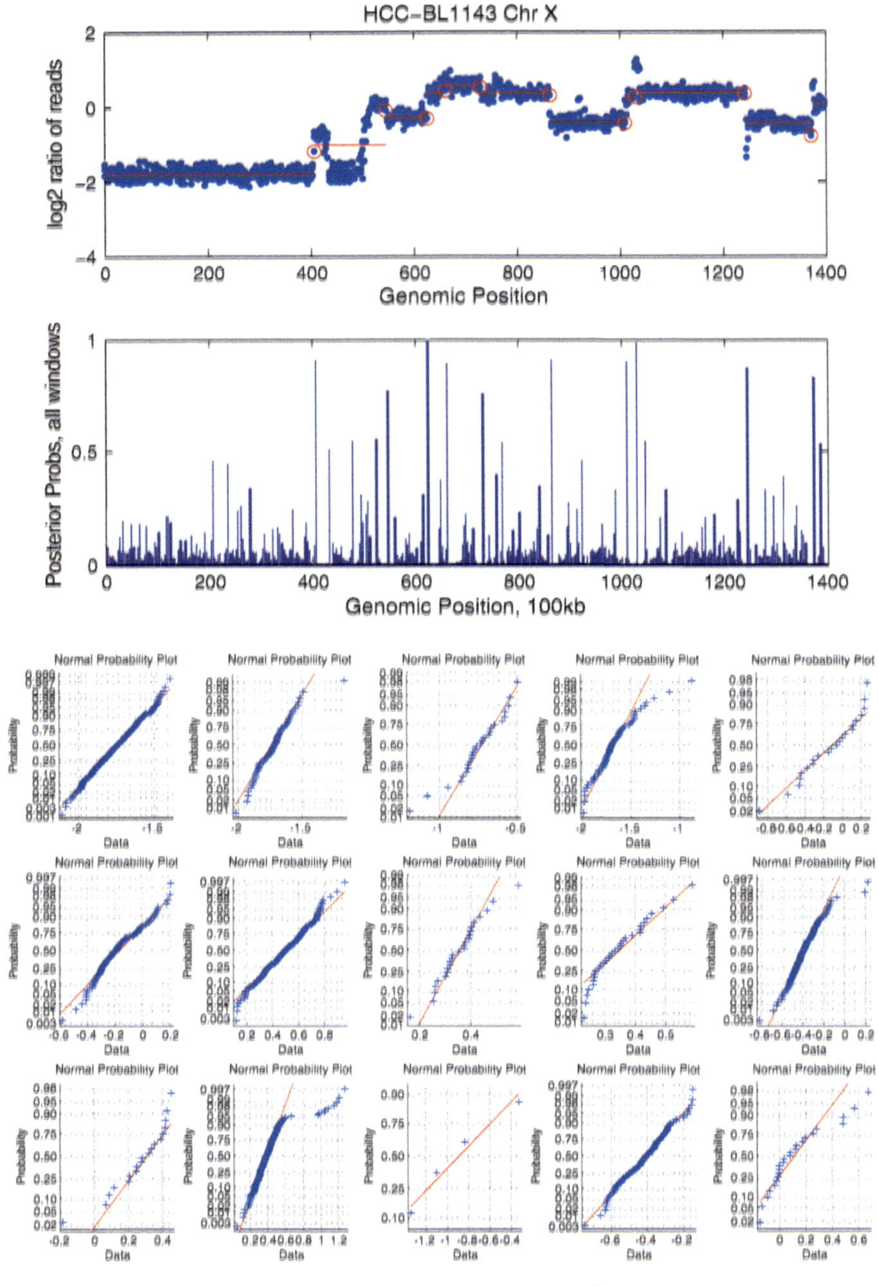

Fig. 4 *Upper panel* breakpoints identified for chromosome X of HCC-BL1143 with window size 26 and threshold of 0.75. There are total of 10 breakpoints. *Lower panel* the normal Q–Q plots of all the 11 segments shown on the *upper panel*

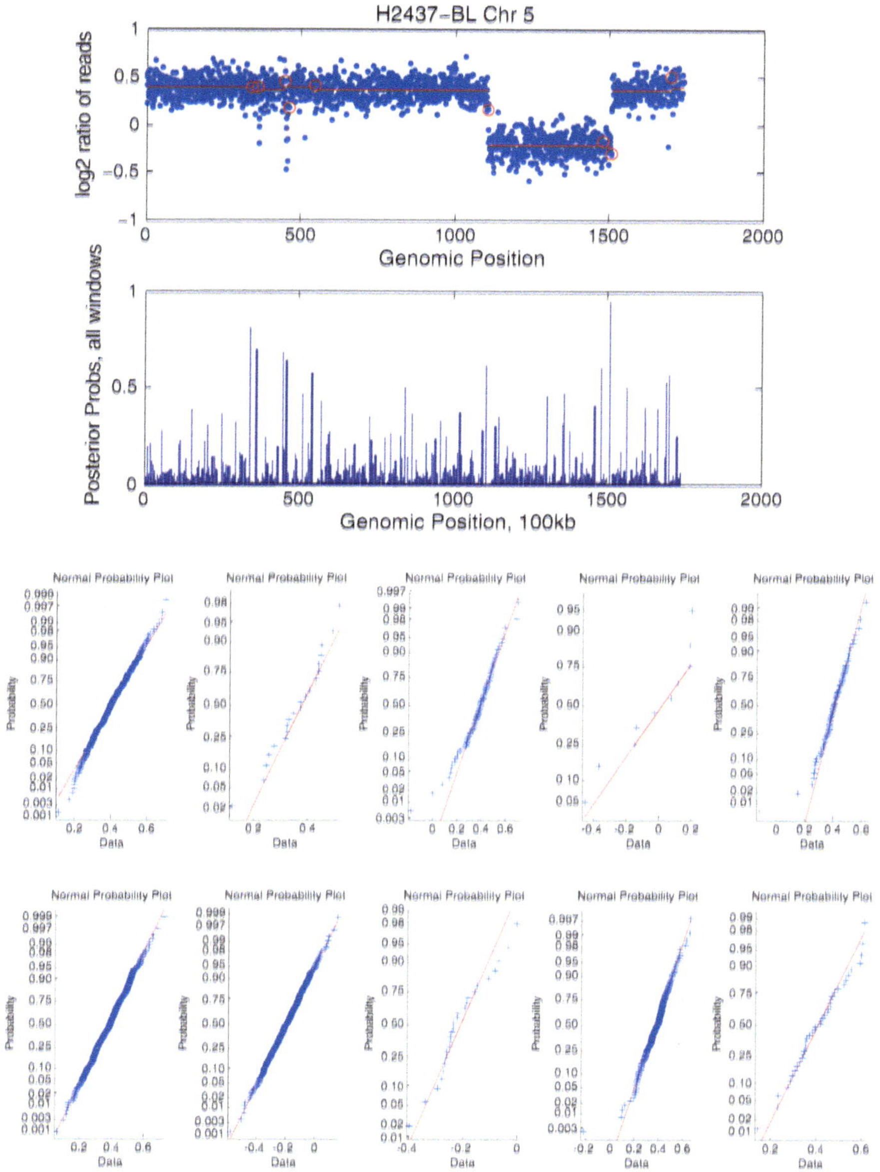

Fig. 5 *Upper panel* breakpoints identified for chromosome 5 of NCI-H2347/BL with window size 26 and threshold of 0.55. There are total of three breakpoints. *Lower panel* the normal Q–Q plots of all the eight segments shown on the *upper panel*

(segment 21), gene TRMT12 amplification results in mRNA overexpression in breast neoplasms, gene KIAA0196 is overexpressed in prostate carcinoma and corresponding gene is amplified in xenografts and hormone-refractory prostate tumors; and mRNA associated with gene TRIB1 is upregulated in ischemic heart disease.

5 Conclusions and Discussions

The next generation sequencing technology has provided ample opportunities for various genetics studies of diseases. It also gives the study of CNVs with a new tool and more insightful information about genetic variations literally at the per base-pair resolution. However, due to the noises inherited from the various steps of the sequencing process, and the availability of computational power, there are still many challenges on how to exploit the information from the NGS data and how to statistically model the data to get best biologically interpretable results. There are not many effective methods available yet. Several methods have been proposed to model the data from different points of views (see the Introduction section). In this paper, we provide a very promising algorithm that has been written into a user friendly R package, SeqBBS. This algorithm can provide fast location of breakpoints with the likelihood of breakpoints being assessed by the posterior probabilities. We applied our algorithm to the reads ratios of all 23 chromosomes of each of the paired cell lines, HCC1954/BL1954, HCC1143/BL1143 and NCI-H2347/2347BL, and we identified all of the breakpoints indicated in [15] plus some extra breakpoints which remains to be biologically validated. Our algorithm has the advantage of fast computing, clear visual presentation with posterior probability indication for possible breakpoints, and confidence interval estimation for copy number ratios of each segment. The method is robust towards the departure of normality assumption. Further work on how to select the threshold value for the posterior probability in consideration of controlling false positive rate needs to be explored.

References

1. R. Redon, S. Ishiwaka, K.R. Fitch, L. Feuk, G.H. Perry, D. Andrews, H. Fiegler, M.H. Shapero, A.R. Carson, W. Chen, E.K. Cho, S. Dallaire, J.L. Freeman, J.R. Gonzalez, M. Gratacos, J. Huang, D. Kalaitzopoulos, D. Komura, J.R. MacDonald, C.R. Marshall, R. Mei, L. Montgomery, K. Nishimura, K. Okamura, F. Shen, M.J. Somerville, J. Tchinda, A. Valsesia, C. Woodwark, F. Yang, J. Zhang, T. Zerjal, J. Zhang, L. Armengol, D.F. Conrad, X. Estivill, C. Tyler-Smith, N.P. Carter, H. Aburatani, C. Lee, K.W. Jones, S.W. Scherer, M.E. Hurles, Global variation in copy number in the human genome. Nature **444**, 444–454 (2006)
2. B. Stranger, M. Forrest, M. Dunning, C. Ingle, C. Beazley, N. Thorne, R. Redon, C. Bird, A. de Grassi, C. Lee, C. Tyler-Smith, N. Carter, S.W. Scherer, S. Tavar, P. Deloukas, M.E. Hurles, E.T. Dermitzakis, Relative impact of nucleotide and copy number variation on gene expression phenotypes. Science **315**, 848 (2007)

3. J. Sebat, B. Lakshmi, D. Malhotra, J. Troge, C. Lese-Martin, T. Walsh, B. Yamrom, S. Yoon, A. Krasnitz, J. Kendall, A. Leotta, D. Pai, R. Zhang, Y.-H. Lee, J. Hicks, S.J. Spence, A.T. Lee, K. Puura, T. Lehtimki, D. Ledbetter, P.K. Gregersen, J. Bregman, J.S. Sutcliffe, V. Jobanputra, W. Chung, D. Warburton, M.-C. King, D. Skuse, D.H. Geschwind, T.C. Gilliam, K. Ye, M. Wigler, Strong association of de novo copy number mutations with autism. Science **316**, 445–449 (2007)
4. P.J. Campbell, P.J. Stephens, E.D. Pleasance, S. O'Meara, H. Li, T. Santarius, L.A. Stebbings, C. Leroy, S. Edkins, C. Hardy, J.W. Teague, A. Menzies, I. Goodhead, D.J. Turner, C.M. Clee, M.A. Quail, A. Cox, C. Brown, R. Durbin, M.E. Hurles, P.A.W. Edwards, G.R. Bignell, M.R. Stratton, P.A. Futreal, Identification of somatically acquired rearrangements in cancer using genome-wide massively parallel paired-end sequencing. Nat. Genet. **40**, 722–729 (2008)
5. H. Stefansson, D. Rujescu, S. Cichon, O.P.H. Pietilinen, A. Ingason, S. Steinberg, R. Fossdal, E. Sigurdsson, T. Sigmundsson, J.E. Buizer-Voskamp, T. Hansen, K.D. Jakobsen, P. Muglia, C. Francks, P.M. Matthews, A. Gylfason, B.V. Halldorsson, D. Gudbjartsson, T.E. Thorgeirsson, A. Sigurdsson, A. Jonasdottir, A. Jonasdottir, A. Bjornsson, S. Mattiasdottir, T. Blondal, M. Haraldsson, B.B. Magnusdottir, I. Giegling, H.-J. Mller, A. Hartmann, K.V. Shianna, D. Ge, A.C. Need, C. Crombie, G. Fraser, N. Walker, J. Lonnqvist, J. Suvisaari, A. Tuulio-Henriksson, T. Paunio, T. Toulopoulou, E. Bramon, M. Di Forti, R. Murray, M. Ruggeri, E. Vassos, S. Tosato, M. Walshe, T. Li, C. Vasilescu, T.W. Mhleisen, A.G. Wang, H. Ullum, S. Djurovic, I. Melle, J. Olesen, L.A. Kiemeney, B. Franke, C. Sabatti, N.B. Freimer, J.R. Gulcher, U. Thorsteinsdottir, A. Kong, O.A. Andreassen, R.A. Ophoff, A. Georgi, M. Rietschel, T. Werge, H. Petursson, D.B. Goldstein, M.M. Nthen, L. Peltonen, D.A. Collier, D. St Clair, K. Stefansson, R.S. Kahn, D.H. Linszen, J. Van Os, D. Wiersma, R. Bruggeman, W. Cahn, L. De Haan, L. Krabbendam, I. Myin-Germeys, Large recurrent microdeletions associated with schizophrenia. Nature **455**, 232–236 (2008)
6. T.-L. Yang, X.-D. Chen, Y. Guo, S.-F. Lei, J.-T. Wang, Q. Zhou, F. Pan, Y. Chen, Z.-X. Zhang, S.-S. Dong, X.-H. Xu, H. Yan, X. Liu, C. Qiu, X.-Z. Zhu, T. Chen, M. Li, H. Zhang, L. Zhang, B.M. Drees, J.J. Hamilton, C.J. Papasian, R.R. Recker, X.-P. Song, J. Cheng, H.-W. Deng, Genome-wide copy-number-variation study identified a susceptibility gene, UGT2B17, for osteoporosis. Am. J. Hum. Genet. **83**(6), 663–674 (2008)
7. A. Rovelet-Lecrux, D. Hannequin, G. Raux, N. Le Meur, A. Laquerrire, A. Vital, C. Dumanchin, S. Feuillette, A. Brice, M. Vercelletto, F. Dubas, T. Frebourg, D. Campion, APP locus duplication causes autosomal dominant early-onset Alzheimer disease with cerebral amyloid angiopathy. Nat. Genet. **38**, 24–26 (2006)
8. S. Moorthie, C.J. Mattocks, C.F. Wright, Review of massively parallel DNA sequencing technologies. Hugo J. **5**, 112 (2001)
9. S. Yoon, Z. Xuan, V. Makarov, K. Ye, J. Sebat, Sensitive and accurate detection of copy number variants using read depth of coverage. Genome Res. **19**, 1586–1592 (2006)
10. C.A. Miller, O. Hampton, C. Coarfa, A. Milosavljevic, ReadDepth: a parallel R package for detecting copy number alterations from short sequencing reads. PLoS One **6**(1), e16327 (2011)
11. A. Magi, L. Tattini, T. Pippucci, F. Torricelli, M. Benelli, Read count approach for DNA copy number variants detection. Bioinformatics **28**, 470–478 (2012)
12. T. Ji, J. Chen, Modeling the next generation sequencing read count data for DNA copy number variant study. Stat. Appl. Genet. Mol. Biol. **14**, 361374 (2015)
13. C. Xie, M.T. Tammi, CNV-seq: a new method to detect copy number variation using high-throughput sequencing. BMC Bioinform. **10**, 80 (2009)
14. A.B. Olshen, E.S. Venkatraman, R. Lucito, M. Wigler, Circular binary segmentation for the analysis of array-based DNA copy number data. Biostatistics **5**(4), 557–572 (2004)
15. D.Y. Chiang, G. Getz, D.B. Jaffe, M.J.T. O'Kelly, X. Zhao, S.L. Carter, C. Russ, C. Nusbaum, M. Meyerson, E.S. Lander, High-resolution mapping of copy-number alterations with massively parallel sequencing. Nat. Methods **6**, 99–103 (2009)
16. T.M. Kim, L.J. Luquette, R. Xi, P.J. Park, rSW-seq: algorithm for detection of copy number alterations in deep sequencing data. BMC Bioinform. **11**(432), 1471–2105 (2010)

17. R. Xi, A.G. Hadjipanayis, L.J. Luquette, T.-M. Kim, E. Lee, J. Zhang, M.D. Johnson, D.M. Muzny, D.A. Wheeler, R.A. Gibbs, R. Kucherlapati, P.J. Park, Copy number variation detection in whole-genome sequencing data using Bayesian information criterion. PNAS **108**, E1128–E1136 (2011)
18. J. Chen, A.K. Gupta, *Parametric Statistical Change Point Analysis - With Applications to Genetics, Medicine, and Finance*, 2nd edn. (Birkhauser, New York, 2012)
19. H. Li, J. Vallandingham, J. Chen, SeqBBS: a change-point model based algorithm and R package for searching CNV regions via the ratio of sequencing reads, in *Proceedings of the 2013 IEEE International Workshop on Genomic Signal Processing and Statistics* (2013), pp. 46–49
20. J. Chen, Y.-P. Wang, A statistical change point model approach for the detection of DNA copy number variations in array CGH data. IEEE/ACM Trans. Comput. Biol. Bioinform. **6**, 529–541 (2009)
21. J. Chen, A. Yiiter, K.-C. Chang, A Bayesian approach to inference about a change point model with application to DNA copy number experimental data. J. Appl. Stat. **38**, 1899–1913 (2011)
22. L.J. Vostrikova, Detecting “disorder” in multidimensional random processes. Sov. Math. Dokl. **2**, 55–59 (1981)
23. R.E. Bellman, S.E. Dreyfus, *Applied Dynamic Programming* (Princeton University Press, Princeton, 1962)
24. www.Biobase-international.com

A Center-Level Approach to Estimating the Effect of Center Characteristics on Center Outcomes

Jennifer Le-Rademacher

Abstract This paper introduces a center-level approach to estimating the effect of center characteristics on outcomes. The proposed method applies to studies where the effect of center characteristics is of primary focus and centers rather than individual patients are entities of interest. Although these studies focus on practices and policies at the center level, it is important to account for the differences in outcomes due to varying patient case-mix. The proposed approach includes two steps. The first step estimates the effect of patient-level characteristics on outcomes so that the variability in patient case-mix can be adjusted prior to estimating the effect of center-level factors. The second step aggregates outcomes (adjusted for patient-level factors) of patients from the same center into a distribution of outcomes representing the response for each center. The outcome distributions are multi-valued responses on which the effects of center-level characteristics are modeled using a symbolic data framework. This method can be used to model the effect of center characteristics on the center-mean outcome as well as the within-center outcome variance. It models the effect of patient characteristics at the patient level and the effect of center characteristics at the center level. The method performs well even when the data come from a classical linear regression model or from a linear mixed effect model. The proposed approach is illustrated using a bone marrow transplant example.

Keywords Center analysis · Center effect · Clustered data · Multi-valued data · Symbolic data

Mathematics Subject Classification 62J99 · 62P10

J. Le-Rademacher (✉)
Department of Health Sciences Research, Mayo Clinic, Rochester, MN 55905, USA
e-mail: Le-Rademacher.Jennifer@mayo.edu

G. Letzter et al. (eds.), *Advances in the Mathematical Sciences*, Association for Women in Mathematics Series 6, DOI 10.1007/978-3-319-34139-2_14

1 Introduction

Medical centers are often interested in the impact of their policies and practices on outcomes. It is especially important for the centers to identify practices associated with poor outcomes. This knowledge may encourage center leadership to change their policies or modify their healthcare delivery models to improve outcomes for patients treated at their center. Loberiza Jr. et al. [20] conducted a study to evaluate the association between transplant center factors and mortality after hematopoietic cell transplantation (HCT). Their analysis included data from 163 transplant centers in the US. Among the 20 center-level factors considered, they found that higher physician caseload (more patients per physician) was associated with lower post-transplant mortality. Their analysis also suggested that early mortality (100 days post-transplant) in patients who received allogeneic transplants were lower at centers where a physician was the initial contact for after office or emergency calls. Another study that was conducted by Eckrich et al. [11] evaluated the association between center characteristics and mortality in pediatric HCT in the US in 2008 and 2009, and it also showed that higher physician caseload was associated with lower mortality. A more recent study of HCT cases in years 2008–2010 from the US conducted by Majhail et al. [21] suggested that higher center volume was associated with lower 100-day and 1-year mortality and having a long-term follow-up or survivorship program was associated with lower one-year mortality. Numerous studies in areas outside of HCT such as Donoghue et al. [10], Sun et al. [27], Trinh et al. [28], Safavi et al. [23], Sen et al. [24], and Sheetz et al. [25] illustrate interest in center characteristics' effect on outcomes across various types of medical areas.

A review of the current medical literature shows that center characteristics' effect is often analyzed using classical (generalized) linear regression [27, 28] or (generalized) linear mixed models [10, 20, 23–25]. These methods simultaneously estimate the effect of patient-level factors and the effect of center-level factors. Moreover, the effect of a covariate, whether measured at the individual patient level or at the center level, is modeled with the individual patient outcome as the response, i.e., patients are the units of observation when estimating the effect of patient-level factors as well as the effect of center-level factors. With widely varied center sizes, outcomes of patients from very large centers may bias the estimates of center characteristics' effect. Furthermore, while the mixed model with a random center effect accounts for potential correlation between outcomes of patients treated from the same center, which is an improvement over the classical linear regression model which assumes that outcomes from all patients are mutually independent, both of these methods assume patient outcomes have a constant variance across centers. Although the constant variance assumption can be relaxed and the association between the dispersion parameter and covariates can be modeled jointly with the mean using the hierarchical (generalized) linear models proposed by Lee and Nelder [15], fitting these models can be challenging in practice. More importantly, this method still models the effect of center-level covariates on individual patient outcome. Again, all these methods are appropriate for studies focusing on the impact of covariates at the individual patient level.

The focus of the method proposed in this paper is on the effect of center-level factors with centers as the units of observation. A simple method that treats centers as the observational units is a two-step approach that first models the effect of patient characteristics, and then the effect of center characteristics are modeled on the average center outcome after adjusting for patient characteristics. We refer to this approach as the center-average model. While the center-average model distinguishes patient-level characteristics from center-level characteristics and treats centers as the units of observation when estimating the effect of center characteristics, using only the center average discards potentially meaningful information such as outcome variability within center. In this paper, we propose a two-step approach that distinguishes patient characteristics from center characteristics and, at the same time, accounts for the variability in center outcomes.

Since patient outcomes are affected by patient-level factors as well as center-level factors, it is essential to account for the differences in outcomes among centers caused by varying patient case-mix in order to accurately identify center practices associated with outcomes. The first step of the proposed method adjusts for varying patient case-mix by first estimating the effect of patient-level characteristics treating individual patients as the units of observation as in step one of the center-average approach. However, motivated by He and Schaubel [12], we propose estimating the effect of patient factors by stratifying by center to avoid bias toward large centers and to preserve the differences among centers.

The second step of the proposed approach evaluates the effect of center-level factors, after adjusting for patient-level characteristics from step one, treating centers as the units of observation. Since this step focuses on the center-level effect on the center's overall outcomes, the entities of interest are the centers rather than the individual patients. Unlike the center-average approach which uses only the average outcome as the response for each center, this method considers the distribution of patient outcomes from each center as the center-level response. The symbolic data framework [8] is used in this step to allow inclusion of outcomes from all patients from each center in the analysis. Unlike a classical observation which takes a single value, a symbolic observation can take a set of values such as an interval or a histogram. Different types of symbolic data have different internal structure. Analysis of symbolic data can be complex due this internal structure. However, using the approach of Le-Rademacher and Billard [16], the outcome distributions can be modeled via their internal parameters using classical regression methods.

The paper is organized as follows. Section 2 describes the proposed method and gives a brief introduction to symbolic data. Section 3 describes the simulation study and shows the simulation results. Section 4 illustrates the proposed method using a bone marrow transplant data example. Discussion about the proposed method is given in Sect. 5.

2 The Proposed Method

The proposed method includes two steps. Step one evaluates the effect of patient-level characteristics on patient outcome using a stratified model. Justification for the use of a stratified model is given in Sect. 2.1. Step two evaluates the effect of center-level characteristics on the center-level outcome after adjusting for patient-level factors using the symbolic data framework. Section 2.2 briefly introduces symbolic data and explains the estimation procedure.

2.1 Adjusting for the Effect of Patient-Level Characteristics

Let Y_{ij} be the outcome (assumed to be a continuous variable) and $\mathbf{Z}_{ij}$ be the patient-level characteristics of patient $j (= 1, \ldots, s_i)$ from center $i (= 1, \ldots, n)$ where s_i is the number of patients from center i and n is the number of centers. The total number of patients from all centers is $N = \sum_{i=1}^{n} s_i$. Let $\boldsymbol{X}_i$ be the center-level characteristics of center i. The proposed modeling process follows.

The first step models the effect of patient-level characteristics $\mathbf{Z}_{ij}$ on the outcome Y_{ij} using the stratified model

$$Y_{ij} = \alpha_i + \mathbf{Z}_{ij}\boldsymbol{\beta} + \varepsilon_{ij}^{s} \tag{1}$$

where ε_{ij}^{s} are assumed to be independent and identically distributed with mean zero and variance σ_s^2. By the stratified model of Eq. (1), $\boldsymbol{\beta}$ represents the within center (or conditional) effect of patient-level characteristics on patient-level outcome and this effect is assumed to be the same across centers. Assuming that patient-factor effect is the same across centers allows separation of the variation caused by patient case-mix from the variation caused by center characteristics. The intercept α_i of Eq. (1) represents the average outcome of center i after adjusting for patient-level characteristics. Then,

$$R_{ij} = Y_{ij} - \mathbf{Z}_{ij}\boldsymbol{\beta} = \alpha_i + \varepsilon_{ij}^{s} \tag{2}$$

represents the outcome of patient j from center i adjusted for his/her patient-level characteristics (henceforth, referred to as the patient adjusted outcome). R_{ij} is the response variable of interest in the following step and R_{ij} can be estimated by $\hat{R}_{ij} = Y_{ij} - \mathbf{Z}_{ij}\hat{\boldsymbol{\beta}}$ where $\hat{\boldsymbol{\beta}}$ are unbiased estimators of $\boldsymbol{\beta}$.

2.2 Estimating the Effect of Center Characteristics

The second part of the analysis evaluates the effect of center-level characteristics on the overall center outcome. The entities of interest here are the centers rather than the individual patients, i.e., the units of observation are the centers. Let R_i be the set of adjusted outcomes of the s_i patients from center i, that is, $R_i = \{R_{ij}, j = 1, \ldots, s_i\}$. Under the classical data framework where each observation can only take a single value, the distribution of values in R_i is often reduced to a single summary value such as the mean, e.g., the center-average approach. If R_{ij} takes the same value for all $j = 1, \ldots, s_i$, the mean can sufficiently represent all values within R_i. Otherwise, using a single value may discard meaningful information about the distribution of values in R_i. To include more information about R_i in the analysis, we propose modeling R_i under the symbolic data framework where classically valued variable is considered a special case. A brief introduction of symbolic data follows. For a comprehensive treatment of the topic, refer to Billard and Diday [8, 9].

In the symbolic data domain, an observation can take multiple values. A common type of symbolic data includes interval data where an observation takes a range of values. For example, suppose R_i is an interval-valued random variable and let $[a_i, b_i]$ be a realization of R_i where $a_i \leq b_i$. It is assumed that all values within $[a_i, b_i]$ are uniformly distributed. Bertrand and Goupil [3] derived the empirical distribution for interval-valued data and defined its symbolic sample mean and its symbolic sample variance as

$$\bar{R}_i = \frac{1}{2n} \sum_{i=1}^{n} (a_i + b_i)$$

$$S^2 = \frac{1}{3n} \sum_{i=1}^{n} [(a_i - \bar{R}_i)^2 + (a_i - \bar{R}_i)(b_i - \bar{R}_i) + (b_i - \bar{R}_i)^2]. \tag{3}$$

Another common type of symbolic data is histogram data where a realization of a histogram-valued variable R_i can be expressed as

$$\left\{[a_i^1, b_i^1), p_i^1; [a_i^2, b_i^2), p_i^2; \ldots; [a_i^{s_i}, b_i^s], p_i^s\right\},$$

where $[a_i^l, b_i^l)$ is called the lth subinterval and p_{ij}^l its associated relative frequency. Billard and Diday [8] and Billard [5] derived the sample mean and the sample variance of histogram-valued R_i as

$$\bar{R}_i = \frac{1}{2n} \sum_{i=1}^{n} \sum_{l=1}^{s} p_i^l (a_i^l + b_i^l),$$

$$S^2 = \frac{1}{3n} \sum_{i=1}^{n} \sum_{l=1}^{s} p_i^l [(a_i - \bar{R}_i)^2 + (a_i - \bar{R}_i)(b_i - \bar{R}_i) + (b_i - \bar{R}_i)^2]. \tag{4}$$

Billard [4, 5] showed that the total sum of squares ($SST = nS^2$), a function of the sample variance S^2 of Eq. (3) for interval data and Eq. (4) for histogram data, can be decomposed into the sum of the internal variation, called the within sum of squares (SSW), and the external variation, called the between sum of squares (SSB). Specifically, $SST = SSW + SSB$. If R_i is a classically valued random variable, S^2 reduces to the classical sample variance where $SSW = 0$, i.e., symbolic data have an internal variance that does not exist in classical data. Extensions of classical statistical methods to symbolic data can be complex due to this internal variance. Methodological development in symbolic data analysis has focused mainly on interval data, including several extensions for linear regression [1, 6, 7, 17–19, 26, 29]. See Noirhomme-Fraiture and Brito [22] for a survey of methodological developments in symbolic data analysis. Some of the challenges encountered when fitting a linear regression model for interval data include adequately accounting for the internal variance, appropriately estimating the predicted response interval, and the lack of theory for inferences. Although inferences are possible using the methods proposed by Silva et al. [26] and Ahn et al. [1], computation for these approaches can be intensive. Furthermore, these methods apply to data where the response variables and the predictors are interval data.

A unique feature of the data in center-characteristic analysis is that the predictors $\boldsymbol{X}_i$ are classically valued while the response variable R_i is symbolic-valued (not necessary interval). Using the internal parameters concept of Le-Rademacher and Billard [16] along with the classical nature of $\boldsymbol{X}_i$, we propose modeling R_i via its internal parameters. Since the internal parameters are also classically-valued, the analysis can then be carried out using existing methods for classical data. This approach can be applied to various types of symbolic data and is not limited to interval data. Interpretations of the effect of center characteristics from this method are intuitive and inferences regarding center-level effect follow from classical linear regression theory. Further details of the proposed approach follow.

Following Le-Rademacher and Billard [16], let $R_i, i = 1, \ldots, n$, be independent symbolic-valued random variables. Suppose R_i take a distribution of values with a density function f with parameter vector $\boldsymbol{\theta}_i$ where the elements of $\boldsymbol{\theta}_i$ is the smallest set of parameters to ensure a one-to-one correspondence with R_i. For example, if R_i is an exponential distribution with mean λ_i then $f(x) = (e^{-x/\lambda_i})/\lambda_i$ and $\boldsymbol{\theta}_i = \lambda_i$, if R_i is an interval $[a_i, b_i]$ then $f(x) = 1/(b_i - a_i)$ and $\boldsymbol{\theta}_i = (a_i, b_i)$, and if R_i is a normal distribution with mean μ_i and variance σ_i^2 then f is the normal density function and $\boldsymbol{\theta}_i = (\mu_i, \sigma_i^2)$. The parameters $\boldsymbol{\theta}_i$ are called the internal parameters of R_i. That is, given a center i, the outcome of patient j, R_{ij}, follows a distribution with density function $f(\boldsymbol{\theta}_i)$ for $j = 1, \ldots, s_i$. Note that $\boldsymbol{\theta}_i$ is a random vector corresponding to the symbolic-valued random variable R_i. The dimension of $\boldsymbol{\theta}_i$ depends on the internal distribution of R_i. Using this internal parameter concept, Le-Rademacher and Billard [16] showed that the sample means and the sample variances in Eqs. (3) and (4) are the maximum likelihood estimators for the overall mean and the overall variance of interval data and histogram data, respectively. They further showed that for a symbolic random variable R_i with internal mean μ_i and internal variance σ_i^2, the overall variance of R_i is a sum of two components, namely, $E(\sigma_i^2) + V(\mu_i)$

which corresponds to the decomposition of $SST = SSW + SSB$ of Billard [4, 5] with $E(\sigma_i^2)$ being the mean internal variation and $V(\mu_i)$ the external variation.

Since $\boldsymbol{\theta}_i$ has a one-to-one relationship with R_i and $\boldsymbol{\theta}_i$ captures the total variance of R_i, the effect of $\boldsymbol{X}_i$ on R_i can be expressed in terms of its effect on $\boldsymbol{\theta}_i$. With $\boldsymbol{\theta}_i$ being a vector of classical values, the effect of $\boldsymbol{X}_i$ on $\boldsymbol{\theta}_i$ can be estimated using well established classical regression methods. Specifically, suppose R_i take a set of values from a distribution with parameters $\boldsymbol{\theta}_i = (\mu_i, \sigma_i^2)$ where $\mu_i = E_i(R_{ij}|R_i)$ and $\sigma_i^2 = V_i(R_{ij}|R_i)$. The effect of $\boldsymbol{X}_i$ on μ_i and σ_i^2 can be modeled by

$$\begin{aligned} \mu_i &= \gamma_{\mu 0} + \boldsymbol{X}_{\mu i}\gamma_{\mu} + \varepsilon_{\mu} \\ \xi_i &= \gamma_{\sigma 0} + \boldsymbol{X}_{\sigma i}\gamma_{\sigma} + \varepsilon_{\sigma} \end{aligned} \tag{5}$$

where $\xi_i = \log(\sigma_i^2)$, $\varepsilon_\mu \sim N(0, \sigma_\mu^2)$, and $\varepsilon_\sigma \sim N(0, \sigma_\xi^2)$. The predictors $\boldsymbol{X}_{\mu i}$ and $\boldsymbol{X}_{\sigma i}$ are subsets of $\boldsymbol{X}_i$ whose effect on μ_i and σ_i is being evaluated. The coefficient γ_μ in Eq. (5) represents the effect of $\boldsymbol{X}_i$ on the center-mean outcome whereas γ_σ represents the effect of these predictors on the within-center outcome variability. Inferences for $\gamma_\mu, \gamma_\sigma$ and model diagnostics for (5) follow Kutner et al. [14].

2.3 Comparison to other Methods

To see the difference between the proposed model compared to classical linear regression model and linear mixed model, let us look at the mean and the variance of the outcome at the patient level. Under classical linear regression model

$$Y_{ij} = \beta_0 + \boldsymbol{Z}_{ij}\beta + \boldsymbol{X}_i\gamma + \varepsilon_{ij} \tag{6}$$

where $\varepsilon_{ij}(j = 1, \ldots, s_i; i = 1, \ldots, n)$ are assumed to be independent and identically distributed (iid) from $N(0, \sigma^2)$; $E(Y_{ij}) = \beta_0 + \boldsymbol{Z}_{ij}\beta + \boldsymbol{X}_i\gamma$ and $V(Y_{ij}) = \sigma^2$. Under the linear mixed model

$$Y_{ij} = \alpha_i^r + \boldsymbol{Z}_{ij}\beta + \boldsymbol{X}_i\gamma + \varepsilon_{ij} \tag{7}$$

where ε_{ij} is defined as in model (6) and α_i^r is the random center effect assumed to be iid under $N(\alpha, \tau^2)$, $E(Y_{ij}) = \alpha + \boldsymbol{Z}_{ij}\beta + \boldsymbol{X}_i\gamma$ and $V(Y_{ij}) = \sigma^2 + \tau^2$.

Under the proposed model, using conditional expectation [16], the expected value of R_{ij} is

$$E(R_{ij}) = E[E(R_{ij}|R_i)] = E(\mu_i) = \gamma_{\mu 0} + \boldsymbol{X}_{\mu i}\gamma_{\mu} \tag{8}$$

and its variance is

$$\begin{aligned} V(R_{ij}) &= E[V(R_{ij}|R_i)] + V[E(R_{ij}|R_i)] \\ &= E(\sigma_i^2) + V(\mu_i) \\ &= E(\exp(\xi_i)) + \sigma_\mu^2 \\ &= \exp(\gamma_{\sigma 0} + \boldsymbol{X}_{\sigma i}\gamma_\sigma + \sigma_\xi^2/2) + \sigma_\mu^2. \end{aligned} \tag{9}$$

Equivalently, $E(Y_{ij}) = \gamma_{\mu 0} + \boldsymbol{Z}_{ij}\beta + \boldsymbol{X}_{\mu i}\gamma_\mu$ and $V(Y_{ij}) = \exp(\gamma_{\sigma 0} + \boldsymbol{X}_{\sigma i}\gamma_\sigma + \sigma_\xi^2/2) + \sigma_\mu^2$.

Although expected patient-level outcome from the proposed model is similar to the other two models, unlike those models which assume constant variance, the variance of Y_{ij} from the proposed model depends on the center-level covariates $\boldsymbol{X}_{\sigma i}$. The analysis of ξ_i, equivalently σ_i^2, provides an important piece of information that classical linear regression model, linear mixed model, and center-average model do not provide. It identifies practices and policies that impact the variability in outcome among patients treated at the same center after adjusting for patient-level risk factors. This information along with the knowledge of center characteristics affecting center-mean outcome (from the analysis of μ_i) can help centers modify their practices to ensure the outcomes of their patients are consistently favorable after adjusting for patient-level risk factors.

The effect of $\boldsymbol{X}_i$ on (μ_i, ξ_i) can be modeled using multivariate method if (μ_i, ξ_i) are assumed to be bivariate normal in situations where the covariance between μ_i and ξ_i cannot be assumed negligible. In the symbolic data framework, classical data are viewed as a special case. The following simulation study shows that the proposed method works as well as linear model and mixed model when applied to classical data generated under classical linear model or linear mixed model scenarios.

3 Simulation Study

A simulation study was conducted to compare the performance of the proposed method to that of classical linear model, mixed model, and the center-average approach. Three scenarios were considered: two under classical data models and one under the proposed symbolic data model.

Data for the first scenario were generated from the linear model of Eq. (6). Data for the second scenario were generated from the mixed model of Eq. (7). Data for the third scenario were generated from the proposed symbolic model of Eq. (5) using the following steps:

i. First, the means μ_i and the log-variances ξ_i were generated from Eq. (5).
ii. The variances $\sigma_i^2 = \exp(\xi_i)$ were computed.
iii. Next, the adjusted patient outcomes R_{ij} for $j = 1, \ldots, s_i$, were generated from $N(\mu_i, \sigma_i^2)$.
iv. Finally, the individual patient outcomes were computed as $Y_{ij} = R_{ij} + \boldsymbol{Z}_{ij}\beta$.

A simulation study with 10,000 replicates was conducted using various combinations of parameters. Results shown in this section represent a specific parameter combination; however, results from other parameter settings lead to the same conclusions. Data for the scenarios shown were generated using the following setting:

- sample size (number of centers), $n = 50$
- number of patients s_i generated from a discrete uniform (10, 100)
- patient-level covariates, $\boldsymbol{Z}_{ij}$, generated from $N(10, 5^2)$, $N(10, 10^2)$, $N(5, 2^2)$
- center-level covariates, $\boldsymbol{X}_i$ of Eqs. (6) and (7) and $\boldsymbol{X}_{\mu i}$ of Eq. (5), generated from $N(10, 5^2)$, $N(10, 10^2)$, $N(5, 2^2)$
- center-level covariates associated with within-center outcome variance, $\boldsymbol{X}_{\sigma i}$ of Eq. (5), generated from $N(1, 1)$ three times
- error term, ε_{ij} from model in Eqs. (6) and (7), generated from $N(0, 5^2)$
- random effect term, α_i^r from model in Eq. (7), generated from $N(2, 2^2)$
- error terms, ε_μ and ε_σ in Eq. (5), generated from $N(0, 1)$
- patient factor effect, $\boldsymbol{\beta}$, was set to $(0, 1, -5)$
- center characteristics effect, γ in Eqs. (6) and (7), and center characteristics effect on the mean center outcomes, γ_μ in Eq. (5), were set to $(0, 1, -.5)$ and
- center-level characteristics effect on the center variance, γ_σ in (5), was set to $(0, .5, -1)$.

Tables 1, 2, and 3 show the estimates, the root mean-squared errors (RMSEs), and the powers from four analysis approaches: classical linear model, the linear mixed model, the center-average model, and the symbolic center-level model. The results shown are for a sample size of 50. Similar conclusions can be drawn from a sample size of 100.

3.1 Results under Classical Linear Model Scenario

Table 1 shows the analysis results of data generated from the classical linear model of Eq. (6). Under this model, all four methods gave unbiased estimates for the effect of patient-level characteristics. However, the RMSEs from the center-average approach are slightly larger than the RMSEs from the other three approaches. The type I error (the probability of rejecting the null hypothesis $\beta_1 = 0$) is close to the 0.05 level and the power (probability of rejecting the hypothesis that $\beta_2 = 0$ or $\beta_3 = 0$) is high across all methods.

All four methods also gave unbiased estimates for the center characteristics effect with the RMSEs from the center-average approach and the symbolic center-level approach slightly larger than those from classical linear approach and the mixed approach. This is due to the fact the RMSEs from classical linear and the mixed approaches were computed based on the total number of patients $N = \sum_{i=1}^{n} s_i$ which is much larger than the number of centers n, the effective sample size in this step for the center-average approach and the symbolic approach. The type I error (the probability of rejecting the hypothesis $\gamma_{\mu_1} = 0$) is close to the 0.05 level for all methods. However, the type I error from the mixed model is lower than the other three approaches. All four methods have good power to detect the effect of center characteristics.

Under this scenario, classical linear model is expected to be the best performer. The simulation results suggest that the mixed method and the symbolic method work equally as well as classical linear regression. This scenario assumes no center characteristics effect on the within-center outcomes variance. The symbolic method correctly estimated $\gamma_\sigma = (0, 0, 0)$ with the type I error at the 0.05 level.

3.2 Results under the Linear Mixed Scenario

Table 2 shows the analysis results under the linear mixed model of Eq. (7). Similar to the results under classical linear model scenario, all four analyses provided unbiased estimates for patient factors effect with the RMSEs from the center-average approach larger than the other three methods. The type I error rate and the power are also similar across all methods.

As expected, with the additional variance due to the random center effect in this scenario, the RMSEs of the center characteristics effect are larger under this scenario than compared to the RMSEs from the classical linear model scenario. Again, all four methods provided unbiased estimates for the effect of center characteristics. The type I errors (probability of rejecting $\gamma_{\mu_1} = 0$) from the linear mixed method, the center-average method, and the symbolic method are close to the 0.05 level. However, the type I error from the classical linear approach is greatly inflated to 0.53. While all four methods have good power to detect center characteristics effect, the power is comparable among the linear mixed approach, the center-average approach, and the symbolic approach while the power from the classical linear approach is again highly inflated.

Under this model, the linear mixed approach is expected to be the best performer. The proposed symbolic approach performs equally as well as the linear mixed approach. The symbolic approach, again, correctly estimated $\gamma_\sigma = (0, 0, 0)$ with the type I error at the 0.05 level under this scenario.

3.3 Results under the Symbolic Center Model Scenario

Table 3 shows the simulation results under the symbolic data model of Eq. (5). Under this scenario, all four methods gave unbiased estimates for patient factors effect. However, the mixed approach and the symbolic approach produced much smaller RMSEs compared to classical linear approach and the center-average approach. This follows from the fact that both the mixed and the symbolic approaches estimated the conditional (within center) effect of patient-level characteristics; the mixed approach treated center as a random effect and the symbolic approach stratified by center. On

Table 1 Simulation results from the classical linear model

Parameter (true value)	Classical model			Mixed model			Center average			Symbolic		
	Estimate	RMSE	Power	Estimate	RMSE	Power	Estimate	RMSE	Power	Estimate	RMSE	Power
$\beta_1(0)$	0.00	0.019	0.048	0.00	0.019	0.048	0.00	0.043	0.049	0.00	0.019	0.049
$\beta_2(1)$	1.00	0.010	1.000	1.00	0.010	1.000	1.00	0.021	1.000	1.00	0.010	1.000
$\beta_3(-5)$	−5.00	0.048	1.000	−5.00	0.048	1.000	−5.00	0.106	1.000	−5.00	0.048	1.000
$\gamma_{\mu_1}(0)$	0.00	0.020	0.052	0.00	0.020	0.038	0.00	0.024	0.048	0.00	0.024	0.047
$\gamma_{\mu_2}(1)$	1.00	0.010	1.000	1.00	0.010	1.000	1.00	0.012	1.000	1.00	0.012	1.000
$\gamma_{\mu_3}(-0.5)$	−0.50	0.052	1.000	−0.50	0.052	1.000	−0.50	0.061	1.000	−0.50	0.061	1.000
γ_{σ_1}										0.00	0.007	0.052
γ_{σ_2}										0.00	0.004	0.049
γ_{σ_3}										0.00	0.018	0.050

Table 2 Simulation results from the linear mixed model

Parameter (true value)	Classical model			Mixed model			Center average			Symbolic		
	Estimate	RMSE	Power	Estimate	RMSE	Power	Estimate	RMSE	Power	Estimate	RMSE	Power
$\beta_1(0)$	0.00	0.020	0.048	0.00	0.019	0.049	0.00	0.043	0.050	0.00	0.019	0.049
$\beta_2(1)$	1.00	0.010	1.000	1.00	0.010	1.000	1.00	0.022	1.000	1.00	0.010	1.000
$\beta_3(-5)$	−5.00	0.051	1.000	−5.00	0.048	1.000	−5.00	0.107	1.000	−5.00	0.048	1.000
$\gamma_{\mu_1}(0)$	0.00	0.069	0.532	0.00	0.063	0.049	0.00	0.064	0.048	0.00	0.064	0.048
$\gamma_{\mu_2}(1)$	1.00	0.035	1.000	1.00	0.032	1.000	1.00	0.032	1.000	1.00	0.032	1.000
$\gamma_{\mu_3}(-0.5)$	−0.50	0.174	0.985	−0.50	0.161	0.862	−0.50	0.162	0.856	−0.50	0.162	0.857
γ_{σ_1}										0.00	0.007	0.052
γ_{σ_2}										0.00	0.004	0.049
γ_{σ_3}										0.00	0.018	0.050

Table 3 Simulation results from the symbolic center effect model

Parameter	Classical model			Mixed model			Center average			Symbolic		
	Estimate	RMSE	Power	Estimate	RMSE	Power	Estimate	RMSE	Power	Estimate	RMSE	Power
$\beta_1(0)$	0.00	0.029	0.049	0.00	0.005	0.053	0.00	0.039	0.051	0.00	0.005	0.053
$\beta_2(1)$	1.00	0.015	1.000	1.00	0.003	1.000	1.00	0.019	1.000	1.00	0.003	1.000
$\beta_3(-5)$	−5.00	0.073	1.000	−5.00	0.013	1.000	−5.00	0.095	1.000	−5.00	0.013	1.000
$\gamma_{\mu_1}(0)$	0.42	0.427	1.000	0.00	0.005	0.068	0.00	0.031	0.051	0.00	0.030	0.049
$\gamma_{\mu_2}(1)$	0.42	0.585	1.000	0.00	1.000	0.065	1.00	0.015	1.000	1.00	0.015	1.000
$\gamma_{\mu_3}(-0.5)$	0.42	0.924	1.000	0.00	0.500	0.065	−0.50	0.077	1.000	−0.50	0.077	1.000
$\gamma_{\sigma_1}(0)$										0.00	0.152	0.051
$\gamma_{\sigma_2}(.5)$										0.50	0.152	0.892
$\gamma_{\sigma_3}(-1)$										−0.99	0.152	1.000

Table 4 Patient-level characteristics associated with one-year survival probability with p-value from the stratified model of Eq. (1)

Patient-level characteristic	p-value
Patient age	<.0001
Karnofsky performance score	0.0002
Disease/status	<.0001
Prior autologous transplant	0.0237
Time from diagnosis to transplant	<.0001
Donor-recipient HLA match	0.0002
Unrelated donor age	0.0068
Sorror comorbidity score	0.0019

Table 5 Frequency of center characteristics considered in bone marrow transplant example

Center characteristic	Category	Number of centers (%)	Number of patients (%)
Allogeneic transplant volume in 2010	≤40	27 (40)	591 (18)
	>40	40 (60)	2729 (82)
Participation in any clinical trials in the past 12 months	No	5 (7)	102 (3)
	Yes	62 (93)	3218 (97)
Center ownership	Government	19 (28)	679 (20)
	Private	48 (72)	2641 (80)

the other hand, the classical linear and the center-average approaches estimated the marginal effect of patient-level characteristics.

Under this scenario, both the classical linear and the mixed model approaches failed to correctly estimate the effect of center-level characteristics. The classical linear approach rejected the null hypothesis almost 100 % of the time although $\gamma_{\mu_1} = 0$ while the power of the linear mixed method is extremely small for $\gamma_{\mu_2} = 1$, $\gamma_{\mu_3} = -.5$. The center-average approach and the symbolic approach both provided unbiased estimates for the center characteristics effect on the mean center outcome with similar RMSEs, type I error, and power. However, the center-average approach cannot estimate γ_σ.

When the data follow the symbolic model of Eq. (5), the center-average approach performed better than classical linear approach and the linear mixed approach, but the RMSEs for patient-level characteristics are much larger compared to the symbolic approach. Although all observations were used in the estimation in the classical linear and the mixed model approaches, both methods assume constant variance across all observations. When the variance depends on factors not included in their respective model, they failed to estimate the effect of center-level characteristics that were

Table 6 Effect of center characteristics on one-year survival after adjusting for patient-level characteristics using the symbolic approach

Outcome	Center characteristic	Estimate	SE	p-value
Mean	Allo volume ≤ 40	−0.072	0.027	0.009
	Participated in clinical trials	−0.013	0.049	0.794
	Privately owned	0.039	0.029	0.179
Variance	Allo volume ≤ 40	0.047	0.037	0.206
	Participated in clinical trials	−0.169	0.067	0.015
	Privately owned	0.005	0.039	0.907

included. Among all methods, only the symbolic approach can provide estimates for the center characteristics effect on within-center outcome variability.

The error terms shown in this example ($\varepsilon_\mu, \varepsilon_\sigma$ of Eq. (5)) were generated from a $N(0, 1)$ distribution. Their variances appear much smaller than the variance of $\varepsilon_{ij} \sim N(0, 5^2)$ of Eqs. (6) and (7) and the variance of the random effect term $\alpha_i^r \sim N(0, 2^2)$ in Eq. (7) used in previous scenarios. However, it is important to note that, by Eq. (2), the variance of Y_{ij} in this model is an exponential function of σ_ξ^2. The magnitude of the variance is further affected by a factor of $\exp(\gamma_{\sigma 0} + \boldsymbol{X}_{\sigma i} \boldsymbol{\gamma}_\sigma)$. Therefore, the variance of Y_{ij} is much larger than first appears and it can increase rapidly with a slight increase in $\boldsymbol{X}_{\sigma i} \boldsymbol{\gamma}_\sigma$. The RSMEs from a scenario in which ε_μ and ε_σ of Eq. (5) were generated from a $N(0, 2^2)$ distribution were larger than those shown in Table 3, but lead to the same conclusions as did simulation results with other parameter combinations. As expected, we observed smaller RMSEs with a larger number of centers ($n = 100$), higher number of patients per center ($s_i \sim Unif(20, 200)$), or with a smaller σ^2 ($\varepsilon_{ij} \sim N(0, 2^2)$).

4 Bone Marrow Transplant Example

The proposed method is illustrated using a subset of data collected by the Center for International Blood and Marrow Transplant Research (CIBMTR). The CIBMTR is comprised of clinical and basic scientists who confidentially share data on their blood and bone marrow transplant patients with the CIBMTR Data Collection Center located at the Medical College of Wisconsin. The CIBMTR is a repository of information about results of transplants at more than 450 transplant centers worldwide. The data used in this example came from a study conducted by Majhail et al. [21] evaluating the effect of transplant center practices on post-transplant survival.

The study included 11,537 patients with malignant and non-malignant diseases who received transplants with graft from related or unrelated donors between 2008 and 2010 from 85 centers. The study identified allogeneic transplant volume in year 2010 as the predictor of one-year overall survival. The one-year overall survival rate was higher at centers that performed 40 or more allogeneic transplants in 2010 compared to the survival rate at centers performed fewer than 40 allogeneic transplants.

A subset of patients (3320 from 67 centers) from the original study with common indications for allogeneic blood and marrow transplant including acute myeloid leukemia (AML), acute lymphoplastic leukemiea (ALL), chronic myeloid leukemia (CML), and myelodysplastic syndrome (MDS) who received bone marrow or peripheral blood stem cell from a sibling donor with matched human leukocyte antigen (HLA) or from an 8/8 or 7/8 HLA-matched unrelated donor was used to illustrate the proposed method. The outcome of interest in this example is the probability of survival at one year computed by the pseudo-values [2], which are the jack-knife estimates of the survival probability at one-year based on the Kaplan-Meier estimator [13]. Without censoring, the pseudo-values are the indicators of the survival status at one year. Since the proposed method applies to continuous response variable, the pseudo-values are used for illustration in place of the indicator of mortality.

Patient-level characteristics considered in this analysis include patient age, Karnofsky performance score, disease type/disease status, whether patient had a prior autologous transplant, time from disease diagnosis to transplant, donor-recipient HLA match, unrelated donor age, and Sorror comorbidity score index. These are patient-level variables known to affect outcomes after allogeneic transplant. Table 4 summarizes the effect of these patient-level characteristics on one-year survival probability from the model (Eq. (1)) stratified by transplant center.

The center-level characteristics considered in this example are center allogeneic transplant volume in 2010 ($\leq$40 vs. >40), whether the center participated in any clinical trials in the past 12 months, and whether the center is owned by the federal government. Table 5 shows the frequencies of these characteristics in terms of the number of centers (third column) as well as the number of patients (fourth column). Note that the number of patients is strongly correlated with allogeneic transplant volume. Although 40 % of centers performed 40 or fewer allogeneic transplant in 2010, patients from these centers only make up 18 % of the total patients in the data set. When estimating the effect of center characteristics, the proposed symbolic approach treats each center as an observation whether 20 or 50 patients came from the center; whereas in classical approaches (classical linear model and the mixed model) patients are the observations. Therefore, many more observations come from very large centers. Outcomes of patients from these large centers may influence the estimates of center characteristics effect.

After adjusting for the patient-level characteristics (shown in Table 4), the outcomes of patients from the same center were combined into a distribution of outcome representing the center. Figure 1 shows the center variance versus the center mean of the adjusted one-year survival probabilities for the 67 centers. The center-mean survival probabilities at one-year post-transplant range from 55–95 %. The variance of one-year survival probabilities for centers range from 13–30 %. Linear regres-

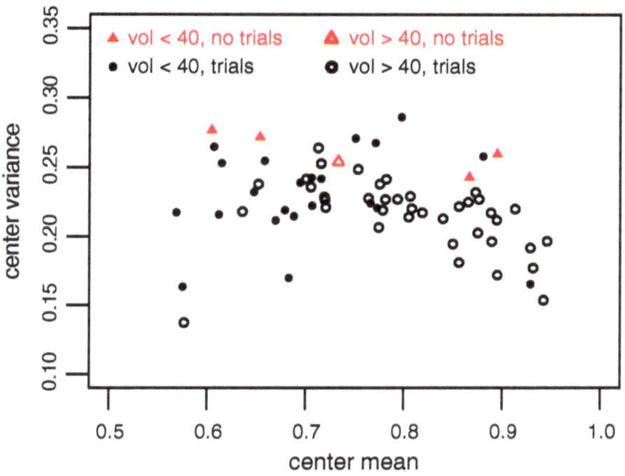

Fig. 1 Center-variance versus center-mean survival probability at one-year

sion models were fit to estimate the effect of center allogeneic transplant volume, participation in clinical trials, and center ownership on the mean survival probability at one year and on the log within-center variance of the survival probability.

Results of the analysis (shown in Table 6) suggest that allogeneic transplant volume in 2010 is a significant predictor of center-mean survival probability with allogeneic transplant volume of 40 or fewer associated with lower survival at one year. This finding is consistent with the result of the original study [21]. Furthermore, participation in clinical trials in the past 12 months is a significant predictor of outcome variability with participation in clinical trials associated with lower variability in one-year survival, i.e., outcomes of patients from centers that participated in any clinical trials in the past 12 months were more consistent than outcomes of patients from centers that did not participate in any clinical trials in the past 12 months. These results are also illustrated in Fig. 1.

For comparison, the data were also analyzed using classical linear model, the mixed model, and the center-average model (results shown in Table 7). Results from these approaches also suggest that allogeneic transplant volume of 40 or less in 2010 was associated with lower survival probability at one year post transplant which is consistent with the conclusions from the symbolic model analysis. However, the effect of participation in clinical trials on outcome consistency cannot be estimated using these methods.

5 Discussion

This paper presents a method to evaluate the effect of center characteristics on overall center outcome. The proposed method focuses on the impact of center-level factors and treats centers as the units of observation. To account for differences in outcomes

Table 7 Effect of center characteristics on one-year survival after adjusting for patient-level characteristics using other approaches

Center characteristic	Classical model			Mixed model			Center average		
	Estimate	SE	p-value	Estimate	SE	p-value	Estimate	SE	p-value
Allo volume ≤40	−0.097	0.023	<.001	−0.091	0.026	<.001	−0.072	0.027	0.009
Participated in clinical trials	0.006	0.049	0.900	0.001	0.054	0.986	−0.012	0.049	0.800
Privately owned	0.030	0.021	0.163	0.030	0.026	0.248	0.037	0.028	0.192

among centers caused by varying patient case-mix, the first step of the method models the effect of patient-level characteristics on individual patient outcome stratified by center. This allows separation between the effect of patient-level factors and the effect of center-level factors.

In contrast with classical linear model and the linear mixed model which estimate the effect of center characteristics on individual patient outcome and the center-average model which estimates the effect on the average center outcome, the proposed method considers the effect of center characteristics on the distribution of outcomes from all patients in the center. Although modeling the variance of a distribution seems a trivial extension, this paper provides theoretical justification for this extension using the symbolic data framework. Using the internal parameter approach of Le-Rademacher and Billard [16] and the fact that center characteristics are classically valued in this type of analysis, well established classical regression models can be used to estimate the effect of center-level factors. This approach accounts for the within-center variation in the outcome distributions while avoiding challenges often encountered in symbolic data regression. Modeling the effect of center characteristics on the internal parameters provide intuitive interpretations. Using classical regression models further allows inferences of model parameters as well as model diagnostics. Most importantly, abundance of software exists for classical regression analysis makes implementation of this approach readily available.

As illustrated in the bone marrow transplant example, the proposed method allows estimation of the effect of center characteristics on more than just the mean. The effect of center characteristics on other measures of center performance such as consistency of center outcomes can be evaluated using this method. Moreover, even when the true effect of center characteristics follow the classical linear model or the linear mixed model, this method performs as well as the method under the true model. On the other hand, classical linear approach and the mixed model approach fail to estimate the effect of center-level factors under the symbolic center effect scenario.

The proposed method is easy to implement. Interpretation of the resulting models from this approach is intuitive. The simulation study and the bone marrow transplant example showed that the method performs well. Validations of the proposed methods to other clinical research areas are ongoing. Some areas of special interest include diabetes research and quality of life (QOL) research where glucose levels or QOL questionnaire data are collected at multiple time points. For these outcomes, the fluctuation, i.e., variance, in glucose levels or QOL scores, can be as clinically relevant as the mean. This method allows identification of factors affecting the within-patient glucose or QOL variance which can lead to better strategies to control fluctuation of sugar levels in diabetics or of quality of life in cancer survivors.

References

1. J. Ahn, M. Peng, C. Park, Y. Jeon, A resampling approach for interval-valued data regression. Stat. Anal. Data Min. **5**, 336–348 (2012)
2. P.K. Andersen, J.P. Klein, J.S. Rosth, Generalized linear models for correlated pseudo-observations with applications to multi-state models. Biometrika **90**, 15–27 (2003)
3. P. Bertrand, F. Goupil, Descriptive statistics for symbolic data, in *Analysis of Symbolic Data: Explanatory Methods for Extracting Statistical Information from Complex Data*, ed. by H.H. Bock, E. Diday (Springer, Berlin, 2000), pp. 106–124
4. L. Billard, Dependencies and variation components of symbolic interval-valued data, in *Selected Contributions in Data Analysis and Classification*, ed. by P. Brito, G. Cucumel, P. Bertrand, F. de Carvalho (Springer, Berlin, 2007), pp. 3–12
5. L. Billard, Sample covariance functions for complex quantitative data, in *Proceedings of the International Association of Statistical Computing Conference 2008* (Yokohama, Japan, 2008)
6. L. Billard, E. Diday, Regression analysis for interval-valued data, in *Data Analysis, Classification, and Related Methods*, ed. by H.A.L. Kiers, J.P. Rassoon, P.J.L. Groenen, M. Schader (Springer, Berlin, 2000), pp. 369–378
7. L. Billard, E. Diday, Symbolic regression analysis, in *Classification, Clustering and Data Analysis: Proceedings of the 8th Conference of the International Federation of Classification Societies (IFCS '02)* (Springer, Poland, 2002), pp. 281–288
8. L. Billard, E. Diday, From the statistics of data to the statistics of knowledge: symbolic data analysis. J. Am. Stat. Assoc. **98**, 470–487 (2003)
9. L. Billard, E. Diday, *Symbolic Data Analysis: Conceptual Statistics and Data Mining* (Wiley, New York, 2006)
10. A.J. Donoghue et al., Effect of hospital characteristics on outcomes from pediatric cardiopulmonary resuscitation: a report from the national registry of cardiopulmonary resuscitation. Pediatrics **118**, 995–1001 (2006)
11. M.J. Eckrich, J. Le-Rademacher, J.D. Rizzo, K.S. Baker, D. Bhatla, D.K. Buchbinder, K.R. Cooke, R. Duerst, A. Gilman, H. Frangoul, N. Kapoor, N.A. Kernan, M. Verneris, M. Eapen, The Effect of Transplant Center Characteristics On Survival After Pediatric Hematopoietic Cell Transplantation. Blood (ASH Annual MeetingAbstracts) **120**, 762 (2012)
12. K. He, D.E. Schaubel, Methods in comparing center-specific survival outcomes using direct standardization. Stat. Med. **33**, 2048–2061 (2014)
13. E.L. Kaplan, P. Meier, Non-parametric estimation from incomplete observations. J. Am. Stat. Assoc. **53**, 457–481 (1958)
14. M.H. Kutner, C.J. Nachtsheim, J. Neter, W. Li, *Applied linear statistical models* (McGraw-Hill, New York, 2005)
15. Y. Lee, J.A. Nelder, Hierarchical generalized linear models. J. Roy. Stat. Soc.: Ser. B (Methodol.). **1**, 619–78 (1996)
16. J. Le-Rademacher, L. Billard, Likelihood functions and some maximum likelihood estimators for symbolic data. J. Stat. Plan Inference **141**, 1593–1602 (2011)
17. N.E. Lima, G. Cordeiro, F. de Carvalho, Bivariate symbolic regression models for interval-valued variables. J. Stat. Comput. Simul. **81**, 1727–1744 (2011)
18. N.E. Lima, F. de Carvalho, Center and range method for fitting a linear regression model to symbolic interval data. Comput. Stat. Data Anal. **54**, 333–347 (2008)
19. N.E. Lima, F. de Carvalho, C.P. Tenorio, Univariate and multivariate linear regression to predict interval-valued features, in *Lecture Notes in Computer Science, AI 2004 Advances in Artificial Intelligence* (Springer, Berlin, 2004), pp. 207–216
20. F.R. Loberiza Jr., M.J. Zhang, S.J. Lee et al., Association of transplant center and physician factors on mortality after hematopoietic stem cell transplantation in the United States. Blood **105**, 2979–2987 (2005)
21. N.S. Majhail, T. Payton, L.W. Mau et al., Provider and center characteristics of US transplant centers and their association with survival after allogeneic hematopoietic cell transplantation

(hct) in adults: results from a national survey conducted by the center for international blood and marrow transplant research (CIBMTR). Blood **122**, 1687 (2013)
22. M. Noirhomme-Fraiture, P. Brito, Far Beyond the classical data models: symbolic data analysis. Stat. Anal. Data Min. **4**, 157–170 (2011)
23. K.V. Safavi, S.X. Li, K. Dharmarajan et al., Hospital variation in the use of noninvasive cardiac imaging and its association with downstream testing, interventions, and outcomes. JAMA Intern. Med. **17**, 546–553 (2014)
24. S. Sen, P.R. Soulos, J. Herrin et al., For-profit hospital ownership status and use of brachytherapy after breast-conserving surgery. Surgery **155**, 776–788 (2014)
25. K.H. Sheetz, S.A. Waits, M.E. Girotti et al., Patient's Perspectives of care and surgical outcomes in michigan - an analysis using the CAHPS hospital survey. Ann. Surg. **260**, 5–9 (2014)
26. A. Silva, N.E. Lima, U. Anjos, A regression model to interval-valued variables based on copula approach, in *Proceedings of the 58th World Statistics Congress of the International Statistical Institute* (Dublin, Ireland, 2011)
27. M. Sun, M. Bianchi, Q.D. Trinh et al., Hospital volume is a determinant of postoperative complications, blood transfusion and length of stay after radical or partial nephrectomy. J. Urol. **187**, 405–410 (2012)
28. Q.D. Trinh, M. Sun, S. Kim et al., The impact of hospital volume, residency, and fellowship training on perioperative outcomes after radical prostatectomy. Urol. Oncol.: Semin. Orig. Invest. **32**, 13–20 (2014)
29. W. Xu, Symbolic Data Analysis: Interval-Valued Data Regression. Ph.D. thesis, University of Georgia (2010)

False Discovery Rate Based on Extreme Values in High Dimension

Junyong Park, DoHwan Park and J. Wade Davis

Abstract In recent years, there has been much work done on high dimensional problems in both theory and applications since high dimensional data are getting more common in broad areas such as microarray data analysis. One important issue in multiple testing problems in high dimensional data is controlling the significance level of large scale simultaneous testing to select significant ones among huge number of genes. In many cases, the true null distribution is assumed to be well-known or a parametric distribution so that p-values can be easily calculated. In practice, the true null distribution may be misspecified or different from the assumed distribution. In this paper, we consider a procedure for a *FDR* based on extreme values which is less sensitive to inaccurate p-values. The normalized forms are assumed to be approximately a standard normal by the central limit theorem (*CLT*). Comparing to the *CLT* approximation, we show that FDR procedure with extreme values achieves a more accurate simultaneous test level under some weaker conditions on sample sizes. We provide simulation studies and a real data example to compare the performance of our proposed procedure and an existing procedure.

Keywords False discovery rate · Extreme value · High dimension · Sparsity

Mathematics Subject Classification Primary · 62H15 · Secondary · 62G32

J. Park · D. Park (✉)
Department of Mathematics and Statistics, University of Maryland
Baltimore County, 1000 Hilltop Circle, Baltimore, MD 21250, USA
e-mail: dhpark@umbc.edu

J. Park
e-mail: junpark@umbc.edu

J.W. Davis
Biostatistics and Research Design Group, University of Missouri,
1 Hospital Dr, Columbia, MO 65212, USA
e-mail: davisjwa@health.missouri.edu

G. Letzter et al. (eds.), *Advances in the Mathematical Sciences*, Association for Women in Mathematics Series 6, DOI 10.1007/978-3-319-34139-2_15

1 Introduction

For the last two decades, high dimensionality is one of the most important issues in statistics especially in the practical use of microarray data. Microarrays can produce from several to more than ten thousand measurements from each of assayed sample simultaneously, but due to the cost, it is common to have a very small sample size. More often, our goal is to identify specific genes to investigate more in details to the individual level on the selected ones.

Let us define the independent random variables, Z_i, $1 \leq i \leq n$. It is of interest to test n null hypothesis simultaneously.

$$H_{0i} : \mu_i = 0 \ \ vs. \ \ H_{1i} : \mu_i > 0, \tag{1}$$

where $\mu_i = E(Z_i)$. Similarly, two sided alternative hypothesis $H_{1i} : \mu_i \neq 0$ can be also considered. A well known procedure to overcome the challenge is Benjamini and Hochberg [1] method (henceforth called the BH-procedure) which is based on the idea of controlling false discovery rate (henceforth FDR). The use of *FDR* goes back to Seeger [12], Simes [12], and Sorić [14], but it became very popular after Benjamini and Hochberg [1] proved a theorem on the control of the false discovery rate which is considered to be the Type I error rate. They also showed that the BH-procedure controls family wise error rate (FWER) under the null hypothesis. The BH-procedure depends on the p-values, denoted by p_is, obtained from Z_is, $1 \leq i \leq n$. The calculation of p-values is based on the assumption of distribution of Z_is or an approximation to Z_is. One of the most typical situations happens when Z_i's have a normalized form such as $Z_i = \frac{\sqrt{m}\bar{X}_i}{\sigma}$, where $\bar{X}_i = \frac{1}{m}\sum_{j=1}^{m} X_{ij}$ with $\mu_i^* = E(X_{ij})$ and $Var(X_{ij}) = \sigma^2$ for some i.i.d. random variables X_{ij}s $1 \leq j \leq m$. Let $\mu_i = \frac{\sqrt{m}\mu_i^*}{\sigma}$; then we have $E(Z_i) = \mu_i$ and $Var(Z_i) = 1$. In such a case, it is natural to consider normal approximation to Z_i by the central limit theorem (*CLT*). However, when the sample size m is not large enough, it is well known that the *CLT* may not provide sufficiently good approximations, especially in the tail. Since small p-values (the tail part) play a major role in the BH-procedure, there is a high risk of misleading results in high dimensional multiple testing problems with small sample m from depending on the *CLT*. As a result, the BH-procedure may fail to control a given level of FDR. Therefore, it is of interest to consider developing a procedure which is less sensitive to inaccurate approximation of p-values than the BH-procedure.

We propose an approach which is based on extreme values and a correction of the selection bias by conditional probability. The dimension reduction is expected from high to moderate dimension by considering only extreme values. Also the conditional probability will remedy bias due to selecting extreme values. This approach of conditional probability has been adopted in many areas, for instance, Greenshtein et al. [8] used conditional likelihood to correct selection bias in high dimensional classification, and Park and Davis [11] considered the conditional maximum likelihood estimation in the high dimensional hypothesis testing problems.

We assume a sparsity condition which means we have a small number of alternative hypotheses to be true. It is a natural and important assumption in high dimensional data. Under the sparsity conditions, most of Z_is observed under alternative hypotheses are expected to be large; so for some given k, $k(<< n)$, extreme values from Z_is are expected to be from alternatives. Of course, some alternative hypotheses could be out of k extreme value; however, we may obtain an advantage of huge dimension reduction if we consider a test procedure based on only k extreme values to detect alternative hypotheses. With this motivation, we shall propose a test procedure that uses only k extreme values.

We demonstrate that the procedure based on k extreme values is less sensitive to the departure from the normality in the sense that the proposed procedure controls *FDR* more reasonably than the BH-procedure when underlying distributions are different from exact normal distribution.

The rest of this paper is organized as follows. In Sect. 2, we review the *FDR* and BH-procedure and demonstrate the motivation of our proposed procedure which is less sensitive to the deviation from normality compared to the BH-procedure. Section 3 provides the derivation of our procedure and Sects. 4 and 5 present simulation studies and a real example of lung cancer data. In Sect. 6, concluding remarks are presented.

2 Overview of *FDR* and BH-procedure

In this section, we briefly review the *FDR* and BH-procedure and point out that the BH-procedure may be sensitive to the deviations from normality when the number of hypothesis is large and the sample size is small.

Let V and R be the number of falsely rejected hypotheses and rejected hypotheses, respectively. FDR is defined by

$$FDR = E\left(\frac{V}{R}\middle| R > 0\right) \tag{2}$$

and FWER is $P(V > 0)$. In high dimensional multiple testing problem, FDR controlling procedure is more popular than FWER controlling procedure since it is well known that the former procedure achieves more power than the latter procedure. One of the *FDR* controlling procedures is the BH-procedure by Benjamini and Hochberg [1] who show that the BH-procedure controls FWER under the null hypotheses when all hypothesis are independent.

To demonstrate the BH-procedure, we first assume that all p-values p_i's are independent. If Z_is are generated from a normal distribution, we obtain $p_i = 1 - \Phi(Z_i) = \bar{\Phi}(Z_i)$. Let $p_{(i)}$ be the order statistics of p_is, $p_{(1)} < p_{(2)} < \cdots < p_{(n)}$ and $H_{(i)}$ be the corresponding hypothesis. Given a level of FDR, namely, α, we take l such that

$$l = \operatorname{argmax}\left\{i : p_{(i)} < \frac{i\alpha}{n}\right\} \tag{3}$$

and reject the first l hypotheses, $H_{(i)}$, $1 \le i \le l$. Benjamini and Hochberg [1] show that the BH-procedure has the property of controlling FWER as well as FDR. From (3), we see that the BH-procedure depends on p-values which are obtained from test statistics Z_is with the knowledge of the distributions of Z_i or approximation to Z_is. Particularly, when Z_i has the normalized form of X_{ij}s for $1 \le j \le m$, it is natural to consider the normal approximation by *CLT* for the calculation of p-value. More specifically, when there are i.i.d. samples X_{ij} for $1 \le j \le m$ for $Var(X_{ij}) = \sigma_i^2$, the normalized form is

$$Z_i = \frac{\sqrt{m}\bar{X}_i}{\sigma_i}, \tag{4}$$

where $\bar{X}_i$ is the sample mean of X_{ij} for $1 \le j \le m$. When F is the true distribution function of Z_i under the null hypothesis, the true p-value, $1 - F(Z_i)$, is not directly available since F is unknown or complicated in computation in practice. Therefore, it is natural to consider approximation of $1 - F(z_i)$ by $1 - \Phi(z_i)$ from the *CLT*. However, the *CLT* does not provide accurate approximation especially in the tail part of Z_i. This is demonstrated as follows by Feller [5, p. 520]:

$$\frac{1 - F(z)}{1 - \Phi(z)} \sim \exp\left(\frac{\xi_3}{6\sigma^3}\frac{z^3}{\sqrt{m}}\right), \tag{5}$$

where $\xi_3 = E(X_{ij}^3)$ and $\sigma^2 = Var(X_{ij})$. Since $z = O(\sqrt{\log n})$ under H_{0i} for all $1 \le i \le n$, $\exp\left(\frac{\xi_3}{6\sigma^3}\frac{x^3}{\sqrt{m}}\right) \sim 1$ if n and m have the relationship $(\log n)^3 = o(m)$. In other words, the normal approximation from the *CLT* is valid when the number of hypotheses n and the sample size m have the relationship $(\log n)^3 = o(m)$. Since the BH-procedure depends on p-values from Z_is, it is expected that the BH-procedure based on p-values from the *CLT* is reliable under $(\log n)^3 = o(m)$; otherwise, the BH-procedure may have misleading results, for instance, failing to control a given level of Type I error rate like the FDR.

The main goal of this paper is to consider a multiple testing procedure which is less sensitive to inaccurate normal approximation by relaxing the condition $(\log n)^3 = o(m)$. The true null distribution belongs to a class of distributions which are approximately normal distribution in the following sense. The tail probability of Z_i has the following property:

$$P(Z_i > z) = \exp\left(-\frac{1}{2}z^2(1 + o(1))\right) \tag{6}$$

when $z \to \infty$ and $z = o(\sqrt{m})$. Note that the tail of a standard normal distribution has also

$$\bar{\Phi}(x) \sim \frac{1}{\sqrt{2\pi}x} \exp\left(-\frac{1}{2}x^2\right) = \exp\left(-\frac{1}{2}x^2(1+o(1))\right). \quad (7)$$

Therefore $o(1)$s in (6) and (7) may differ. Even when the $o(1)$ terms are slightly different, such differences are expected to cause sensitivity in high dimensional testing problems. The level of a simultaneous test such as the BH-procedure, is accurate when $\log n = o(m^{1/3})$ from (5). This relationship may break down if the sample size (m) is too small or the dimension (n) is too large.

In the next section, we propose a multiple testing procedure based on k extreme values for some k and conditional probabilities. We demonstrate that the BH-procedure based on extreme values is less sensitive to inaccuracy of p-values from normal approximation than the BH-procedure under some class of true null distributions. We discuss how to choose k depending on n as well based on our asymptotic results.

3 FDR Based on Extreme Values

As demonstrated in the previous section, it is well-known that the BH-procedure controls a given level of *FDR* α when all hypotheses are independent. In practice, p-values are obtained under some distributional assumption of test statistics Z_is. However, it is often hard to identify the distribution of Z_is, therefore it is natural to consider an approximation such as normal approximation to the distribution of Z_is. From the approximation to Z_is, it is inevitable that one obtains approximate p-values. Therefore the BH-procedure based on inaccurate p-values may give misleading test results. It is of interest to develop a "robust" procedure in high dimensional multiple testing in the sense that the procedure is less sensitive to the errors by approximation of p-values. From the view point of FDR, we define a "robust" procedure as follows.

Definition 1 Given approximate p-values, a procedure for hypothesis (1) is robust if the procedure with its $FDR_n = E(V_n/R_nI(R_n > 0))$ controls a given level of *FDR* α; that is,

$$FDR_n \leq \alpha \text{ as } n \to \infty. \quad (8)$$

Let F be the true distribution function of Z_i. Then the p-value is $1 - F(Z_i) = \bar{F}(Z_i)$ which could be either impossible to identify or very difficult to calculate. If F is exactly normal, the BH-procedure is robust, otherwise the BH-may not control right α level of *FDR*. Throughout this paper, we are interested in approximating the normal distribution especially in the tail. So we consider a class of distribution such that the behavior of the tail probability of Z_i under H_0 has

$$P(Z_i > x) = \bar{F}(x) \sim \alpha(x)\bar{\Phi}(x) \text{ as } x \to \infty. \quad (9)$$

for some function $\alpha(x)$ satisfying $\log \alpha(x) = o(x^2)$ as $x \to \infty$. Here, $a(x) \sim b(x)$ means for two different functions $a(x)$ and $b(x)$, $a(x)/b(x) \to 1$ as $x \to \infty$.

For example, if $\alpha(x) = 1$, the tail probability of Z_i is exactly the same as that of a standard normal distribution, otherwise $1 - F(x)$ and $1 - \Phi(x)$ have similar forms, like (6) and (7) except $o(1)$ term. To summarize, we define the following class of distribution functions F, denoted by $\mathcal{F}$

$$\mathcal{F} = \left\{ F : \bar{F}(x) \sim \alpha(x)\bar{\Phi}(x), \quad \frac{d}{dx}\log \alpha(x) = o(x), \log \alpha(x) = o(x^2) \right\}. \quad (10)$$

Depending on the behavior of $\alpha(x)$, an approximate p-value obtained from normal approximation may be seriously different from the true p-value, $\bar{F}(x)$. For instance, when $\alpha(x) = x^3$, then $\frac{d}{dx}\log \alpha(x) = o(x)$; so the corresponding distribution function is in $\mathcal{F}$. However, as $x \to \infty$, $\frac{\bar{F}(x)}{\bar{\Phi}(x)} \to \infty$. Therefore, an approximate p-value $\bar{\Phi}(x)$ could be much different from $\bar{F}(x)$ depending on $\alpha(x)$ for the large value of x.

In addition to the class of F, we consider another important structural assumption called sparsity. The sparsity condition is a general assumption in high dimensional data in which only a small number of features or hypotheses is significant and the others are just noise or from null hypotheses. In the context of the multiple testing problem in (1), large values of test statistics Z_is are believed to be generated from alternative hypotheses without further information. More formally, let us define the order statistics $Z_{(1)} \leq Z_{(2)} \leq \cdots \leq Z_{(n)}$ of $Z_1, Z_2, \ldots, Z_n$ and

$$(Y_0, Y_1, \ldots, Y_k) \equiv (Z_{(n-k)}, Z_{(n-k+1)}, \ldots, Z_{(n)}) \quad (11)$$

be the largest $k + 1$ test statistics for some $k << n$. Similarly, we can define hypothesis $H_{(i)}$ corresponding to $Z_{(i)}$. One advantage of considering $Y_0, \ldots, Y_k$ instead of $Z_1, \ldots, Z_n$ is reducing dimension of hypotheses from n to k. Based on the k largest test statistics or equivalently k smallest p-values, we propose a more robust procedure which is less sensitive to approximate p-values than the BH-procedure.

Using k extreme values, we actually test the following random hypothesis which is modification of (1): for $1 \leq i \leq k$,

$$H'_{0i} : \mu'_i = 0 \quad vs. \quad H'_{1i} : \mu'_i > 0, \quad (12)$$

where μ'_i is the mean value corresponding to Y_i. Therefore, we focus on $(\mu'_1, \ldots, \mu'_k)$ corresponding to $(Y_1, \ldots, Y_k)$ instead of the original $(\mu_1, \ldots, \mu_n)$. Note that μ'_is may not be ordered although Y_i's are.

Modification (12) defines an uncommon hypothesis, called a random hypothesis. This is because the identity of the tested hypotheses is random; it depends on the (random) selection of the k largest statistics. In data mining, the practice is also known as "data snooping" or "fishing". However this selection procedure can lead to a very misleading statistical inference due to selection bias. In our context, we correct selection bias by considering conditional probabilities. Conditional probabilities have been used to correct selection bias in some previous studies; classification in Greenshtein et al. [8], estimation in Greenshtein et al. [9], Park and Davis [11] and Park [10]; and hypothesis testing in Greenshtein et al. [7] and Park and Davis [11].

Based on the selected k extreme values and corresponding hypotheses, we develop a procedure which controls *FDR* more reasonably than the BH-procedure. Of course, since the proposed procedure tests only a subset of hypotheses, it could be conservative in the sense that procedures for testing (12) tend to reject a smaller number of hypotheses than the BH-procedure. However, the BH-procedure can be too liberal to control a given level of *FDR*, while the proposed procedure based on extreme values is expected to be more reliable.

We now describe our proposed procedure. Let Z_i be standardized as in (4). We assume that n and m increase simultaneously, where m depends on n, and adopt the notation of $F_{m,n}$ for the distribution of Z_i under the null hypotheses. Similarly, we define $\alpha_{n,m}(z)$ and

$$\mathcal{F}_{n,m} = \left\{ F_{n,m} : \bar{F}_{n,m}(z) = \alpha_{n,m}(z)\bar{\Phi}(z),\ \frac{d}{dz}\log\alpha(x) = o(z^2),\ \log\alpha_{n,m}(z) = o(z^2) \right\}.$$

However, for notational simplicity, we suppress all subscripts n and m.

Lemma 1 *If Z_i has the form of (4) for $1 \le i \le n$, then for $z = o(m^{1/2})$, $F \in \mathcal{F}$.*

Proof From (5), let $\alpha(z) = \exp\left(\frac{\xi_3}{6\sigma^3}\frac{z^3}{\sqrt{m}}\right)$. Then $\log\alpha(z) = \frac{\xi_3}{6\sigma^3}\frac{z^3}{\sqrt{m}} = o(z^2)$ due to $z = o(\sqrt{m})$. □

Our proposed procedure is simply the BH-procedure with k extreme values. More formally, we apply the BH-procedure to the k p-values $(p_{(1)}, \ldots, p_{(k)})$ corresponding to the k extreme values $(Y_1, Y_2, \ldots, Y_k)$. For this purpose, we want $(p_{(1)}, \ldots, p_{(k)})$ to be uniformly distributed under the null hypotheses. However, this is not true since there exists selection bias, i.e., $(p_{(1)}, \ldots, p_{(k)})$ are obtained from k largest values among n values where $n >> k$. To consider the BH-procedure only for k p-values, we need to correct the bias. One way is to consider conditional probabilities. Given $Y_0 = Z_{(n-k)}$ as in (11), define the conditional probability of $Y_i = Z_{(n-k+i)}$, $p^{c*}_{(i)} \equiv \bar{F}(Y_i)/\bar{F}(Y_0)$. Then $(p^{c*}_{(1)}, \ldots, p^{c*}_{(k)})$ are uniformly distributed under the null hypotheses. Instead of $p^{c*}_{(i)}$, we use approximate probabilities based on a normal approximation, denoted by $p^{c}_{(i)}$:

$$p^{c}_{(i)} = \frac{\bar{\Phi}(Y_i)}{\bar{\Phi}(Y_0)}. \tag{13}$$

For $F \in \mathcal{F}$, we already observed that approximate p-values may not be accurate due to $\bar{F}(x)/\bar{\Phi}(x) \sim \alpha(x)$ which may go to ∞ or 0 depending on $\alpha(x)$. On the other hand, conditional probabilities from normal approximation can be close to the true conditional probability if, for $y > c$,

$$\frac{\bar{F}(y)/\bar{F}(c)}{\bar{\Phi}(y)/\bar{\Phi}(c)} \sim \frac{\alpha(y)}{\alpha(c)} = \frac{\exp\left(\frac{\mu_3}{6\sigma^3}\frac{y^3}{\sqrt{m}}\right)}{\exp\left(\frac{\mu_3}{6\sigma^3}\frac{c^3}{\sqrt{m}}\right)} = \exp\left(\frac{\mu_3}{6\sigma^3}\frac{y^3 - c^3}{\sqrt{m}}\right) \tag{14}$$

is close to 1. We shall show that the above at most ratio converges to 1 uniformly in $1 \leq i \leq k_n$ for some choice of k_n.

We first present the following lemma which is used in proving the theorem.

Lemma 2 *When* $k_n/n \to 0$ *and* $k_n \to \infty$, *then* $X_{(n)} - X_{(n-k_n)} = \frac{\log k_n}{\sqrt{2\log n}}(1 + o_p(1))$.

Proof: See Appendix. □

From Lemma 2, we obtain the following theorem which states that a test based on the conditional probability for k_n largest values is expected to be more robust than the BH-procedure.

Theorem 1 *If the following conditions hold;* $(\log k_n)^2 \log n = o(m)$, $k_n \to \infty$ *and* $y - c = O(\frac{\log k_n}{2\sqrt{\log n}})$ *for* $y > c$, *then* $\frac{\bar{F}(y)/\bar{F}(c)}{\bar{\Phi}(y)/\bar{\Phi}(c)} \to 1$.

Proof: Since $\bar{F}(y)/\bar{\Phi}(y) \sim \exp(\frac{\mu_3}{6\sigma^3}\frac{y^3}{\sqrt{m}})$,

$$\begin{aligned}
\frac{\bar{F}(y)/\bar{F}(c)}{\bar{\Phi}(y)/\bar{\Phi}(c)} &\sim \exp\left(\frac{y^3 - c^3}{\sqrt{m}}\right) = \exp\left(\frac{(y-c)(y^2 + cy + c^2)}{\sqrt{m}}\right) \\
&\sim \exp\left(O\left(\frac{2\log k_n}{\sqrt{\log n}}\right) 3\frac{2\log n}{\sqrt{m}}\right) \\
&\sim \exp\left(O\left(\frac{\log k_n \sqrt{\log n}}{\sqrt{m}}\right)\right) = \exp(o(1)) \sim 1
\end{aligned}$$

from $(\log k_n)^2 \log n = o(m)$. □

Theorem 1 states that conditional probabilities from a normal approximation are uniformly close to the true conditional probabilities while unconditional p-values from normal approximation could be different from the true p-values. Therefore, the BH-procedure based on approximate p-values may produce misleading test results while the procedure based on the approximate conditional probabilities is not affected seriously from normal approximation. We evaluate the BH-procedure and our procedure from the view point of FDR in later sections with simulations and real data example.

To summarize, our procedure, called *eFDR* procedure, based on k_n extreme values is as follows.

1. Select $k_n + 1$ largest observations, $(Y_0, Y_1, \ldots, Y_{k_n}) \equiv (X_{(n-k_n)}, X_{(n-k_n+1)}, \ldots, X_{(n)})$.
2. Calculate conditional p-values of Y_i, $1 \leq i \leq k_n$:

$$p^c_{(i)} = \frac{1 - \Phi(Y_k)}{1 - \Phi(Y_0)}$$

3. Apply the BH-procedure to $p^c_{(1)} < p^c_{(2)} < \cdots < p^c_{(k_n)}$.

Theorem 1 uses the condition on k_n, which is $(\log k_n)^2 \log n = o(m)$, while the FDR procedure based on p-values requires $(\log n)^3 = o(m)$ from the explanation after Eq. (5). We recommend taking $k_n = (\log n)^{(\log n)^\tau}$ for $\tau = \frac{1}{2} - \varepsilon$. The motivation for this choice is that we have $o(m) = (\log n)^{2\tau - \varepsilon}(\log \log n)^2 = O((\log n)^2)$ which means the proposed procedure reduces a $\log n$ term compared to $(\log n)^3 = o(m)$ from directly using p-values. Of course, k_n can be either larger or smaller than our recommended k_n. In general, there is some risk for large or small values of k_n. If k_n is too large or close to n, we don't have advantage compared to FDR based on p-value. On the other hand, if k_n is too small, then we will lose lots of true alternatives. We will leave the choice of k_n for future work.

4 Simulations

In order to evaluate the performance of the proposed procedure, we present simulation studies. We compare our proposed procedure with the BH-procedure from the view point of FDR and V. It is expected that V obtained from the BH-procedure is larger than that from the proposed procedure for heavier tailed distribution than normal. However, our procedure is expected to preserve a given level of *FDR* more reasonably than the BH-procedure. We consider various configurations based on combinations of p, m, and l (the number of alternatives). We use $n = 2 \times 10^4, 5 \times 10^4, 10^5$, $m = 20, 30, 50$. For different combinations of (n, m), we also consider several values of l from 200 to 2000. We consider the following formulation. Define $Z_i = \frac{\sqrt{m}\bar{X}_{ij}}{\sigma} + \mu_i$ where X_{ij}'s are i.i.d. with $E(X_{ij}) = 0$ and $\sigma^2 = Var(X_{ij})$; l μ_i's are generated from Uniform (2, 4) and the other $p - l$ μ_i's are 0. Then our hypothesis is $H_{0i} : \mu_i = 0$ and $H_{1i} : \mu_i \neq 0$ for $1 \leq i \leq p$. For distributions of X_{ij}, we consider the normal, t and centered exponential distribution. Our interest is to find alternatives from p hypothesis while controlling FDR. In our simulations, we approximate *FDR* using the false discovery proportion (FDP): Note

$$E\left(\frac{R}{V}\middle| R > 0\right) \approx \frac{1}{L}\sum_{i=1}^{L}\frac{R_i}{V_i}I(V_i > 0),$$

where $\frac{R_i}{V_i}$ is the FDP obtained from the ith simulated data and $L = 1000$ simulations. Throughout the simulations, we consider the level of *FDR* =0.05. Table 1 shows the simulations when X_{ij}'s are generated from $N(0, 1)$. The BH-procedure and proposed procedure control a given level of FDR, in particular the BH-procedure seems controls more closely to 0.05 while the proposed procedure is a bit more conservative.

Table 2 displays the results when X_{ij}'s are generated from the centered exponential distribution. In other words, $X_{ij} = Z_{ij} - 1$ where $Z_{ij} \sim exp(1)$. We consider different configurations of (n, l, m) and the results show that the *FDR* of the BH-

Table 1 $Z_i = X_i + \mu_i$ where X_is are generated from N(0, 1)

	eFDR			BH		
(n, l)	*FDR*	$E(V)$	$E(R)$	*FDR*	$E(V)$	$E(R)$
$(2 \times 10^4, 200)$	0.036	51.3	1.9	0.047	57.9	2.8
$(2 \times 10^4, 500)$	0.019	150.7	2.9	0.048	209.8	10.2
$(5 \times 10^4, 500)$	0.032	139.3	4.6	0.050	166.3	8.4
$(5 \times 10^4, 1000)$	0.018	260.8	4.7	0.049	388.7	19.2
$(10^5, 1000)$	0.027	252.3	6.8	0.049	322.4	16.0
$(10^5, 2000)$	0.011	448.2	5.0	0.048	793.3	38.7

μ_i for $1 \leq i \leq l$ are generated from uniform [2, 4] and $\mu_i = 0$ for $l+1 \leq i \leq n$

Table 2 $Z_i = \frac{\sqrt{m}(\bar{X}_i - 1)}{\sigma} + \mu_i$ where X_{ij}s are generated from exponential distribution with 1

	eFDR			BH		
(n, l, m)	*FDR*	$E(V)$	$E(R)$	*FDR*	$E(V)$	$E(R)$
$(2 \times 10^4, 200, 20)$	0.268	64.0	17.3	0.348	98.6	34.6
$(2 \times 10^4, 200, 30)$	0.214	59.1	12.9	0.286	85.9	24.9
$(2 \times 10^4, 200, 50)$	0.163	58.3	9.6	0.221	79.6	17.9
$(2 \times 10^4, 500, 20)$	0.133	132.3	17.7	0.235	255.8	60.4
$(2 \times 10^4, 500, 30)$	0.101	140.6	14.4	0.193	252.0	49.0
$(2 \times 10^4, 500, 50)$	0.075	145.9	11.1	0.155	247.4	38.6
$(5 \times 10^4, 500, 20)$	0.241	136.4	33.0	0.349	247.7	86.9
$(5 \times 10^4, 500, 30)$	0.193	127.1	24.7	0.289	214.4	62.2
$(5 \times 10^4, 500, 50)$	0.138	135.0	18.8	0.217	207.3	45.3
$(5 \times 10^4, 1000, 20)$	0.129	229.1	29.8	0.256	521.5	133.7
$(5 \times 10^4, 1000, 20)$	0.102	234.2	24.0	0.213	488.6	104.6
$(5 \times 10^4, 1000, 20)$	0.073	245.4	18.0	0.167	474.1	79.6
$(10^5, 10^3, 20)$	0.217	233.1	50.9	0.350	495.2	173.9
$(10^5, 10^3, 30)$	0.170	225.6	38.5	0.285	437.2	125.0
$(10^5, 10^3, 50)$	0.124	232.1	28.9	0.220	407.5	90.0
$(10^5, 2 \times 10^3, 20)$	0.107	374.3	40.3	0.255	1056.6	270.5
$(10^5, 2 \times 10^3, 20)$	0.083	370.9	30.8	0.213	959.8	205.3
$(10^5, 2 \times 10^3, 20)$	0.058	390.3	22.7	0.169	931.3	157.9

μ_i for $1 \leq i \leq l$ are generated from uniform [2, 4] and $\mu_i = 0$ for $l+1 \leq i \leq n$

procedure exceeds 0.05 more seriously than the proposed procedure. For instance, when $(n, l, m) = (2 \times 10^4, 500, 20)$, *FDR* from the BH-procedure is 0.167, while the FDR from the proposed procedure is 0.073. Of course, as we pointed out, the number of rejected hypothesis from the BH-procedure is larger than those from the proposed procedure. However, a higher proportion of discoveries is falsely rejected hypothesis in the BH-procedure.

Similarly, Table 3 shows the results when X_{ij}s are generated from t distribution with degrees of freedom 5 for different configurations of (n, l, m). Similar to Table 2,

Table 3 $Z_i = \frac{\sqrt{m}\bar{X}_i}{\sigma} + \mu_i$ where X_{ij}s are generated from t_5

	eFDR			BH		
(n, l, m)	FDR	$E(V)$	$E(R)$	FDR	$E(V)$	$E(R)$
$(2 \times 10^4, 200, 20)$	0.113	63.2	7.2	0.131	72.6	9.6
$(2 \times 10^4, 200, 30)$	0.087	62.0	5.5	0.106	71.2	7.6
$(2 \times 10^4, 200, 50)$	0.068	60.8	4.2	0.084	69.0	5.9
$(2 \times 10^4, 500, 20)$	0.053	160.5	8.5	0.090	230.1	20.9
$(2 \times 10^4, 500, 30)$	0.043	158.0	6.8	0.078	222.9	17.6
$(2 \times 10^4, 500, 50)$	0.033	160.7	5.4	0.068	226.2	15.5
$(5 \times 10^4, 500, 20)$	0.103	139.7	14.6	0.132	175.6	23.3
$(5 \times 10^4, 500, 30)$	0.079	142.7	11.3	0.105	175.9	18.6
$(5 \times 10^4, 500, 50)$	0.061	132.2	8.2	0.085	162.5	13.9
$(5 \times 10^4, 1000, 20)$	0.054	260.3	14.2	0.099	418.5	41.9
$(5 \times 10^4, 1000, 30)$	0.041	269.8	11.2	0.083	418.6	34.9
$(5 \times 10^4, 1000, 50)$	0.032	272.6	8.8	0.071	416.9	29.7
$(10^5, 10^3, 20)$	0.092	265.4	24.6	0.130	365.4	47.9
$(10^5, 10^3, 30)$	0.073	251.1	18.4	0.106	340.9	36.3
$(10^5, 10^3, 50)$	0.056	252.4	14.2	0.085	334.7	28.7
$(10^5, 2 \times 10^3, 20)$	0.044	425.2	19.0	0.099	839.2	83.7
$(10^5, 2 \times 10^3, 30)$	0.032	439.0	14.1	0.083	849.2	71.2
$(10^5, 2 \times 10^3, 50)$	0.023	435.9	10.4	0.071	815.8	58.2

μ_i for $1 \leq i \leq l$ are generated from uniform [2, 4] and $\mu_i = 0$ for $l + 1 \leq i \leq n$

Table 3 also shows that the BH-procedure produces larger *FDR*s than the proposed procedure, while the number of rejected hypothesis from the BH-procedure is larger than from the proposed procedure.

Figure 1 show the trace plots of FDP_l and V_l for 1000 repetition in simulations. FDP_ls from *eFDR* are smaller than those from the BH-procedure so that the *eFDR* procedure provides more reliable FDP values while *eFDR* rejects more conservatively than the BH-procedure.

To summarize, when underlying distribution is exactly normal, then the BH-procedure is more accurate in controlling a given level of FDR than *eFDR*. However, when the underlying distribution is not guaranteed to be exactly normal or heavy tailed distribution, then *eFDR* gives more reasonable sizes of *FDR*; while the BH-procedure obtains inflated FDR values leading to greater proportion of falsely rejected hypothesis than the proposed procedure.

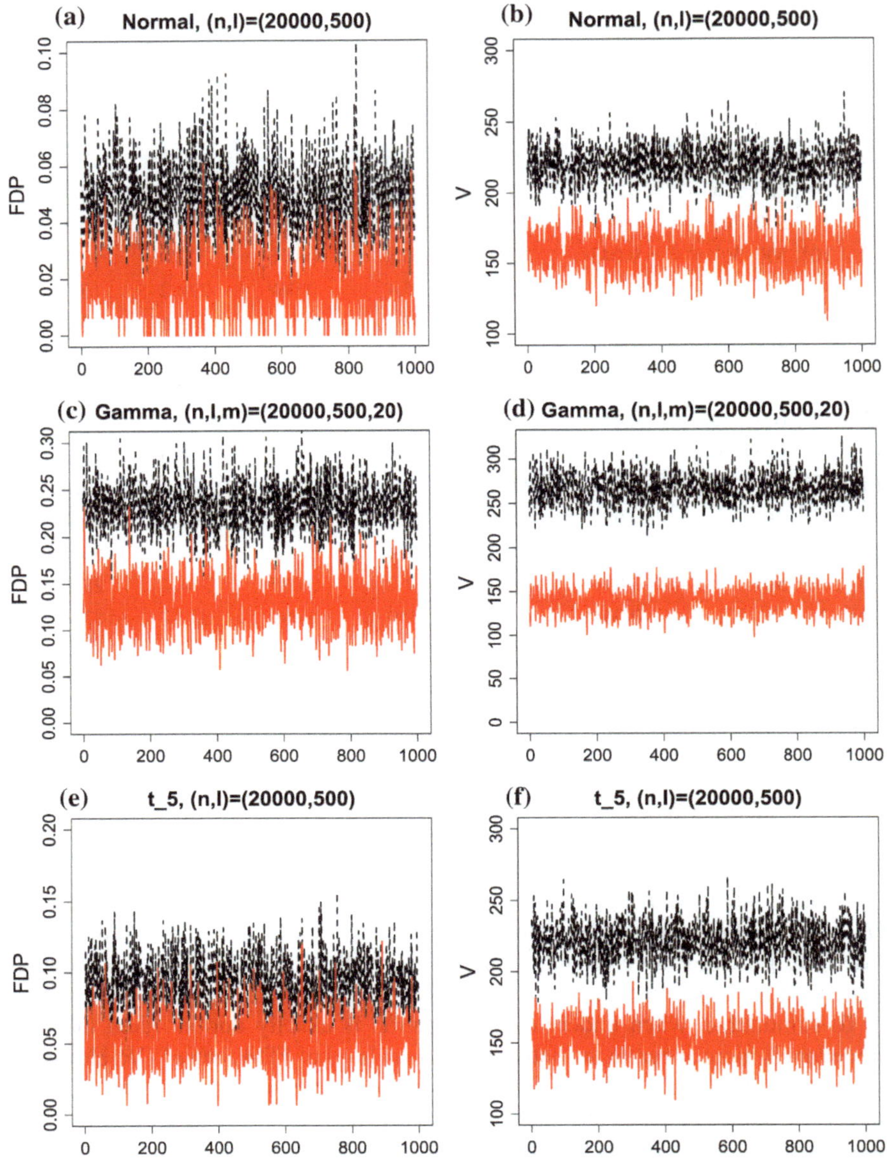

Fig. 1 *Left panels* show the plots of $FDP_i = \frac{R_i}{V_i}$ for $1 \leq i \leq 1000$. *Right panels* show the plots of V_i for $1 \leq i \leq 1000$. *Dotted lines* and *solid lines* are for BH-procedure and eFDR procedure, respectively

5 Real Example

Now we consider lung cancer data which were previously analyzed by Gordon et al. [6]. The data is available at http://www.chestsurg.org. There are $p = 12{,}533$ genes and 181 samples coming from two classes, $n_1 = 29$ samples from malignant pleural mesothelioma (MPM) and $n_2 = 150$ from adenocarcinoma (ADCA).

Since the variance is unknown, for each gene, we calculate two sample t-statistics for ith gene, say t_i, with pooled sample variance, $z_i = \Phi^{-1}(F_{n_1+n_2-2}(t_i))$. As pointed out by Efron [4], the null distribution may not be a standard normal distribution, $N(\mu_0, \sigma_0^2)$. Therefore, it is reasonable to estimate the null distribution empirically and adjust the p-values. Efron [4] uses quadratic approximation at $z = 0$ and obtains $(\hat{\mu}_0, \hat{\sigma}_0)$. For more details, see Efron [4]. Once we have estimators of μ_0 and σ_0, we calculate p-values from $N(\hat{\mu}_0, \hat{\sigma}_0^2)$ or equivalently, use $z_i^\times = \frac{z_i - \hat{\mu}_0}{\hat{\sigma}_0}$ under $N(0, 1)$. For $n = 12,533$, $k = (\log n)^{(\log n)^{0.5}} = 987$. So approximating, we use $k = 900$.

Given $\alpha = 0.05$, the BH-procedure selects 171 significant genes while *eFDR* selects 142 significant genes. If the null distribution is fairly close to normal distribution, the BH-procedure with estimation of the null distribution in Efron [4] would be ideal and our proposed procedure considered conservative. However, if the true null distribution seriously deviates from normality, then our proposed would be more reliable in the sense that FDR is controlled more reasonably while the BH-procedure has more falsely rejected hypothesis than a given level of FDR.

6 Concluding Remark

In this paper, we investigated the sensitivity of the BH-procedure to deviation of the null distribution from normality and propose a *FDR* procedure based on extreme values of observations. We demonstrated that considering k extreme values first reduces dimension from n to k as well as improve accuracy of p-values corresponding to k extreme values. Simulation studies show that our proposed procedure controls *FDR* more reasonably than the BH-procedure. This means the BH-procedure discovers more alternatives. However, the BH-procedure has a larger proportion of false discoveries. In real data example of microarray data, the BH-procedure obtains more discoveries, but the obtained discoveries may not be reliable in case the true null distribution of gene expressions are not close to normal distribution. Therefore, although the proposed procedure selects fewer genes, those selected genes are considered to be more reliable than those selected from the BH-procedure from the view point of controlling a given *FDR*.

Appendix

Proof of Lemma

Define $a_{n,n}$ and a_{n,k_n} satisfying $\bar{F}(a_{n,n}) = \frac{1}{n}$ and $\bar{F}(a_{n,k_n}) = \frac{k_n}{n}$, respectively. Each of them is upper $1/n$ or k_n/n quantile of the distribution F. Using $\log \alpha(x) = o(x^2)$ and Mill's ratio $1 - \Phi(x) \sim \frac{\phi(x)}{x}$ for $x \to \infty$, we have

$$1 - F(x) = \alpha(x)(1 - \Phi(d_n)) \sim \frac{\alpha(x)}{x}\phi(x) \sim \exp\left(-\frac{1}{2}x^2(1 + o(1))\right) \quad (15)$$

we can derive $a_{n,n} \sim \sqrt{2\log n}$ and $a_{n,k_n} \sim \sqrt{2\log \frac{n}{k_n}}$. $b_n(X_{n,n} - a_{n,n})$ and $c_n(X_{n,k_n} - a_{n,k_n})$ have nondegenerating asymptotic distribution (but not necessarily normal distribution) for some increasing sequences b_n and c_n. See more details in Chibisov [2]. From this, we have $X_{n,n} = \sqrt{2\log n} + O_p(b_n^{-1}) = \sqrt{2\log n} + o_p(1)$ and $X_{n,n} = \sqrt{2\log \frac{n}{k_n}} + O_p(c_n^{-1}) = \sqrt{2\log \frac{n}{k_n}} + o_p(1)$. Therefore,

$$\begin{aligned} X_{(n)} - X_{(n-k_n)} &= \sqrt{2\log n} - \sqrt{2\log(n/k_n)} + o_p(1) \\ &= \frac{2\log n - 2\log(n/k_n)}{\sqrt{2\log n} + \sqrt{2\log(n/k_n)}} + o_p(1) \\ &= \frac{2\log k_n}{\sqrt{2\log n}} + o_p(1). \end{aligned}$$

□

References

1. Y. Benjamini, Y. Hochberg, Controlling the false discovery rate : a practical and powerful approach to multiple testing. J. Roy. Stat. Soc. Ser. B **57**, 1165–1188 (1995)
2. D.M. Chibisov, On limit distributions for order statistics. Theory Probab. Appl. **9**, 142–148 (1964)
3. S. Dudoit, M.J. van der Laan, *Multiple Testing Procedures with Applications to Genomics* (Springer, Heidelberg, 2008)
4. B. Efron, Large-Scale simultaneous hypothesis testing: the choice of an null hypothesis. J. Am. Stat. Assoc. **99**, 96–104 (2004)
5. W. Feller, *An Introduction to Probability and its Applications* (Wiley, New York, 1966)
6. G.J. Gordon, R.V. Jensen, L.L. Hsiao, S.R. Gullans, J.E. Blumenstock, S. Ramaswamy, W.G. Richards, D.J. Sugarbaker, R. Bueno, Translation of microarray data into clinically relevant cancer diagnostic tests using gene expression ratios in lung cancer and mesothelioma. Cancer Res. **62**, 4963–4967 (2002)
7. E. Greensthien, J. Park, Robust test for detecting a signal in a high dimensional sparse normal vector. J. Stat. Plan. Inference **142**, 1445–1456 (2012)

8. E. Greenshtein, J. Park, G. Lebanon, Regularization through variable selection and conditional MLE with application to classification in high dimensions. J. Stat. Plan. Inference **139**, 385–395 (2009)
9. E. Greensthein, J. Park, Y. Ritov, Estimating the mean of high valued observations in high dimension. J. Stat. Theory Pract. **2**, 407–418 (2008)
10. J. Park, Shrinkage estimator in normal mean vector estimation based on conditional maximum likelihood estimators. Stat. Probab. Lett. **93**, 1–6 (2014)
11. J. Park, J.W. Davis, Estimating sums of means of high valued observations in high dimensional multivariate binary data and its application. J. Stat. Plan. Inference. **141**, 10211030 (2011)
12. P. Seeger, A note on a method for the analysis of significances en mass. Technometrics **10**, 586–593 (1968)
13. R.J. Simes, An improved Bonferroni procedure for multiple tests of significance. Biometrika **73**, 586–593 (1986)
14. B. Sorić, Statistical "discoveries" and effect size estimation. J. Am. Stat. Assoc. **84**, 608–610 (1989)

Part VI
Differential Equations

Asymptotic and Oscillatory Behavior of Dynamic Equations on Time Scales

Raegan Higgins

Abstract In this study, the behavior of solutions to certain second-order nonlinear dynamic equations on unbounded time scales are considered. Our objective is to obtain conditions for the existence of solutions of this dynamic equation. Results from the theory of lower and upper solutions for related dynamic equations and results from calculus will be used to reach our goal.

Keywords Oscillation · Asymptotic behavior · Dynamic equations · Time scales

Mathematics Subject Classification 34N05 · 39A10 · 39A21

1 Introduction

Following Stefan Hilger's milestone paper [12], a rapidly growing body of literature has sought to unify, extend, and generalize ideas from the continuous calculus, discrete calculus, and quantum calculus to arbitrary time-scale calculus. Since the introduction of time scales, many authors have expounded on various aspects of this new theory. A book on the subject by Bohner and Peterson [6] summarizes and organizes much of the time scale calculus; we also refer to [5] by Bohner and Peterson for advances in dynamic equations on time scales.

For over a decade, there has been significant interest in studying the asymptotic behavior solutions of dynamic equations on a time scale. This has lead to many attempts to harmonize the oscillation theory for the continuous and the discrete cases, to include them in one comprehensive theory, and to extend the results to

Raegan presented this work at the special session "Research from the Cutting EDGE." Raegan is the seventeenth member of the EDGE (Enhancing Diversity in Graduate Education) Program to receive her doctorate in Mathematics. The goal of EDGE is to strengthen the ability of women to successfully complete their graduate programs in the Mathematical Sciences. Please see the preface for more information about the EDGE Program and its founders.

R. Higgins (✉)
Texas Tech University, 2500 Broadway Ave, Lubbock, TX 79409, USA
e-mail: raegan.higgins@ttu.edu

G. Letzter et al. (eds.), *Advances in the Mathematical Sciences*, Association for Women in Mathematics Series 6, DOI 10.1007/978-3-319-34139-2_16

more general time scales. Some references are [1, 4, 7, 10, 11], and those cited therein.

Throughout this paper, we assume t_0 is nonnegative and belongs to the time scale $\mathbb{T}$ which is unbounded above. We define the time scale interval $[t_0, \infty)_{\mathbb{T}}$ by

$$[t_0, \infty)_{\mathbb{T}} := [t_0, \infty) \cap \mathbb{T}.$$

We are concerned with the asymptotic and oscillatory behavior of the solutions of the second-order nonlinear dynamic equation

$$y^{\Delta\Delta}(t) + f(t, y^{\sigma})g(y^{\Delta}) = 0, \tag{1}$$

where $\sup \mathbb{T} = \infty$. We assume
(A_0) $f, f_y : \mathbb{T} \times \mathbb{R} \to \mathbb{R}$ are continuous in y and right-dense continuous in t and $g : \mathbb{R} \to \mathbb{R}$ is continuous.
(A_1) $f(t, 0) = 0,\ t \in [t_0, \infty)_{\mathbb{T}}$.
(A_2) $f_y(t, y) \geq 0$ and is nondecreasing in y for $t \in [t_0, \infty)_{\mathbb{T}}$ and $y \geq 0$.
(A_3) $g(v) > 0$ for all $v \in \mathbb{R}$.

By considering

$$y^{\Delta\Delta}(t) + f_y(t, \alpha)y = 0,$$

where $\alpha \in \mathbb{R}$ depends on the solutions of (1), we will study (1). In this paper, we intend to use the method of upper and lower solutions to obtain oscillation criteria for (1) under certain conditions.

2 A Time-Scale Introduction

A time scale $\mathbb{T}$ is an arbitrary nonempty closed subset of the real numbers; see [5, 6]. Within that set, define the forward jump operator $\sigma : \mathbb{T} \to \mathbb{T}$ as

$$\sigma(t) = \inf\{s \in \mathbb{T} : s > t\}$$

and the backward jump operator $\rho : \mathbb{T} \to \mathbb{T}$ by

$$\rho(t) = \sup\{s \in \mathbb{T} : s < t\}$$

where $\inf \emptyset := \sup \mathbb{T}$ and $\sup \emptyset := \inf \mathbb{T}$. A point $t \in \mathbb{T}$ is right-dense, right-scattered, left-dense, left-scattered if $\sigma(t) = t,\ \sigma(t) > t,\ \rho(t) = t,\ \rho(t) < t$, respectively. The graininess functions $\mu, \nu : \mathbb{T} \to [0, \infty)$ are defined by $\mu(t) = \sigma(t) - t$ and $\nu(t) = t - \rho(t)$. Define the set $\mathbb{T}^{\kappa}$ by $\mathbb{T}^{\kappa} = \mathbb{T} \setminus \{M\}$ if $\mathbb{T}$ has a left-scattered maximum M and $\mathbb{T}^{\kappa} = \mathbb{T}$ otherwise.

Definition 1 Assume $f : \mathbb{T} \to \mathbb{R}$ is a function and let $t \in \mathbb{T}^{\kappa}$. Define $f^{\Delta}(t)$ to be the number (provided it exists) with the property that given any $\varepsilon > 0$, there is a neighborhood U of t such that

$$|[f(\sigma(t)) - f(s)] - f^{\Delta}(t)[\sigma(t) - s]| \leq \varepsilon|\sigma(t) - s| \quad \forall s \in U.$$

The function $f^{\Delta}(t)$ is called the delta derivative of f at t.

When $\mathbb{T} = \mathbb{R}$, $f^{\Delta}(t) = f'(t)$, and when $\mathbb{T} = \mathbb{Z}$, $f^{\Delta}(t) = f(t+1) - f(t)$. In the case that $\mathbb{T} = q^{\mathbb{Z}} \cup \{0\}$ for $q > 1$, we have

$$f^{\Delta}(t) = \begin{cases} \dfrac{f(\sigma(t)) - f(t)}{\mu(t)} = \dfrac{f(qt) - f(t)}{(q-1)t}, & \text{if } t \in \mathbb{T} \setminus \{0\}; \\ \lim\limits_{s \to 0} \dfrac{f(0) - f(s)}{0 - s} = \lim\limits_{s \to 0} \dfrac{f(s) - f(0)}{s}, & \text{otherwise} \end{cases}$$

because 0 is a right-dense minimum and every other point in $\mathbb{T}$ is both right-scattered and left-scattered, i.e., isolated. A function $f : \mathbb{T} \to \mathbb{R}$ is right-dense continuous provided it is continuous at right-dense points in $\mathbb{T}$ and its left-sided limits are finite at left-dense points in $\mathbb{T}$.

Definition 2 Let $f : \mathbb{T} \to \mathbb{R}$ be a function and let $a,\ b \in \mathbb{T}$. If there is a function $F : \mathbb{T} \to \mathbb{R}$ such that $F^{\Delta}(t) = f(t)$ for all $t \in \mathbb{T}^{\kappa}$, then F is a delta antiderivative of f. In this instance, the integral is given by

$$\int_a^b f(t)\, \Delta t = F(b) - F(a) \quad \text{for all} \quad a, b \in \mathbb{T}.$$

All right-dense continuous functions are delta integrable; see [6, Theorem 1.74]. In the case the time scale $[a,b]_{\mathbb{T}}$ consists of only isolated points and f is right-dense continuous,

$$\int_a^b f(t)\, \Delta t = \begin{cases} \sum\limits_{t=a}^{\rho(b)} \mu(t) f(t), & \text{if } a < b; \\ 0, & \text{if } a = b; \\ -\sum\limits_{t=b}^{\rho(a)} \mu(t) f(t), & \text{if } a > b. \end{cases}$$

3 Preliminary Results

In this section, we provide fundamental results necessary to prove our main results. We begin with the following generalization of [15, Theorem 3].

Theorem 1 *Let $f(t, y)$ be a continuous function of $t \geq t_0$ and $|y(t)| < \infty$. Assume that for all positive t and nonzero y, $yf(t, y) > 0$, and for each fixed t, $f(t, y)$ is nondecreasing in y for $y > 0$. Then a necessary condition for*

$$y^{\Delta\Delta}(t) + f(t, y^{\sigma}) = 0, \quad t \geq t_0 \tag{2}$$

to have a bounded nonoscillatory solution is that

$$\int^{\infty} tf(t, c)\Delta t < \infty \tag{3}$$

for some constant $c > 0$.

Proof Suppose y is a bounded eventually positive solution of (2). So there exists $T \in [t_0, \infty)_{\mathbb{T}}$ such that $y(t) > 0$ for $t \geq T$. As $f(t, y) > 0$ for all $y > 0$, $y^{\Delta\Delta}$ is eventually negative. So $y^{\Delta}(t)$ is decreasing and tends to a limit L that is either positive, zero, negative, or $-\infty$. If $L < 0$ or if $L = -\infty$, $y(t)$ would be eventually negative. Hence $0 \leq L < \infty$. In fact, $L = 0$ since if $L > 0$, then $y(t)$ would be unbounded. Integrating (2) from s to T_1, we obtain

$$y^{\Delta}(s) = y^{\Delta}(T_1) + \int_s^{T_1} f(r, y^{\sigma}(r))\Delta r.$$

Letting T_1 approach infinity yields

$$y^{\Delta}(s) = \int_s^{\infty} f(r, y^{\sigma}(r))\Delta r$$

since $0 = L < \infty$. Integrating again from t_1 to t, we have

$$\begin{aligned}
y(t) - y(t_1) &= \int_{t_1}^{t} y^{\Delta}(s)\Delta s \\
&= \int_{t_1}^{t}\int_s^{\infty} f(r, y^{\sigma}(r))\Delta r\Delta s \\
&= \int_{t_1}^{t}\int_{t_1}^{r} f(r, y^{\sigma}(r))\Delta s\Delta r + \int_t^{\infty}\int_{t_1}^{t} f(r, y^{\sigma}(r))\Delta s\Delta r \\
&= \int_{t_1}^{t} (r - t_1)f(r, y^{\sigma}(r))\Delta r + \int_t^{\infty} (t - t_1)f(r, y^{\sigma}(r))\Delta r \\
&\geq \int_{t_1}^{t} (r - t_1)f(r, y^{\sigma}(r))\Delta r,
\end{aligned}$$

and so

$$M \geq y(t) > \int_{t_1}^{t} (r - t_1)f(r, y^{\sigma}(r))\Delta r$$

for some $M > 0$. Since $\int_{t_1}^{t} (r - t_1) f(r, y^\sigma(r)) \Delta r$ is an increasing function of t, we have

$$\int_{t_1}^{\infty} (r - t_1) f(r, y^\sigma(r)) \Delta r < \infty.$$

By the monotonicity of f, we have

$$\int_{t_1}^{\infty} (r - t_1) f(r, y(t_1)) \Delta r < \infty.$$

Hence (3) is necessary.

To prove our main results, we need a method to study boundary value problems (BVPs). Specifically, we will define functions called upper and lower solutions that, not only imply the existence of a BVP but also provide bounds of a solution. Consider the second-order equation

$$y^{\Delta\Delta} = f(t, y^\sigma) \tag{4}$$

where f is continuous on $[a, b]_{\mathbb{T}} \times \mathbb{R}$.

Definition 3 [6, *Definition 6.53*] We say that $\alpha \in C_{rd}^2$ is a *lower solution* of (4) on $[a, \sigma^2(b)]_{\mathbb{T}}$ provided

$$\alpha^{\Delta\Delta}(t) \geq f(t, \alpha^\sigma(t)) \quad \text{for all} \quad t \in [a, b]_{\mathbb{T}}.$$

Similarly, $\beta \in C_{rd}^2$ is called an *upper solution* of (4) on $[a, \sigma^2(b)]_{\mathbb{T}}$ provided

$$\beta^{\Delta\Delta}(t) \leq f(t, \beta^\sigma(t)) \quad \text{for all} \quad t \in [a, b]_{\mathbb{T}}.$$

Theorem 2 [6, Theorem 6.54] *Let f be continuous on $[a, b]_{\mathbb{T}} \times \mathbb{R}$. Assume that there exist a lower solution α and an upper solution β of (4) with*

$$\alpha(a) \leq A \leq \beta(a) \quad \textit{and} \quad \alpha(\sigma^2(b)) \leq B \leq \beta(\sigma^2(b))$$

such that

$$\alpha(t) \leq \beta(t) \quad \textit{for all} \quad t \in [a, \sigma^2(b)]_{\mathbb{T}}.$$

Then the BVP

$$y^{\Delta\Delta} = f(t, y^\sigma) \textit{ on } [a, b]_{\mathbb{T}}, \quad y(a) = A, \quad y(\sigma^2(b)) = B$$

has a solution y with

$$\alpha(t) \leq y(t) \leq \beta(t) \quad \textit{for all} \quad t \in [a, \sigma^2(b)]_{\mathbb{T}}.$$

The following is a generalization of Theorem 7.4 of [13].

Theorem 3 *Let f be continuous on $[a, b]_{\mathbb{T}} \times \mathbb{R}$. Assume that there exist a lower solution α and an upper solution β of (4) with $\alpha(t) \leq \beta(t)$ for all $t \in [a, \infty)_{\mathbb{T}}$. Then for any $\alpha(a) \leq c \leq \beta(a)$ the BVP*

$$y^{\Delta\Delta} = f(t, y^{\sigma}), \quad y(a) = c \tag{5}$$

has a solution y with

$$\alpha(t) \leq y(t) \leq \beta(t) \quad \textit{for all} \quad t \in [a, \infty)_{\mathbb{T}}.$$

Proof It follows from Theorem 2 that for each $n \geq 1$ there is a solution $y_n(t)$ of $[a, a+n]_{\mathbb{T}}$ with $y_n(a) = c$, $y_n(a+n) = \beta(a+n)$ and $\alpha(t) \leq y_n(t) \leq \beta(t)$ on $[a, a+n]_{\mathbb{T}}$. Thus, for any fixed $n \geq 1$, $y_m(t)$ is a solution on $[a, a+n]_{\mathbb{T}}$ satisfying $\alpha(t) \leq y_m(t) \leq \beta(t)$ for all $m \geq n$. Hence, for $m \geq n$, the sequence $\{y_m(t)\}$ is pointwise bounded on $[a, a+n]_{\mathbb{T}}$.

We claim that $\{y_m(t)\}$ is equicontinuous on $[a, a+n]_{\mathbb{T}}$ for any fixed $n \geq 1$. Since f is continuous and $y_m(t) \leq \beta(t)$ for all $t \in [a, a+n]_{\mathbb{T}}$, there is constant $K > 0$ such that $|y_m^{\Delta\Delta}(t)| = |f(t, y_m^{\sigma}(t))| \leq K$ for all $t \in [a, a+n]_{\mathbb{T}}$. It follows that

$$\begin{aligned} y_m^{\Delta}(t) - y_m^{\Delta}(a) &= \int_a^t y_m^{\Delta\Delta}(s)\, \Delta s \\ &\leq \int_a^t K\, \Delta s \\ &= K(t-a) \\ &\leq K(a+n-a) \\ &= Kn \end{aligned}$$

which gives that

$$|y_m^{\Delta}(t)| \leq |y_m^{\Delta}(a)| + |Kn| =: L.$$

Consequently,

$$|y_m(t) - y_m(s)| = \left| \int_s^t y_m^{\Delta}\, \Delta s \right| \leq L|t-s| < \varepsilon$$

for all $t, s \in [a, a+n]_{\mathbb{T}}$ provided $|t-s| < \delta = \dfrac{\varepsilon}{L}$. Hence the claim holds.

So by the Ascoli–Arzela Theorem and a standard diagonalization argument, $\{y_m(t)\}$ contains a subsequence which converges uniformly on all compact subintervals $[a, a+n]_{\mathbb{T}}$ of $[a, \infty)_{\mathbb{T}}$ to a solution $y(t)$, which is the desired solution of the (5) that satisfies $\alpha(t) \leq y(t) \leq \beta(t)$ for all $t \in [a, \infty)_{\mathbb{T}}$.

4 Main Results

We now establish necessary and sufficient conditions for the existence of certain types of solutions of (1).

Theorem 4 *Assume* (A_0)-(A_3) *hold and let* $\alpha_0 > 0$. *Additionally, assume* $\sigma(t)/t$ *is bounded. Then the following statements are equivalent:*

1. *For each* $0 < \alpha < \alpha_0$ *there is a solution* $u_\alpha(t)$ *of* $u^{\Delta\Delta}(t) + f(t, u^\sigma)g(u^\Delta) = 0$ *satisfying* $\lim\limits_{t\to\infty} u_\alpha(t) = \alpha$.
2. $\int^\infty \sigma(t) f_y(t, \alpha)\Delta t < \infty$ *for* $0 < \alpha < \alpha_0$.

Proof Assume $\int^\infty \sigma(t) f_y(t, \alpha_1)\Delta t = \infty$ for some $0 < \alpha_1 < \alpha_0$ and let $\alpha_1 < \beta < \alpha_0$. Let $u_\beta(t)$ be the corresponding solution of $u_\beta^{\Delta\Delta} + f(t, u_\beta^\sigma)g(u_\beta^\Delta) = 0$ with $\lim\limits_{t\to\infty} u_\beta(t) = \beta$. Choose $\delta > 0$ such that $\alpha_1 + \delta < \beta$ and let $T \geq 0$ be such that $u_\beta^\sigma(t) \geq \alpha_1 + \delta$ for all $t \geq T$. Then for $t \geq T$

$$u_\beta^{\Delta\Delta} = -f(t, u_\beta^\sigma)g(u_\beta^\Delta) \leq 0.$$

Hence $u_\beta^\Delta > 0$ and decreases to a limit, and this limit must be zero since u_β is bounded. Therefore, $u_\beta(t) \leq \beta$ for $t \geq T$. By applying the Mean Value Theorem, we obtain

$$\frac{f(t, u_\beta^\sigma(t)) - f(t, \alpha_1)}{u_\beta^\sigma(t) - \alpha_1} = f_y(t, \eta(t)) \quad \text{for some } \eta(t) \in (\alpha_1, u(\beta^\sigma(t))).$$

Now by the monotonicity of f_y, we have

$$\begin{aligned} f_y(t, \alpha_1) &\leq f_y(t, \eta(t)) \\ &\leq \frac{f(t, u_\beta^\sigma(t)) - f(t, \alpha_1)}{u_\beta^\sigma(t)) - \alpha_1} \\ &\leq \frac{u_\beta^\sigma(t)}{u_\beta^\sigma(t) - \alpha_1} \frac{f(t, u_\beta^\sigma(t)}{u_\beta^\sigma(t)} \\ &\leq \frac{\beta}{\delta} \frac{f(t, u_\beta^\sigma(t))}{u_\beta^\sigma(t)} \end{aligned}$$

for $t \geq T$. Since $\lim\limits_{t\to\infty} u^\Delta(t) = 0$, there exists $T_1 \geq T$ such that $g(u_\beta^\Delta(t)) \geq \dfrac{g(0)}{2} > 0$ for all $t \geq T_1$. Hence, for $t \geq T_1$, we have

$$
\begin{aligned}
u_\beta^{\Delta\Delta}(t) &= -f(t, u_\beta^\sigma(t))g(u_\beta^\Delta(t)) \\
&\le -\frac{f_y(t,\alpha_1)}{\beta}\delta u_\beta^\sigma(t)\frac{g(0)}{2} \\
&= -kf_y(t,\alpha_1)u_\beta^\sigma(t)
\end{aligned}
$$

where $k = g(0)\dfrac{\delta}{2\beta}$. Also, $\alpha_1^{\Delta\Delta} = 0 \ge -kf_y(t,\alpha_1)\alpha_1$. Hence, by Theorem 3, there is a solution $z(t)$ of $z^{\Delta\Delta} + kf_y(t,\alpha_1)z^\sigma = 0$ with $0 < \alpha_1 \le z(t) \le u_\beta(t) \le \beta$ on $[T,\infty)_{\mathbb{T}}$. By Theorem 1, it follows that

$$
\int^\infty kctf_y(t,\alpha_1)\Delta t < \infty
$$

for some $c > 0$. Since $\sigma(t)/t$ is bounded, we have

$$
\int^\infty \sigma(t)f_y(t,\alpha_1)\Delta t < \infty,
$$

which is the desired contradiction.

Conversely, let $0 < \alpha < \alpha_0$ be such that

$$
\int^\infty \sigma(t)f_y(t,\alpha)\Delta t < \infty
$$

and let

$$
M = \max\{g(v) : 0 \le v \le \alpha\}.
$$

Choose $T \ge 0$ such that

$$
\int_T^\infty (\sigma(s) - T)f_y(s,\alpha)\Delta s < \frac{1}{M} \quad \text{and} \quad \int_T^\infty f_y(s,\alpha)\Delta s < \frac{1}{M}.
$$

We shall now define a sequence of functions on $[T,\infty)_{\mathbb{T}}$ in the following manner:

Let $y_0(t) = \alpha,\ t \ge T$. Now for $t \ge T$

$$
\begin{aligned}
0 &\le \int_t^\infty (\sigma(s) - t)f(s,\alpha)g(0)\Delta s \\
&= \int_t^\infty (\sigma(s) - t)[f(s,\alpha) - f(s,0)]g(0)\Delta s \\
&= \int_t^\infty (\sigma(s) - t)\alpha f_y(s,\eta(s))g(0)\Delta s, \quad \eta(s) \in (0,\alpha) \\
&\le \int_t^\infty (\sigma(s) - t)\alpha f_y(s,\alpha)g(0)\Delta s
\end{aligned}
$$

$$
\begin{aligned}
&\leq \alpha M \int_t^\infty (\sigma(s) - t) f_y(s, \alpha) \Delta s \\
&\leq \alpha M \int_T^\infty (\sigma(s) - T) f_y(s, \alpha) \Delta s \\
&< \alpha.
\end{aligned}
$$

By defining $y_1(t) := \alpha - \int_t^\infty (\sigma(s) - t) f(s, \alpha) g(0) \Delta s$, $t \geq T$, we have $0 \leq y_1(t) < \alpha$. Differentiating y_1, we obtain

$$
\begin{aligned}
y_1^\Delta(t) &= 0 - \left[\int_t^\infty - f(s, \alpha) g(0) \Delta s + (\sigma(t) - \sigma(t)) f(t, \alpha) g(0) \right] \\
&= \int_t^\infty f(s, \alpha) g(0) \Delta s \\
&\leq M \int_t^\infty f(s, \alpha) \Delta s \\
&= M \int_t^\infty [f(s, \alpha) - f(s, 0)] \Delta s \\
&= \alpha M \int_t^\infty f_y(s, \eta(s)) \Delta s, \quad \eta(s) \in (0, \alpha) \\
&\leq \alpha M \int_T^\infty f_y(s, \alpha) \Delta s \\
&< \alpha.
\end{aligned}
$$

So $0 \leq y_1^\Delta(t) < \alpha$ for $t \geq T$. Proceeding inductively, we define for all $m \geq 1$

$$
y_{m+1}(t) := \alpha - \int_t^\infty (\sigma(s) - t) f(s, y_m^\sigma(s)) g(y_m^\Delta(s)) \Delta s, \quad t \geq T, \tag{6}
$$

and obtain $0 \leq y_m(t),\ y_m^\Delta(t) \leq \alpha$ for all $m \geq 1$. Hence the sequence $\{y_m(t)\}_{m=0}^\infty$ is uniformly bounded and equicontinuous. The Ascoli-Arzela Theorem along with a standard diagonalization argument yields a uniformly convergent subsequence $\{y_{m_k}(t)\}$ on compact subintervals of $[T, \infty)_\mathbb{T}$. Let

$$
u_\alpha(t) := \lim_{k \to \infty} y_{m_k}(t),
$$

for $t \in [T, \infty)$. It follows that

$$
\lim_{k \to \infty} f(t, y_{m_k}(t)) g(y_{m_k}^\Delta(t)) = f(t, u_\alpha(t)) g(u_\alpha^\Delta(t))
$$

uniformly on compact subintervals of $[T, \infty)_\mathbb{T}$. Replacing m in (6) by m_k and letting $k \to \infty$, we get

$$u_\alpha(t) = \alpha - \int_t^\infty (\sigma(s) - t) f(s, u_\alpha^\sigma(s)) g(u_\alpha^\Delta(s)) \Delta s$$

on $[T, \infty)_\mathbb{T}$. Consequently, $u_\alpha(t)$ is a solution of $u_\alpha^{\Delta\Delta}(t) + f(t, u_\alpha^\sigma)(t) g(u_\alpha^\Delta) = 0$. As $\lim_{t\to\infty} u_\alpha(t) = \alpha$, the proof is complete.

Remark 1 If g is positive and continuous on $\mathbb{R}$ and $f(t, y) = -f(t, -y)$, then $\int^\infty \sigma(t) f_y(t, \alpha) \Delta t < \infty$ for $0 < |\alpha| < \alpha_0$ if and only if for each $0 < |\alpha| < \alpha_0$ there is a solution $u_\alpha(t)$ of $u^{\Delta\Delta} + f(t, u^\sigma) g(u^\Delta) = 0$ with $\lim_{t\to\infty} u_\alpha(t) = \alpha$.

Corollary 1 *There is a solution $u_\alpha(t)$ of $u^{\Delta\Delta} + f(t, u) g(u^\Delta) = 0$ with $\lim_{t\to\infty} u_\alpha(t) = \alpha$ for all $\alpha > 0$ if and only if $\int^\infty \sigma(t) f_y(t, \alpha) \Delta t < \infty$ for all $\alpha > 0$.*

We continue with an example that shows how Theorem 4 can be applied.

Example 1 Consider the dynamic equation

$$y^{\Delta\Delta}(t) + \frac{y^2(\sigma(t))}{t^5} \left(y^\Delta(t)\right)^2 = 0 \tag{7}$$

on $\mathbb{T} = 3^{\mathbb{N}_0}$. Here $f(t, y) = \frac{y^2}{t^5}$ and $g(v) = v^2$. Let $\alpha_0 > 0$ be given.

Immediately we have that (A_3) holds. By the choice of f, (A_1) holds and f is continuous in y and right-dense continuous in t. Then $f_y(t, y) = \frac{2y}{t^5}$ is continuous in y and right-dense continuous in t, and so (A_0) holds. Also, for $y \geq 0$ and $t \in \mathbb{T}$, $f_y(t, y) \geq 0$ and $f_y(t, y)$ is strictly increasing since $f_{yy}(t, y) = \frac{2}{t^5} > 0$. Thus (A_2) holds. Next, observe that for $t = 3^n$, $n \in \mathbb{N}_0$,

$$\frac{\sigma(t)}{t} = \frac{3t}{t} = 3 < \infty.$$

We now have that all assumptions of Theorem 4 are satisfied. Additionally, for all $\alpha > 0$, if $b = 3^m$ and $t = 3^n$, $m, n \in \mathbb{N}_0$, we have

$$\begin{aligned}
\int_1^\infty \sigma(t) f_y(t, \alpha)\, \Delta t &= \lim_{b\to\infty} \int_1^b \sigma(t) f_y(t, \alpha)\, \Delta t \\
&= \lim_{m\to\infty} \sum_{t=1}^{\rho(3^m)} 3t \frac{2\alpha}{t^5} (3t - t) \\
&< 12\alpha \lim_{m\to\infty} \sum_{t=1}^{3^{m-1}} \frac{1}{t^3} \\
&< \infty.
\end{aligned}$$

Since condition (ii) of Theorem 4 holds, we conclude that there is a solution satisfying (i) for all $0 < \alpha < \alpha_0$.

In [14] it is shown that $y'' + a(t)y^{2n+1} = 0,\ n \geq 0$, where $a(t) \geq 0$ for $t \geq 0$ and $g(v) = 1$ for all v, has solutions for which

$$\lim_{t\to\infty} \frac{y(t)}{t} = \alpha > 0$$

if and only if

$$\int^{\infty} t^{2n+1} a(t)dt < \infty.$$

We will show that an analogous result is true for the dynamic equation (1) provided $f(t, y)$ satisfies the following additional condition.

(A_4) There exist positive real numbers c and λ such that

$$\liminf_{v\to\infty} \frac{f(t, v)}{v f_v(t, cv)} \geq \lambda$$

for all sufficiently large t.

Note that in the case of $y'' + a(t)y^{2n+1} = 0$, c and λ may be any positive real numbers with $\lambda c^{2n} \leq 1/(2n + 1)$. We continue by establishing the following result.

Theorem 5 *Assume (A_0)–(A_3) hold and let there be a positive real number β with*

$$\int^{\infty} \sigma(t) f_y(t, \beta\sigma(t))\Delta t < \infty.$$

Then there exists a solution $y(t)$ of $y^{\Delta\Delta} + f(t, y^{\sigma}(t))g(y^{\Delta}) = 0$ such that $\lim_{t\to\infty} \frac{y(t)}{t}$ exists and is positive.

Proof Let $T > 0$ be such that

$$\int_T^{\infty} \sigma(t) f_y(t, \beta\sigma(t))\Delta t < \frac{1}{2M},$$

where $M = \max\{g(v) : 0 \leq v \leq \beta\}$. We define a solution of $u^{\Delta\Delta} + f(t, u^{\sigma})g(u^{\Delta}) = 0$ by

$$u(T) = 0, \quad u^{\Delta}(T) = \beta,$$

and we assert that the solution satisfies $u^{\Delta}(t) \geq \frac{\beta}{2}$ for $t \geq T$. Observe that $u(t) > 0$ and $u^{\Delta}(t) > 0$ for some $t > T$. Assume, for the sake of contradiction, that there is a $\delta > 0$ with $\delta < \frac{\beta}{2}$ and a $t_1 > T$ with $u^{\Delta}(t_1) = \delta$ and $u(t) > 0$ on $(T, t_1]_{\mathbb{T}}$. Then for $T \le t \le t_1$ we have

$$u^{\Delta}(T) = u^{\Delta}(t) + \int_T^t f(s, u^{\sigma}(s))g(u^{\Delta}(s))\Delta s. \tag{8}$$

Since $u^{\Delta\Delta}(t) \le 0$ on $(T, t_1]_{\mathbb{T}}$ and $u(t)$ is decreasing on $(T, t_1]_{\mathbb{T}}$, we have

$$u^{\Delta}(t) \le \beta \text{ on } (T, t_1)_{\mathbb{T}} \quad \text{and} \quad u(t) \le \beta(t - T) \text{ on } (T, t_1)_{\mathbb{T}}.$$

By applying the Mean Value Theorem to (8) and the monotonicity of f_y we have

$$\begin{aligned}
\beta = u^{\Delta}(T) &= u^{\Delta}(t) + \int_T^t f(s, u^{\sigma}(s))g(u^{\Delta}(s))\Delta s \\
&\le u^{\Delta}(t) + M \int_T^t f(s, u^{\sigma}(s))\Delta s \\
&= u^{\Delta}(t) + M \int_T^t [f(s, u^{\sigma}(s)) - f(s, 0)]\Delta s \\
&= u^{\Delta}(t) + M \int_T^t u^{\sigma}(s) f_y(s, \eta(s))\Delta s, \quad 0 < \eta(s) < u^{\sigma}(s) \\
&\le u^{\Delta}(t) + M\beta \int_T^t (\sigma(s) - T) f_y(s, u^{\sigma}(s))\Delta s \\
&\le u^{\Delta}(t) + M\beta \int_T^t \sigma(s) f_y(s, \beta\sigma(s))\Delta s \\
&< u^{\Delta}(t) + M\beta \frac{1}{2M} \\
&= u^{\Delta}(t) + \frac{\beta}{2}.
\end{aligned}$$

Hence, $u^{\Delta}(t_1) > \frac{\beta}{2}$, a contradiction. Thus, $u^{\Delta}(t) \ge \frac{\beta}{2}$ on $[T, \infty)_{\mathbb{T}}$ and $\lim_{t\to\infty} u^{\Delta}(t)$ exists and is positive. By L'Hôpital's Rule [6, Theorem 1.120], we have $\lim_{t\to\infty} \frac{u(t)}{t}$ exists and is positive.

If we assume condition (A_4), then we may establish the converse of Theorem 5.

Theorem 6 *Assume conditions (A_0)–(A_4) hold. Then* (1) *has a solution,* $y(t)$, *such that* $\lim_{t\to\infty} \frac{y(t)}{t}$ *exists and is positive if and only if*

$$\int^{\infty} \sigma(t) f_y(t, \beta\sigma(t))\Delta t < \infty \textit{ for some } \beta > 0.$$

Proof Let $\alpha > 0$ and let $y(t)$ be a solution of (1) with

$$\lim_{t\to\infty} \frac{y(t)}{t} = \alpha.$$

Let $T \geq 0$ be such that $y(t) \geq \alpha t/2$ for $t \geq T$ and let

$$m := \min\{g(v) : 0 \leq v \leq y^{\Delta}(T)\}.$$

By condition (A_4), there is a $T_1 \geq T$ such that

$$f(t, y^{\sigma}(t)) \geq \lambda y^{\sigma}(t) f_y(t, cy^{\sigma}(t)) \geq \lambda \frac{\alpha\sigma(t)}{2} f_y\left(t, \frac{\alpha\sigma(t)}{2}\right) = k\sigma(t) f_y\left(t, \frac{\alpha\sigma(t)}{2}\right)$$

for $t \geq T_1$, where $k = \dfrac{\lambda\alpha}{2}$. Since $0 < y^{\Delta}(t) \leq y^{\Delta}(T)$ for $t \geq T$, we have

$$f(t, y^{\sigma}(t)) g(y^{\Delta}(t)) \geq mk\sigma(t) f_y\left(t, \frac{c\alpha\sigma(t)}{2}\right), \quad t \geq T_1.$$

Therefore,

$$\begin{aligned} y^{\Delta}(T_1) &= y^{\Delta}(t) + \int_{T_1}^{t} f(s, y^{\sigma}(s)) g(y^{\Delta}(s)) \Delta s \\ &\geq y^{\Delta}(t) + \int_{T_1}^{t} mk\sigma(s) f_y\left(s, \frac{c\alpha\sigma(s)}{2}\right). \end{aligned}$$

Since $\lim_{t\to\infty} y^{\Delta}(t) \geq 0$,

$$\int_{T_1}^{\infty} \sigma(s) f_y\left(s, \frac{c\alpha\sigma(s)}{2}\right) < \infty,$$

and this proves the theorem.

Example 2 Consider the dynamic equation

$$y^{\Delta\Delta}(t) + \frac{y^2(\sigma(t))}{t^5} \left(y^{\Delta}(t)\right)^2 = 0 \tag{7}$$

on $\mathbb{T} = 3^{\mathbb{N}_0}$. As shown in Example 1, conditions (A_0)–(A_4) hold. If we choose $c = \frac{1}{8}$ and $\lambda = 2$, then

$$\liminf_{v\to\infty} \frac{f(t, v)}{v f_v(t, cv)} = \liminf_{v\to\infty} \frac{t^{-4}v^2}{\frac{1}{4}t^{-4}v^2} = 4 > 2.$$

Hence all assumptions of Theorem 6 are satisfied. Furthermore, if $b = 3^m$ and $t = 3^n$, $m,\ n \in \mathbb{N}_0$, for $\beta = \frac{1}{4}$ we have

$$\begin{aligned}
\int_1^\infty \sigma(t) f_y(t, \beta\sigma(t))\, \Delta t &= \lim_{b\to\infty} \int_1^b \sigma(t) f_y(t, \beta\sigma(t))\, \Delta t \\
&= \lim_{m\to\infty} \sum_{t=1}^{\rho(3^m)} 3t \frac{2\frac{1}{4}3t}{t^5}(3t - t) \\
&< 9 \lim_{m\to\infty} \sum_{t=1}^{3^{m-1}} \frac{1}{t^2} \\
&< \infty.
\end{aligned}$$

Therefore, we conclude that there is a solution y of (7) such that $\lim_{t\to\infty} \frac{y(t)}{t}$ exists and is positive.

5 Conclusion

We have obtained conditions for the existence of a bounded nonoscillatory solution with prescribed limit at infinity and a nonoscillatory solution whose derivative has positive limit at infinity to

$$y^{\Delta\Delta}(t) + f(t, y^\sigma) g(y^\Delta) = 0.$$

These results were attained using the method of upper and lower solutions and applying the Mean Value Theorem and L'Hôpital's Rule.

References

1. E. Akin-Bohner, M. Bohner, S. Djebali, T. Moussaoui, On the asymptotic integration of nonlinear dynamic equations, Adv. Difference Equ. (2008). (Art. ID 739602, 17p)
2. F.V. Atkinson, On second order non-linear oscillations. Pac. J. Math. **5**, 643–647 (1955)
3. M. Bohner, G. Guseinov, Improper integrals on time scales. Dyn. Syst. Appl. **12**(1–2), 45–65 (2003). (Special issue: dynamic equations on time scales)
4. M. Bohner, D.A. Lutz, Asymptotic behavior of dynamic equations on time scales. J. Differ. Equ. Appl. **7**(1) 21–50 (2001). (Special issue in memory of W. A. Harris, Jr)
5. M. Bohner, A. Peterson (eds.), *'Advances in Dynamic Equations on Time Scales,* 1st (Birkhäuser, Boston, 2003)
6. M. Bohner, A. Peterson, Dynamic Equations on Time Scales: An Introduction with Applications, 1st (Birkhäuser, Boston, 2001)
7. M. Bohner, S. Stević, Asymptotic behavior of second-order dynamic equations. Appl. Math. Comput. **188**(2), 1503–1512 (2007)

8. L.H. Erbe, Nonoscillatory solutions of second order nonlinear differential equations. Pac. J. Math. **28**(1), 77–85 (1969)
9. L. Erbe, A. Peterson, Boundedness and oscillation for nonlinear dynamic equations on a time scale. Proc. Am. Math. Soc. **132**, 735–744 (2004)
10. L. Erbe, A. Peterson, S.H. Saker, Oscillation criteria for second-order nonlinear dynamic equations on time scales. J. Lond. Math. Soc. (2), **67**(3), 701–714 (2003)
11. R. Higgins, Some oscillation results for second-order functional dynamic equations. Adv. Dyn. Syst. Appl. **5**(1), 87–105 (2010)
12. S. Hilger, Analysis on measure chains–a unified approach to continuous and discrete calculus. Results Math. **18**(1–2), 18–56 (1990)
13. L.K. Jackson, Subfunctions and second-order ordinary differential inequalites. Adv. Math. **2**, 307–363 (1968)
14. R.A. Moore, Z. Nehari, Nonoscillation theorems for a class of nonlinear differential equations. Trans. Am. Math. Soc. **93**, 30–52 (1959)
15. J.S.W. Wong, On second order nonlinear oscillation. Funkcial. Ekvac. **11**, 207–234 (1969)

Part VII
Sharing the Joy: Engaging Undergraduate Students in Mathematics

Using Applications to Motivate the Learning of Differential Equations

Karen M. Bliss and Jessica M. Libertini

Abstract The field of differential equations is rich with applications that can be used to motivate and facilitate learning. This paper presents a variety of ways a modeling-focus can be adopted, based on the desired learning outcomes of an individual course. We offer an overview of the implementation of this approach within a specific course and identify several modeling scenarios that can be used either to introduce and motivate a lesson topic or to allow students to apply recently acquired skills to a meaningful problem. While using applications has clear benefits for our students, many of whom go on to pursue engineering degrees, adding these components to a course can be challenging, so this paper also addresses some logistical approaches to folding these applications into a course with success. Student feedback is also provided as evidence of the value of adopting such an approach.

Keywords Differential equations · Teaching · Applications · Modeling

Mathematics Subject Classification 97M06 · 97D06 · 34

1 Motivation

It is not uncommon for students to arrive to college with the misconception that studying mathematics is a pointless exercise, especially now that computers and calculators are so capable. Many students have had experiences with "word problems" that are not realistic and have been contrived to make the math they are learning seem relevant. These experiences may have led even science and engineering students to underestimate the value of mathematics and question its role in their education.

Instructors who lead with relevant applications can get students excited about answering questions and learning the relevant mathematical skills [1, 2]. For exam-

K.M. Bliss · J.M. Libertini (✉)
Department of Applied Mathematics, Virginia Military Institute, Lexington, VA 24450, USA
e-mail: libertinijm@vmi.edu

K.M. Bliss
e-mail: blisskm@vmi.edu

G. Letzter et al. (eds.), *Advances in the Mathematical Sciences*, Association for Women in Mathematics Series 6, DOI 10.1007/978-3-319-34139-2_17

ple, the context of a disease outbreak can incite natural student curiosity about the efficacy of different policies designed to combat the spread of the disease. Similarly, in the context of a murder mystery, students' curiosity can lead them to develop Newton's Law of Cooling. Students are more likely to deepen and persevere in their learning of mathematics if they know that they need to learn new concepts in order to answer the questions at hand [3–6]. Including applications can also promote development in many important non-mathematical areas as well; students can enhance more general problem-solving skills, teamwork, and communication. Furthermore student confidence in all of these areas can increase through this approach [7].

Differential equations classes often serve students majoring in engineering and the sciences, many of whom envision careers in industry. There have been recent outcries from the engineering educational community challenging "both the pedagogical practice of teaching non-contextualized mathematics and the lack of transparency regarding the significance of mathematics to engineering" [8]. Our engineering students benefit from a class that incorporates applications as this builds their ability to distill a "messy" problem into a mathematical form, identifying the most important inputs and outputs needed to answer the question at hand. They are well-served in a course that continually forces them to reconcile the abstract mathematics with the real-world problem, using their knowledge from each to inform and increase their knowledge in the other.

For mathematics and math education majors, exposure to applications is equally important. The field of mathematics suffers from a negative public image, and all mathematics majors should be prepared to be ambassadors for the field. Even students who prefer pure mathematics should be able to articulate that mathematics can be a powerful tool for addressing issues that are relevant and whose answers are highly valued. Likewise, future mathematics teachers need to be able to provide examples of how math has been leveraged to solve problems across the spectrum of disciplines.

Driven by the motivating factors above, in this article, we present the flexibility of a modeling first approach, details of a specific course implementation, an acknowledgement of challenges along with mollifiers, and student feedback.

2 Flexibility in Adoption

Despite increasing awareness of best practices in STEM education, many faculty do not adopt these practices due to a variety of barriers [9–11]. Specifically, differential equations courses are often subject to content guidelines to meet accreditation requirements and/or to satisfy prerequisites for follow-on courses. Additionally, some faculty are subject to assessment constraints, such as departmental exam policies or restrictions on projects in lieu of exams. Given these pressures, it is understandable that instructors may feel that they do not have the time or freedom to adopt a modeling-based approach for differential equations. Contrary to this opinion, we have found that a modeling-based approach is highly adaptable. In addition to working flexibly within constraints, such an approach allows instructors to tailor their courses to work towards a variety of learning objectives, both mathematical and nonmathematical. In the following section, we provide a variety of factors that can influence the structure of a differential equations course as well as present ways those factors are compatible with a modeling-based approach.

Instructors face a variety of constraints. Some are immediately obvious, but others may not become apparent until the course has already begun. Some of these constraints are curricular; there may be specific content and learning objectives that must be addressed in the course. In addition to external curricular constraints, instructors may have their own personal learning objectives for their students. We broadly classify any other constraints as environmental.

2.1 Curricular Constraints and Learning Objectives

As determined by a combination of external factors, such as accreditation requirements and instructor choice, each differential equation course addresses a unique set of learning objectives spanning both mathematical and non-mathematical domains.

The mathematical and quantitative goals may broadly include the following.

- *Practicing analytic solution techniques.* One misconception is that incorporating realistic applications necessitates the introduction of complicated models whose solutions can only be computed numerically. Although many real-world applications can lead to complex models, there exist many applications which can be modeled using equations that are readily solvable by classical analytical techniques. Additionally, even for those applications with cumbersome models, this provides an opportunity to address the important skill of reducing a model to one that is tractable.
- *Solving via computational methods.* Many differential equations courses address numerical approaches in a variety of technological settings such as spreadsheets, computer algebra systems, and programming environments. Within each of these settings, the instructor can vary the role that technology plays; for example, while one instructor may ask students to modify an existing file, others may require students to produce their own files from scratch. Regardless of the specific role of technology in a particular course, the inclusion of it allows for rich explorations of increasingly complex mathematical models.
- *Developing models and interpreting results in context.* The inclusion of meaningful applications can entice students to dive deeper into a model, allowing them to discover new concepts, such as parameter sensitivity and solution stability, and even pose their own research questions. Based on the goals of the course, these explorations can be framed quantitatively and/or qualitatively.

  The introduction of both a problem and a stakeholder presents an opportunity to practice the skills of mathematical abstractions and interpreting mathematical results in context. Through this step of interpreting results, students can reflect on the reasonableness of their results, which can help them identify possible errors in their mathematical solutions. In this way, the inclusion of mathematical modeling provides a support structure for the more traditional course goals of differential equations.

In industry, questions are rarely well-posed and easily isolated. While many colleges would like their students to be able to distill "messy" problems into something tractable, students rarely have the opportunity to experience that in a more traditional differential equations course. Applications and modeling also provide opportunities for students to build a desirable set of quantitative skills that translate broadly.

While it can be easy to identify mathematical objectives, employers emphasize the need for college graduates to have skills beyond subject matter proficiency. In addition to strong problem-solving skills, lists of desirable attributes in job candidates frequently include strong technical communication skills and the ability to work collaboratively [17–19]. The inclusion of a modeling-based approach can offer a platform to develop broad non-mathematical skills, including the following.

- *Communication.* The inclusion of applications makes it easy to emphasize a variety of communication skills. Through thoughtful assignment development, instructors can emphasize skills such as data visualization, oral presentation, and writing proficiency. Regardless of the communication medium, in their future careers, students will undoubtedly encounter situations where they need to communicate their work either to a technical audience or to a general audience. To target either of these goals, instructors can introduce the application through the lens of a stakeholder, asking their students to frame their response for this audience. For example, an instructor who wants to focus attention on technical writing skills might ask for a project write-up to be written in the form of an internal memo, with students writing to their boss, who is knowledgable in their field, about how they found a solution and implications of their results. If, on the other hand, the instructor wants to focus attention on communication to a broad audience, she might give the same project, but ask the students to have their write-up be in the form of an article that might appear in a newspaper (where the audience is not knowledgable of the field). Additionally, students can be prompted to advocate for a position (explaining why something needs to be designed differently), which is different than just reporting results. When instructors intentionally force the students to think about their audience and to argue for a particular position, students are more likely to perceive the assignments as important and relevant; this can break the cycle of students simply generating a product in the hopes of meeting some minimum requirements *for* their instructor.
- *Collaboration.* By encouraging group work, not only can students take on more challenging problems, but they are able to practice important skills, such as teamwork and leadership. Additionally, longer projects can give students the opportunity to practice various aspects of management, including time management, project management, and personnel management. The honing of these collaborative skills is best realized through intentional decisions by the instructor in the types and frequency of deliverables.

2.2 Environmental Constraints

Every academic institution has its own environment defined by both the physical spaces and the culture. As instructors include applications, it is important to consider the following about their local learning environment:

- the types of technology available to their students as well as their students' comfort level in using such technology;
- the types of equipment and spaces available for students to run experiments;
- the physical classroom space (moveable versus fixed seating, chalkboards versus whiteboards, etc.);
- the number of students in a class;
- the cultural norms of the students.

The authors have experimented with folding mathematical modeling into their courses at multiple institutions, including the United States Military Academy at West Point, Quinnipiac University, the University of Rhode Island, and Virginia Military Institute. Independent of environmental factors, we have found that students benefit from the inclusion of modeling. The positive outcomes from these early classroom explorations culminated in the development of a full implementation of a modeling-based differential equations course, the details of which are presented in the next section.

3 Classroom Implementation Example

In the spring of 2015, we implemented a model-based approach in two sections of differential equations at Virginia Military Institute. We recognize that the environment at Virginia Military Institute is quite different from most institutions, but we argue that every school offers a unique environment, and this implementation discussion is intended to serve as an example of how this teaching approach was adapted to one specific environment. To help frame the decisions behind this implementation, we briefly explain the curricular and environmental constraints and how those shaped the course. We then go on to discuss resources and list some of the modeling scenarios used in this implementation.

3.1 Shaping the Course

As the majority of our differential equations students are engineering majors, we opted to keep the existing (traditional) list of course topics, which has been carefully developed to address the needs of our engineering majors. To be consistent with other sections of the course, we maintained a schedule of four exams (each worth between

10–15% of the course grade) plus a final exam (worth 30% of the course grade). The modeling component was then added to this existing framework. With the exception of exam weeks, each Friday the students were presented with an application or scenario that, once modeled, would either be solvable using techniques they had already learned or would motivate the development of new techniques. These Friday classes were known as Fun Fridays, as they offered a change of pace from the rest of the week in which skills and techniques were presented and practiced in a format that toggled between mini-lectures and group work.

Our mathematics curriculum includes a mathematical software course. However, most students take this course after differential equations, so we felt that it made sense to limit the use of technology in this particular differential equations course. As a result, the mathematical goals of this course focused primarily on the development and use of analytic solution techniques as well as qualitative and quantitative reasoning in the context of the problem.

As approximately 50% of our students commission into the military, we also have an institutional emphasis on communication, leadership, and teamwork. Therefore, we had students work in groups in class, but they submitted individually written reports each week.

Our classes are small, and we have chalkboards that nearly encircle the classroom. While the students spent each Friday working on these modeling scenarios in their classroom, many used the boards. This allowed us to quickly identify serious missteps, and if desired, an opportunity to intervene, either through Socratic inquiry or more directed assistance.

3.2 The Modeling Scenarios

The course included nine application scenarios; Table 1 highlights a few of these, with citations where available. These scenarios were selected to align with the techniques learned in class, following a traditional list and sequence of differential equations topics. As indicated in the last line of Table 1, at the end of the course, the students were challenged with developing their own scenarios.

Table 1 Select modeling scenarios

Mathematical topic	Modeling scenario
Separation of variables	Time of death/Newton's law of cooling [15]
Integrating factors	Growth of an oil slick [16]
Exact equations	Fluid potential lines and streamlines
Homogeneous linear systems	Paracetamol absorption
Laplace transforms	Student-developed scenarios

4 Overcoming Challenges

Change can be uncomfortable and can present challenges. Here we present some challenges that we've faced in adopting the modeling-forward approach, as well as how we have addressed them.

4.1 Where Can I Find Scenarios?

Adding applications to a differential equations course can be intimidating. However, there are several resources available for faculty seeking differential equations applications. Specifically, there are repositories available online through CODEE, COMAP, and SIMIODE [12–14]. The CODEE website offers a variety of materials that "promote the teaching and learning of ordinary differential equations" [12]. Some of the pieces are examples of models without explanation of how one could adapt it for use in a classroom, while others (mainly under the Projects heading) offer examples of resources one might use with a class. COMAP offers materials for instructors considering incorporating modeling into differential equations. However, instructors must be COMAP members (and pay an annual membership fee) in order to access those materials. The SIMIODE repository includes scenarios ranging from short demonstrations to longer projects. Each SIMIODE scenario includes editable files, e.g. TeX, for student versions and teacher versions, as well as insights for what an instructor can expect if used "as is," and many offer suggestions for modifications, allowing the reader to adapt the scenario to a specific classroom implementation. Scenarios written for a lesson involving a particular use of technology also include annotated versions of the relevant files, such as Mathematica notebooks or Excel spreadsheets. Faculty need to register on the SIMIODE website in order to have access to the teacher versions of the materials, but membership is free.

In addition to online resources, the world immediately around us offers countless examples of practical applications. Whether it is the strength and sweetness of sweet tea or the population of fish in a nearby pond, one need not be a disciplinary expert to identify scenarios where a quantity is related to a rate of change. However, if you happen to have discipline-specific expertise, scenarios culled from your experiences are often valuable, as you can provide students with both enthusiasm and informed insight. Another valuable resource can be faculty from other STEM fields and contacts from local businesses who may work on problems that are represented with differential equations.

Lastly, towards the end of the course, we found that many of our students were ready to design their own scenarios.

4.2 So Much Grading!

Depending on how you choose to assess your students' work, this approach can lead to significant amounts of grading. Although as mathematicians we may feel more comfortable with clear point-based guidelines and rubrics, we have found that using a more holistic approach reduces grading time while still providing students with valuable feedback. For example, we typically create feedback checklists with broad topics, such as clarity, organization, mathematical accuracy, and translation between the mathematical work and the modeling scenario. After providing the feedback, each paper receives an overall score based on whether it was strong (A), average (C), insufficient (F), or somewhere in between (B or D). Also, if a student turns in a very poor paper, we may tell him/her that we are not willing to put in the effort to give feedback until they give us a paper that is ready to receive feedback; in other words, we have them to resubmit in order to get a grade other than a zero.

Also, if you are clear in your expectations, then the quality of the work is generally higher. By providing actionable feedback to students early in the semester and/or providing revision opportunities, the grading of reports can actually become enjoyable by the end of the semester.

4.3 Is This Really Your Work?

As professional academics, we understand collaboration and academic integrity, as we are immersed in a culture of crediting others as we present our work. However, our students have typically not had any opportunity to distinguish between cheating and working together. Since the problems we propose are challenging by design, it is important that students feel comfortable working together. We believe it is also important that they understand how to give credit to these collaborations, especially in an environment with a strict student honor code. Therefore, we dedicate time to explain our expectations, policies, and procedures before the first assignment, and we provide practice opportunities through homework assignments that require students to seek out, use, and thusly reference resources.

4.4 I Have Too Much Content to Cover; I Don't Have Time for This

Learning takes time, and the deeper learning promoted through this practice takes even more time. We have found that it helps to have students struggle through the hardest parts of the project together in the classroom; this means that there is less time to present material through lecture. It can be helpful to prepare students for a modeling scenario by providing a pre-lab homework assignment, not unlike the prelabs in chemistry and physics; students read some background material and answer a few questions so that they are ready to tackle the challenges in class. To make the best use of class time, it helps to be flexible on which elements of the project are completed in class versus which are completed out of class. Our best experiences with teaching this way have resulted from basing decisions about class time on a prioritized list of course objectives.

4.5 *My Students Don't Want to Think*

Some students see the differential equations course as a box that must be checked in order to complete their major; they fail to see how the ideas are connected to their discipline and/or life goals. As a result, they may just seek algorithmic proficiency, versus concept ownership. Students may initially be resistant and even disgruntled about the difficulty and ambiguity of modeling scenarios. In an era when student evaluations are often tied to tenure and promotion decisions, it can seem unwise to adopt a practice that cause students to react negatively. Both of the authors have faced this challenge, but we have successfully implemented a few methods to address this concern.

- Make the problem realistic, and introduce a stakeholder who will value their results.
- Openly acknowledge to students that the problems are a bit like a mountain; it can be challenging to make the climb, but once they do, they can appreciate the view from the top.
- Let your students know that it is normal to struggle, get things wrong, and become stuck. Remind them that the real world will not present them with neatly packaged and labeled challenges, and that the skills and perseverance they are gaining now will help them going forward.
- Stick with it. It can take up to three-quarters of the semester to have your students realize the value of these experiences. However, often the most frustrated students at the beginning of the semester are the most appreciative at the end of the semester. In the next section, we present student comments to this effect.

5 Student Feedback

Although we believed that this approach would be beneficial to students, it has been reassuring to receive student feedback sharing this view. In end-of-course evaluations, the section of the course that had been most resistant to the modeling scenarios had over 70 % of students explicitly say that the modeling scenarios were the most intellectually stimulating element of the course, with many of the remaining 30 % making other positive comments about this approach elsewhere on the evaluation form. Students also had the opportunity to write essays about their experiences in the course. In this more open-ended setting, students shared both their initial frustrations and their ultimate sense of accomplishment. Below are some excerpts from these essays.

> Unlike other math classes where I just learn equations and forget it later, differential equations can be applied to nearly everything. For instance, our last assignment was to create our own fun Friday. At the beginning of the semester I thought I would never be able to create an application to determine, for example, the spread of an oil slick, spread of a life threatening disease, or absorption of drugs. However, I was wrong, I managed to create an application using a Laplace transform function that can calculate the volume or flow rate of vehicles on a highway. Though I did not finish the application, it was the principle and how I was able to understand and create an application that is relevant to the real world. I think its easy for me to say that I have more confidence in tackling problems after taking this course. ... If we did not have fun Friday applications then I would not have been nearly as successful, and my overall performance in this course would have been much worse.

> After completing the fun Friday where tanks were used to predict the amount of a drug in the human body, I was amazed at how a simple technique can be used to model such a complex system. ... Previously I had thought that math was a theoretical subject but now I know how wrong I was. Differential equations has broadened my level of thinking. I am now more confident in my problem-solving skills by being able to approach problems in different ways than before. I have a better appreciation for math and see the relationship that it has with many different aspects of life. I know that I will be able to use this class as a foundation to guide me along in other classes that I take, and eventually in my future endeavors.

> This made my nine years of math worthwhile. After nine years of math, I never really knew when we were going to put it all together. I knew that I had to know my trig and algebra for most of my civil engineering. Now once I head into my advanced courses, I know that I would need to retain math within differential equations. Additionally, I feel like my translation of math into English has gotten better over the course. It has become easier to explain my method of madness to other people and you.

> This is the first time in any math course that I have taken at college that I actually feel like I can apply it to the real world and I think it is because of this that I have enjoyed taking this class so much (all admittedly it did take a while). In solving a lot of the problems we did in class (especially the modeling ones) it required quite a bit of leg work which at times was difficult to work through. By the end of the semester I definitely feel more confident in being able to look at a lengthy problem, break it up into a bunch of smaller tasks and tackle it, something I was not very good at the start of the semester. As an engineer it's not very likely that the problems I will be tasked with will be easily solvable and as such working on the basic skills needed will go a long way into making me more confident in my mathematical abilities. With the direct application into the real world with the Fun Friday's we got a genuine look at how the skills we learned in class apply to modeling things. For me, this was the most stimulating and after getting a couple under my belt and figuring out what was expected of me I actually began to have some fun with them.

> I expected a cut and dry math class similar to many of the other math classes I have taken. I have enjoyed them but they were not always applicable to subjects I was studying in other classes. This semester, however, really showed me the application of all the math concepts we have been using since the beginning. Not only did I see many applications through the Fun Friday's, but I was able to directly apply it to my engineering classes. ... I feel as though I could see a problem outside of class and at least set up the general outline of the math necessary to solve it. I have seen myself begin to think of things through mathematical modeling.

This last essay went on to describe how that student was already using what he had learned in the course to help in his leadership job of determining how the freshmen class of cadets will efficiently file into and out of the dining hall at meal times.

The above quotations capture the sentiment of the students in the course. Many students initially wrestled with the stark difference between this approach and the more traditional approach of their previous mathematics courses. However, as evidenced through both the open-ended feedback opportunities and the course evaluations, even the most resistant students eventually embraced the value of this approach, acknowledging how it had promoted their mathematical development, their problem-solving skills, their communication skills, and their confidence in all of these areas.

Acknowledgments Funding for the authors' presentation at the symposium was generously provided by the Association for Women in Mathematics (AWM). The authors thank Brian Winkel and the SIMIODE community for guidance and professional support. The authors also thank their students for sharing their insights about the course and their learning.

References

1. D. Perin, Facilitating Student Learning Through Contextualization. Community College Research Center (CRCC) Working Paper **29**. http://files.eric.ed.gov/fulltext/ED516783.pdf
2. S. Bell, Project-Based Learning for the 21st Century: Skills for the Future. The Clearing House: J. Educ. Strateg. Issues Ideas (2010). doi:10.1080/00098650903505415
3. R. Geier et al., Standardized test outcomes for students engaged in inquiry-based curricula in the context of urban reform. J. Res. Sci. Teach. **45**(8), 922–939 (2008)
4. J.W. Thomas, A Review of Research on PBL (2000), http://www.bobpearlman.org/BestPractices/PBL_Research.pdf. Cited 16 Jul 2015
5. M. Gultekin, The effect of project based learning on learning outcomes in the 5th grade social studies course in primary education. Educ. Sci. Theory Pract. **5**(2), 548–556 (2005)
6. J. Boaler, Mathematics for the moment, or the millennium?, in *Education Week* (1999), http://www.edweek.org/ew/articles/1999/03/31/29boaler.h18.html. Cited 15 Jul 2015
7. Y. Doppelt, Implementing and assessment of project-based learning in a flexible environment. Int. J. Technol. Des. Educ. **13**, 55–272 (2003)
8. D. Harris et al., Mathematics and its value for engineering students: what are the implications for teaching? Int. J. Math. Educ. Sci. Technol. (2014). doi:10.1080/0020739X.2014.979893
9. S.E. Brownell, K.D. Tanner, Barriers to faculty pedagogical change: lack of training, incentives, and ... tensions with professional identity? Life Sci. Educ. **11**, 339–346 (2012)
10. Beach Henderson, Facilitating change in undergraduate STEM practices: an analystic review of the literature. J. Res. Sci. Teach. Finkelstein **48**(8), 952–984 (2011)
11. C. Henderson, M. Dancy, Impact of physics education research on the teaching of introductory quantitative physics in the United States. Phys. Rev. ST Phys. Educ. Res. **5**, 020107 (2011)
12. Community of Ordinary Differential Equations Educators (CODEE) mainpage. http://www.codee.org. Cited 15 Jul 2015
13. Consortium for Mathematics and Its Applications (COMAP) mainpage. http://www.comap.com. Cited 15 Jul 2015
14. Systemic Initiative for Modeling Investigations and Opportunities with Differential Equations (SIMIODE) mainpage. https://www.simiode.org. Cited 15 Jul 2015
15. B. Winkel, Time of death, in *SIMIODE* (2015) Available via SIMIODE. https://www.simiode.org/resources/393. Cited 15 Jul 2015

16. K. Bliss, Spread of an oil slick, in *SIMIODE* (2015) Available via SIMIODE. https://www.simiode.org/resources/198. Cited 15 Jul 2015
17. Hart Research Associates: Falling Short? College Learning and Career Success, in *Association of American Colleges and Universities* (2015) http://www.aacu.org/leap/public-opinion-research/2015-survey-results. Cited 16 Jul 2015
18. National Association of Colleges and Employers: Job Outlook 2015 (2015), https://www.naceweb.org/surveys/job-outlook.aspx. Cited 16 Jul 2015
19. Maguire Associates, Inc.: The role of higher education in career development: employer perceptions, in *Chronicle of Higher Education* (2012), https://chronicle.com/items/biz/pdf/Employers%20Survey.pdf. Cited 16 Jul 2015

What Is a Good Question?

Brigitte Servatius

Abstract Antoine Gombaud had one and Pascal answered it. Fermat had one and Wiles answered it. Erdős had many and Carl Pomerance, at JMM15, shared the story of his collaboration with Erdős. Good questions can get many people hooked. Good questions are not always formulated by an experienced mathematician, sometimes they come from students. We discuss how we use good questions to get students interested in mathematics and how requesting students to formulate mathematical questions promotes learning.

Keywords Proofs course · Transition course · Mathematical writing · Proof writing

Mathematics Subject Classification Primary 97D50

Antoine Gombaud, a.k.a. Chevalier de Méré, asked Pascal in 1654: Which is more likely, to get at least one six rolling a die four times, or to get at least one double six rolling two dice twenty-four times. This question led to a correspondence between Pascal and Fermat, and 1654 is considered the birth year of probability theory. The story is mentioned among some historical examples in [2], see also [1] for more history.

Paul Erdős is fondly remembered by many a mathematician for asking many questions whose estimated difficulty he indicated by offering a dollar amount for a solution. While it is unknown how much he spent on answers, his success is evident by other numbers. Erdős is cited 13148 times by 7259 authors in the MR Citation Database, he is author of 1426 publications.

In his talk entitled *Letters from the master: My correspondence with Paul Erdős* given at the Joint Mathematics Meetings in San Antonio on January 10, 2015, see [4], Carl Pomerance relates the following story:

> I began my career at the University of Georgia in 1972. In the spring of my second year there a fortuitous event happened: On April 8, 1974, Hank Aaron of the Atlanta Braves hit his 715th career homerun, thus finally eclipsing the supposedly unbeatable total of 714 set some

B. Servatius (✉)
Mathematical Sciences Department, Worcester Polytechnic Institute,
100 Institute Rd, Worcester, MA 01609, USA
e-mail: bservat@wpi.edu

G. Letzter et al. (eds.), *Advances in the Mathematical Sciences*, Association for Women in Mathematics Series 6, DOI 10.1007/978-3-319-34139-2_18

> 40 years earlier by Babe Ruth. I was watching the game on television, and I noticed that $714 \times 715 = 2 \times 3 \times 5 \times 7 \times 11 \times 13 \times 17$ so we have two consecutive integers whose product is also the product of the first k primes for some k. It seemed to me that this was not likely to occur ever again.
>
> The next day I challenged my colleague David Penney to find an interesting property of 714 and 715, and he quickly saw the same thing. He asked a numerical analysis class he was teaching, and one of the students came up with another property: The sum of prime divisors of 714 equals the sum of prime divisors of 715.
>
> With Penney and a student, Carol Nelson, I quickly wrote a paper for the Journal of Recreational Mathematics, which was accepted by return mail and was published that same spring.

The point to be made here is that an interesting observation got turned into a question. At first a question from mathematician to mathematician, then a question posed in class—and a student came up with a new property. Hence Carl Pomerance's story is a perfect example on how to use good questions to get students interested in math. More about the impact of this question appears in [3].

In our *Bridge to Higher Mathematics* course, a sophomore course that teaches proof techniques, we study some famous theorems (=good questions) and discuss several proofs to each one. To get started, we ask students to split into two groups, a question group and an answer group. The question group is required to come up with a good question for the answer group to solve, with the caveat that the question group should be able to judge the validity and correctness of the proposed solution by the answer group. Strangely enough there usually are far fewer volunteers to pose questions than volunteers for answering. Why is that so? Students are rarely asked to formulate questions, they find it difficult to come up with a question. For the few theorems of which they remember the full statement, they usually do not remember a proof. They hesitate to ask a routine question, because they are unsure whether or not to accept the mere answer as the solution. For example if the question asks for the derivative of a function, they are unsure whether or not the answer should involve the limit definition of derivative. In short, they are unsure of what a good question is and they are unsure of what a solution is. Evolving question- and answer-group discussions reveal this uncertainty to the students without any instructor comment.

What causes this uncertainty in quality judgement of question and answer? We are probably not spending enough time in stressing the importance of looking at a question and understanding a question before solving it. In the typical freshman calculus courses drill exercises are formulated in an abbreviated fashion such as: In Exercises 10–25 compute the derivative of the given function and evaluate it at the given point. While these problems are necessary and valuable many students solve them without thinking about the concepts. They just practice a skill, they do not think about a question/solution process. If an electronic system is used for homework submission and grading, the students only type in an answer. Collecting these homework problems in written form reveals that many students produce a correct answer by some mysterious string of expressions without the use of a single equality symbol. When asked to compute the rate of change of $f(x)$ with respect to x at $x = b$ they do not know what to do, even if they could solve Exercises 10–20. The so-called "word" problems are considered difficult.

After mastering the routine exercises and acquisition of the skill, students should be made aware of the problems they now can tackle. Letting the student formulate a problem that classmates and the problem poser have to solve leads to much deeper student engagement in the subject matter. Students are critically examining the problem statement before diving into a solution process.

Here is my favorite example from my teaching experience at Worcester Polytechnic Institute:

In a Calculus I course I assigned one of the typical textbook problems: *Given* $f(x) = ax^3 + bx^2 + cx + d$ *and the point* Q *with coordinates* (x_1, y_1), *find the equation of the tangent line to* f *that contains* Q.

One student complained:

> There is only one answer in the book.

At first I was puzzled by the complaint. It turned out that the student did not know how to start the problem and turned to the answer in the back of the book for a hint. The student also looked up similar problems involving tangents to a circle and these had two answers—a good observation, I thought. I asked this student to turn his complaint into a mathematical question.

After many office hours and interesting discussions we arrived at the following problem formulation which the student asked in the Bridge class a year later as a member of the question group.

Question *Given the graph of a cubic* $f(x)$, *determine the sets* P_k *of all points* p_k *with the property that exactly k tangents to* f *intersect in* p_k. *How many (nonempty) regions are there? What are the possible values of k?*

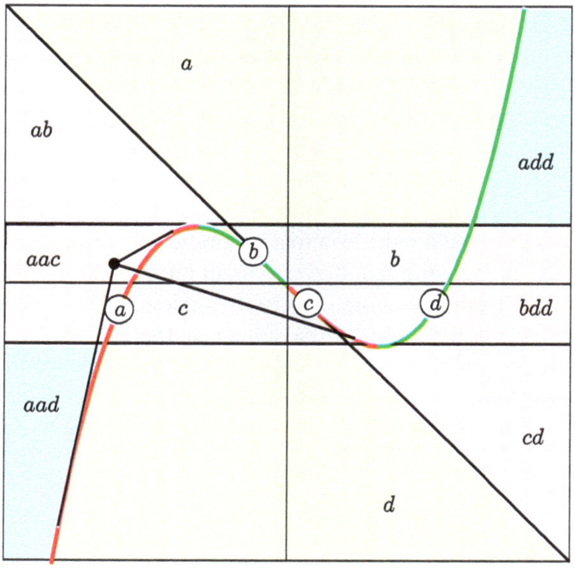

Fig. 1 The multiplicity of symbols in a region label indicates the number of tangents to the arcs of the cubic with the same label. The tiny *yellow regions* should be labeled *bb* and *cc*

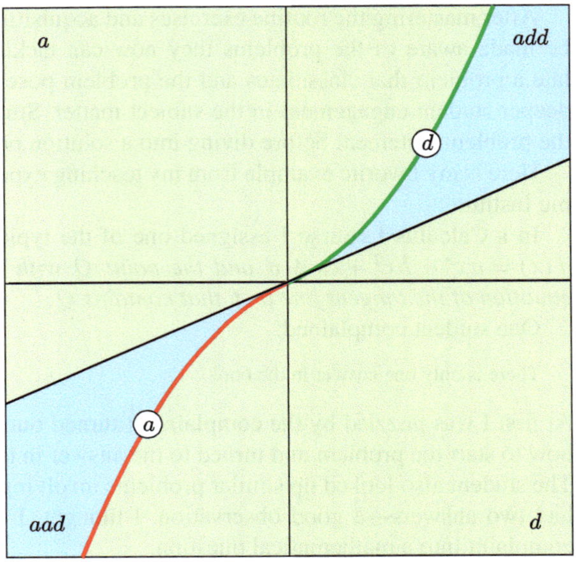

Fig. 2 The multiplicity of the numbers in a region label indicates the number of tangents to the arc of the cubic with the same label

The solution, namely that k ranges from one to three, is in Figs. 1 and 2. Maximum, minimum, and inflection point separate the graph of the cubic into four parts (Fig. 1). It is clear that to each of these four parts there are at most two tangents from any point in the plane. If the cubic only has an inflection point but no maximum or minimum the situation is as in Fig. 2. The graph of the cubic together with the tangents to the critical points divide the plane into regions with the property that the number of tangents from a point in the region to the cubic is constant. The regions are labeled using the symbols by which the arcs of the cubic are labeled. The multiplicity of a symbol in a region label indicates the number of tangents to the arc of the cubic with the same label. It should be pointed out that for the constants used in the calculus book problem there was indeed a unique solution, the point Q was indeed in a green region.

It was delightful to listen to deliberations of the answer group. They did not stop after picking a (in their opinion) best solution, they also formulated generalizations—and checked a reduction to a quadratic. Many students testified that this was the first time in their life that they *did* math rather than just learnt it.

To foster this kind of activity students should be encouraged to regularly solve problems posed in journals such as the *Pi Mu Epsilon Journal* or the *American Mathematical Monthly*. Soon they will experience that a good question is one that gets them hooked. Getting hooked makes students advance from problem solvers to problem posers.

References

1. K. Devlin, *The Unfinished Game: Pascal, Fermat, and the Seventeenth-Century Letter that Made the World Modern* (Basic Books, New York, 2008)
2. A. Engel, *Wahrscheinlichkeitsrechnung und Statistik* (Ernst Klett Verlag, Stuttgart, 1973)
3. C. Pomerance, Ruth-Aaron numbers revisited, in *Paul Erdős and His Mathematics, I (Budapest, 1999)*, vol. 11, Bolyai Society Mathematical Studies (János Bolyai Mathematical Society, Budapest, 2002), pp. 567–579
4. C. Pomerance, Letters from the master: my correspondence with Paul Erdős (2015). https://math.dartmouth.edu/~carlp/homtalk.pdf

Part VIII
Discrete Math and Theoretical Computer Science

Information Measures of Frequency Distributions with an Application to Labeled Graphs

Cliff Joslyn and Emilie Purvine

Abstract The problem of describing the distribution of labels over a set of objects is common in many domains. Cyber security, social media, and protein interactions all care about the manner in which labels are distributed among different objects. In this paper we present three interacting statistical measures on label distributions, thought of as integer partitions, inspired by entropy and information theory. Of central concern to us is how the open- versus closed-world semantics of one's problem leads to different ways that information about the support of a distribution is accounted for. In particular, we can consider the number of labels seen in a particular data set in relation to both the number of items and the number of labels available, if known. This will lead us to consider both two alternate entropy normalizations, and a new measure specifically of support size, based not on entropy but on nonspecificity measures as used in nontraditional information theory. The entropy- and nonspecificity-based measures are related in their ability to index integer partitions within Young's lattice. Labeled graphs are discussed as a specific case of labels distributed over a set of edges. We describe a use case in cyber security using a labeled directed multigraph of IPFLOW. Finally, we show how these measures respond when labels are updated in certain ways corresponding to particular changes of the Young's diagram of an integer partition.

Keywords Information measures · Distributions · Entropy · Nonspecificity · Labeled graph

Mathematics Subject Classification 94A17 · 05C90 · 05A17

C. Joslyn · E. Purvine (✉)
Pacific Northwest National Laboratory, Seattle, WA 99352, USA
e-mail: Emilie.Purvine@pnnl.gov

C. Joslyn
e-mail: Cliff.Joslyn@pnnl.gov

G. Letzter et al. (eds.), *Advances in the Mathematical Sciences*, Association for Women in Mathematics Series 6, DOI 10.1007/978-3-319-34139-2_19

1 Introduction

Given a nonempty collection of labeled entities, how are we to measure the distribution of their labels as drawn from a set of available discrete attributes? Of course, such a basic question should, and does, have many answers already. In particular, entropy measures are commonly used to measure the spread and shape of a distribution, but are defined only on probability distributions. And while probability distributions, as relative frequencies, are easily and uniquely derived from counts, nonetheless this transformation loses information about the support relative to the original count distribution.

We have sought in the literature measures which completely characterize the size and shape of the distribution of labels of such a collection, but were surprised not to find them. We present candidates here, incorporating two forms of uncertainty-based information measures. Entropy and normalized entropy are retained as standard probabilistic approaches over a fixed support, which we introduce as a measure of the "smoothness" of a count distribution. But we also introduce a "dispersion" measure of the degree of support itself relative to the total count. This is a kind of normalized nonspecificity measure, a non-probabilistic uncertainty-based information measure.

In Sect. 2 we introduce integer partitions, different ways to represent them, and our inspiration from collections of labeled items. Additionally, in this section we introduce entropy measures on discrete probability distributions. Next, in Sect. 3 we introduce our two candidate functions measuring the *smoothness* and *dispersion* of a distribution of labels. Then in Sect. 4 we narrow to the case of integer partitions derived from labeled degree distribution where each part counts the number of edges with a particular label. We additionally give a use case related to cyber security. Finally, in Sect. 5 we explore the theoretical aspects of our two functions as they relate to integer partitions and Young diagrams.

2 Preliminaries

In our Preliminaries section we introduce three separate concepts: frequency distributions, integer partitions, and information measures. These three somewhat disjoint concepts will be more strongly related in Sect. 3 and beyond.

2.1 Frequency Distributions

We begin by considering a finite set $X = \{x_j\}_{j=1}^{n}$ of n items, each with a label $l \in \Lambda$, drawn from a set of $\mu = |\Lambda|$ labels. Our goal in this paper is to study the distribution of these μ labels over the set X. We begin then, by forming a frequency vector $\gamma = \langle \gamma_k \rangle_{k=1}^{\mu}$, where $\gamma_k \in \mathbb{N}$ is the number of items in X that are labeled l_k. Notice here that although we have μ labels available to us we may not use all of the labels, so there might be some $\gamma_k = 0$ in our frequency distribution. In fact, it may be the case that $\mu > n$ so that there *must* be zeros in our distribution as it's impossible to use more than n labels on a set of n objects.

2.2 *Integer Partitions*

An **integer partition**, or simply a **partition**, of a positive integer n is a list of strictly positive integers, $\lambda = \langle\lambda_1, \lambda_2, \ldots, \lambda_m\rangle$ such that $\lambda_i \geq \lambda_{i+1}$ and $\sum_i \lambda_i = n$. Each λ_i is called a **part**, and we denote the number of parts in λ by $m := |\lambda|$. For example, $\lambda = \langle 5, 5, 3, 1\rangle$ is a partition of 14 with $m = 4$ parts.

There are many areas in which integer partitions may arise, but we are concerned with those induced by a collection of labeled items. Consider again a collection $X = \{x_j\}_{j=1}^n$ of n items, each with a label drawn from a set of labels Λ. When we introduced frequency distributions in the previous section we did our counting "label-wise". Here, to form an integer partition, we will count "item-wise". The labels l_i naturally partition X into $m \leq n$ disjoint blocks $X_i \subseteq X$ where for each $1 \leq i \leq m$, all the labels for the items in X_i are the same. Let $\lambda_i := |X_i|$, $1 \leq i \leq m$ be the number of items in block i, so that $1 \leq \lambda_i \leq n$. Then sort them down, so that $\lambda_i \geq \lambda_{i+1}$. Finally, let $\lambda := \langle\lambda_i\rangle_{i=1}^m$. λ is now an integer partition representing the counts of the m labels actually seen, amongst the n items in question.

The item-wise label frequency distribution λ created from X and Λ essentially counts the same thing as the label-wise distribution γ, so let's consider the relationship of γ to λ. Clearly $\mu \geq m$, so $\lambda \subseteq \gamma$ in the sense that each $\lambda_i \in \lambda$ has a corresponding $\gamma_k \in \gamma$, but γ must have $\mu - m$ additional $\gamma_k = 0$. Basically, γ is λ padded with zero counts for any unused labels. Note that if $\mu = m$ then $\gamma = \lambda$ will be a proper integer partition. Therefore, we may refer to γ in the general case below, and λ when we rely on having an integer partition. We will return later to a discussion of γ versus λ in the case of information measures on label distributions, but for now we return to our discussion on integer partitions.

There are various ways of representing partitions either as decreasing lists of numbers as already stated, or as diagrams. A **Young diagram** of a partition λ is an array of boxes which is top and left aligned in which each row represents a single part of the partition. An example is shown in Fig. 1. We will often refer to both the Young diagram of a given partition, and the partition itself, by the same name, e.g., λ. Sometimes these are also called **Ferrers diagrams**, though Ferrers diagrams are often drawn with dots instead of boxes. We will come back to Young diagrams in Sect. 5.

There is a great deal of research in the theory of partitions. Most prominent are enumerative combinatorics questions of the form "How many partitions are there that have a specific property, as a function of n?" Many properties are explored including restricting the number of parts, type of parts (e.g., even parts, distinct parts), and other more complicated types of restrictions.

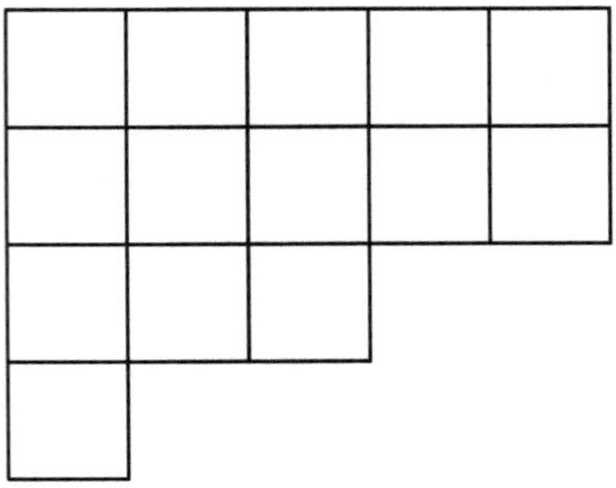

Fig. 1 The Young diagram corresponding to partition $\langle 5, 5, 3, 1\rangle$ of 14

Our research is looking not at these enumerative questions, but instead at calculating information measures, to be introduced in the following sections, over the set of integer partitions, and studying the distribution of these statistics as n, m, and λ_1 (the largest part) vary. Specifically we are interested in characterizing these statistics over the integer partition poset,[1] P_n, referenced in [1, 9]. P_n is the poset of integer partitions of n ordered by refinement. So, if $\eta \leq \lambda$ in P_n then λ has fewer parts than η, and each part in λ is split into multiple parts in η. For example, $\eta = \langle 5, 4, 2, 1, 1, 1\rangle$ is a refinement of $\lambda = \langle 5, 5, 3, 1\rangle$ since one of the 5 s is split into 4, 1, the 3 is split into 2, 1, and everything else remains the same. The elements of P_n can be grouped by their number of parts m, making P_n a *graded* poset [1, 6]. An example of the Hasse diagram of the integer partition poset P_6 is shown in Fig. 2. A Hasse diagram is a visual representation of the cover relations in a poset. In general, each element $p \in P$ from the poset $\mathscr{P} = (P, \leq)$ is shown, and an edge from p to q with p below q indicates that $p \prec q$. We call each group of partitions with the same number of parts a *rank* of P_n. Notice that the ranks of P_n are separated onto different vertical levels of the Hasse diagram.

Finally, note that the elements of each integer partition poset P_n sit within Young's lattice, a lattice of all integer partitions ordered by containment of their Young diagrams. One Young diagram, η, is contained within another, λ, if every box in η occurs in λ. In other words, if we label each box with two integer coordinates, the x coordinate increasing along columns and the y increasing along rows (as in labeling matrix entries), then every label that occurs in η will also appear in λ. The Young's lattice for $n \leq 6$ is shown in Fig. 3, the dashed edges indicate partition refinements in Young's lattice produced by adding or deleting boxes in a Young diagram λ, and thus a change in n. Solid edges indicate partition refinements within each of the P_n produced by splitting or merging blocks, and thus a change in m, but with no change in n, these solid edges are not included in Young's lattice. The significance of the G notations at the bottom of Fig. 3 will be discussed later in Sect. 3. Notice that Young's lattice is also graded, with each rank being the set of partitions in one P_i.

[1]The integer partition poset should not be confused with the *set* partition *lattice* [1].

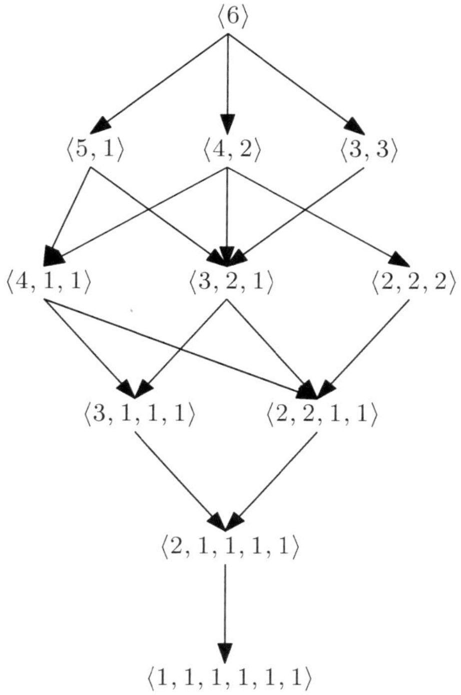

Fig. 2 The Hasse diagram of the integer partition poset for partitions of $n = 6$, P_6

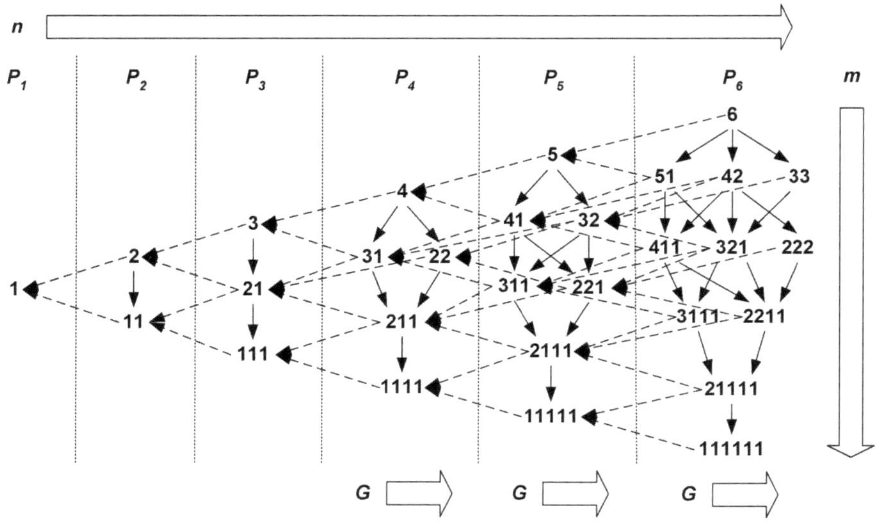

Fig. 3 Partition posets P_i (*solid arrows*) embedded as ranks i within Young's lattice (*dashed arrows*)

2.3 Information Measures

Let $\mathbf{p} = \langle p_1, p_2, \ldots, p_\mu \rangle$ be a discrete probability distribution, where the $p_k := p(y_k) \in [0, 1]$ are the probabilities of a random variable Y taking values in the set $\{y_1, \ldots, y_\mu\}$, so that $\sum_{k=1}^{\mu} p_k = 1$. The **entropy** of the distribution $\mathbf{p}$, denoted $H(\mathbf{p})$, is given by the following equation:

$$H(\mathbf{p}) = -\sum_{k=1}^{\mu} p_k \log_2 p_k.$$

In the case where any $p_k = 0$ we define $0 \log_2(0) := 0$. This is a logical definition since $\lim_{x \to 0} x \log_2(x) = 0$. There are continuous analogs of entropy where the sum is replaced by an integral, but we only deal with the discrete version in this paper.

Entropy can be thought of as a measure of uncertainty in a probability distribution. If the distribution is uniform, $\mathbf{p}_u = \left\langle \frac{1}{\mu}, \ldots, \frac{1}{\mu} \right\rangle$, then the entropy is $H(\mathbf{p}_u) = -\log_2(\frac{1}{\mu}) = \log_2(\mu)$. This represents maximum uncertainty, i.e., any outcome is equally likely. Contrast that with the fully skewed distribution $\mathbf{p}_s = \langle 0, \ldots, 0, 1 \rangle$ which has entropy $H(\mathbf{p}_s) = -\log_2(1) = 0$. This represents no uncertainty since the outcome is determined with all of the probability mass sitting on only one possibility.

Normalizing entropy by its maximum, $\log_2(\mu)$, is a standard approach to effectively measure the shape of the probability distribution. We call this normalized entropy **p-smoothness** of a probability distribution $\mathbf{p}$ and denote it by $\widetilde{G}(\mathbf{p})$:

$$\widetilde{G}(\mathbf{p}) = \frac{H(\mathbf{p})}{\log_2(\mu)} = \frac{-\sum_{k=1}^{\mu} p_k \log_2 p_k}{\log_2(\mu)} \in [0, 1]. \tag{1}$$

Just as in the definition for entropy, we must define $\widetilde{G}(\mathbf{p})$ in a special case. This time we treat $\mu = 1$, which would require dividing by $\log_2(1) = 0$. In this case we let $\widetilde{G}(\langle 1 \rangle) := 1$ which agrees with $\lim_{x \to 1} \frac{x \log_2(x)}{\log_2(x)} = 1$. We are using the word "smooth" here as a synonym for "close to uniform". We know that a uniform distribution has maximum entropy and therefore, when normalized, $\widetilde{G}(\mathbf{p}) = 1$. So, a highly smooth distribution is one that is close to uniform. On the other hand, a very skewed distribution, one that looks "lumpy", will have low entropy (low uncertainty) and therefore $\widetilde{G}(\mathbf{p})$ smaller. So, the closer the distribution is to perfectly smooth, or uniform, the higher $\widetilde{G}(\mathbf{p})$ will be.

3 Information Measures on Label Distributions

In this paper we are concerned with measuring the shape of a distribution of labels over a set of items. Specifically, given an arbitrary collection of n labeled items $X = \{x_j\}_{j=1}^{n}$, how can we best characterize the distribution of their labels Λ? Now, this distribution is not a probability distribution but instead it is an *absolute* frequency distribution. Classically, the first thing done is to transform the associated label-wise distribution, γ, described in Sect. 2.2, which we recall may be a proper integer partition, λ, into a *relative* frequency distribution $\mathbf{f}$ as

$$\mathbf{f}(\gamma) := \gamma/n = \langle f_k \rangle_{k=1}^{\mu} = \langle \gamma_k/n \rangle_{k=1}^{\mu}$$

so that now $f_k \in [0, 1]$ and $\sum_{k=1}^{\mu} f_k = 1$. Relative frequency distributions $\mathbf{f}$ are typically interpreted as (discrete) probability distributions $\mathbf{p}$ by considering f_k as the probability of choosing an item labeled l_k when picking an item randomly from the set of n labeled items, $X = \{x_j\}$. Therefore, we can calculate the entropy and $\mathbf{p}$-smoothness as introduced in the previous section. With slight abuse of notation here we define

$$\widetilde{G}(\gamma) := \widetilde{G}(\mathbf{f}(\gamma)) = \frac{H(\mathbf{f}(\gamma))}{\log_2(\mu)} = \frac{-\sum_{k=1}^{\mu} \frac{\gamma_k}{n} \log_2 \frac{\gamma_k}{n}}{\log_2(\mu)}.$$

Now, consider our set X with items labeled from Λ and create both the integer partition λ item-wise and the distribution γ label-wise. We can treat λ similarly, taking

$$\mathbf{f}(\lambda) := \lambda/n = \langle f_i \rangle_{i=1}^{m} = \langle \lambda_i/n \rangle_{i=1}^{m} .$$

$\mathbf{f}(\lambda)$ is basically $\mathbf{f}(\gamma)$ stripped of its $\mu - m$ trailing zeros. And then we have

$$\widetilde{G}(\lambda) := \widetilde{G}(\mathbf{f}(\lambda)) = \frac{H(\mathbf{f}(\lambda))}{\log_2(m)} = \frac{-\sum_{i=1}^{m} \frac{\lambda_i}{n} \log_2 \frac{\lambda_i}{n}}{\log_2(m)}.$$

We observe that if $m < \mu$ then $\widetilde{G}(\gamma) \neq \widetilde{G}(\lambda)$ despite being calculated from the same sets X and Λ. Non-normalized entropy H, the numerator of $\widetilde{G}$, cannot distinguish the relative frequency distributions of λ and γ as it is blind to zero-padding, but the normalization factor is different in each case, $\log_2(\mu)$ on the left-hand side and $\log_2(m)$ on the right.

The question of whether or not these measures *should* be equal is not for this paper to decide. However, we present an additional measure which is blind to zero-padding on absolute frequency distributions and later will show that there can be advantages in using one or the other. We define **smoothness** as

$$G(\gamma) = G(\mathbf{f}(\gamma)) := \frac{H(\mathbf{f}(\gamma))}{\log_2(m)} = \frac{-\sum_{i=1}^{\mu} \frac{\gamma_k}{n} \log_2 \frac{\gamma_k}{n}}{\log_2(m)}.$$

Notice the difference being that here we normalize by the number of nonzero entries in γ, and now $G(\gamma) = G(\lambda) = \widetilde{G}(\lambda)$. So, G is only sensitive to the *absolute* support, m, of λ or γ, that is, the number of labels actually *seen*, as opposed to the number μ of labels which are *available*. Another way of saying this is that $\widetilde{G}$, as opposed to G, makes a *closed-world* assumption that we know in advance the universe of discourse Λ of available labels. In open-world situations, Λ may be unknown, unspecified, or so large as to be meaningless to the problem being modeled. The open world assumption of G is thus independent of any implicit assumptions about the space of labels.

While the open-world smoothness G provides an important alternative to the traditional closed-world normalized entropy $\widetilde{G}$, still for our question of characterizing counts of labels in the context of integer partitions λ, or more general label distributions γ, both G and $\widetilde{G}$ have flaws. Both relative frequency distributions $\mathbf{f}(\gamma), \mathbf{f}(\lambda)$ lose information relative to absolute frequency distributions γ, λ. In particular, consider two label frequency distributions γ and γ' of n and n' objects. Let $\gamma' = \alpha \cdot \gamma$ for some $\alpha \in \mathbb{N}$, so that, incidentally, $n' = \alpha n$. Then additionally $\mathbf{f}(\gamma) = \mathbf{f}(\gamma')$, and of course, both $G(\gamma) = G(\gamma')$ and $\widetilde{G}(\gamma) = \widetilde{G}(\gamma')$. In reducing to relative frequencies, we have lost connection to the interpretation of the partitions as counts with respect to a total number of items n, n', and thus neither smoothness G nor $\mathbf{p}$-smoothness $\widetilde{G}$ can distinguish these cases.

We have observed now that both G and $\widetilde{G}$ are insensitive to the total number of items that we are labeling, $n = |X|$, and additionally that $\widetilde{G}$ distinguishes between the distributions item-wise (λ an integer partition) and label-wise (γ an arbitrary non-negative integer distribution) when we have more labels available, μ, than observed, m, whereas G does not. It is not our goal in this paper to decide which of $\widetilde{G}$ and G is more reasonable to use, indeed it often depends on the application. In some cases, and often in those that we are concerned with, e.g., cyber networks, either the number of possible labels is very large compared to the number of labels seen, or we simply do not know how many labels are possible. In these cases of large μ normalizing with respect to μ gives little to no information as all $\widetilde{G}$ values will then be very small. It is for these reasons that we generally turn our attention in this paper to the use of G over $\widetilde{G}$, or when we do use $\widetilde{G}$ we will often make the assumption that $\mu = n$. We note that we will contrast G and $\widetilde{G}$ again later using this assumption on μ when we introduce a third measure called κ. Additionally since $G(\lambda) = G(\gamma)$ we will proceed working with only integer partitions.

As evidenced by the prior discussion on G versus $\widetilde{G}$ it's clear that there is an important relationship between the three quantities n, m, and μ. While the number of items n and the number of available labels μ themselves need not be related, we do have $m \leq \min(n, \mu)$; that is, there can be as many labels, m, in X as the number of items n, unless $\mu < n$, in which case there are not enough labels to go around, and some items x_i must "double up." So, there are two edge cases:

$m = n$: Here all items have a distinct label, so that $\lambda = \langle 1, 1, \ldots, 1 \rangle$. Also $n \leq \mu$, and our label-wise frequency distribution is $\gamma = \langle 1, 1, \ldots, 1, 0, 0, \ldots 0 \rangle$, with m ones and $\mu - m$ zeroes.

$m = \mu$: Here all available labels are used, so that $n \geq \mu$, and $\lambda = \gamma$ with all the $\lambda_k \geq 1$.

Consider P_6 as shown in Fig. 4. The integer partitions $\lambda \in P_6$ are shown adorned with the values for G, and m is shown on the right. The κ values on the right of this

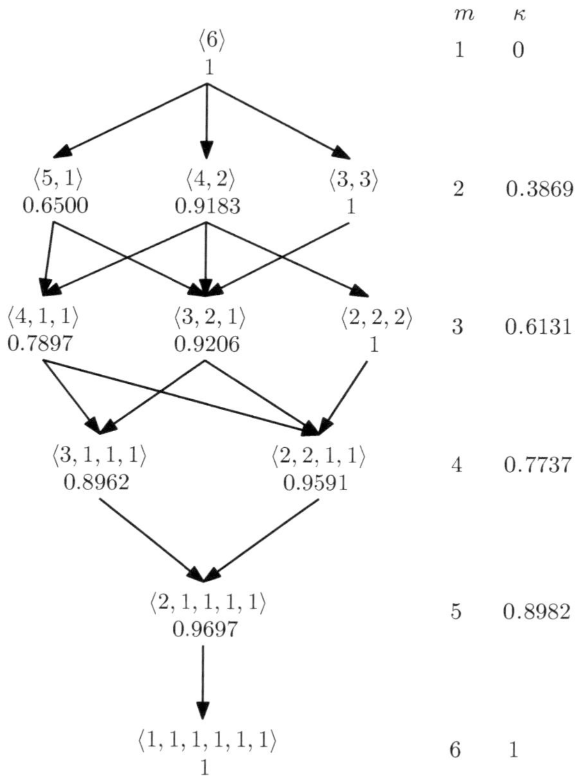

Fig. 4 The integer partition P_6 adorned with smoothness $G(\lambda)$

figure will be defined later. As a general matter, for a given n, each level of P_n includes partitions with the same number of parts. $G(\lambda)$ then orders the λ within each level, with $G(\lambda) = 1$ iff λ is uniform, and $G(\lambda) \to 0$ for $\lambda = \langle n-k, 1, \ldots, 1\rangle$ for each $1 \leq k \leq n-1$ as $n \to \infty$. Note that in some cases we have $G(\lambda_1) = G(\lambda_2)$ when $\lambda_1 \neq \lambda_2$, e.g. when $n = 20$ and $m = 6$ we have $G([8, 3, 3, 2, 2, 2]) = G([6, 4, 4, 4, 1, 1])$. We will order these partitions according to lexicographic ordering when they occur.

The role of G to order partitions is illustrated in Fig. 3 in Young's lattice, with G operating within each P_n. So more generally, for any partition λ of any n, we seek measures to place it within Young's lattice in terms of:

- Its rank within Young's lattice, which is clearly just n.
- Within each rank of Young's lattice (that is, within P_n), its "horizontal" placement. This is its smoothness $G(\lambda)$.
- And again within each P_n, its "vertical" level within P_n.

So concerning this final quantity, we are left with the question, is there a way to order the integer partitions based on how many parts they have? The number of parts, m, is the obvious answer, as together with G this ordering would essentially create a coordinate system on the integer partition poset P_n.

We could simply measure the number of parts m relative to the number of possible parts n, whether or not some absolute number of labels μ set an upper bound on m. For a partition of n with m parts this would simply seem to be $\frac{m}{n}$. However, the distribution of the number of partitions of n with a given number of parts has a long tail, i.e., there are relatively few partitions of n with large numbers of parts, and we seek a measure which has larger gaps when there are many partitions, when m is smaller, and smaller gaps when there are fewer partitions, rather than something linear in the number of parts. Additionally, while working with information measures like G, there are great advantages of working with log functions, due to additivity and other properties. We therefore use a log ratio and define what we call **dispersion**, denoted by $\kappa(\lambda)$, as

$$\kappa(\lambda) = \frac{\log_2(m)}{\log_2(n)} \in [0, 1].$$

Figure 4 also shows the κ value for each level of P_6. Like G, we also have $\kappa(\lambda) \in [0, 1]$, but now $\kappa = 0$ if and only if $\lambda = \langle n \rangle$ which is the case only when $m = 1$. Additionally, $\kappa = 1$ if and only if $\lambda = \langle 1, 1, \ldots, 1 \rangle$ when $m = n$. Notice that $\kappa = 1$ implies $G = 1$, but we can have $G = 1$ with small κ, as in $\lambda = \langle 3, 3 \rangle \in P_6$.

Note that unlike G, the value of κ does *not* depend on the actual *values* of the $\lambda_i \in \lambda$, but only on m, the *number* of parts into which λ is divided, and n the total sum of λ. It is effectively a measure of the *support* of the partition λ of the integer n *relative* to n itself. And while G, as an entropy, is a measure of the information content of the partition λ, interpreted as a relative frequency (that is, discrete probability) distribution, $\mathbf{f}(\lambda)$, so κ is *also* a measure of information, although not an entropy, but rather a **Hartley** measure [5].

In the context of **generalized information theory**, a Hartley measure is an information measure called a **nonspecificity N**. Given a collection of m choices, then their nonspecificity is simply $\mathbf{N}(m) = \log_2(m)$. Note that this quantity $\log_2(m)$ is also the maximal entropy $H(\mathbf{p})$ when $\mathbf{p}$ is uniform, and in this particular case these measures coincide. But like entropy, in more general cases nonspecificities can take on more complex forms, and are fully-fledged information measures in that they satisfy basic axioms of additivity, monotonicity, and normalization as they quantify the amount of uncertainty present in a collection of choices which are not probabilistically weighted.

In our case, our $\kappa = \log_2(m)/\log_2(n)$ is thereby a **normalized nonspecificity**. As a general matter, nonspecificities are defined and used in the context of possibilistic information theory, or possibility theory [2, 3, 8]. Although exploring possibilistic measures of support is not the purpose of this paper, here it is sufficient to observe that as entropy measures the probabilistic constraint placed on a collection of m choices by the probabilities f_i, so κ measures the non-probabilistic constraint placed on a collection of n choices by the selection of m of them.

We have already observed that we can use κ and G as two "coordinates" within the Hasse diagram of the integer partition poset. We use κ to tell the vertical level of a partition, a y coordinate in the Hasse diagram, and G as an x coordinate telling how far to the right (closer to uniform) a partition is. This is illustrated in Fig. 5(a) where κ (y-axis) and G (x-axis) are calculated on all partitions of $n = 20$. We see clear levels corresponding to each m value yielding different κ values. The G values then go up to a maximum of nearly 1 depending on if m divides 20 or not.

Another important observation is as follows. In our "typical case" of $n = \mu$ so that we have no more labels than there are items, we actually have that

$$\widetilde{G} = G \cdot \kappa.$$

This, together with $\kappa \in [0, 1]$, shows that $\widetilde{G}$ can be interpreted as both a weighting down of G by κ, and simply as a quantity reflecting both smoothness and dispersion through the product, capturing information from both G and κ. So if $\widetilde{G}$ captures information from both κ and G, the question is: can we use it to determine both the x and y coordinates of a partition? In order for this to be the case we would need to be able to break up the range of $\widetilde{G}$, [0, 1], into disjoint intervals, one for each level of the poset (m value). Unfortunately, we can only do this for $n \leq 7$. In the integer partition poset for $n = 8$ we cannot decompose the interval [0, 1] into disjoint intervals for each m value. For $m = 3$ the $\widetilde{G}$ values range between 0.3538 and 0.5204, and for $m = 4$ the values range between 0.5163 and 0.6667. There is a clear overlap between the two ranges, so we cannot in general use $\widetilde{G}$ to decide which level a partition is on. The plot in Fig. 5b shows the plot of κ now against $\widetilde{G}$ (with $n = \mu$) to show the overlap in $\widetilde{G}$ ranges as it is broken up into its κ levels.

Considering the relation between our three measures G, $\tilde{G}$, and κ, we can conclude that it is more enlightening to use both κ and G separately to give information about

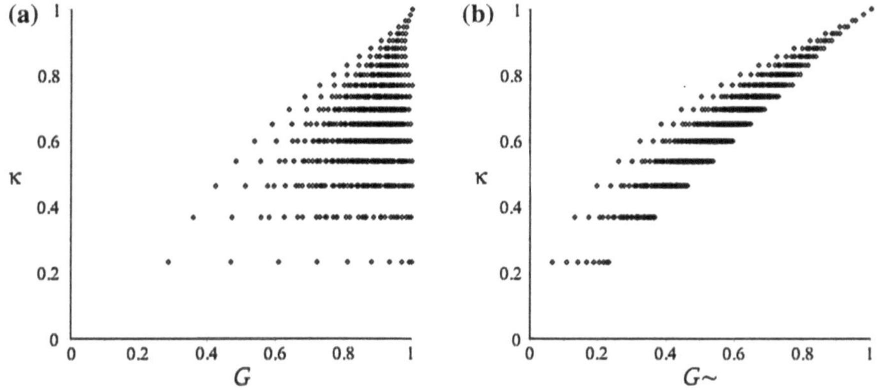

Fig. 5 Plots showing how G, κ, and $\widetilde{G}$ interact for all partitions of $n = 20$. **a** κ versus G. **b** κ versus $\widetilde{G}$ with $\mu = n$

a partition λ. But at the same time, it might be convenient to use $\widetilde{G}$ as an alternate entropy normalization, a discounting of G by κ, or a scalar combining G and κ. Effectively, $\widetilde{G}$ provides a single, scalar quantity reflecting both the horizontal and vertical position within P_n. Together with n itself, we can use them to uniquely characterize any integer partition λ within Young's lattice.

4 Application to Labeled Degree Distributions

So far in this paper we have described dispersion (κ) and smoothness (G) as measures on integer partitions, λ, or more specifically on frequency distributions of a set of labels, Λ, on a set of objects X. Now we come to the application of labeled graphs. Consider a directed graph or multigraph, $G = (V, E)$, with edge label function $f : E \to \Lambda$. Theoretically, any collection of edges can be considered as our set of objects, X, but we will restrict to the case where the set of edges has a common source vertex or target vertex. Given a vertex, $v \in V$, let $S_v = \{e = e_1e_2 \in E : e_1 = v\}$ be those edges which have v as their source vertex, and $T_v = \{e = e_1e_2 \in E : e_2 = v\}$ be those which have v as their target vertex. We may treat S_v and T_v separately as base sets, or take their union and consider the full set of edges incident on v. For example, see Fig. 6.

Given these sets of edges, S_v and T_v, we consider not just the size of the sets, as traditionally considered when investigating degree distribution, but also their dispersion and smoothness with respect to the labeling function f and label set Λ. Referring back to Fig. 6 the integer partition corresponding to T_v is $\lambda_{T_v} := \langle 2, 1, 1 \rangle$ where the labels are in the order C, A, D, and integer partition for S_v is $\lambda_{S_v} := \langle 2, 1, 1, 1 \rangle$ for labels B, A, C, E. If we wish to take the full set of edges into account we have $\lambda_v := \langle 3, 2, 2, 1, 1 \rangle$ for labels C, A, B, D, E. Figure 7 shows the Young diagrams for these three example partitions. Then we can calculate the dispersion, smoothness, and λ-smoothness of these distributions. These values are found in Table 1. As we might expect, λ_{S_v} has the highest smoothness among the three partitions and also the highest dispersion. But, these three partitions are not very diverse and so we get very similar G, $\widetilde{G}$, and κ values.

Fig. 6 An example vertex in a labeled directed graph with 4 in-edges and 5 out-edges

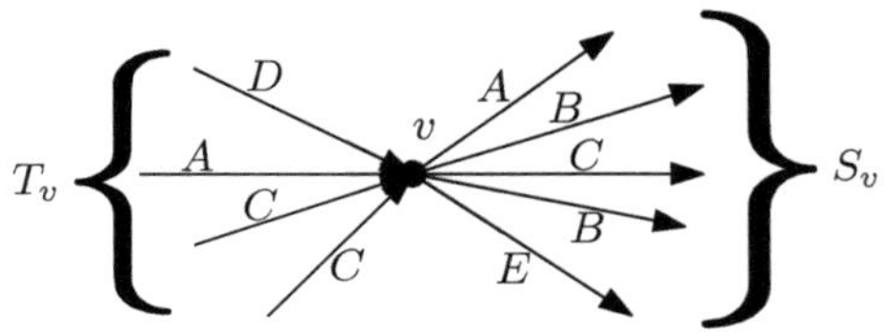

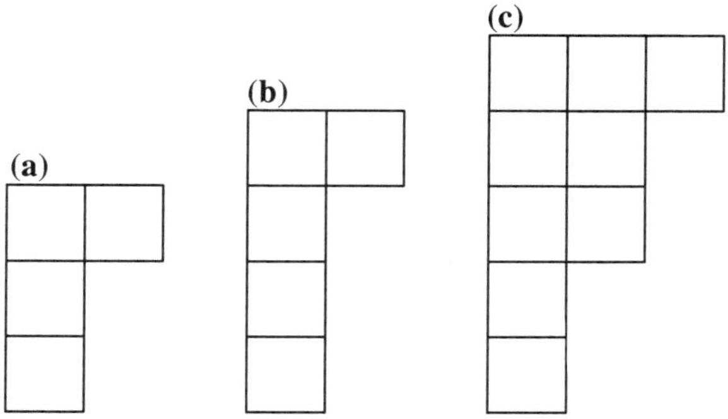

Fig. 7 Young diagrams for our three example partitions. **a** λ_{T_v}. **b** λ_{S_v}. **c** λ_v

Table 1 Smoothness, λ-smoothness, and dispersion values for three example partitions

	λ_{T_v}	λ_{S_v}	λ_v
G	0.9464	0.9610	0.9463
$\widetilde{G}$	0.7500	0.8277	0.6931
κ	0.7925	0.8614	0.7325

4.1 Cyber Security Use Case

This question of assessing the shape of labeled degree distributions, or more generally sets of labeled objects or integer partitions, through information measures was originally motivated by a problem in cyber security. Securing cyber systems has become more and more necessary since attacks to large companies and governments have become increasingly common. Detecting different types of attacks while maintaining system resiliency, i.e., being able to complete missions in the face of attack, is a major focus in cyber security.

We focus on the use of a specific type of cyber data called NetFlow, or more generally IPFLOW, which is IP communication traffic collected at routers and switches throughout the network. A single IPFLOW record contains a source IP and port, a destination IP and port, as well as other data about the information being sent including start and end time, number of packets, and number of bytes. We can study IPFLOW data by transforming it into an IPFLOW multigraph in which vertices are IP addresses, or IP:port pairs, and edges indicate a flow of information from one IP to another. Figure 8 shows an example IPFLOW graph where IPs have been reduced down to their last two octets (e.g., a.x instead of $\alpha.\beta$.a.x), and edges are labeled with a flow ID.

Many common attack types have signatures for the way that they are carried out that can be seen in the IPFLOW graph. Our observation was that these signatures can manifest as extreme shifts in smoothness and dispersion values. For example, consider a denial of service (DoS) or distributed denial of service (DDoS) attack. This occurs when an adversary, or a distributed group of adversaries, floods a server with external communication requests. This overloads the server so that it cannot respond

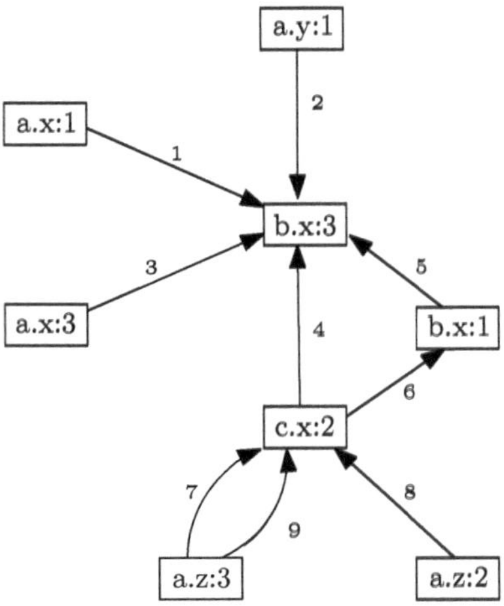

Fig. 8 An example IPFLOW graph with vertices being IP:port pairs and edges labeled with a flow ID

to all of the requests in a timely manner, effectively making the server unavailable to legitimate requests.

If we are looking at a single attacker, IP_A, a DoS attack manifests as a large out-degree with the majority of edges having the same destination. Consider the set of edges with source IP_A and label each edge with its destination IP address. The distribution of these labels will have a single, or very few, labels with high count and any others with very low count. The smoothness of this distribution will likely depend on the number of victims and non-victims being contacted by IP_A. If IP_A only contacts victims then smoothness is likely to be high, but if they are also contacting others (e.g., fellow attackers or a controller in a DDoS attack) then smoothness will be low. However, we can say that dispersion will be very low since the number of communications (the number of edges or items) will be much larger than the number of IPs that are contacted (the number of labels).

In Fig. 9 we show the outgoing smoothness versus outgoing dispersion for vertices in an IPFLOW graph when edges are labeled as described above, by the destination IP address. We used NetFlow from the 2013 Visual Analytics Science and Technology (VAST) Challenge data set which contains synthetic NetFlow with well-documented ground truth attacks [7]. The blue data points are external IP addresses and those large blue points on the left that are labeled are known to be attackers. The cluster of blue points on the upper right, with the IP 10.0.0.5 singled out, are virtual websites. Their high smoothness and dispersion mean that they send information fairly uniformly to other IP addresses, and they do not send very much to each, which is expected behavior for a website. The size of each circle indicates how many flows were sent. We have more results on this data set using these dispersion and smoothness measures as well as other analysis in [4].

Other types of attacks have similar characterizations through smoothness and dispersion. Port scans, for example, where one or many external IPs contact a single vertex through many ports, can be seen again in the IPFLOW graph. The port scanners

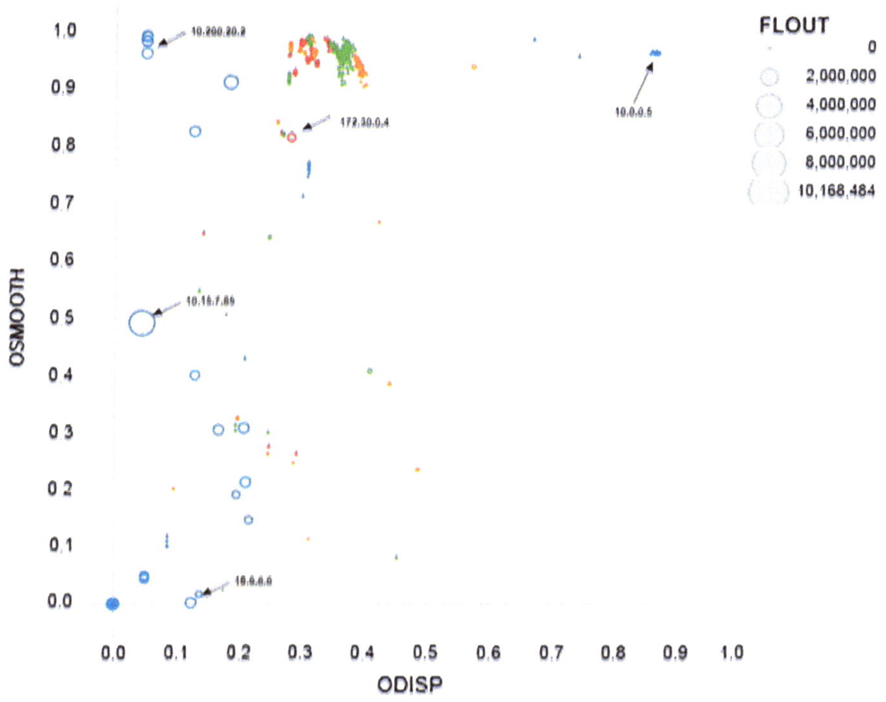

Fig. 9 Outgoing smoothness (y-axis) versus outgoing dispersion (x-axis) for vertices in an IPFLOW graph created by synthetic IPFLOW data around the time of a DDoS attack

will have both high outgoing dispersion and high outgoing smoothness when edges are labeled with destination IP and port. Each scanner contacts each port on a given IP address a small number of times, and does so fairly uniformly.

Though we have only worked in the cyber security use case we can see the value of using these measures on vertices in other domains to enrich the degree distribution. Labeled graphs have a richer set of information than unlabeled graphs and so we should be using that extra information to perform analysis. Other possible application areas could be social network graphs where edge labels are type of communication or type of relationship between people, or protein interaction graphs where distributions come from weights on edges indicating magnitude of interaction.

5 Relation to Young Diagrams

Although we were inspired to create the dispersion and smoothness functions to measure the shape of labeled degree distributions in directed, labeled graphs, we realize that they are also interesting mathematical functions on the set of integer partitions. In this section we will give a few of our observations about how G and κ change when we change an integer partition in two specific ways.

Recall that a Young diagram is one way to pictorially represent an integer partition as m rows of boxes, the ith row containing λ_i boxes. We were interested in how G and κ vary as we transform the Young diagram. Two transformations will be considered, (1) moving one box from row j to row i, in the case of labeled items this corresponds to changing one label to something else already seen, and (2) taking the conjugate of the diagram.

5.1 Moving Boxes

We will first consider moving one box from row j to row i. Consider two partitions, λ and η, where we form η from λ by switching one box as described. In this case we can write η in terms of λ

$$\eta_\ell = \begin{cases} \lambda_i + 1 & \ell = i \\ \lambda_j - 1 & \ell = j \\ \lambda_\ell & \textit{else}. \end{cases}$$

We are interested in how G and κ change as we go from λ to η. If we make the stipulation that $\lambda_j \geq 2$ then we are never removing an entire part from λ, therefore $\kappa(\lambda) = \kappa(\eta)$. However, G will change as long as $\lambda_i + 1 \neq \lambda_j$.

$$\begin{aligned}
\log_2(m)\left[G(\mathbf{f}(\lambda)) - G(\mathbf{f}(\eta))\right] &= -\sum_{\ell=1}^{m} \frac{\lambda_\ell}{n}\log_2\frac{\lambda_\ell}{n} + \sum_{\ell=1}^{m}\frac{\eta_\ell}{n}\log_2\frac{\eta_\ell}{n} \\
&= -\frac{\lambda_i}{n}\log_2\frac{\lambda_i}{n} - \frac{\lambda_j}{n}\log_2\frac{\lambda_j}{n} + \\
&\quad + \frac{\lambda_i+1}{n}\log_2\frac{\lambda_i+1}{n} + \frac{\lambda_j-1}{n}\log_2\frac{\lambda_j-1}{n} \\
&= -\frac{\lambda_i}{n}\log_2\frac{\lambda_i}{n} + \frac{\lambda_i}{n}\log_2\frac{\lambda_i+1}{n} + \frac{1}{n}\log_2\frac{\lambda_i+1}{n} + \\
&\quad - \frac{\lambda_j}{n}\log_2\frac{\lambda_j}{n} + \frac{\lambda_j}{n}\log_2\frac{\lambda_j-1}{n} - \frac{1}{n}\log_2\frac{\lambda_j-1}{n} \\
&= \frac{\lambda_i}{n}\left[\log_2\frac{\lambda_i+1}{\lambda_i}\right] + \frac{\lambda_j}{n}\left[\log_2\frac{\lambda_j-1}{\lambda_j}\right] + \\
&\quad + \frac{1}{n}\left[\log_2\frac{\lambda_i+1}{\lambda_j-1}\right]
\end{aligned}$$

We will bound each of these terms independently to see what kind of change in G we can have. First, notice that λ_i cannot be bigger than $n-1$ since we are adding 1 to it, and λ_j cannot be less than 2 since we are subtracting 1 from it, it also cannot be more than $n-1$ or there would be nowhere else to move a box. Therefore, $1 \leq \lambda_i \leq n-1$ and $2 \leq \lambda_j \leq n-1$.

Now, let's focus on upper and lower bounds for the first term, $\frac{\lambda_i}{n}\left[\log_2 \frac{\lambda_i+1}{\lambda_i}\right]$. It is not difficult to check that $\frac{d}{dx}\left(\frac{x}{n}\log_2 \frac{x+1}{x}\right)$ is positive for $x \geq 1$, so this term must be increasing. Therefore, we can get upper and lower bounds for the term by plugging in the maximum and minimum values for λ_i, respectively.

$$\frac{1}{n} = \frac{1}{n}\log_2 \frac{1+1}{1} \leq \frac{\lambda_i}{n}\left[\log_2 \frac{\lambda_i+1}{\lambda_i}\right] \leq \frac{n-1}{n}\log_2 \frac{n-1+1}{n-1} = \frac{n-1}{n}\log_2 \frac{n}{n-1}.$$

Next, we look at bounds for the second term, $\frac{\lambda_j}{n}\left[\log_2 \frac{\lambda_j-1}{\lambda_j}\right]$. Again, we can see that the derivative $\frac{d}{dx}\left(\frac{x}{n}\log_2 \frac{x-1}{x}\right)$ is positive for $x \geq 2$ and so we have an increasing function of λ_j. We again plug in maximum and minimum values for λ_j to get upper and lower bounds.

$$-\frac{2}{n} = \frac{2}{n}\log_2 \frac{1}{2} = \frac{2}{n}\log_2 \frac{2-1}{2} \leq \frac{\lambda_j}{n}\left[\log_2 \frac{\lambda_j-1}{\lambda_j}\right] \leq \frac{n-1}{n}\log_2 \frac{n-2}{n} < 0.$$

Finally, we need to get bounds for the third term, $\frac{1}{n}\left[\log_2 \frac{\lambda_i+1}{\lambda_j-1}\right]$. In this case we have a function of both λ_i and λ_j so using the derivative to tell if it is increasing or decreasing will not work. Instead, we bound $\frac{\lambda_i+1}{\lambda_j-1}$ and then take the $\log_2$ of those bounds.

$$\frac{2}{n-2} = \frac{1+1}{n-1-1} \leq \frac{\lambda_i+1}{\lambda_j-1} \leq \frac{(n-1)+1}{2-1} = n.$$

Then, our bounds for $\frac{1}{n}\left[\log_2 \frac{\lambda_i+1}{\lambda_j-1}\right]$ are:

$$\frac{1}{n}\log_2 \frac{2}{n-2} \leq \frac{1}{n}\left[\log_2 \frac{\lambda_i+1}{\lambda_j-1}\right] \leq \frac{1}{n}\log_2 n$$

Putting everything together we can bound our original quantity, $\log_2(m)\left[G\left(\frac{\lambda}{n}\right) - G\left(\frac{\eta}{n}\right)\right]$. The upper bound is given by:

$$\log_2(m)\left[G\left(\frac{\lambda}{n}\right) - G\left(\frac{\eta}{n}\right)\right] \leq \frac{n-1}{n}\log_2 \frac{n}{n-1} + 0 + \frac{1}{n}\log_2 n = \frac{1}{n}\log_2\left(\frac{n^n}{(n-1)^{n-1}}\right).$$

The lower bound is

$$\log_2(m)\left[G\left(\frac{\lambda}{n}\right) - G\left(\frac{\eta}{n}\right)\right] \geq \frac{1}{n} - \frac{2}{n} + \frac{1}{n}\log_2 \frac{2}{n-2} = \frac{1}{n}\log_2 \frac{1}{n-2}$$

Both of these bounds tend to 0 as $n \to \infty$, so that means that in the long run if we just change one bit in the integer partition we don't change G by very much. This makes sense because intuitively as n gets larger, moving one bit, or relabeling one element of X, should impact the smoothness less and less as n gets larger.

5.2 Conjugation

The second way we can transform a Young diagram is by conjugating it. This is flipping it over its main diagonal, as in Fig. 10. The conjugate of λ is written as λ^*. Because κ depends entirely on the number of parts (along with the value n being partitioned), and the number of parts of λ^* is equal to the largest part of λ, we can prove sharp bounds on $\kappa(\lambda^*)$ in terms of $\kappa(\lambda)$.

Proposition 1 *Let λ be an integer partition of n with m parts, and λ^* be its conjugate, with m^* parts. Then*

$$1 - \kappa(\lambda) \leq \kappa(\lambda^*) \leq \frac{\log_2(n - n^{\kappa(\lambda)} + 1)}{\log_2(n)}, \tag{2}$$

and these bounds are sharp.

Proof We have already mentioned the fact that the number of parts of λ^* is equal to the largest part of λ, or λ_1. Therefore, we can write $\kappa(\lambda^*)$ in terms of the largest part of λ as

$$\kappa(\lambda^*) = \frac{\log_2(m^*)}{\log_2(n)} = \frac{\log_2(\lambda_1)}{\log_2(n)}. \tag{3}$$

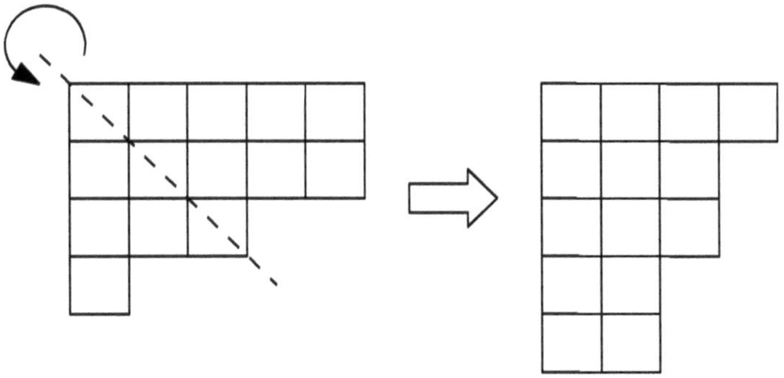

Fig. 10 The conjugate of a partition is created by flipping the associated Young diagram over the dotted diagonal line. For example, the conjugate of $\lambda = \langle 5, 5, 3, 1 \rangle$ is $\lambda^* = \langle 4, 3, 3, 2, 2 \rangle$

Now, observe that we can bound λ_1 in terms of m. We claim an upper bound $\lambda_1 \leq n - m + 1$. If λ_1 were strictly larger than $n - m + 1$ and all other parts were equal to 1 (the smallest they can be) then we would have

$$n = \sum_{i=1}^{m} \lambda_i = \lambda_1 + (m-1) \cdot 1 > (n - m + 1) + m - 1 = n$$

which is a contradiction. This upper bound is sharp which can be seen when $\lambda_1 = n - m + 1$ and all other parts are equal to 1.

Then, we claim a lower bound of $\lambda_1 \geq \frac{n}{m}$. Here if instead $\lambda_1 < \frac{n}{m}$ then we would also have $\lambda_i < \frac{n}{m}$ for all i since the parts are in decreasing order. Then in this case

$$n = \sum_{i=1}^{m} \lambda_i < m \cdot \frac{n}{m} = n$$

again a contradiction. The lower bound is also sharp whenever m is a factor of n by allowing $\lambda_i = \frac{n}{m}$ for all $1 \leq i \leq m$. We can now use these bounds to prove the inequalities in (2).

We substitute $\lambda_1 \geq \frac{n}{m}$ in (3) to prove the first inequality:

$$\begin{aligned}\kappa(\lambda^*) &= \frac{\log_2(\lambda_1)}{\log_2(n)} \\ &\geq \frac{\log_2\left(\frac{n}{m}\right)}{\log_2(n)} = \frac{\log_2(n) - \log_2(m)}{\log_2(n)} \\ &= 1 - \kappa(\lambda).\end{aligned}$$

Next, we can substitute $\lambda_1 \leq n - m + 1$ again into (3) to prove the second inequality:

$$\begin{aligned}\kappa(\lambda^*) &= \frac{\log_2(\lambda_1)}{\log_2(n)} \\ &\leq \frac{\log_2(n - m + 1)}{\log_2(n)}.\end{aligned}$$

We cannot break up the log in this case, but we can write m as a function of $\kappa(\lambda)$ by inverting κ.

$$\begin{aligned}\kappa(\lambda) &= \frac{\log_2(m)}{\log_2(n)} \\ \log_2(n)\kappa(\lambda) &= \log_2(m) \\ 2^{\log_2(n)\kappa(\lambda)} &= 2^{\log_2(m)} \\ n^{\kappa(\lambda)} &= m\end{aligned}$$

Substituting this back in to finish our bound we indeed see that

$$\kappa(\lambda^*) \leq \frac{\log_2(n - m + 1)}{\log_2(n)} = \frac{\log_2(n - n^{\kappa(\lambda)} + 1)}{\log_2(n)}.$$

To see that these bounds are sharp we simply need to provide a $[\lambda, \lambda^*]$ pair for each inequality which makes it into an equality. For the upper bound this is quite easy. Given any $q < n$ we have a partition $\lambda = \langle q, 1, \ldots, 1\rangle$ and then $\lambda^* = \langle m, 1, \ldots, 1\rangle$. Since this achieves the upper bound on λ_1 that we used in proving the upper bound on $\kappa(\lambda^*)$ we can turn the one inequality into an equality. For the lower bound we do the same thing. First assume that n is not prime, so we can write it as $n = n_1 \cdot n_2$ for some $n_1, n_2 < n$ and $n_1, n_2 \in \mathbb{Z}$. Then we let $\lambda = \langle n_1, n_1, \ldots, n_1\rangle$ where we have n_2 copies of n_1, and $\lambda^* = \langle n_2, n_2, \ldots, n_2\rangle$ with n_1 copies of n_2. Again, because this achieves the lower bound on λ_1 in terms of m we can turn our inequality into an equality. Now, in the case that n is prime we still achieve our lower bound for the trivial partition $\lambda = n$ where $\kappa(\lambda) = 0$. Then $\lambda^* = \langle 1, 1, \ldots, 1\rangle$ and we have $\kappa(\lambda^*) = 1 = 1 - \kappa(\lambda)$.

In order to illustrate the bounds in Proposition 1 see Fig. 11. The x axis is $\kappa(\lambda)$ and the y axis is $\kappa(\lambda^*)$. The line and curve are the bounds, and you can see many instances of points sitting on the bounds. This picture is for $n = 20$.

Given that there are sharp bounds on $\kappa(\lambda^*)$ in terms of $\kappa(\lambda)$ we asked whether or not we can do the same for $G(\lambda^*)$. In fact we can get a lower bound, but the form is much more complicated. And an upper bound seems nonexistent from looking at the points themselves. In Fig. 12 we have a picture similar to that in Fig. 11, again

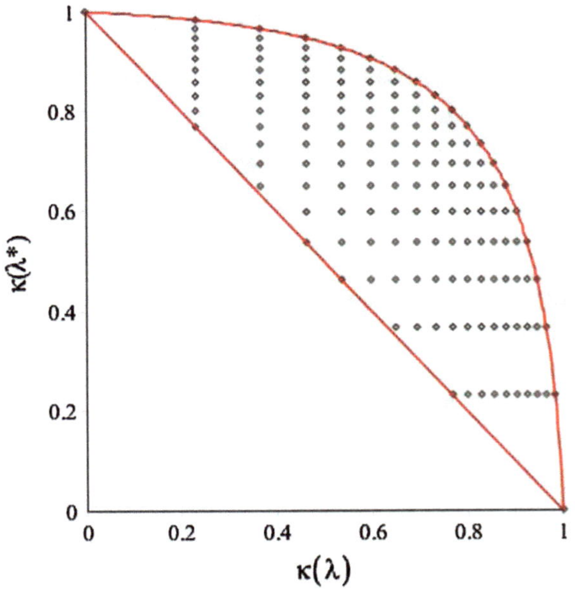

Fig. 11 Plot of $\kappa(\lambda^*)$ versus $x = \kappa(\lambda)$ for $n = 20$

Fig. 12 Plot of $G(\lambda^*)$ versus $x = G(\lambda)$ for $n = 20$

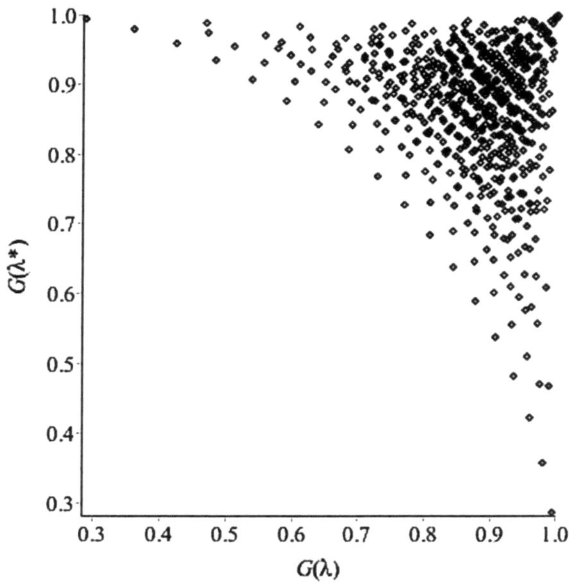

for $n = 20$, but this time we give the G values. Notice that there is a clear lower bound curve traced out by the points. These lowest elements correspond to the case where $\lambda = \langle \lambda_1, 1, 1, \ldots, 1\rangle$ and then $\lambda^* = \langle m, 1, 1, \ldots, 1\rangle$ since partitions of this form have the lowest G value. However, the form of G is much more complicated than that of κ and we are not able to invert G to get a function for m in terms of $G(\lambda)$. Therefore, the solution is more of a numerical approximation than a closed form function.

6 Conclusion

In this paper we introduced two functions, dispersion (κ) and smoothness (G), which measure the shape of frequency distributions in two different ways. First, dispersion assesses how many bins, or labels, there are in the distribution versus how many objects. If there are a similar number of objects and bins then the distribution is very dispersed, but if there are many more objects than bins then the distribution is very narrow. Secondly, smoothness uses a normalized entropy to measure how close the distribution is to uniform. We showed how these measures function on directed labeled graphs and more specifically to the cyber security use case. Finally, we explored how G and κ change when we make specific changes to the partition through Young diagram manipulation.

We believe that these two functions can help discover changes in evolving data and characterize labeled directed multigraphs in much of the same way as the degree distribution characterizes unlabeled graphs.

Acknowledgments The research described in this paper is part of the Asymmetric Resilient Cybersecurity Initiative at Pacific Northwest National Laboratory. It was conducted under the Laboratory Directed Research and Development Program at PNNL, a multi-program national laboratory operated by Battelle for the U.S. Department of Energy.

References

1. G. Birkhoff, *Lattice Theory*, vol. 25, 3rd edn. (American Mathematical Society, Providence, 1940)
2. G. de Cooman, D. Ruan, E. Kerre (eds.), *Foundations and Applications of Possibility Theory* (World Scientific, Singapore, 1995)
3. D. Dubois, H. Prade, *Possibility Theory* (Plenum Press, New York, 1988)
4. C. Joslyn, W. Cowley, E. Hogan, B. Olsen, Discrete mathematical approaches to graph-based traffic analysis, in *2014 International Workshop on Engineering Cyber Security and Resilience (ECSaR14)* (2014)
5. G. Klir, *Uncertainty and Information: Foundations of Generalized Information Theory* (Wiley, Hoboken, 2006)
6. R.P. Stanley, *Enumerative Combinatorics*, vol. 1 (Cambridge UP, Cambridge, 1997)
7. Visual Analytics Science and Technology (VAST) Challenge (2013). http://vacommunity.org/VAST+Challenge+2013
8. O. Wolkenhauer, *Possibility Theory with Applications to Data Analysis* (Wiley, New York, 1998)
9. G.M. Ziegler, On the poset of partitions of an integer. J. Comb. Theory, Ser. A **42**(2), 215–222 (1986)

Integrating and Sampling Cuts in Bounded Treewidth Graphs

Ivona Bezáková, Erin W. Chambers and Kyle Fox

Abstract In this paper, we consider the problem of evaluating (s, t)-cuts in a bounded treewidth graph. In particular, we show how to compute the partition function for weighted cuts of the graph, i.e., the total weight of all (s, t)-cuts where the weight of a single cut is the product of its edge weights. This method can also easily be adapted to work with additive weights for the cost of a cut. We also present a method for sampling a cut proportional to its weight in linear time. Computing the partition function is #P-hard for general graphs, and our sampling algorithm is simple enough to prove useful is several application areas. Finally, we discuss an alternative method for sampling cuts that uses Markov chains and show that, in the worst case, its mixing time is exponential in the size of the graph even when the graph has bounded treewidth.

Keywords Treewidth · Minimum cuts · Algorithms

Mathematics Subject Classification 05C83 · 05C85 · 97P20

Ivona Bezáková, partially supported by NSF, Award No. CCF-1319987. Portions of this research took place while Kyle Fox was a student at the University of Illinois at Urbana-Champaign and while the author was a postdoctoral fellow at the Institute for Computational and Experimental Research in Mathematics, Brown University. Erin W. Chambers, partially supported by NSF, Grants No. CCF-1054779 and IIS-1319573. Kyle Fox, partially supported by the Stutzke Dissertation Completion Fellowship from the University of Illinois at Urbana-Champaign.

I. Bezáková
Department of Computer Science, Rochester Institute of Technology,
Rochester, NY 14623, USA
e-mail: ib@cs.rit.edu

E.W. Chambers (✉)
Department of Computer Science and Mathematics, Saint Louis University,
St. Louis, MO 63103, USA
e-mail: echambe5@slu.edu

K. Fox
Department of Computer Science, Duke University, Durham, NC 27708, USA
e-mail: kylefox@cs.duke.ed

G. Letzter et al. (eds.), *Advances in the Mathematical Sciences*, Association for Women in Mathematics Series 6, DOI 10.1007/978-3-319-34139-2_20

1 Introduction

Given a directed graph $G = (V, E)$ of bounded treewidth with edge weights $w(\cdot)$ and two designated vertices s and t, we describe a linear time algorithm for computing the partition function for weighted cuts in G. We assume a model of computation where pairs of values can be added, multiplied, and compared in constant time; requiring time proportional to the bit-complexity of numbers only increases our running times by a near-linear factor. After running our algorithm, it becomes possible to sample cuts proportionally to the product of edge weights in linear time. We emphasize that the probability of a cut being chosen in an individual sample is *exactly* as described above and not approximate.

For an (s, t)-cut $C \subseteq V$, $s \in C$, $t \notin C$, its weight $w(C)$ is the *product* of weights for edges crossing the cut, that is:

$$w(C) = \prod_{(x,y):x\in C, y\notin C, (x,y)\in E} w(x, y).$$

Weights corresponding to products are used in many applications. For example, in physics and biology, the weights often represent probabilities or energies of individual events. In practice, the actual probabilities are unknown and are replaced by energies that are proportional to the probabilities. However, for certain applications including the maximum likelihood principle, one needs a close estimate on the probability of an event, not just its energy. To compute this probability, one needs to scale the energy of the event by the sum of the energies of all events. This scaling quantity is known as the *partition function*. Our algorithms for computing the partition function and sampling cuts appear in Sects. 3 and 4, respectively.

In Sect. 5, we extend the above results to work with multiple sources and sinks. The extension is surprisingly simple; however, this is the first result we are aware of on evaluating cuts in the multiple sources and sinks setting. In Sect. 6, we describe how to modify the above techniques to count and sample (s, t)-cuts that have minimum summed edge weight. While the minimum weight (s, t)-cut counting procedure is nearly the same as that given for computing the partition function, we feel it is different enough to be of independent interest.

As further motivation for our approach, we conclude this report with a discussion on Markov chain methods for (approximately) sampling (s, t)-cuts by multiplicative weight (Sect. 7). Markov chains have been used successfully for a variety of difficult counting problems, either to approximately count the number of solutions or to provide an approximately uniformly random solution [13]. A common and natural approach is to use a Glauber dynamics type chain that in each transition modifies the current state by a constant number of sites. For cuts, this means adding or removing a constant number of vertices to/from the current cut. We show that this approach may need exponential time to converge to an approximately uniform distribution on cuts, even in graphs of treewidth 2 or less. Therefore, for essentially any graph class Glauber dynamics Markov chains will not yield polynomial-time approximations of the partition function for weighted (s, t)-cuts. In contrast, our algorithms handle an interesting and widely studied class of graphs in (fast) polynomial time, and provide motivation for considering extensions to more general classes of graphs such as minor free graph families.

1.1 Related Work

For (s, t)-cuts the partition function Z is defined as the sum of the weights of all (s, t)-cuts:

$$Z := \sum_{C:C\subseteq V, s\in C, t\notin C} w(C).$$

Notice that for a cut C, its actual probability when sampling proportionally to its weight is $w(C)/Z$. Computing the partition function is #P-hard. This fact can be shown via a reduction from the problem of counting minimum cardinality (s, t)-cuts, which is #P-complete [17]. Assigning weight $1/2^n$ to all edges in the graph, where n is the number of vertices, means the minimum cardinality cuts will dominate the partition function.

This problem of counting minimum weight (s, t)-cuts in general graphs is #P-complete [17] even for unit weights, and can be reduced to the problem of counting maximal antichains in a poset [2]. Ball and Provan first considered the problem of counting minimum cuts and gave a polynomial time algorithm to compute the number of minimum cardinality (s, t)-cuts in an (s, t)-planar graph (where the source and sink are on the same face) [2]. Later, Bezáková and Friedlander generalized the algorithm for arbitrary locations of s and t in a planar graph [3] while also allowing arbitrary edge capacities. Chambers, Fox, and Nayyeri further generalized the algorithm for directed graphs embedded on orientable surfaces of bounded genus [7].

The problem of counting minimum (cardinality) cuts was originally motivated by questions in network reliability [2, 8, 14, 16]. In particular, the problem is closely related to the probabilistic connectness of stochastic graphs, where edges may fail with known probabilities [2]. More recently, counting minimum cuts has been studied for its applications to problems in computer vision. In these applications, the pixels of an image are interpreted as vertices in a graph with edges between the vertices describing the similarity between pixels. Minimum cuts provide a high quality way to segment the pixels of the image [6]. Counting minimum cuts is closely related to sampling these cuts, allowing for a varied selection of high quality segmentations.

1.2 Courcelle's Theorem for Bounded Treewidth Graphs

Courcelle showed in 1990 that any graph property describable in counting monadic second order logic can be decided in linear time if the input graph has bounded treewidth [9]. There are often practicality concerns with using this particular metatheorem, however, since many direct applications of it lead to hidden constants that are doubly or triply exponential in the treewidth [11].

Courcelle's theorem has been extended in a variety of ways. One extension particularly relevant to our work is as follows. Weight the edges of a graph G and fix a monadic second order logic formula ϕ with one free set variable. For any set of edges A, the weight of A can be defined as either the sum or product of its edge weights. If G has bounded treewidth, it is possible to sum the weight of all sets A that satisfy ϕ in linear time [10]. In particular, this result implies we can compute the partition function for weighted cuts of a graph G in linear time in bounded treewidth graphs, one of our main results.

The main advantage of our partition function algorithm over the meta-theorems mentioned above is that our algorithm is very simple, and the dependence on the treewidth of G is only singly exponential. As stated earlier, standard applications of Courcelle's theorem often have doubly or triply exponential dependence on treewidth.

Perhaps of greater interest, our algorithm also provides a simple way to randomly sample cuts, one of the key motivations behind the study of partition functions; to the best of our knowledge, no such sampling is known under the general framework of Courcelle's theorem.

2 Tree Decompositions and Treewidth

A *tree decomposition* $\mathcal{T}$ of a graph $G = (V, E)$ is a pair $(T, \mathcal{X})$ where T is a tree and $\mathcal{X}$ is a family of subsets (or *bags*) of V such that:

- Each node u of T has a corresponding subset $X_u \in \mathcal{X}$ and $\cup_{X \in \mathcal{X}} X = V$;
- For every edge $uv \in E$ there is a bag $X \in \mathcal{X}$ such that $u, v \in X$.
- For any three nodes $u, v, w \in T$ such that v is on the u-to-w path in T, $X_u \cap X_w \subseteq X_v$.

In this paper, we will refer to the *bags or nodes* of T and the *vertices* of G to avoid confusion.

The *width* of a tree decomposition $(T, \mathcal{X})$ is $\max_{X \in \mathcal{X}} |X| - 1$. The *treewidth* of a graph is the minimum possible width of a tree decomposition of the graph. Any graph of treewidth k has a tree decomposition with at most $n - k + 1$ nodes [5].

Tree decompositions were originally introduced by Halin [12] and were rediscovered (and popularized) by Robertson and Seymour [18]. While it is NP-complete to decide if any graph has treewidth at most k [1], a tree decomposition can be constructed in linear time if k is a constant (the dependance on k is exponential) [4].

We only consider tree decompositions $\mathcal{T} = (T, \mathcal{X})$ with $O(n)$ nodes where T is rooted at some node r and every node of T has at most two children. We can modify any tree decomposition of width k and $O(n)$ nodes to meet the last assumption in $O(kn)$ time without increasing the width of the decomposition by replacing any node v with d children by $d - 1$ nodes, each with 2 children and the same bag as v.

3 Computing the Partition Function

In this section we show how to compute the partition function for weighted cuts of G. Let $G = (V, E, w)$ be a positively edge-weighted graph (directed or not) with treewidth k. Let $s, t \in V$ be the source and the sink, respectively, $s \neq t$. For an (s, t)-cut $C \subseteq V, s \in C, t \notin C$, its weight $w(C)$ is the *product* of weights for edges crossing the cut, that is:

$$w(C) = \prod_{(x,y):x \in C, y \notin C, (x,y) \in E} w(x, y).$$

We wish to compute the total weight, that is, the sum of weights over all (s, t)-cuts.

Let $\mathcal{T}$ be a tree decomposition of G with width k. For every edge $e = (x, y) \in E$, choose exactly one bag X_u with $x, y \in X_u$ as its *designated bag*. For a bag X_u, we refer to the set E_u of edges $e = (x, y) \in E$ such that $x, y \in X_u$ and X_u is the designated bag for e, as the *designated edges* for X_u.

We are now ready to present the algorithm. The idea is to compute, for each bag and for each of its subsets, the total weight of the cuts that are consistent with the subset, where the weight takes into account only the edges that are designated to this bag or to one of its descendents. Algorithm 1 contains pseudocode for the algorithm. The correctness of the algorithm is chiefly explained by Lemma 1.

We first analyze the running time of the algorithm. It does a single pass through the tree T, where in each node it goes through $2^{O(k)}$ operations. There are $O(n)$ nodes total in T, so we get an $2^{O(k)}n$ running time. More precisely, each node actually has $O(2^{3k}k)$ operations since there are 2^{k+1} choices for C, 2^{2k+2} choices for C_1 and C_2, and it takes $O(k)$ time to verify that for each pair C_1 and C_2, we have $C_1 \cap X_v = C \cap X_{v_1}$ and $C_2 \cap X_v = C \cap X_{v_2}$. We get a total running time of $O(2^{3k}kn)$.

Next we prove the correctness of our algorithm.

Lemma 1 *Let v be a node of T and let $P(v)$ be the set of all descendants of v in T (including v). Let $\tilde{V}_v = \cup_{u \in P(v)} X_u$ be the union of all the bags corresponding to $P(v)$, and let $\tilde{E}_v = \cup_{u \in P(v)} E_u$ be the union of all the edges that are designated for those bags. At the time when a node v is marked as done, the following holds for each $C \subseteq X_v$:*

$$weightDP_v[C] = \sum_{\tilde{C} \subseteq \tilde{V}_v, s \notin (\tilde{V}_v \setminus \tilde{C}), t \notin \tilde{C}, \tilde{C} \cap X_v = C} \left(\prod_{x,y:x \in \tilde{C}, y \notin \tilde{C}, (x,y) \in \tilde{E}_v} w(x, y) \right).$$

Proof We will prove the lemma by induction on the number of descendants of v. For the base case, v is a leaf node of T. The summation goes through a single $\tilde{C}$ as $\tilde{v} = X_v$ and $\tilde{C} \cap X_v = C$ implies that $\tilde{C} = C$. The claim follows directly from steps 4 and 7 of the algorithm.

For the inductive case, suppose that v is not a leaf and that the claim holds for all nodes with fewer descendants; in particular, the claim holds for v's children. Let $\tilde{C} \subseteq \tilde{V}_v$ be such that $s \notin (\tilde{V}_v \setminus \tilde{C})$, $t \notin \tilde{C}$, and $\tilde{C} \cap X_v = C$. We will show that

Algorithm 1: Computing the partition function for weighted (s,t)-cuts in a graph G with treewidth k

Require: A graph G, a corresponding tree decomposition $\mathcal{T} = (T, \mathcal{X})$ of width k with T rooted at node r, and distinct vertices s, t

Mark all nodes of T as not done.
for every leaf node u of T **do**
  for every subset $C \subseteq X_u$ such that $s \notin (X_u \setminus C)$ and $t \notin C$ **do**
    Let $\mathrm{weightDP}_u[C]$ be the product of all the designated edges for X_u that are cut by C: $\mathrm{weightDP}_u[C] := \prod_{(x,y)\in E_u, x\in C, y\notin C} w(x,y)$.
  end for
  for all other subsets $C \subseteq X_u$ **do**
    Let $\mathrm{weightDP}_u[C] := 0$.
  end for
  Mark u as done.
end for
for every non-leaf node v of T such that all its children are done **do**
  Let v_1 be v's child and v_2 be the other child (if it exists).
  for every subset $C \subseteq X_v$ such that $s \notin (X_v \setminus C)$ and $t \notin C$ **do**
    Let $\mathrm{weightDP}_v[C] := 0$.
    for every subset $C_1 \subseteq X_{v_1}$ such that $C_1 \cap X_v = C \cap X_{v_1}$ and every subset $C_2 \subseteq X_{v_2}$ such that $C_2 \cap X_v = C \cap X_{v_2}$ (if applicable) **do**
      Add $\mathrm{weightDP}_{v_1}[C_1] \cdot \mathrm{weightDP}_{v_2}[C_2]$ to $\mathrm{weightDP}_v[C]$. (If there is no v_2, take $\mathrm{weightDP}_{v_2}[C_2] = 1$.)
    end for
    Set $\mathrm{weightDP}_v[C] := \mathrm{weightDP}_v[C] \cdot \prod_{(x,y)\in E_v, x\in C, y\notin C} w(x,y)$.
  end for
  for all other subsets $C \subseteq X_v$ **do**
    Let $\mathrm{weightDP}_v[C] := 0$.
  end for
  Mark v as done.
end for
return $\sum_{C\subseteq X_r} \mathrm{weightDP}_r[C]$.

the algorithm includes the weight $\prod_{x,y: x\in\tilde{C}, y\notin\tilde{C}, (x,y)\in\tilde{E}_v} w(x,y)$ in the computation of $\mathrm{weightDP}_v[C]$.

Recall that v_1 and v_2 (if it exists) are v's children in T. Let $\tilde{C}_i = \tilde{C} \cap \tilde{V}_{v_i}$ be the restriction of the cut $\tilde{C}$ to the descendants of v_i and let $C_i = \tilde{C} \cap X_{v_i}$ be its restriction to the bag X_{v_i}. Notice that $C_i \cap X_v = C \cap X_{v_i}$. Let $\pi(C) = \prod_{(x,y)\in E_v, x\in C, y\notin C} w(x,y)$. The sum of weights for all $\tilde{C}$ as described above is

$$\sum_{\tilde{C}} \left[\pi(C) \cdot \prod_{(x,y)\in\tilde{E}_{v_1}, x\in\tilde{C}_1, y\notin\tilde{C}_1} w(x,y) \cdot \prod_{(x,y)\in\tilde{E}_{v_2}, x\in\tilde{C}_2, y\notin\tilde{C}_2} w(x,y) \right]$$
$$= \pi(C) \sum_{\tilde{C}} \left[\prod_{(x,y)\in\tilde{E}_{v_1}, x\in\tilde{C}_1, y\notin\tilde{C}_1} w(x,y) \cdot \prod_{(x,y)\in\tilde{E}_{v_2}, x\in\tilde{C}_2, y\notin\tilde{C}_2} w(x,y) \right].$$

By the definition of tree decompositions, any pair of choices for $\tilde{C}_1$ and $\tilde{C}_2$ as allowed above will include or exclude the same members of $\tilde{V}$. By this fact and distribution, we see the expression above is equal to

$$\pi(C) \cdot \text{weightDP}_v[C_1] \cdot \text{weightDP}_v[C_2].$$

The lemma immediately yields the main theorem of this section.

Theorem 1 *Algorithm 1 correctly computes the partition function for all (s, t)-cuts. The running time is $O(2^{3k}kn)$.*

Proof Apply the lemma to $v = r$. Then we get $\tilde{V}_r = V$, $\tilde{E}_r = E$. An (s, t)-cut $\tilde{C}$ will be accounted for by weightDP$_r[C]$ for $C = \tilde{C} \cap X_r$. Summing across all C's we account for all (s, t)-cuts.

4 Sampling Cuts

One interesting application of our dynamic programming formulation is that it can be easily modified to aid in repeatedly sampling cuts. Recall that in many applications, these weights are probabilities of some individual event's occurrence, and may be estimates or energies that correspond to probabilities but lack the scaling factor. Therefore, sampling provides a method via which events can be repeatedly selected from the given probability distribution.

Algorithm 1 given above builds a dynamic programming table that can be used to randomly sample cuts proportionally to their weight. For a node v, the subset of vertices $\tilde{V}_v$ as described in Lemma 1, and a subset of vertices $C \subseteq X_v$, we need a way to sample a subset of vertices $\tilde{C} \subseteq \tilde{V}_v$ such that $s \notin (\tilde{V}_v \setminus \tilde{C})$, $t \notin \tilde{C}$, and $\tilde{C} \cap X_v = C$ proportionally to $\prod_{x,y:x \in \tilde{C}, y \notin \tilde{C}, (x,y) \in \tilde{E}_v} w(x, y)$. Let v_1 and v_2 be the children of v in T (assuming v_2 exists). In order to sample this subset, our algorithm randomly selects two subsets of vertices $C_1 \subseteq X_{v_1}$ and $C_2 \subseteq X_{v_2}$ such that $C_1 \cap X_v = C \cap X_{v_1}$ and $C_2 \cap X_v = C \cap X_{v_2}$. It does so proportionally to weightDP$_{v_1}[C_1] \cdot$ weightDP$_{v_2}[C_2]$. Similar to before, if v_2 does not exist, then our algorithm only selects C_1 and it assumes weightDP$_{v_2}[C_2] = 1$ when giving weights to the subsets. Once C_1 and C_2 are selected in $O(k)$ time, it recursively selects subsets from $\tilde{V}_{v_1}$ and $\tilde{V}_{v_2}$ using the same procedure.

In order to select an (s, t)-cut, it selects a set $C_r \subseteq X_r$ from the root bag proportionally to weightDP$_r[C_r]$ in $O(k)$ time. It then uses the above procedure to select the whole cut $\tilde{C}$. A random sample is performed once per tree node, so the entire procedure takes $O(kn)$ time.

Theorem 2 *There exists an algorithm to sample (s, t)-cuts proportionally to their weight in $O(kn)$ time per sample after running Algorithm 1 once.*

5 Multiple Source-Sink Cuts

We now describe an extension to our previous algorithms to handle computing the partition function when there are multiple sources and multiple sinks in the input graph G. In essence, this is a simple modification to our previous algorithms, but as far as we are aware it is the first result on evaluating cuts in the multiple sources and sinks setting.

The extension works as follows. Let S be the set of source vertices and T be the set of sink vertices. Our algorithms modify the graph G by adding two vertices s^* and t^*. They then add an edge from s^* to every member of S and an edge from every member of T to t^*. Vertex s^* is set as the only source and vertex t^* is set as the only sink.

These edges are given weight 0 so that any (s^*, t^*)-cut that divides S or T will have weight 0. We can add s^* and t^* to every bag in any tree decomposition of G, increasing the width of the decomposition by 2 while still maintaining it as a valid tree decomposition after modifying G. The partition function and number of minimum weight (S_1, S_2)-cuts can still be computed in $O(2^{3k}kn)$ time.

6 Minimum (s, t)-cuts

In this section, we describe how to count and sample (s, t)-cuts that have minimum summed edge weight. Our algorithm is very similar to the one used for computing the partition function. The pseudocode appears in Algorithm 2. The key idea behind our algorithm for counting minimum cuts is that our dynamic programming procedure takes a subset of vertices for a bag and returns *two* values, the *weight* of any minimum (s, t)-cut consistent with that subset, and the *number* of these minimum weight cuts. When computing the two values for a node v's bag, it enumerates all consistent subsets for the children of v. The children's subsets only contribute to v's number variable if the sum of their weight variables is minimum. The proof of correctness is nearly the same as that given earlier for computing the partition function.

Theorem 3 *Algorithm 2 correctly computes the number of minimum weight (s, t)-cuts. The running time is $O(2^{3k}kn)$.*

Similar to before, our algorithm for counting minimum (s, t)-cuts builds a dynamic programming table that can be used to sample minimum weight (s, t)-cuts uniformly at random. The procedure is the same as the one given for sampling cuts proportionally to multiplicative weight, except the sampling algorithm will pick subsets of vertices C_1 and C_2 for each node v's children proportionally to the product of C_1 and C_2's number variables. The algorithm only considers subsets C_1 and C_2 where the sum of their weight variables is minimum.

Theorem 4 *There exists an algorithm to sample minimum weight (s, t)-cuts uniformly at random in $O(kn)$ time per sample after running Algorithm 2 once.*

Algorithm 2: Counting the minimum weight (s, t)-cuts in a graph G with treewidth k

Require: A graph G, a corresponding tree decomposition $\mathcal{T} = (T, \mathcal{X})$ of width k with T rooted at node r, and distinct vertices s, t

Mark all nodes of T as not done.
for every leaf node u of T **do**
  for every subset $C \subseteq X_u$ such that $s \notin (X_u \setminus C)$ and $t \notin C$ **do**
    Let $\text{countDP}_u[C]$ be the total weight and number of cuts designated by C: $\text{countDP}_u[C] := \left(\sum_{(x,y)\in E_u, x\in C, y\notin C} w(x, y), 1\right)$.
  end for
  for all other subsets $C \subseteq X_u$ **do**
    Let $\text{countDP}_u[C] := (\infty, 0)$.
  end for
  Mark u as done.
end for
for every non-leaf node v of T such that all its children are done **do**
  Let v_1 be v's child and v_2 be the other child (if it exists).
  for every subset $C \subseteq X_v$ such that $s \notin (X_v \setminus C)$ and $t \notin C$ **do**
    Let (minWeight, cutCount) $:= (\infty, 0)$.
    for every subset $C_1 \subseteq X_{v_1}$ such that $C_1 \cap X_v = C \cap X_{v_1}$ and every subset $C_2 \subseteq X_{v_2}$ such that $C_2 \cap X_v = C \cap X_{v_2}$ (if applicable) **do**
      $(\text{subWeight}_i, \text{subCount}_i) := \text{countDP}_{v_i}[C_i]$ for $i \in \{1, 2\}$. (If there is no v_2, take $\text{countDP}_{v_2}[C_2] = (0, 1)$.)
      if $\text{subWeight}_1 + \text{subWeight}_2 = \text{minWeight}$ **then**
        Set (minWeight, cutCount) $:=$ (minWeight, cutCount $+ \text{subCount}_1 \cdot \text{subCount}_2$).
      end if
      if $\text{subWeight}_1 + \text{subWeight}_2 < \text{minWeight}$ **then**
        Set (minWeight, cutCount) $:= (\text{subWeight}_1 + \text{subWeight}_2, \text{subCount}_1 \cdot \text{subCount}_2)$.
      end if
    end for
    Set $\text{countDP}_v[C] := (\text{minWeight} + \sum_{(x,y)\in E_u, x\in C, y\notin C} w(x, y), \text{cutCount})$.
  end for
  for all other subsets $C \subseteq X_v$ **do**
    Let $\text{countDP}_u[C] := (\infty, 0)$.
  end for
  Mark v as done.
end for
Let (minWeight, cutCount) $:= (\infty, 0)$.
for every subset $C \subseteq X_r$ **do**
  (subWeight, subCount) $:= \text{countDP}_r[C]$.
  if subWeight $=$ minWeight **then**
    Set (minWeight, cutCount) $:=$ (minWeight, cutCount $+$ subCount).
  end if
  if subWeight $<$ minWeight **then**
    Set (minWeight, cutCount) $:=$ (subWeight, subCount).
  end if
end for
return cutCount.

7 Markov Chain Techniques: Slow Mixing

In this section we discuss using Markov chains to generate a random (s, t)-cut approximately proportional to its multiplicative weight. In particular, we provide a simple undirected graph with bounded treewidth for which Markov chains that modify only a constant portion of the cut need exponential time to get close to the stationary distribution. We begin with a refresher on Markov chains before getting into our results.

7.1 Markov Chain Preliminaries

A *Markov chain* is a pair (Ω, P), where Ω is a set of *states* and P is a (right) stochastic matrix of size $|\Omega| \times |\Omega|$ that specifies the probabilities $P(x, y)$ of *transitioning* from state x to state y. A distribution π on states Ω is *stationary* if $\pi(y) = \sum_{x \in \Omega} \pi(x) P(x, y)$ for all $x, y \in \Omega$; in other words, if starting from a state chosen according to the distribution π, after one step of the Markov chain the states are distributed according to π. Notice that $P^t(x, y)$ is the probability of transitioning from x to y in t steps. A Markov chain is irreducible if for every x, y there is a t such that $P^t(x, y) > 0$; it is aperiodic if $\gcd\{t : P^t(x, y) > 0\} = 1$ for every x, y. An irreducible and aperiodic Markov chain has a unique stationary distribution; moreover, if the transition matrix is symmetric (that is $P(x, y) = P(y, x)$ for every x, y), then the stationary distribution is uniform (that is $\pi(x) = 1/|\Omega|$). The Metropolis–Hastings technique can be used to modify the transition probabilities of a symmetric Markov chain to achieve a desired stationary distribution σ. In particular, for an irreducible, aperiodic, and symmetric Markov chain $M = (\Omega, P)$, we can construct a Markov chain $M_\sigma = (\Omega, P_\sigma)$ such that $P_\sigma(x, y) = P(x, y) \min\{\sigma(y)/\sigma(x), 1\}$ for $x \neq y$.

The *mixing time* $\tau(\varepsilon) := \max_{x \in \Omega} \min\{t : d_{tv}(P^t(x, .), \pi) < \varepsilon\}$ is the time needed to get ε-close to stationarity when starting from an arbitrary state x. The *total variation distance* $d_{tv}(\mu, \pi) := \sum_{x \in \Omega} (\mu(x) - \pi(x))/2$ measures the closeness of two distributions μ and π. For any $A \subset \Omega$, let $\pi(A) := \sum_{x \in A} \pi(x)$. A quantity known as *conductance*

$$\Phi := \min_{A \subset \Omega, \pi(A) \leq 1/2} \frac{\sum_{x \in A, y \in \Omega - A} \pi(x) P(x, y)}{\pi(A)} \tag{1}$$

can be used to bound the mixing time of an ergodic Markov chain (from above and below). In particular, for a Markov chain with $P(u, u) \geq 1/2$, for every $u \in \Omega$,

$$\frac{1}{2}\left(\frac{1}{2\Phi} - 1\right) \log\left(\frac{1}{2\varepsilon}\right) \leq \tau(\varepsilon) \leq \frac{2}{\Phi^2} \log\left(\frac{1}{\pi_{\min}\varepsilon}\right), \tag{2}$$

where $\pi_{\min} = \min_{x \in \Omega} \pi(x)$ [15, 19]. The requirement on $P(u, u) \geq 1/2$ is technical, typically used to guarantee that a chain is aperiodic. For every Markov chain

$M = (\Omega, P)$ there exists a so-called *lazy* Markov chain $M_{\text{lazy}} = (\Omega, P_{\text{lazy}})$ that with probability 1/2 stays in the current state, otherwise it follows transitions of M; formally, $P_{\text{lazy}} = 1/2(I + P)$ where I is the identity matrix. The stationary distribution of M_{lazy} is the same as that of M. Intuitively, a lazy Markov chain takes about twice as long to mix compared to the original chain.

7.2 Glauber Dynamics Markov Chains for Cuts

We discuss Markov chains for all (s, t)-cuts sampled proportionally to their multiplicative weight, as well as Markov chains for sampling just minimum (s, t)-cuts. We mentioned that our earlier dynamic programming results for bounded treewidth graphs can be easily modified to use additive weights and/or to be restricted to only minimum cuts. However, for Markov chain based sampling the situation is different and we present slow mixing examples for both scenarios.

Let Ω be the set of all minimum (s, t)-cuts. Consider a Glauber dynamics Markov chain that tries to modify a single site in each transition. Then, the transition from a current state C is as follows:

1. choose a random vertex $v \in V - \{s, t\}$,
2. let $C' := C \oplus \{v\}$, the symmetric difference of C and $\{v\}$,
3. if $C' \in \Omega$, then C' is the next state; otherwise, the chain stays in C.

As this chain is symmetric, its stationary distribution is uniform. The chain moves from C to C' with probability $1/(n-2)$ if the chain is not lazy and with probability $1/(2(n-2))$ for its lazy version. More general Glauber dynamics chains attempt to modify more sites in one transition. For c modified sites the transition probabilities are $\Theta(1/n^c)$ and the chain is increasingly more likely to reject a move in step 3 due to $C' \notin \Omega$. As such, Markov chains that modify the current state *locally*, in other words, by changing only a constant-size part of the state, are generally preferred.

For weighted cuts, let Ω be the set of all (s, t)-cuts. The desired stationary distribution is $\pi_w(C) = w(C)/Z$, where $Z = \sum_{x \in \Omega} w(x)$ is the normalization factor, i.e., the partition function. The Metropolis–Hastings variant of the Glauber dynamics chain redefines step 3 as follows: if $C' \in \Omega$, then with probability $\min\{\pi(C')/\pi(C), 1\}$ state C' becomes the next state; otherwise, the chain stays in C. Notice that we do not need to know the (generally difficult to compute) normalization factor Z, since $\pi_w(C')/\pi_w(C) = w(C')/w(C)$.

7.3 Slow Mixing for All Weighted Cuts

We present a simple family of graphs for which the lazy Metropolis–Hastings variant of the above Markov chain needs exponential time to mix.

Consider the following undirected weighted graph $G = (V, E, w)$, where $V = \{u_1, u_2, \ldots, u_n\}$ with edges $E = \{(u_i, u_{i+1}) \mid i \in [n-1]\}$ of weights $w(u_1, u_2) = w(u_{n-1}, u_n) = 1$ and $w(u_i, u_{i+1}) = 1/2^n$ for $2 \le i \le n-2$. Let $s = u_1$ and $t = u_n$. Graph G is a path and therefore has treewidth 1.

In this case $\Omega := \{\{s\} \cup S \mid S \subseteq V - \{s,t\}\}$. Notice that there are only two (s,t)-cuts with weight 1, namely, $\{u_1\}$ and $\{u_1, \ldots, u_{n-1}\}$, and that the (multiplicative) weight of any other (s,t)-cut is at most $1/2^n$. Therefore, $Z \le 1 + 1 + (2^{n-2} - 2)/2^n < 3$.

Let $A = \{\{s\}\}$. Then, $1/3 < \pi_w(A) = w(A)/Z = 1/Z < 1/2$. The probability of moving from cut $\{s\}$ to another cut $\{s, u_i\}$ is at most $1/(2(n-2)2^n)$ since we choose u_i with probability $1/(2(n-2))$ and accept the move with probability $w(\{s, u_i\})/w(\{s\}) \le 1/2^n$ (more precisely, the acceptance probability is $1/2^n$ if $i = 2$ and $(1/2^n)^2$ otherwise).

We claim that the conductance out of A is exponentially small, see (1):

$$\Phi \le \frac{\sum_{x \in A, y \in \Omega - A} \pi_w(x) P_w(x,y)}{\pi_w(A)} = \frac{\sum_{y \in \Omega - A} \pi_w(\{s\}) P_w(\{s\}, y)}{\pi_w(\{s\})}$$
$$= \sum_{y \in \Omega - A} P_w(\{s\}, y) \le (n-2)\frac{1}{2(n-2)2^n} = \frac{1}{2^{n+1}}.$$

Therefore, we can bound the mixing time, see (2):

$$\tau(\varepsilon) \ge \frac{1}{2}\left(\frac{1}{2\Phi} - 1\right)\log\left(\frac{1}{2\varepsilon}\right) \ge \frac{1}{2}(2^n - 1)\log\left(\frac{1}{2\varepsilon}\right).$$

The mixing time is exponential in n. If we instead want to bound the mixing time in terms of input size, then note there are $n-1$ edge weights, each of up to n bits. Therefore, the size of the input is $\Theta(n^2)$ and the mixing time is still super polynomial, as it is exponential in the square root of the size of the input.

7.4 Slow Mixing for Minimum (s,t)-cuts

We conclude this paper with a family of graphs for which the Markov chain for minimum (s,t)-cuts need exponential time to mix. For simplicity we assume additive weights, as is standard for minimum (s,t)-cuts due to their correspondence to maximum s-t flows. We note that the same example with edge weights $1/2$ yields slow mixing arguments in case of multiplicative weights.

For any integer $\ell \ge 1$, consider the following undirected unweighted graph $G = (V, E)$, where

$$V = \{s, t, u, a_1, a_2, \ldots, a_\ell, b_1, b_2, \ldots, b_\ell\}$$

and

$$E = \{(s, a_i), (a_i, u), (u, b_i), (b_i, t) \mid i \in [\ell]\}.$$

Graph G is series-parallel and therefore has treewidth at most 2. It has $n = 2\ell + 3$ vertices. Any (s, t)-cut with value ℓ is minimum (since the value of the maximum s-t flow is ℓ). Therefore, $A := \{\{s\} \cup C_a \mid C_a \subseteq \{a_1, \ldots, a_\ell\}\}$ and $B := \{\{s, a_1, \ldots, a_\ell, u\} \cup C_b \mid C_b \subseteq \{b_1, \ldots, b_\ell\}\}$ are sets of minimum (s, t)-cuts. We claim that there are no other minimum (s, t)-cuts.

Lemma 2 *If C is a minimum (s, t)-cut, then $C \in A$ or $C \in B$.*

Proof Suppose that $u \notin C$. Then, for every i, either (s, a_i) or (a_i, u) is cut; for total cut cost ℓ. If there is a $b_j \in C$, then (b_j, t) increases the cut cost beyond ℓ, a contradiction. In this case, $C \in A$.

Suppose that $u \in C$. Then, for every j, either (u, b_j) or (b_j, t) is cut; for total cut cost ℓ. If there is an $a_i \notin C$, then (s, a_i) increases the cut cost beyond ℓ, a contradiction. In this case, $C \in B$.

Therefore, $\Omega = A \cup B$. Notice that to move from A to B one has to pass through the state $y' := \{s, a_1, \ldots, a_\ell, u\}$; however, there is a single state $x' := \{s, a_1, \ldots, a_\ell\}$ in A that can move to y'. Since $|A| = |B| = 2^\ell$, we have $\pi(A) = 1/2$ and $\pi(x) = 1/2^{\ell+1}$ for any $x \in \Omega$. Therefore, we can bound the conductance of the lazy chain as follows:

$$\Phi \leq \frac{\sum_{x \in A, y \in \Omega - A} \pi(x) P(x, y)}{\pi(A)} = \frac{1/(2^{\ell+1}) 1/(2(n-2))}{1/2} = \frac{1}{2^{(n-1)/2}(n-2)}.$$

Then, the mixing time is bounded by:

$$\tau(\varepsilon) > \frac{1}{2}\left(\frac{1}{2\Phi} - 1\right) \log\left(\frac{1}{2\varepsilon}\right) \geq (2^{(n-5)/2}(n-2) - 1) \log\left(\frac{1}{2\varepsilon}\right).$$

Thus, we need exponential mixing time in n to get ε-close to the uniform distribution even if ε is a constant. The computation can be adapted to show exponential mixing time for Glauber dynamics Markov chains that change c vertices at a time for any constant c.

8 Conclusions

In this paper, we presented a simple dynamic programming algorithm to compute the partition function for weighted cuts of a bounded treewidth graph. This algorithm easily extends to multiple source multiple sink cuts as well. We also provided an algorithm to sample cuts under our framework in the same amount of time, and demonstrated that Markov chain techniques to generate cuts require exponential time to converge in our setting.

We remark that in many computer vision applications the graph is a grid graph with two extra vertices, the source and the sink, that are each connected to a set of grid vertices. This situation arises, for example, for the Random Markov Field model. When using maximum likelihood to determine the best parameters for the model, one needs to compute the partition function across all weighted cuts. Unfortunately, this graph does not have bounded treewidth. We leave the study of evaluating cut problems for planar graphs with two apices, the source and the sink, for future work.

Acknowledgments Portions of this research took place during the authors' participation in Dagstuhl Seminar 13421: Algorithms for Optimization Problems in Planar Graphs.

References

1. S. Arnborg, D. Corneil, A. Proskurowski, Complexity of finding embeddings in a k-tree. SIAM J. Algebr. Discret. Methods **8**(2), 277–284 (1987). http://epubs.siam.org/doi/abs/10.1137/0608024
2. M.O. Ball, S.J. Provan, Calculating bounds on reachability and connectedness in stochastic networks. Networks **13**, 253–278 (1983)
3. I. Bezáková, A.J. Friedlander, Counting and sampling minimum (s, t)-cuts in weighted planar graphs in polynomial time. Theoret. Comput. Sci. **417**, 2–11 (2012)
4. H. Bodlaender, A linear-time algorithm for finding tree-decompositions of small treewidth. SIAM J. Comput. **25**(6), 1305–1317 (1996). http://epubs.siam.org/doi/abs/10.1137/S0097539793251219
5. H.L. Bodlaender, Dynamic programming on graphs with bounded treewidth, in *Automata, Languages and Programming*, vol. 317, Lecture Notes in Computer Science, ed. by T. Lepistö, A. Salomaa (Springer, Berlin, 1988), pp. 105–118
6. Y. Boykov, O. Veksler, Graph cuts in vision and graphics: theories and applications, in *Handbook of Mathematical Models in Computer Vision*, ed. by N. Paragios, Y. Chen, O. Faugeras (Springer, New York, 2006), pp. 79–96
7. E.W. Chambers, K. Fox, A. Nayyeri, Counting and sampling minimum cuts in genus g graphs, in *Proceedings of the Twenty-ninth Annual Symposium on Computational Geometry, SoCG '13*, ACM, New York, NY, USA (2013), pp. 249–258. http://doi.acm.org/10.1145/2462356.2462366
8. C.J. Colbourn, Combinatorial aspects of network reliability. Ann. Oper. Res. **33**, 1–15 (1991)
9. B. Courcelle, The monadic second-order logic of graphs I. Recognizable sets of finite graphs. Inf. Comput. **85**, 12–75 (1990)
10. B. Courcelle, J. Makowsky, U. Rotics, On the fixed parameter complexity of graph enumeration problems definable in monadic second-order logic. Discret. Appl. Math. **108**(1–2), 23–52 (2001). http://www.sciencedirect.com/science/article/pii/S0166218X00002213 (Workshop on Graph Theoretic Concepts in Computer Science)
11. M. Grohe, Algorithmic meta theorems, in *Graph-Theoretic Concepts in Computer Science*, vol. 5344, Lecture Notes in Computer Science, ed. by H. Broersma, T. Erlebach, T. Friedetzky, D. Paulusma (Springer, Berlin, 2008), pp. 30–30
12. R. Halin, S-functions for graphs. J. Geom. **8**(1–2), 171–186 (1976). http://dx.doi.org/10.1007/BF01917434
13. M. Jerrum, *Counting, Sampling and Integrating: Algorithms and Complexity*. Lectures in Mathematics. ETH Zürich. (Springer, New York, 2003). http://books.google.com/books?id=aLINQMsDQQ0C

14. D.R. Karger, A randomized fully polynomial time approximation scheme for the all terminal network reliability problem, *Proceedings of the Twenty-Seventh Annual ACM Symposium on Theory of Computing, STOC '95*, ACM, New York, NY, USA, (1995), pp. 11–17
15. G. Lawler, A. Sokal, Bounds on the l2 spectrum for Markov chains and markov processes: a generalization of Cheeger's inequality. Trans. Am. Math. Soc. **309**, 557–580 (1988)
16. H. Nagamoch, Z. Sun, T. Ibaraki, Counting the number of minimum cuts in undirected multigraphs. IEEE Trans. Reliab. **40**, 610–614 (1991)
17. S.J. Provan, M.O. Ball, The complexity of counting cuts and of computing the probability that a graph is connected. SIAM J. Comput. **12**, 777–788 (1983)
18. N. Robertson, P. Seymour, Graph minors. III. Planar tree-width. J. Comb. Theory Ser. B **36**(1), 49–64 (1984). http://www.sciencedirect.com/science/article/pii/0095895684900133
19. A. Sinclair, M. Jerrum, Approximate counting, uniform generation and rapidly mixing markov chains. Inf. Comput. **82**(1), 93–133 (1989)

Considerations on the Implementation and Use of Anderson Acceleration on Distributed Memory and GPU-based Parallel Computers

John Loffeld and Carol S. Woodward

Abstract Recent work suggests that Anderson acceleration can be used as an accelerator to the fixed-point iterative method. To improve the viability of the algorithm, we seek to improve its computational efficiency on parallel machines. The primary kernel of the method is a least-squares minimization within the main loop. We consider two approaches to reduce its cost. The first is to use a communication-avoiding QR factorization, and the second is to employ a GMRES-like restarting procedure. On problems using 1,000 processors or less, we find the amount of communication too low to justify communication avoidance. The restarting procedure also proves not to be better than current approaches unless the cost of the function evaluation is very small. In order to begin taking advantage of current trends in machine architecture, we also studied a first-attempt single-node GPU implementation of Anderson acceleration. Performance results show that for sufficiently large problems a GPU implementation can provide a significant performance increase over CPU versions due to the GPU's higher memory bandwidth.

Keywords Anderson acceleration · Nonlinear solvers · Fixed-point iteration · TSQR

Mathematics Subject Classification 65B99 · 65N22 · 65H10

1 Introduction

Nonlinear root finding problems of the form $f(u) = 0$ are common in computational science problems and especially when computing the solution of discretized PDEs. For large-scale systems, the Newton–Krylov method is commonly used due to the

J. Loffeld · C.S. Woodward (✉)
Lawrence Livermore National Laboratory, Center for Applied Scientific Computing, Livermore, CA 94551, USA
e-mail: woodward6@llnl.gov

J. Loffeld
e-mail: loffeld1@llnl.gov

G. Letzter et al. (eds.), *Advances in the Mathematical Sciences*, Association for Women in Mathematics Series 6, DOI 10.1007/978-3-319-34139-2_21

fast, often quadratic, convergence of the Newton iteration [3, 11] and the scalability of linear Krylov methods. However, the need to solve a large linear system each iteration involving the Jacobian of f results in high computational cost per step. Furthermore, for complex problems, finding an analytical expression for the Jacobian can be nontrivial or evaluation of the Jacobian may be expensive (see e.g., [9]). Numerical approximations of the action of the Jacobian times a vector can be used instead, but doing so can compromise the rate of convergence of the Krylov iteration. In either case, a scalable preconditioner is often required for good performance when solving the linear systems, and the preconditioner adds considerable complexity to the solution process.

The nonlinear problem can be posed as a stationary problem $u = g(u)$ through the relation $f(u) = u - g(u) = 0$. Fixed-point iteration can then be applied to this system. Compared to methods based on the Newton iteration, the fixed-point method is simple to implement and has a lower computational cost per step, as it does not require use of the derivative. Unfortunately, the iteration does not always converge since the function, $g(u)$, must be a contraction map [10]. When the iteration does converge, the rate of convergence is often slow, typically only linear. However, recent work has shown that the rate of convergence of fixed-point iteration can be improved through Anderson acceleration (AA) [1, 16]. This work raises the possibility of using Anderson-accelerated fixed-point iteration as an alternative to Newton–Krylov in cases where determining the Jacobian is difficult, evaluating the Jacobian is expensive, or the rate of convergence of the Krylov iteration is slow due to a lack of a good preconditioner.

AA improves the rate of convergence of the fixed-point iteration by utilizing information from more than just the most previous iterate. For each iterate, it chooses a linear combination over m prior iterates that minimizes the fixed-point residual in the least-squares sense. This approach of maximizing rate of convergence through a residual minimizing choice of next iterate is similar to the idea behind the generalized minimum residual method (GMRES) iterative linear solver. Indeed, on linear problems, a mathematical equivalence in the rate of convergence between AA and GMRES has been shown in [16]. The size of the least-squares problem that must be solved each AA iteration is $n \times m$, where n is the number of unknowns in the problem. Solving a large least-squares problem each iteration makes AA more expensive per step than basic fixed-point iteration. If $g(u)$ is not a dominant cost, minimizing the cost of the least-squares problem is the key for making the method efficient. In this paper, we consider two approaches to implement AA that are aimed at lowering the cost of solving the least-squares problem. Both cases compromise the rate of convergence of the iteration; so any net benefit from either approach is a matter of balancing cost per iteration and the number of iterations that must be computed. We found the trade-off in one case not to be favorable for the sizes of problems we tested, but the other approach to be modestly favorable in some instances.

The first approach is aimed at large-scale problems implemented on distributed-memory machines with at least thousands of processors and problem sizes with tens of thousands or more unknowns per processor. We consider whether it is possible, or even necessary, to lower the MPI communication cost of the least-squares problem.

In Anderson acceleration, the least-squares problem can be solved in a variety of ways. QR factorization is a good choice due to a balance between efficiency and accuracy [6, 16]. With some approaches for the QR problem, the factorization can be incrementally updated each iteration without requiring a full factorization [7, 15]. When computing the QR decomposition on distributed-memory machines, use of panel factorization, blocking and tiling optimizations, such as those employed in the ScaLAPACK library [2, 4], can give a significant performance benefit over unoptimized algorithms. Recent communication-avoiding QR algorithms use such techniques to minimize interprocessor communication cost, which may be of particular benefit on large-scale machines [5]. Such algorithms minimize communication of the same matrix elements by performing operations on sub-matrices of size greater than rank one. Such techniques are generally unsuitable for per-iteration update of the factorization. Employing those algorithms in the context of AA would require updating the QR factorization only every k vectors, limiting the degree of acceleration in the fixed-point iteration. Nevertheless, use of such algorithms might be beneficial if the computational savings outweigh the cost of computing additional iterations.

The second approach, restarting, is applicable to all sizes of problems, as well as both serial and parallel implementations. In this paper, we tested it on small-scale parallel problems. Restarting is a commonly used technique for reducing cost in GMRES [12]. For linear problems, a mathematical equivalence in the rate of convergence between a truncated-and-restarted AA and a restarted GMRES has been shown [16]. Despite the equivalence, the underlying operations computed by each method are different, i.e., they have a different computational cost structure. However, costs are comparable enough between them that one would expect the trade-off between rate of convergence and computational savings to balance similarly between the two methods on linear problems. However, in AA, rather than periodically truncating the iteration and restarting with the last iterate, a factorization over a sliding window of k previous iterates can be maintained [16]. Thus, rather than periodically fully discarding all previous iterates and starting over with a single vector, only the most stale iterate is deleted each iteration. This contrasting approach, which we refer to as "sliding," allows a less severe reduction in the rate of convergence, at the cost of applying the delete procedure each iteration. We tested the overhead to determine whether its additional cost was worth the better acceleration. We found that avoiding the computational cost of the delete operation through restarting is modestly beneficial in some cases but not all.

Finally, we tested the performance of a GPU implementation of Anderson acceleration as a whole. High Performance Computing machines increasingly employ accelerators such as GPUs, due to their high concurrency. For an algorithm to be well-suited for current and future machines, a significant amount of its parallelism must be captured within each node through the accelerator. Parallelism through MPI alone is no longer sufficient. As a step towards an implementation of Anderson acceleration suitable for current and future supercomputers, we developed a single-node GPU implementation of Anderson acceleration that is based on the GPU-optimized BLAS library, CuBLAS. GPUs are balanced differently than CPUs and require a higher degree of concurrency to operate efficiently. On the machines we tested, we found that the GPU version was quite inferior to the CPU version when the number of unknowns in a problem was about 10,000 or less. When the number was greater, the GPU version was able to greatly outperform the CPU version due to the higher memory bandwidth on the device.

This paper is organized as follows. Section 2 describes the MPI-based implementation of AA used for the experiments. Section 3 describes some performance measurements of the implementation to determine the balance of local computation and interprocessor communication. Section 4 details performance comparisons between a communication-avoiding implementation of AA versus the base implementation. In Sect. 5 we give a performance comparison of a restarted version of AA versus the base implementation, which uses a sliding window of past iterates. We describe and give performance results for the GPU implementation in Sect. 6. Finally, in Sect. 7, we make some final conclusions and describe some possible future work.

2 Anderson Acceleration

In this section, we describe the baseline implementation of AA used in our numerical experiments. The implementation is part of the C language KINSOL package of solvers for nonlinear algebraic equations from the SUNDIALS suite of codes [8, 14]. Methods in SUNDIALS are written on top of an abstracted vector API so that they are independent of whether and how parallelism is used. The SUNDIALS distribution is equipped with a number of packages that include implementations of the vector kernels for serial, thread-parallel, and distributed memory parallel (with MPI) vectors, although users can supply their own. The library abstracts away details about how the data is mapped on to processors and how communication is handled between processors when computing operations on the vectors. As such, we specify the implementation of AA below in a "Matlab-like" manner, only specifying details about parallelization when needed.

Our goal is to solve fixed-point problems of the form: Given $g : \mathbb{R}^n \rightarrow \mathbb{R}^n$, find u such that $u = g(u)$. The Anderson-accelerated fixed-point method is given in Algorithm 1.

In practical implementation, the constrained least-squares problem is often formulated as the following equivalent unconstrained least-squares problem [6, 16]: find $\gamma^{(i)} = (\gamma_0^{(i)}, \ldots, \gamma_{m_i-1}^{(i)})^T$ such that $\min_\gamma \|f_i - F_i\gamma\|_2$, where $F_i \equiv \left[\Delta f_{i-m_i}, \ldots, \Delta f_{i-1}\right]$ and $\Delta f_j = f_{j+1} - f_j$. The least-squares coefficient vectors α and γ are related by $\alpha_0 = \gamma_0, \alpha_j = \gamma_j - \gamma_{j-1}$ for $1 \leq j \leq m_i - 1$, and $\alpha_{m_1} = 1 - \gamma_{m_i-1}$. The next iterate then becomes $u_{i+1} = g(u_i) - \sum_{j=0}^{m_i-1} \gamma_j^{(i)} \left[g(u_{i-m_i+j+1} - g(u_{i-m_i+j})\right]$.

The KINSOL implementation of AA follows the approach described by Walker in [15]. The least-squares problem is solved by performing the QR factorization of F_i and using backward substitution to solve the upper triangular system $R_i\gamma = Q_i^T f_i$.

Algorithm 1: Anderson acceleration

Input: u_0, $m \geq 1$, and ϵ

$u_1 \leftarrow g(u_0)$
for $i = 1, 2, \ldots,$ *until* $\|u_{i+1} - u_i\| < \epsilon$ **do**
  $m_i \leftarrow \min\{m, i\}$
  $H_i \leftarrow [f_{i-m_i}, \ldots, f_i]$, where $f_j = g(u_j) - u_j$
  Solve the constrained least-squares problem for $\alpha^{(i)} = (\alpha_0^{(i)}, \ldots, \alpha_{m_i}^{(i)})^T$ s.t.
    $\min_\alpha \|H_i\alpha\|_2$ s.t. $\sum_{j=0}^{m_i} \alpha_j = 1$
  $u_{i+1} \leftarrow \sum_{j=0}^{m_i} \alpha_j^{(i)} g(u_{i-m_i+j})$
end
Output u_i

Note that when $i < m$, the size of F is $n \times i$, and a new vector is added to the right of F_i in each iteration. After the mth iteration, F_i remains of fixed size, $n \times m$, but in each iteration a column vector is removed from the left of F_i while a new vector is added to the right. It is inefficient to refactorize F_i anew each step, so two helper procedures are used to in-place update Q_i and R_i based on the previous factorization. QRAdd updates the factorization to account for the addition of a vector to the right of F_i, while QRDelete updates for the removal of a vector from the left.

QRAdd uses modified Gram–Schmidt to orthonormalize each new Δf_{i-1} against the previous columns of Q_{i-1}. The resulting vector becomes the new rightmost column of Q_i. Algorithm 2 gives pseudo-code for the procedure using Matlab notation.

Algorithm 2: QRAdd

Input : $Q \in \mathbb{R}^{n \times m_i}$, $R \in \mathbb{R}^{m_i \times m_i}$, and Δf_{i-1}
Output: $Q \in \mathbb{R}^{n \times m_i+1}$ and $R \in \mathbb{R}^{m_i+1 \times m_i+1}$

for $j = 1$ *to* $m_i - 1$ **do**
  $R(j, m) \leftarrow Q(:, j)^T * \Delta f_{m-1}$
  $\Delta f_{i-1} \leftarrow \Delta f_{i-1} - R(j, m) * Q(:, j)$
end
$Q(:, m) \leftarrow \Delta f_{i-1}/\|\Delta f_{i-1}\|_2$ and $R(m, m) \leftarrow \|\Delta f_{i-1}\|_2$.

On a distributed-memory machine with p processors, Q_i is represented as a set of column vectors, with each processor receiving n/p rows of each vector. Communication between processors is incurred by the dot products $Q(:, j)^T * \Delta f_{m-1}$ and when computing the norm $\|\Delta f_{i-1}\|_2$ (implemented with a dot product). The results of the dot products are broadcast to all processors, resulting in a copy of R_i on each processor.

QRDelete uses Givens rotations to update the factorization when a vector is removed from F. The procedure is based on the observation that if $F_{k-1} = Q * R$ then $F_{k-1}(:, 2 : m) = Q * R(:, 2 : m)$, where $R(:, 2 : m)$ is upper Hessenberg. Note that Q and $R(:, 2 : m)$ do not constitute a QR factorization of $F_{k-1}(:, 2 : m)$. They can be updated to be one by using Givens rotations to return $R(:, 2 : m)$ to upper triangular form and then applying the inverse of those rotations to Q. Specifically, if we determine Givens rotations $J_1, \cdots, J_{m-1}$ such that $J_{m-1} * \cdots * J_1 * R(:, 2 : m)$ is upper triangular, then

$$F_{k-1}(:, 2 : m) = Q * R(:, 2 : m) = Q * J1' * \cdots * J'_{m-1} * J_{m-1} * \ldots * J1 * R(:, 2 : m),$$

and setting $Q = Q * J'_1 * \cdots * J'_{m-1}$ and $R = J_{m-1} * \ldots * J1 * R(:, 2 : m)$ gives a QR factorization for F_{k-1}. The pseudo-code for QRDelete is shown in Algorithm 3.

Algorithm 3: QRDelete

Input : $Q \in \mathbb{R}^{n \times m}$ and $R \in \mathbb{R}^{m \times m}$
Output: $Q \in \mathbb{R}^{n \times m-1}$ and $R \in \mathbb{R}^{m-1 \times m-1}$

for $i = 1$ *to* $m - 1$ **do**
  $b \leftarrow \sqrt{R(i, i+1)^2 + R(i+1, i+1)^2}$
  $c \leftarrow R(i, i+1)/b$ and $s \leftarrow R(i+1, i+1)/b$
  $R(i, i+1) \leftarrow d$ and $R(i+1, i+1) \leftarrow 0$
  if $i < m - 1$ **then**
    for $j = i + 2$ *to* m **do**
      $d \leftarrow c * R(i, j) + s * R(i+1, j)$
      $R(i+1, j) \leftarrow -s * R(i, j) + c * R(i+1, j)$ and $R(i, j) \leftarrow d$
    end
  end
  $V \leftarrow c * Q(:, i) + s * Q(:, i+1)$
  $Q(:, i+1) \leftarrow -s * Q(:, i) + c * Q(:, i+1)$ and $Q(:, i) \leftarrow V$
end
$Q \leftarrow Q(:, 1 : m - 1)$ and $R \leftarrow R(1 : m - 1, 2 : m)$

We are interested in the balance between MPI communication and local on-node cost. The former can be broken down into a bandwidth and latency cost. The latter can be further broken down into the cost of floating-point operations and the cost of data transfers between the processor and memory.

In the case of the SUNDIALS implementation, all MPI communication comes from the dot products found in QRAdd and backwards substitution. On each processor, the dot product kernel sums over the local portion of the vectors itself and only calls MPI's Allreduce routine for the summed value. As a result, the reduction is done only over a single number, and the time spent within MPI is nearly completely a latency cost, with very little bandwidth cost. Only synchronous communication is used, so the processor remains idle during the reduction.

The three main kernels in AA, QRAdd, QRDelete, and backwards substitution, are all comprised solely of vector–vector operations. As such, the ratio of floating-point cost to data transfer cost within the kernels (the arithmetic intensity) is very low, well less than one flop per byte. As a result the on-node cost is dominated by the time spent in streaming data between memory and the processor.

Therefore, in AA, the comparison between local-node cost and MPI communication essentially reduces to a balance between local memory transfer cost and the latency costs for the reductions. The TSQR algorithm reduces communication cost by reducing latency on distributed-memory machines, so it targets the main form of communication cost found in AA.

Both the on-node and MPI costs of QRAdd and backwards substitution are very similar. Within each AA iteration, QRAdd performs $m_i - 1$ dot products, performs $m_i - 1$ linear sums, and incurs the data transfer costs of streaming over the involved m_i vectors. Backwards substitution must perform m_i dot products to determine $\gamma^{(i)}$ and then perform m_i linear sums to produce the next iterate vector. It also incurs the data transfer costs of operating on those $m_i + 1$ vectors.

QRDelete is not invoked unless $i > m$, in which case $m_i = m$ and the sizes of the matrices remain fixed at $Q \in \mathbb{R}^{n \times m}$ and $R \in \mathbb{R}^{m \times m}$. In that case, the local cost of QRDelete is on par with the other two kernels in that it applies Givens rotations, implemented as a pair of linear sums, to m vectors (as well as to the very small matrix R). No MPI communication is required. If the number of total iterations is considerably greater than m, the local cost of QRDelete is a substantial portion of the overall on-node cost of the whole method. If the history of iterates is flushed every m iterations and the Anderson iteration restarted with $u_{i+1} = g(u_i) - \sum_{j=0}^{m_i-1} \gamma_j^{(i)}(g(u_{i-m_i+j+1} - g(u_{i-m_i+j})$ then QRDelete is not needed and the cost per step would be reduced. However, the rate of convergence of the iteration would also be hindered. We test whether it is favorable to make this trade-off in Sect. 5.

3 Balance of Communication Versus Computation

In this section, we discuss some performance measurements of the ratio of interprocessor communication versus local-node computation for KINSOL's AA implementation. The tests were conducted on two machines: The Blue Gene Q system "Vulcan," and the Intel Xeon-based system "Cab." Vulcan is composed of 24,576 16-core PowerPC A2 processors running at 1.6 GHz. The processors are connected by a high-speed, low-latency network configured as a 5D torus. IBM's BG/Q MPI library, based on MPICH2 1.4, was used as the MPI library for the tests on Vulcan. Cab is a cluster of Intel Xeon E5-2670 processors, with two 8-core CPUs on each node in a shared-memory configuration, and a total of 1,296 nodes. The nodes are connected via an InfiniBand QDR network in a two-stage federated fat-tree.

On both Vulcan and Cab, the implementation was instrumented with MPI's Wtime function. On Vulcan, the timer has sub-nanosecond resolution, whereas on Cab the timer has microsecond resolution. In both cases, timer accuracy was not a limitation. SUNDIALS only uses blocking communication, so the time spent on communication and computation could be measured independently.

On Vulcan, KINSOL was configured to compute a fixed-point problem using AA for four iterations, with a window size of $m = 4$, and for 16 iterations with a window size of $m = 16$ on a problem with 1, 10, 100, 1,000, 10,000, and 100,000 unknowns per processor. The problem was run on 100 processors and 1,000 processors, using only a single core per processor to ensure that all MPI communication was done over the network and not through shared memory. The $g(u)$ problem was a dummy function that returned random values for the vector elements. Note that the cost of AA depends only on the number of iterations computed, the value for m, the number of elements in a vector per processor, and the number of processors, but does not depend on the contents of the solution vector. The choice of $g(u)$ is irrelevant for cost measurement as long as the number of iterations does not change. In our timings, we ignored the cost of computing $g(u)$; the cost measured is for just the Anderson algorithm itself.

The outcome on Vulcan was very similar in all cases so we only show the largest problem with 1,000 processors and with 16 iterations. The results are shown in Table 1a. The rightmost column gives the percentage of overall cost that was spent on MPI communication. While communication dominates when the number of unknowns per processor is small, it becomes negligible for 10,000 unknowns per processor or greater. For most large-scale problems, 10,000 unknowns per processor is quite lean, so the cases of 10,000 unknowns or greater are the most relevant. For problems where the number of processors is 1,000 or less, communication is not a major cost for AA on Vulcan.

On Cab, the problem setup was the same except for two differences. The problem was run on 100 processors and 256 processors, as 256 was the limit of our access. Furthermore, the problem was additionally run with 1,000,000 unknowns to better show how communication falls off in importance relative to computation. As on Vulcan, the outcomes were quite similar regardless of the number of processors or iterations, so we display only the 16 iteration case with 256 processors in Table 1b. We see that communication on Cab is a greater proportion of overall cost compared to Vulcan, which is not surprising considering the less capable network on Cab. However, the communication cost is still overtaken by the cost of local computation as the number of unknowns increases, and the cost becomes minor even for lean problems.

Overall, we conclude that communication is not a major cost in AA for the scale of problems we have considered. It may be of greater importance when using a larger number of processors.

4 TSQR Versus Modified Gram–Schmidt

In this section, we consider whether the communication-avoiding Tall Skinny QR (TSQR) algorithm might give better performance due to reduced MPI latency compared to the modified Gram–Schmidt implementation in KINSOL (QRAdd). We tested with the distributed-memory TSQR implementation from NuLAB [13]. The details of the algorithm can be found in [5]. We employed the variant where communication is done using a binary reduction tree. The local QR solves on sub-blocks

Table 1 Local node and MPI time costs in seconds for AA in KINSOL when computing for 16 iterations with $m = 16$. On both machines, the MPI cost becomes minor when the number of unknowns per processor is modest or larger

Unknowns per processor	Total	Local	MPI	% MPI
(a) Run times on Vulcan				
1	4.8E-03	4.0E-04	4.4E-03	91.7 %
10	4.8E-03	4.6E-04	4.3E-03	90.4 %
100	5.7E-03	1.2E-03	4.5E-03	79.0 %
1,000	1.4E-02	9.3E-03	5.0E-03	34.9 %
10,000	9.3E-02	8.8E-02	5.0E-03	5.4 %
100,000	8.8E-01	8.8E-01	5.0E-03	0.6 %
(b) Run times on Cab				
1	6.2E-03	3.4E-05	6.1E-03	99.55 %
10	6.1E-03	5.9E-05	6.1E-03	99.0 %
100	9.8E-03	6.6E-05	9.7E-03	99.3 %
1,000	8.5E-03	3.3E-04	8.1E-03	96.1 %
10,000	1.6E-02	3.5E-03	1.3E-02	78.5 %
100,000	6.2E-02	4.3E-02	1.9E-02	30.8 %
1,000,000	5.6E-01	5.1E-01	4.6E-02	8.2 %

were computed using a LAPACK library optimized for the respective machine, which in the case of Vulcan was IBM's ESSL library and on Cab was Intel's Math Kernel Library (MKL).

The matrices were the same size and used the same partitioning over processors as those in Sect. 3, allowing us to directly compare the performance of TSQR versus modified Gram–Schmidt. The matrices were filled with random data. Note that the amount of computation and communication does not depend on the content of the matrices, only their dimensions.

The relative performance of TSQR compared to a non-communication-avoiding algorithm improves with the width of the matrix. In particular, TSQR performs better relative to KINSOL for the case when the matrix is 16 columns wide instead of 4 so we only display that case in Table 2. The Vulcan measurements were done on 1,000 processors and the ones on Cab were done using 256 processors.

The far right column of the table shows the overall performance of TSQR relative to modified Gram–Schmidt. We see that on Vulcan the overall cost is significantly reduced. However, the percentage of overall cost that is communication is trivial on Vulcan for problems with even a modest number of unknowns per processor. The cost reduction is from savings in computation not MPI communication. The lower computational cost of TSQR is in part because it uses a tuned library, ESSL, while KINSOL is untuned. For QR factorization, the performance gain of tuned libraries comes primarily through panelization and tiling, which requires operations to be performed on sub-matrices wider than a single column. This requirement is at odds with in-place updating the QR factorization one column at a time, as is done in KINSOL and required by the algorithm as written above. To exploit tuned libraries fully, AA would need to perform the QR factorization over k vectors at a time, where the performance gain would increase with k up to some saturation point. That would mean acceleration could be applied only every kth iteration, and ordinary fixed-point would need to be used for the iterations in between. To prevent the rate of convergence from being disastrously reduced, k would need to remain small, putting the needs of convergence at odds with the needs of QR algorithm optimizations. It is possible that for some problems there is a balance that results in a net reduction in overall cost, but we have not yet found such a case on Vulcan.

On Cab, the performance of TSQR is unexpectedly poor. As seen in Table 2b, both the communication and computational cost are increased in TSQR over KINSOL by orders of magnitude. We initially believed this was a mistake in our problem setup, but after much investigation we have not found anything particular. The approach taken by TSQR to solving the QR problem is quite different from that of KINSOL, and it appears to balance unfavorably on Cab's architecture. We will continue to investigate the underlying cause of this. In any case, in light of the results of Sect. 3, communication avoidance is not expected to be helpful on the scale of problems we have tested.

In conjunction with the tests in Sect. 3, we conclude that communication avoidance is not generally helpful for problems computed on 1,000 processors or less. For larger scale problems, communication cost may increase relative to computation to the point where avoiding communication becomes important, but that possibility is not tested by our measurements. However, current trends in supercomputer architecture are moving away from large node counts and moving more parallelism to within each node. Current supercomputers such as Sequoia at Lawrence Livermore National Laboratory and Titan at Oak Ridge National Laboratory have large node counts of about 100,000 and 20,000, respectively. The replacement machines are planned to only have several thousand nodes in each case, with most of the parallelism coming from GPUs. This makes the case for use of communication avoidance in Anderson acceleration at the MPI level uncompelling.

5 Restarting

Since communication is not an important cost for the cases we tested, we consider a possible approach for reducing computational cost. As discussed in Sect. 2, we can restart the iteration every m iterations using only the most recent iterate. As with restarting in GMRES, doing so mitigates the quadratic increase in cost per iteration, but may also reduce the rate of convergence. However, unlike GMRES, AA can also control quadratic growth in cost by limiting the number of past iterates, m, and updating the QR factorization in-place, as discussed in Sect. 2. This practice is used in the current KINSOL implementation. For easy comparison with "restarting", we

Table 2 Local node and MPI time costs (in seconds) for the QR factorization kernel in AA when computed using TSQR, for 16 iterations with a window size of 16. On Vulcan, the communication cost is nontrivially reduced compared with the KINSOL case, but communication is too small a percentage of the overall cost for this to matter. In Cab, the communication cost is actually greatly increased

(a) Run times on Vulcan

Unknowns per processor	Total	Local	MPI	% MPI	Overall % GS
1	2.6E-05	2.2E-05	4.0E-06	15.4 %	21.3 %
10	3.2E-04	3.2E-04	3.0E-06	0.93 %	39.3 %
100	1.0E-03	7.1E-04	3.2E-04	31.1 %	44.2 %
1,000	1.8E-03	1.5E-03	3.2E-04	17.6 %	23.0 %
10,000	2.8E-02	2.7E-02	3.3E-04	1.2 %	32.5 %
100,000	6.9E-02	6.8E-02	8.9E-04	1.3 %	8.5 %

(b) Run times on Cab

Unknowns per processor	Total	Local	MPI	% MPI	Overall times GS
1	1.3E-00	6.4E-01	6.6E-01	50.7 %	437x
10	2.8E-00	3.9E-01	2.4E-00	85.8 %	1045x
100	3.2E-00	3.8E-01	2.8E-00	88.0 %	593x
1,000	3.7E-00	8.2E-01	2.9E-00	77.8 %	793x
10,000	4.0E-00	8.4E-01	3.2E-00	79.1 %	450x
100,000	3.6E-00	1.1E-00	2.5E-00	69.0 %	100x

will label this case as "sliding", since the window of past iterates slides forward each iteration. By limiting m through sliding, the rate of convergence is also reduced, but not as severely as by restarting. The trade-off is that QRDelete must be called each iteration past the mth one, which increases the cost per iteration.

We tested on Vulcan and Cab whether restarting gives better performance than sliding on a restricted-additive-Schwarz (RAS) iteration applied to the 2D Poisson problem. Details about the RAS problem can be found in [16], where tests of Anderson acceleration compared to fixed-point iteration.

$$\Delta u + 20u + 20u_x + 20u_y = f \quad \text{in } D = [0,1] \times [0,1], \quad \text{where } u = 0 \text{ on } \delta D.$$

The problem was discretized using centered differences discretization on a 128^2 node grid, with $f = -10$. The domain was divided into four sub-domains per direction, for a total of 16 sub-domains, with three grid lines of overlap between neighbors.

The linear sub-domain problems were solved with a direct solver, but the computational cost of $g(u)$ was ignored. Only the cost of the operations in AA itself was measured. Of course in practice the cost of $g(u)$ may matter greatly, but the complexity varies widely between problems. The measurements are therefore an optimistic bound. Compared to the normal implementation, restarting increases the number of

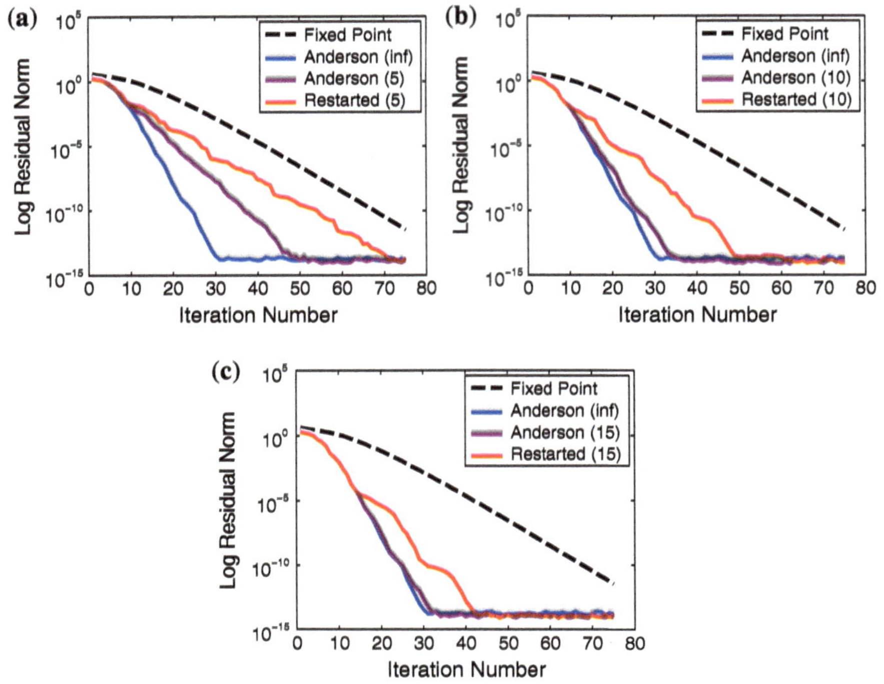

Fig. 1 Comparison of the rate of convergence of restarted Anderson acceleration versus full Anderson acceleration on a 2D Poisson problem. The running times are listed in Table 3. (**a**) $m = 5$. (**b**) $m = 10$. (**c**) $m = 15$

iterations that must be computed, so if the trade-off is not worthwhile when $g(u)$ has zero cost, it will not be worthwhile when the cost of $g(u)$ is included.

Before considering computational efficiency, we first look at how restarting affects convergence compared to sliding. Restarting was compared with sliding on the RAS problem with the following parameters. The problem was run when restarting every 5, 10, and 15 iterations, when sliding with m limited to 5, 10, and 15, and without restriction on m. In what follows, we label that last case the baseline case. For comparison, the problem was also computed using fixed-point iteration with no acceleration. The convergence plots are shown in Fig. 1. We see that AA generally converges much more quickly than the fixed-point iteration. The rate of convergence is reduced for both restarting and sliding, although less so for sliding. As m increases, the rates of convergence improve for both sliding and restarting. When $m = 10$, the rate of convergence for sliding is almost the same as for the baseline case. However, even when $m = 15$, the rate of convergence for restarting is still significantly reduced compared to the baseline case.

We now turn to computational efficiency. On both Vulcan and Cab, restarting was tested using 16 processors using the same parameters as the previous paragraph. The costs, not including that of evaluating $g(u)$, are shown in Table 3a, b for the sizes of

m and number of iterations that corresponds to how long it takes to reach a tolerance of 10^{-14}. For example, as can be seen in the figures, it takes about 30 iterations for the baseline iteration to reach the limit of precision. Table 3 shows that it takes 0.072 seconds on Vulcan to compute those 30 iterations. Any case on that machine with lower times to reach machine precision is an improvement on the baseline case.

On Vulcan, we see that sliding and restarting with $m = 5$ give an improvement over the baseline case, and sliding with $m = 10$ gives a negligible improvement. For sliding, even though the cost of QRAdd is reduced due to the smaller window size, QRDelete is also called. We can see this call adds significant overhead. For example, when $m = 15$, sliding needs about the same number of iterations as the baseline case, but the cost of QRAdd is smaller due to the restricted size of m. However, the overall cost is still higher due to the overhead of QRDelete. In contrast, restarting avoids the overhead of QRDelete, but it must compute a larger number of iterations than sliding. The balance results in a cost on Vulcan that is similar between restarting and sliding. When $m = 10$, restarting requires slightly more time than sliding, while when $m = 15$, restarting takes slightly less time.

Restarting is more favorable on Cab, with the most time-consuming case of restarting using less time than the least expensive case of sliding. A window size of $m = 5$ gives the best improvement over the baseline case for both sliding and restarting. In that case, sliding costs about 82 % of the baseline case while restarting incurs 62 % of the time, which makes restarting 75 % of the cost of sliding. The most expensive case for restarting was with $m = 15$, where it incurred 74 % of the cost of the baseline case while sliding was 109 %.

We see that the cost of QRDelete is modest to the point that restarting gives no benefit over sliding on Vulcan, even when $g(u)$ has no cost, but still expensive enough that avoiding it gives an improvement on a different machine architecture. This test is only for one problem, and the effect on the rate of convergence of sliding and

Table 3 Cost of Anderson acceleration when using restarting versus sliding the QR factorization using QRDelete. The number of iterations corresponds to those needed by each iteration in Fig. 1. Restarting is less efficient on Vulcan but more efficient on Cab

Type	m	Iters	Time (s)	Type	m	Iters	Time (s)
(a) Vulcan							
Sliding	5	47	0.058	Restarted	5	71	0.058
Sliding	10	33	0.071	Restarted	10	49	0.073
Sliding	15	32	0.089	Restarted	15	42	0.085
Baseline	30	30	0.072				
(b) Cab							
Sliding	5	47	0.028	Restarted	5	71	0.021
Sliding	10	33	0.031	Restarted	10	49	0.022
Sliding	15	32	0.037	Restarted	15	42	0.025
Baseline	30	30	0.034				

restarting varies from problem to problem. It can be expected, though, that restarting will require a significantly higher number of iterations over sliding in general. Even on a machine like Cab, the increased number of iterations could be harmful if the cost of $g(u)$ were nontrivial. However, restarting requires almost no additional complexity in the implementation on top of sliding and coexists with it easily. Like with most implementations of GMRES, the option can be left to the user and might be valuable on some problems.

6 GPU Implementation

On current high performance computing machines, the majority of the compute capacity on each node now comes from accelerators such as GPUs and the Intel Phi line of processors. A well-balanced algorithm for modern supercomputers not only has to be efficient at MPI communication between nodes, but also must make good use of the local accelerators. The architectures of such systems are balanced differently than pure CPU ones, so algorithms must be adapted to take full advantage of them. Along the path for developing an implementation of Anderson acceleration well-suited for modern machines, we have begun work on implementations that make use of accelerators. In this section, we describe a first step effort to implement Anderson acceleration on GPUs. The implementation is currently only for a single node and is not yet fully optimized for the GPU architecture, but still shows a considerable performance increase over a CPU implementation. Based on lessons learned from this initial effort, a better optimized and MPI-capable implementation will be developed in future work.

Compared to traditional CPUs, GPUs are characterized by a much higher level of single instruction multiple thread (SIMT) parallelism. For the purposes of this paper, they can be thought of as vector processors, where thousands of vector or matrix elements can be processed in parallel simultaneously using the same instructions. The high SIMT concurrency gives such processors up to an order of magnitude higher peak flops rate than CPUs. GPUs on HPC class machines also have their own RAM, which generally has five to ten times higher bandwidth than the main memory of CPUs. As a trade-off, their caching systems are comparatively limited, and the latency to RAM is also many times higher. Instead of using large low-latency caches to minimize the performance penalty of accessing RAM, as done in CPUs, GPUs instead attempt to hide the latency behind a much higher degree of parallelism. Even if some vector elements are stalled waiting for data to transfer from memory, the high level of concurrency ensures some other elements are likely to have their data requests satisfied and are ready to continue, thus keeping the processor busy.

Algorithms in scientific computing fall within a spectrum between those that are compute bound and those that are memory bound. Compute-bound algorithms require a large number of floating-point operations to be performed per byte loaded from memory. After each chunk of data is loaded, the processor remains busy for a long while and the memory system must wait for the processor to finish before transferring the next group of data. As a result, the speed of the processor itself is the rate limiting factor for the performance of the algorithm. Matrix–matrix operations such as those found in Level 3 of the Basic Linear Algebra Subprograms (BLAS) library are generally compute bound. On the other hand, the situation is reversed in memory-bound algorithms. The processor performs only a limited number of operations per byte, so the processor largely remains idle waiting for data transfers to complete, and the performance of the algorithm is now determined by how fast the data can be transferred from memory. Vector–vector operations such as those in Level 1 BLAS, and matrix–vector operations such as those in Level 2 BLAS are both memory bound on most architectures.

Our current GPU implementation of Anderson acceleration is based on the GPU-optimized BLAS library CuBLAS from Nvidia. The form of the algorithm still follows the structure listed in Algorithms 1 through 3, and the main loop still runs on the CPU. However, except for the small R matrix, the data for the algorithm resides in the GPU RAM, and each vector operation is performed by calling Level 1 BLAS operations on the GPU. For example, each dot product in 2 is done using the CuBLAS function cublasDdot. Because the implementation is based on vector–vector operations, the algorithm is expected to be highly memory bound. Therefore, the better flops rate of the GPU is not expected to be helpful, but the much higher memory bandwidth should still give the GPU implementation a performance advantage over a CPU implementation if the bandwidth is well utilized. Unfortunately, invoking an operation on the GPU has a high overhead and each call to a BLAS operation on the GPU incurs about $10\,\mu s$ of latency. Furthermore, the CPU blocks during the execution of each BLAS call (host pointer mode was set) and only the default CUDA stream was used. Therefore, there is no overlap of work between the CPU and GPU, nor between GPU kernels, to hide the overhead of the BLAS calls. To make the cost of each call worthwhile, that overhead must be amortized over a sufficiently large amount of work, i.e., over a sufficiently large vector. We can expect the GPU implementation to perform poorly for small vector sizes, but to perform well if the vector length is sufficiently large to amortize the overhead and allow the high bandwidth to be exploited over many vector elements.

For comparison, a single-node CPU implementation was also developed that keeps the data within the CPU RAM and uses standard BLAS instead of CuBLAS. The bandwidth of the CPU RAM is lower than that of the GPU RAM, but the caching system is superior and the overhead to invoke a BLAS call is comparatively negligible. As such, we can expect the CPU version to outperform the GPU version for small vector lengths, as the cache will be able to hold most of the data and the overhead per BLAS call will dominate the GPU implementation. However, for sufficiently large vector lengths, the data will no longer fit within CPU cache and the overhead per BLAS call will be well amortized on the GPU. The higher bandwidth of the GPU should then allow for higher performance over the CPU version.

The CPU implementation was linked with Intel's optimized BLAS routines provided in the Intel Math Kernel Library (MKL). The code was tested against both single-threaded and multi-threaded versions of the library. Multi-threading allows a greater number of operations to access the memory system at the same time, provid-

Table 4 Configuration of the two machines on which the GPU implementation was tested

	Primary	Secondary
CPU	*Intel Xeon E5-2670*	*Intel i5-3570K*
Cores/socket	8	4
# sockets	2	1
Clock rate	2.6 GHz	3.4 GHz
L1 Cache	32 KB	32 KB
L2 Cache	256 KB	256 KB
L3 Cache	20 MB	6 MB
RAM	256 GB DDR3	16 GB DDR3
Memory bandwidth	102.4 GB/s	21.0 GB/s
GPU	*Tesla K40m*	*GeForce GTX 680*
Architecture	Kepler	Kepler
Clock rate	745 MHz	1.18 GHz
L2 Cache	1.5 MB	512 KB
RAM	12 GB GDDR5	4 GB GDDR5
Memory bandwidth	288 GB/s	192 GB/s

ing better memory bandwidth utilization at the cost of some thread synchronization overhead.

The implementations were tested on two sets of machines. The primary is representative of a node on a current HPC-grade machine and was used for the performance timings. The secondary machine has a consumer-grade GPU and CPU. It was not used to gather the primary results, but rather to supplement our understanding of the performance through low-level profiling. Our access rights to the machine allowed use of hardware performance counters that could measure bandwidth usage, which was not possible on the main machine. The configuration of both machines is specified in Table 4.

The implementations were tested on the primary machine with four sets of experiments. The first set was run for four Anderson iterations with a window size of $m = 4$ (i.e., without QRDelete) for vector lengths ranging from one to ten million, increasing in factors of ten. The remaining experiments were run with 16 Anderson iterations over the same range of vector lengths, but with window sizes of $m = 4$, $m = 8$, and $m = 16$. Note that in the last case, QRDelete is also not used. Besides running on the GPU, all four sets were run on the CPU using both one thread and 16 threads. Other numbers of threads were tested but gave results intermediate to the one and sixteen thread cases. As with previous experiments, a dummy function that returned random values was used and the cost of the function was not included in the timings. The results are shown in Fig. 2.

The outcome in all four cases is very similar. For vector lengths below ten thousand, the CPU versions require significantly less time than the GPU version in all four experiments. The cost of the CPU implementation remains approximately con-

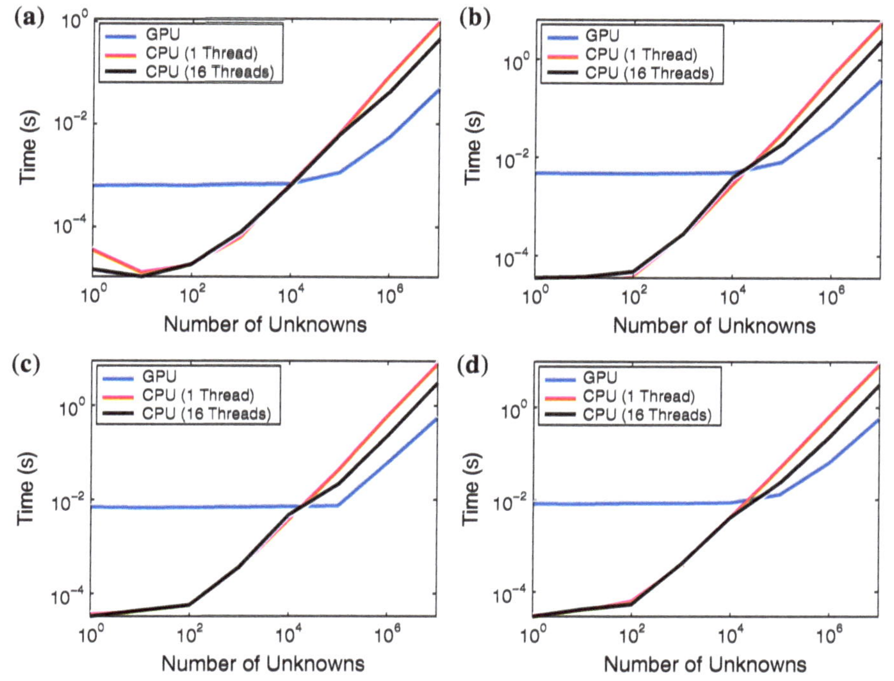

Fig. 2 Performance of a GPU implementation versus a CPU implementation using one and four threads on the primary GPU machine. For sufficiently large vector lengths, the GPU version outperforms the CPU version due to the higher memory bandwidth on the GPU. For smaller vector lengths, the high overhead of invoking BLAS routines prevents the GPU implementation from being competitive. (**a**) 4 iterations, $m = 4$. (**b**) 16 iterations, $m = 4$. (**c**) 16 iterations, $m = 8$. (**d**) 16 iterations, $m = 16$

stant until the vector size reaches a hundred elements, after which the cost begins to converge to a linear increase in cost with vector length. For the GPU version, the cost remains constant until about ten thousand elements per vector due to the high overhead of invoking BLAS calls. Note that in each of the four experiments, the number of vector operations is constant and independent of the vector length. That is why the GPU cost remains constant until the vector length is increased enough for the amount of work per vector to dominate over the overhead per vector operation. Beyond ten thousand elements per vector, the cost slowly approaches a linear increase in cost with length. When the vectors are large enough that both the CPU and GPU costs have linear scaling, we expect the ratio in performance to be roughly equal to the ratio in effective memory bandwidth. For the case of four iterations with $m = 4$, the run time for the GPU at ten million unknowns per vector is 5.4×10^{-2} s, while it is 4.5×10^{-1} s for the multi-threaded CPU case, giving a ratio of about 8.5. For the case of 16 iterations with a window size of $m = 16$, for a vector length of ten million the GPU run time was 6.5×10^{-1} s, while it was 3.1 s in the multi-threaded CPU case, resulting in a ratio of only 4.8, implying the CPU bandwidth is not fully utilized. For the 16 iteration cases with $m = 4$ and $m = 8$, the performance ratios were 5.7 times and 4.9 times, respectively, compared to the multi-threaded times.

To verify that the higher bandwidth of the GPU is the primary cause for its performance advantage, the CPU implementation was profiled on the 4-core machine using the Intel VTune profiler, and the GPU implementation was profiled on the corresponding GeForce GTX 680 using the Nvidia Visual Profiler. Both profilers are able to measure the memory bandwidth usage directly using low-level hardware performance counters. The qualitative results of the experiments on the second machine were similar to those of Fig. 2, except that the difference between the single and multi-threaded CPU cases was much less. For the largest vector size, when computing four iterations with $m = 4$, the difference in run time between the GPU and multi-threaded CPU case was a ratio of 10.0, equal to the peak bandwidth ratio for the machine. For the 16 iteration cases, the ratios were 8.2 for $m = 4$, 7.8 for $m = 8$, and 7.6 for $m = 16$. Measuring these cases in VTune, for both CPU versions the bandwidth usage over time had the profile of extended periods of high bandwidth during the BLAS calls interspersed with shorter periods of low bandwidth usage between the calls. The peak bandwidth in the multi-threaded case reached over 20 GB/s, which is very close to the peak bandwidth of the machine, while the average bandwidth was 16.1 GB/s. In the single-threaded case, the peak bandwidth reached about 19.5 GB/s, but the profile was considerably less uniform. The average bandwidth was 15.4 GB/s. The lower and less consistent bandwidth was due to only having a single thread access the memory bus, preventing the bandwidth from being consistently held high. Despite the lower bandwidth in the single-threaded case, the run time was always nearly identical to the multi-threaded case on the 4-core machine. This is due to the OpenMP synchronization overhead in the multi-threaded case, resulting in the performance balancing to about the same overall cost. We assume that for the 16-core machine, the overall bandwidth utilization using multiple threads was even better than for the 4-core case, giving a significant net win over a single thread. For the GPU with the largest size of vector, the bandwidth usage within the BLAS calls was generally about 155 GB/s and between the calls the bandwidth was near zero. Almost all of the time was spent within the BLAS calls instead of between. For shorter vector lengths, the percentage of time spent between calls increased, reflecting less amortization of the overhead of invoking the routines, and the bandwidth utilization also fell within the BLAS calls due to the lower amount of concurrency utilizing the memory system. For example, when the vector length was 10,000, the bandwidth within the BLAS calls fell to only several hundred MB/s.

We conclude that GPUs can provide a significant performance increase over CPU implementations as long as the number of unknowns in the problem is sufficiently high to allow the bandwidth to be exploited. Improvements to the performance of the implementation could come in two forms. The first is to achieve better memory bandwidth utilization. The current implementation has good memory efficiency when the vector lengths are high, but there is some room for improvement. Perhaps a more fruitful approach would be to design an implementation that is less memory bound. Transfers of data are a form of communication and communication-avoiding algorithms were in fact first designed to minimize memory cost. Attempting to trade rate of convergence for a reduction in communication cost was not effective in the MPI case because the amount of inter-node communication was too low to make doing so worthwhile. However, in the on-node case, for large enough vector sizes, almost all of the cost is memory communication. It might be that a GPU implementation of communication-avoiding QR factorization or some other communication minimizing approach would give a net performance advantage. We will explore this possibility in future work.

7 Conclusions and Future Work

In this paper, we considered whether communication-avoiding QR algorithms in AA could increase efficiency on distributed-memory machines. We found that on 1,000 processors, communication was not significant enough to require communication avoidance. In future work, we will test whether communication becomes more significant when utilizing more processors.

The Anderson iteration can be restarted in a manner similar to GMRES, which mitigates the quadratic growth in cost and memory from an increasing set of past iterates. However, AA can also do an in-place update of the QR factorization to achieve a similar benefit. The latter approach limits the rate of convergence less than the former, but has higher overhead. We tested on only a single problem, but the results suggest that the overhead from the in-place update is high enough that restarting can be modestly beneficial in some cases. We will test on a larger set of problems to see how the balance varies between problems.

Implementation of AA for GPUs can give a sizable performance increase over CPU implementations when the number of unknowns is sufficiently large due to the higher memory bandwidth of GPU memory. We did not find a benefit from MPI-level communication avoidance, but the highly memory-bound nature of our current GPU implementation suggests communication avoidance may be useful at the GPU level. We will investigate this in future work.

Acknowledgments This work was performed under the auspices of the U.S. Department of Energy by Lawrence Livermore National Laboratory under Contract DE-AC52-07NA27344. LLNL-PROC-675918.

References

1. D.G. Anderson, Iterative procedures for nonlinear integral equations. J. Assoc. Comput. Mach. **12**, 547–560 (1965)
2. L.S. Blackford, J. Choi, A. Cleary, E. D'Azevedo, J. Demmel, I. Dhillon, J. Dongarra, S. Hammarling, G. Henry, A. Petitet et al., *ScaLAPACK Users' Guide*, vol. 4 (SIAM, Philadelphia, 1997)

3. P.N. Brown, Y. Saad, Hybrid Krylov methods for nonlinear systems of equations. SIAM J. Sci. Statist. Comput. **11**, 450–481 (1990)
4. J. Choi, J. Demmel, I. Dhillon, J. Dongarra, S. Ostrouchov, A. Petitet, K. Stanley, D. Walker, R.C. Whaley, ScaLAPACK: a portable linear algebra library for distributed memory computer design issues and performance, *Applied Parallel Computing Computations in Physics, Chemistry and Engineering Science* (Springer, Heidelberg, 1996), pp. 95–106
5. J. Demmel, L. Grigori, M. Hoemmen, J. Langou, Communication-optimal parallel and sequential QR and LU factorizations. SIAM J. Sci. Comput. **34**(1), 206–239 (2012). http://dx.doi.org/10.1137/080731992
6. H. Fang, Y. Saad, Two classes of multisecant methods for nonlinear acceleration. Numer. Linear Algebra Appl. **16**, 197–221 (2009)
7. Hammarling, S., Lucas, C.: Updating the QR factorization and the least squares problem. Tech. rep., The University of Manchester (2008). http://citeseerx.ist.psu.edu/viewdoc/summary?doi=10.1.1.142.2571
8. A.C. Hindmarsh, P.N. Brown, K.E. Grant, S.L. Lee, R. Serban, D.E. Shumaker, C.S. Woodward, SUNDIALS: suite of nonlinear and differential/algebraic equation solvers. ACM Trans. Math. Softw. **31**(3), 363–396 (2005). http://doi.acm.org/10.1145/1089014.1089020
9. J.E. Jones, C.S. Woodward, Preconditioning Newton–Krylov methods for variably saturated flow, in *Computational Methods in Water Resources*, vol. 1, ed. by L.R. Bentley, J.F. Sykes, C. Brebbia, W. Gray, G.F. Pinder (Balkema, Rotterdam, 2000), pp. 101–106
10. C. Kelley, *Iterative Methods for Linear and Nonlinear Equations, Frontiers in Applied Mathematics*, vol. 16 (SIAM, Philadelphia, 1995)
11. D.A. Knoll, D.E. Keyes, Jacobian-free Newton–Krylov methods: a survey of approaches and applications. J. Comp. Phys. **193**, 357–397 (2004)
12. Y. Saad, M.H. Schultz, GMRES: a generalized minimal residual algorithm for solving nonsymmetric linear systems. SIAM J. Sci. Stat. Comput. **7**(3), 856–869 (1986)
13. E. Solomonik, G. Ballard, N. Knight, M. Jacquelin, P. Koanantakool, E. Georganas, D. Matthews, NuLAB. https://github.com/solomonik/NuLAB/
14. SUNDIALS (SUite of Nonlinear and DIfferential/ALgebraic Solvers), http://www.llnl.gov/casc/sundials
15. H. Walker, Anderson acceleration: Algorithms and implementations. Tech. Rep. MS-9-21-45, Worcester Polytechnic Institute (2011)
16. H.F. Walker, P. Ni, Anderson acceleration for fixed-point iterations. SIAM J. Numer. Anal. **49**(4), 1715–1735 (2011). http://dx.doi.org/10.1137/10078356X

MIX
Papier aus verantwortungsvollen Quellen
Paper from responsible sources
FSC® C105338
FSC
www.fsc.org

If you have any concerns about our products,
you can contact us on
ProductSafety@springernature.com

In case Publisher is established outside the EU,
the EU authorized representative is:
Springer Nature Customer Service Center GmbH
Europaplatz 3, 69115 Heidelberg, Germany

Printed by Libri Plureos GmbH
in Hamburg, Germany